On the Applications of EMG Sensors and Signals

On the Applications of EMG Sensors and Signals

Editor

Ernest N. Kamavuako

MDPI • Basel • Beijing • Wuhan • Barcelona • Belgrade • Manchester • Tokyo • Cluj • Tianjin

Editor
Ernest N. Kamavuako
King's College London
UK

Editorial Office
MDPI
St. Alban-Anlage 66
4052 Basel, Switzerland

This is a reprint of articles from the Special Issue published online in the open access journal *Sensors* (ISSN 1424-8220) (available at: https://www.mdpi.com/journal/sensors/special_issues/EMG_Sensors).

For citation purposes, cite each article independently as indicated on the article page online and as indicated below:

LastName, A.A.; LastName, B.B.; LastName, C.C. Article Title. *Journal Name* **Year**, *Volume Number*, Page Range.

ISBN 978-3-0365-5989-6 (Hbk)
ISBN 978-3-0365-5990-2 (PDF)

Contents

About the Editor

Ernest N. Kamavuako

Ernest N. Kamavuako has been a Senior Lecturer in the Department of Engineering at King's College London since October 2017 and Professor at the University of Kindu since August 2022. He received his Master's and PhD degrees in Biomedical Engineering from Aalborg University, Denmark, in 2006 and 2010, where he was an Assistant Professor (2010-2014) and Associate Professor (2014-2017), with excellent teaching and supervision skills. From 2012 to 2013, he was a Visiting Postdoctoral Fellow at the Institute of Biomedical Engineering at the University of New Brunswick. Between 2017 and 2021, he was appointed Adjunct Professor in the Department of Electrical and Computer Engineering at the University of New Brunswick, Canada. Between February and September 2017, he was an Academic Visitor at the Department of Bioengineering, Imperial College London, United Kingdom. He has a good publication record with primary research interests related to using EMG recordings to control upper limb prostheses and quantifying fluid intake. Dr Kamavuako was Associate Editor for IEEE Transactions on Neural Systems and Rehabilitation Engineering between 2016 and 2021. He has Guest-Edited the Special Issue "On the Applications of EMG Sensors and Signals" in Sensors, MDPI. He is now an Associate Editor for Neuroprosthetics within Frontiers in Neuroscience and Rehabilitation Engineering within Frontiers in Rehabilitation Sciences.

sensors

Editorial

On the Applications of EMG Sensors and Signals

Ernest N. Kamavuako [1,2]

[1] Department of Engineering, King's College London, London WC2R 2LS, UK; ernest.kamavuako@kcl.ac.uk; Tel.: +44-207-848-8666

[2] Faculté de Médecine, Université de Kindu, Kindu, Maniema, Democratic Republic of the Congo

Citation: Kamavuako, E.N. On the Applications of EMG Sensors and Signals. *Sensors* **2022**, *22*, 7966. https://doi.org/10.3390/s22207966

Received: 9 October 2022
Accepted: 17 October 2022
Published: 19 October 2022

1. Introduction

The ability to execute limb motions derives from composite command signals (or efferent signals) that stem from the central nervous system through the highway of the spinal cord and peripheral nerves to the muscles that drive the joints. The brain encodes information about a given movement using electrical impulses, referred to as action potentials. The modulation of commands is possible via the recruitment principle and motor unit firing frequency. Because of direct access to the brain, spinal cord and peripheral nerves pose a significant challenge requiring mainly invasive approaches, and muscles provide excellent access to study motor control. Electromyography (EMG) is a technique used for evaluating and recording electrical activity produced by muscles. EMG signals can be harvested on the skin's surface, under the skin, and inside the muscle, providing different levels of information. It is regarded as the biological amplifier of nerve impulses, providing an improved signal-to-noise ratio. Over the last few decades, there have been considerable technological advances, including sensor miniaturization and advanced signal processing algorithms. The development has empowered EMG sensors and the associated EMG signals to find applications in many areas. In electrodiagnostic (EDX) medicine, EMG has often been used to cover the entire spectrum of EDX techniques using the needle electrode to record muscle electrical activity. In robotic and rehabilitation applications, the EMG signal captures human intention, creating a communication channel between humans and robots. The EMG is also used to assess swallowing function in motor neuron disorders and estimate fluid intake. Motion analysis, feedback, and handwriting modelling are other important areas which involve the application of EMG sensors and signals. EMG sensors and signals are also used in many research laboratories involved in biomechanics, movement disorders, motor control, postural control, neuromuscular physiology, and physical therapy. This Special Issue attempts to capture the latest advances in EMG sensor development, EMG sensor application, and EMG signal conditioning from theoretical and experimental approaches. Nineteen papers (two reviews and seventeen research papers) have been published, providing useful information on the application of EMG sensors and signals. The 17 research papers addressed several exciting themes: swallowing, motion detection and prosthesis control, muscle synergies, robotic exoskeleton, driver behavior, signal conditioning, and muscle assessment.

2. Overview of Contribution

The first review paper [1] presented an in-depth study of 28 articles on TMR-based prosthesis control strategies, consolidating the field's current knowledge and outlining the limits of these strategies. We learned that there is a lack of accepted reference standard performance evaluation due to several evaluation tests based on different metrics. The diversity of these evaluation tests makes it difficult to define common guidelines used for understanding the potential of the proposed control systems. The second review [2] studied 42 articles. Exactly half of the articles were associated with muscle activity during activities of daily living, and the other half were related to a reduction in the synergy-based dimensionality of EMG. The review identified fundamental challenges and issues relevant to comprehensively understanding human hand behavior, providing more intuitive control of prostheses, and achieving realistic biomechanical models.

Within the swallowing theme, Ye Lin et al. used conditional Granger causality from surface EMG signals to analyze the directed functional coordination between different swallowing muscles in healthy and dysphagia patients swallowing water, saliva, and yoghurt [3]. The study indicated that the analysis of functional coordination supplied relevant information for evaluating motor control synergy and provided a step towards identifying new and robust biomarkers for the early detection of dysphagia. A second paper hinted at the potentialities of surface EMGs to differentiate between liquid and non-liquid swallows and to estimate fluid intake volume [4]. Nevertheless, the performance is dependent on EMG features. Further research is needed to explore this potential, including research in real-life environments. Such a novel system is intended for care homes as a step forward in reducing the rate of dehydration in older adults and improving the quality of care in healthcare settings. Using EMG to monitor diet and fluid intake is viable because there are no privacy issues compared with video- or sound-based monitoring.

Motion detection and prostheses were the topics of seven papers in the Special Issue. A proposed model classifies hand motions and transforms the classification sequence into virtual movement with the Opensim environment [5]. The model was developed using an online database. The model has turned out to be an excellent tool for the practical design of hand prostheses or human-computer based on hand motions. Hagengruber et al. proposed a new labelling approach for proportional EMG-based control that efficiently maps muscular activity to proportional control input with good accuracy [6]. Another study [7] addressed the performance issues caused by electrode shifts, feature vectors, and posture groups. They proposed adding more electrode shift training sessions. Furthermore, the Pearson correlation coefficient helped to select the feature vector. These findings might contribute to the optimization of sEMG-based pattern recognition algorithms.

Commercial prostheses devices use electrodes that can be costly for low-income settings. As a result, low-cost custom-made embroidered EMG electrodes have proven to perform similarly to conventional gel-based EMG electrodes in an online experiment [8], paving the way for low-cost myoelectric sensors. Noor et al. explored the potential use of surface EMG as a controlling mechanism for developing a home-based lower limb rehabilitative device for stroke patients [9]. Experimental data showed a moderate positive correlation between the Fugl–Meyer assessment (FMA) scores and classification accuracies. Using ablation analysis, the authors of [10] investigated the feasibility of using wearable sensors and machine learning to differentiate between standing, walking, running, and sprinting. Results demonstrated that using a reduced sensor set on the lower legs performed similarly to using all sensors. A study effectively summarized the movement detection theme by introducing a system that offers the possibility of longitudinal experiments with advanced prostheses based on a cost-effective Arduino system platform and coupled it with wearable EMG sensors [11]. Home trials with people with limb loss will be carried out in future works.

Muscle synergy was addressed in two studies. One study [12] tested the hypothesis that different subsets of muscle synergies are used in various movement and postural tasks. The approach includes extracting muscle synergies and examining whether these synergies explain each motor task. Their results may support the notion of low dimensionality in motor outputs. The low-dimensionality principle states that the central nervous system flexibly recruits fundamental muscle synergies to execute diverse human behaviors. Another study on muscle synergies [13] proposed a novel muscle synergy extraction method based on multivariate curve resolution–alternating least squares (MCR-ALS). This approach overcomes the limitation of the non-negative matrix factorization (NMF) method used for extracting non-sparse muscle synergy and provides higher repeatability and intra-subject consistency.

Benito de Pedro et al. compare the immediate effectiveness in latent myofascial trigger points between ischemic compression techniques and deep dry needling using surface EMG activity in the lateral and medial gastrocnemius of triathletes [14]. Authors recommended that deep dry needling could be advisable for triathletes who train at speeds lower than 1 m/s. Ischemic compression could be more advisable for training or competitions at speeds greater than 1.5 m/s. Lozano García et al. proposed two noninvasive indices (neuromechanical coupling and mechanical efficiency) of parasternal intercostal muscles [15]. The two indices are used for the regular assessment of patients with disordered respiratory mechanics using noninvasive wearable and wireless devices. The last assessment study [16] presented and validated a framework for continuously assessing fatigue using a system-based monitoring paradigm. Ultimately, monitoring and assessing fatigue has important implications for preventing neuromuscular injury, optimizing training loads, and guiding effective and individualized treatment strategies for rehabilitation.

Robotic exoskeleton driven by EMG has received significant attention in the literature. One study in the Special Issue [17] proposed an artificial neural network-trained adaptive controller mechanism to navigate a wheelchair-mounted upper limb robotic exoskeleton. The EMG from upper limb muscles informs users' intentions, and thus are the input sources to the system. The authors claimed that the system could be tested on people with muscular dystrophy and neurodegenerative diseases.

One article on driver distraction [18] recommended the use of haptic guidance with adaptive authority for both attentive and distracted drivers because it resulted in a lower driver workload than manual driving and fixed authority. Furthermore, haptic guidance with adaptive authority could reduce lane departure risk.

An interesting paper on statistical analysis [19] proposed a wave train electrical activity method for exploratory data analysis based on 2D and 3D area under curve (AUC) diagrams. The technique was designed to analyze, among many biosignals, EMG data collected from patients with Parkinson's disease. The aim was to treat the EMG signal as a combination of wave trains and address the generalized characteristics of the EMG signal based on local time–frequency changes in the signal. The proposed method and AUC diagrams could reveal new regularities associated with the high-accuracy diagnosis of Parkinson's disease.

3. Conclusions

This Special Issue demonstrated the breadth of applications to which EMG sensors and signals make extensive contributions. Within the area of swallowing functions, the specific domain of bolus type and fluid volume estimation from EMG has received less attention. Diet and nutrition are key determinants of healthy ageing, within which optimal hydration is important, as a balance between fluids and electrolytes is necessary if cells are to survive and function normally. However, most papers on swallowing functions have aimed to quantify the changes in swallowing biomechanics due to disorders such as dysphagia or stroke. I want to encourage researchers within the field of EMG sensors and applications to contribute to the urgent development of validated methods of assessing fluid intake to support clinical practice and research interventions, and to prevent dehydration, a costly recurrent issue in older adults. The scale of EMG applications provides strong evidence that whenever physiologically appropriate muscles are accessible, EMG sensors should be sought before any other methods, such as access to peripheral nerves.

Funding: This research received no external funding.

Institutional Review Board Statement: Not applicable.

Informed Consent Statement: Not applicable.

Data Availability Statement: Not applicable.

Conflicts of Interest: The author declares no conflict of interest.

References

1. Mereu, F.; Leone, F.; Gentile, C.; Cordella, F.; Gruppioni, E.; Zollo, L. Control Strategies and Performance Assessment of Upper-Limb TMR Prostheses: A Review. *Sensors* **2021**, *21*, 1953. [CrossRef]
2. Jarque-Bou, N.J.; Sancho-Bru, J.L.; Vergara, M. A Systematic Review of EMG Applications for the Characterisation of Forearm and Hand Muscle Activity during Activities of Daily Living: Results, Challenges, and Open Issues. *Sensors* **2021**, *21*, 3035. [CrossRef]
3. Ye-Lin, Y.; Prats-Boluda, G.; Galiano-Botella, M.; Roldan-Vasco, S.; Orozco-Duque, A.; Garcia-Casado, J. Directed Functional Coordination Analysis of Swallowing Muscles in Healthy and Dysphagic Subjects by Surface Electromyography. *Sensors* **2022**, *22*, 4513. [CrossRef] [PubMed]
4. Malvuccio, C.; Kamavuako, E.N. The Effect of EMG Features on the Classification of Swallowing Events and the Estimation of Fluid Intake Volume. *Sensors* **2022**, *22*, 3380. [CrossRef] [PubMed]
5. Amezquita-Garcia, J.; Bravo-Zanoguera, M.; Gonzalez-Navarro, F.F.; Lopez-Avitia, R.; Reyna, M.A. Applying Machine Learning to Finger Movements Using Electromyography and Visualization in Opensim. *Sensors* **2022**, *22*, 3737. [CrossRef] [PubMed]
6. Hagengruber, A.; Leipscher, U.; Eskofier, B.M.; Vogel, J. A New Labeling Approach for Proportional Electromyographic Control. *Sensors* **2022**, *22*, 1368. [CrossRef]
7. Kim, J.; Koo, B.; Nam, Y.; Kim, Y. sEMG-Based Hand Posture Recognition Considering Electrode Shift, Feature Vectors, and Posture Groups. *Sensors* **2021**, *21*, 7681. [CrossRef]
8. Kamavuako, E.N.; Brown, M.; Bao, X.; Chihi, I.; Pitou, S.; Howard, M. Affordable Embroidered EMG Electrodes for Myoelectric Control of Prostheses: A Pilot Study. *Sensors* **2021**, *21*, 5245. [CrossRef]
9. Noor, A.; Waris, A.; Gilani, S.O.; Kashif, A.S.; Jochumsen, M.; Iqbal, J.; Niazi, I.K. Decoding of Ankle Joint Movements in Stroke Patients Using Surface Electromyography. *Sensors* **2021**, *21*, 1575. [CrossRef] [PubMed]
10. Gonzalez, S.; Stegall, P.; Edwards, H.; Stirling, L.; Siu, H.C. Ablation Analysis to Select Wearable Sensors for Classifying Standing, Walking, and Running. *Sensors* **2021**, *21*, 194. [CrossRef] [PubMed]
11. Wu, H.; Dyson, M.; Nazarpour, K. Arduino-Based Myoelectric Control: Towards Longitudinal Study of Prosthesis Use. *Sensors* **2021**, *21*, 763. [CrossRef] [PubMed]
12. Saito, H.; Yokoyama, H.; Sasaki, A.; Kato, T.; Nakazawa, K. Flexible Recruitments of Fundamental Muscle Synergies in the Trunk and Lower Limbs for Highly Variable Movements and Postures. *Sensors* **2021**, *21*, 6186. [CrossRef]
13. Ma, Y.; Shi, C.; Xu, J.; Ye, S.; Zhou, H.; Zuo, G. A Novel Muscle Synergy Extraction Method Used for Motor Function Evaluation of Stroke Patients: A Pilot Study. *Sensors* **2021**, *21*, 3833. [CrossRef] [PubMed]
14. Benito-de-Pedro, M.; Calvo-Lobo, C.; López-López, D.; Benito-de-Pedro, A.I.; Romero-Morales, C.; San-Antolín, M.; Vicente-Campos, D.; Rodríguez-Sanz, D. Electromyographic Assessment of the Efficacy of Deep Dry Needling versus the Ischemic Compression Technique in Gastrocnemius of Medium-Distance Triathletes. *Sensors* **2021**, *21*, 2906. [CrossRef] [PubMed]
15. Lozano-García, M.; Estrada-Petrocelli, L.; Torres, A.; Rafferty, G.F.; Moxham, J.; Jolley, C.J.; Jané, R. Noninvasive Assessment of Neuromechanical Coupling and Mechanical Efficiency of Parasternal Intercostal Muscle during Inspiratory Threshold Loading. *Sensors* **2021**, *21*, 1781. [CrossRef] [PubMed]
16. Madden, K.E.; Djurdjanovic, D.; Deshpande, A.D. Using a System-Based Monitoring Paradigm to Assess Fatigue during Submaximal Static Exercise of the Elbow Extensor Muscles. *Sensors* **2021**, *21*, 1024. [CrossRef] [PubMed]
17. Schabron, B.; Desai, J.; Yihun, Y. Wheelchair-Mounted Upper Limb Robotic Exoskeleton with Adaptive Controller for Activities of Daily Living. *Sensors* **2021**, *21*, 5738. [CrossRef] [PubMed]
18. Wang, Z.; Suga, S.; Nacpil, E.J.C.; Yang, B.; Nakano, K. Effect of Fixed and sEMG-Based Adaptive Shared Steering Control on Distracted Driver Behavior. *Sensors* **2021**, *21*, 7691. [CrossRef] [PubMed]
19. Sushkova, O.S.; Morozov, A.A.; Gabova, A.V.; Karabanov, A.V.; Illarioshkin, S.N. A Statistical Method for Exploratory Data Analysis Based on 2D and 3D Area under Curve Diagrams: Parkinson's Disease Investigation. *Sensors* **2021**, *21*, 4700. [CrossRef] [PubMed]

sensors

MDPI

Article

The Effect of EMG Features on the Classification of Swallowing Events and the Estimation of Fluid Intake Volume

Carlotta Malvuccio [1] and Ernest N. Kamavuako [1,2,*]

[1] Department of Engineering, King's College London, London WC2R 2LS, UK; carlotta.1.malvuccio@kcl.ac.uk
[2] Faculté de Médecine, Université de Kindu, Site de Lwama II, Kindu, Maniema, Congo
* Correspondence: ernest.kamavuako@kcl.ac.uk; Tel.: +44-2-078-488-666

Abstract: Nowadays, society is experiencing an increase in the number of adults aged 65 and over, and it is projected that the older adult population will triple in the coming decades. As older adults are prone to becoming dehydrated, which can significantly impact healthcare costs and staff, it is necessary to advance healthcare technologies to cater to such needs. However, there has not been an extensive research effort to implement a device that can autonomously track fluid intake. In particular, the ability of surface electromyographic sensors (sEMG) to monitor fluid intake has not been investigated in depth. Our previous study demonstrated a reasonable classification and estimation ability of sEMG using four features. This study aimed to examine if classification and estimation could be potentiated by combining an optimal subset of features from a library of forty-six time and frequency-domain features extracted from the data recorded using eleven subjects. Results demonstrated a classification accuracy of 95.94 ± 2.76% and an f-score of 94.93 ± 3.51% in differentiating between liquid swallows from non-liquid swallowing events using five features only, and a volume estimation RMSE of 2.80 ± 1.22 mL per sip and an average estimation error of 15.43 ± 8.64% using two features only. These results are encouraging and prove that sEMG could be a potential candidate for monitoring fluid intake.

Keywords: surface electromyography; swallowing events; geriatrics; hydration; fluid intake

Citation: Malvuccio, C.; Kamavuako, E.N. The Effect of EMG Features on the Classification of Swallowing Events and the Estimation of Fluid Intake Volume. *Sensors* **2022**, 22, 3380. https://doi.org/10.3390/s22093380

Academic Editor: Andrea Facchinetti

Received: 20 March 2022
Accepted: 26 April 2022
Published: 28 April 2022

1. Introduction

The indisputable fact that our society faces nowadays is the increase in the population aged 65 and over. According to the United Nations [1], by 2050, one in six people will be aged over 65. The number of persons aged 80 years and over is projected to triple. This increase in the number of older adults is primarily due to societal improvements in lifestyle and advances in technology and healthcare. Nevertheless, it will soon pose new challenges if action is not taken to advance the current monitoring technology, as older adults necessitate more care and medications. One challenge in our healthcare system is dehydration, which is a recurrent issue in older adults [2,3]. Dehydration is mainly provoked by a diminished thirst sensation with ageing and mobility or memory impairments [4,5]. Moreover, light hydration in older adults increases the chances of falls and hence the possibility of bone fractures, other than potentially leading to drug intoxication, confusion, and delirium in severe cases [6–8]. Furthermore, it has been recently suggested that chronic suboptimal hydration might be a risk factor for increased mortality in COVID-19 patients on top of risk factors such as male gender and older age [9].

There are no gold standards for identifying dehydration [10,11]. Moreover, signs of dehydration, such as skin turgor, are often confounded by the age of the patients, and assessing dehydration is made difficult by a shortage of staff, absence of appropriate laboratory analyses and memory impairment of senile patients [12,13]. Thus, the best option is to ensure that older adults intake the recommended daily fluid volume [14,15]. To keep track of service users' fluid intake, nurses must fill manually fluid monitoring charts, contributing to a substantial increase in their workload. This method is inaccurate and prone to misreporting, as nurses mainly rely on patient information and relatives rather than direct observation of the patient [16]. Thus, maintaining adequate oral hydration for older residents is an ongoing challenge for staff members, bringing the prevalence of dehydration in nursing homes residents at 37% in the United Kingdom (UK) alone [17], adding strain to healthcare costs. For example, Xiao et al. [18] reported that the average length of stay for dehydration in the Healthcare Cost and Utilisation Project (HCUP) data was 4.6 days, amounting to a total hospitalisation charge per person of $7442. Therefore, there is an urgent need to develop accurate methods to monitor fluid intake in older adults to ease the burden on healthcare staff and healthcare finances.

Only a handful of studies in the literature have tried to develop fluid intake monitoring systems using signals harvested directly from the subjects. Most of these used microphones only [19,20], and data collection was performed on a very limited number of participants. Amft and Tröster [21] combined surface electromyography (sEMG) with microphones to perform classification between solid and liquid foods. The reported classification accuracy was less than 84%. Apart from this paper, to the best of our knowledge, there is no other record in the literature of sEMG used with the specific scope of quantifying fluid intake. Indeed, most of the documentation found in the literature refers to sEMG being used to study dysphagia, although some of these studies hinted at the possibility that certain neck sEMG features such as peak amplitudes change according to the swallowed volume [22,23]. Therefore, we conducted a preliminary study on swallowing events classification and volume estimation using four predefined time-domain features. The aim was to assess if neck sEMG could distinguish between saliva and liquid swallows and then if it was possible to perform volume estimation. The results of our published study [24] showed a mean classification accuracy between saliva and liquid swallows of 86.69 $\pm$ 5.52% using the k-Nearest Neighbour classifier. The average Root Mean Square Error was 2.01 $\pm$ 1.39 mL per swallowed sip using an Artificial Neural Network (ANN) with two hidden layers, each with fifteen neurons.

While these results were encouraging, it was necessary to investigate further whether classification and volume estimation could be improved using a more comprehensive set of features and studying what could be the ideal combination to obtain the best performance possible. Again, the need to carry out this investigation stemmed from the fact that there is no record in the literature about an extensive analysis of neck sEMG features. Hence, we extracted forty-six single features, of which thirteen were frequency-domain, using the list of features detailed in the paper from Phyniomark et al. [25], and we applied these on the sEMG of neck muscles involved during the swallowing events.

2. Materials and Methods

The Research Ethics Committee of King's College London (LRS-18/19-10877) approved this study. We recruited eleven healthy participants (3 F, 8 M, age range from 20 to 67 years) with no known underlying medical conditions. All participants agreed voluntarily to participate in the experiment, and written informed consent was obtained.

Participants sat comfortably, and the skin around the neck was prepared using alcohol wipes. Two Delsys Trigno sensors (Delsys Incorporated, Natick, MA, USA) were set as sEMG and placed on the belly of the sternohyoid muscles (infrahyoid group), as shown in Figure 1. The sternohyoid muscles were chosen, as these are the most superficial muscles of the infrahyoid group. The correct anatomical placement of the sensors was identified via palpation of the neck muscles of the participants, as the distancing and position were highly dependent on the sex and body conformation of participants. We collected data collection using two pieces of equipment. A TREE KHR502 electronic scale (resolution of 0.01 g and a capacity of 500 g) was used to measure the volume of water ingested by the subjects during each task. The Delsys Trigno wireless EMG system (gain: 42 V; bandwidth: 20–450 Hz, sampling rate: 1 kHz) collected the sEMG signals produced during the swallowing process. The length of each recording was manually set as 10 s.

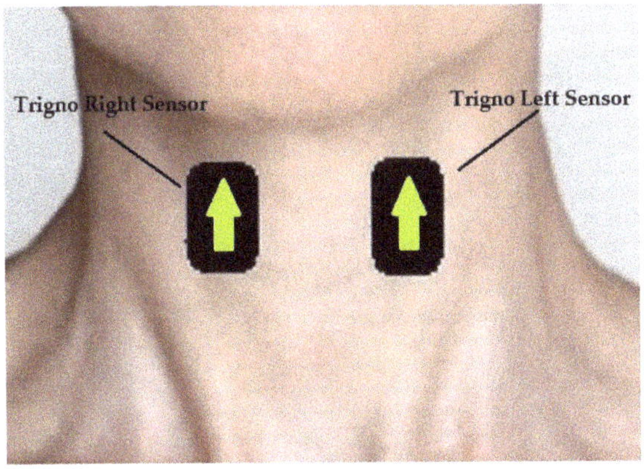

Figure 1. The figure shows the anatomical position on which sensors were placed. The arrows indicate the use of Delsys trigno system.

Once sensors were placed on the correct anatomical positions, subjects were asked to perform seven tasks, with six including water at room temperature. The subject was instructed to swallow for each task following a cue. The time between cues was at the participant's discretion, and the recording was initiated two seconds prior to each cue. The first task consisted of performing one saliva swallow, which was repeated five times. Tasks II, III, and IV consisted of ingesting water from the administered container five times. For each of these three tasks, the container used by the subject was randomly changed in a cup (Task II), bottle (Task III) and straw (Task IV), respectively, and subjects were instructed to sip as they would normally do in real-life scenarios. The volume of each sip was calculated using the laboratory scale. The selected fluid container was filled and placed on the scale, and we noted the initial weight of the container. We cued each subject to take a single sip as typically as possible and place the container back on the scale. We then subtracted the final weight from the initial weight to note the swallowed volume. Thus, we did not impose specific volumes, and the only condition was to take a single sip. Task V consisted of the participant being administered a total liquid volume corresponding to the highest volume ingested in Task II plus 5 mL and was performed once. This task had the scope of improving linear regression and volume estimation. Adding 5 mL to the highest ingested volume would ideally produce an observation corresponding to the maximum swallowing capacity of the subject.

The collected EMG signals were processed using the MATLAB R2020b version. First, a Kaiser window FIR bandpass filter with a bandpass frequency range of 20 to 400 Hz, transition band steepness of 0.85, and stopband attenuation of 60 dB was applied to the left and right sternohyoid signals. In order to perform burst extraction and identify the burst region, the signal was smoothed using a moving RMS envelope with a window length of 1000 ms. These two steps are illustrated by the graphs shown in Figure 2. The burst region was identified by locating the peak with the highest value of the smoothed signal and including the 750 data points (0.75 s) to the left and right of the highest peak, thus resulting in a total burst duration of 1500 ms. Once the burst location was identified, the burst was extracted from the raw bandpass filtered signal. Finally, baseline noise was extracted by taking the last 1.5 s of the recorded signal that contained no swallowing information based on throughout visual inspection of the signals. A total of forty-six single features, thirty-three time-domain and thirteen frequency-domain features were computed from each 1500 ms window (burst and the baseline noise). The features used were selected based on the paper from Phinyomark et al. [25] and presented in Table 1.

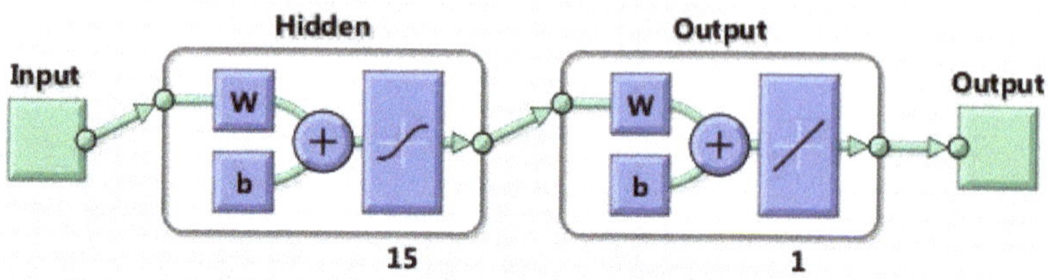

Figure 2. The figure shows the diagram of the ANN model used in this study.

Table 1. This table presents the forty-six single features included in this study. The features are presented in the order that these were computed.

Feature Full Name	Abbreviation	Parameters
Integrated EMG	IEMG	-
Mean Absolute Value	MAV	-
Mean Absolute Value 1	MAV 1	-
Mean Absolute Value 2	MAV 2	-
Simple Squared Integral	SSI	-
Variance of EMG	VAR	-
Root Mean Square	RMS	-
Second V-Order	V2	v = 2
Third V-Order	V3	v = 3
Log Detector	LOG	-
Waveform Length	WL	-
Average Amplitude Change	AAC	-
Difference Absolute Standard Deviation Value	DASDV	-
Maximum Fractal Length	MFL	-
Myopulse Percentage Rate	MYOP	threshold = 5.5 μ
Willinson Amplitude	WAMP	threshold = 0.3 × σ (noise)
Modified Mean Absolute Value	MMAV	-
Zero Crossing	ZC	threshold = 0.3 × σ (noise)
Slope Sign Change	SSC	-
Abs. val. of Third Temporal Moment	TM3	order = 3
Abs. val. of Fourth Temporal Moment	TM4	order = 4
Abs. val. of Fifth Temporal Moment	TM5	order = 5
Abs value of the Summation of Square Root	ASS	-
Mean Value of Square Root	MSR	-
Absolute value of the Summation of the expth root of the given signal and its Mean	ASM	-
Kurtosis	Kurt	-
Skewness	Skew	-
Amplitude of the First burst	AFB	-

Table 1. *Cont.*

Feature Full Name	Abbreviation	Parameters
Mean Power	MNP	-
Total Power	TTP	-
Median Frequency	MDF	-
Mean Frequency	MNF	-
Peak Frequency	PKF	-
First Spectral Moment	SM1	order = 1
Second Spectral Moment	SM2	order = 2
Third Spectral Moment	SM3	order = 3
Frequency Ratio	FR	lc < MNF; hc > MNF
Mean Power Density	MPD	-
Power Spectrum Deformation	PSDd	-
Variance of Central Frequency	VCF	-
Higuchi Fractal Dimension	HFD	k = 128
Sample Entropy	SaEn	m = 2, r = 0.2 σ
Approximate Entropy	ApEn	m = 2, r = 0.2 σ
Maximum to Minimum Drop in Power Density Ratio	dPDR	-
Power Spectrum Ratio	PSR	n = 20
Area Under the Curve	AUC	-

Features were calculated per 1500 ms window to represent a single sip. As this study aimed to verify which features combination performs best for our scope, stepwise forward selection was used. Each problem (classification or estimation) was performed using single features. The feature with the highest performance metric was retained as the best feature and then combined with each of the remaining 45 features and tested in pairs. The pair with the highest performance metric was then combined with each of the remaining 44 features, and then, the procedure was repeated until the performance parameters reached a plateau or did not improve further. A one-way analysis of variance (ANOVA) test was used to test if the resulting metrics significantly differed with an increasing number of features.

The classification was modelled per subject as a two-class problem, with one class containing the baseline noise and saliva swallows' data together (including 11 noise observations and five saliva bursts) against the class containing liquid swallows (a total of 16 observations). The classifiers employed were Linear Discriminant Analysis (LDA) and k-Nearest Neighbour with k = 1 (KNN). To estimate the performance of the classifiers, the Leave-One-Out Cross-Validation (LOOCV) method was used with the following performance metrics: Accuracy, Sensitivity, Specificity, Precision, and F-score, as presented in Equations (1)–(5). Accuracy quantifies the ability of the model to assign the observation to the correct class. Sensitivity measures the ability of the model to predict actual liquid swallows. Specificity measures the power of the model to correctly predict non-liquid swallowing events. Precision aims to quantify the proportion of liquid swallows classified as liquid swallows. Then, the F-score was calculated to assess if the model trade-off between precision and sensitivity was acceptable.

$$\text{Accuracy} = \frac{\text{true liquid swallows} + \text{true non liquid events}}{\text{number of obs}} \tag{1}$$

$$\text{Sensitivity} = \frac{\text{true liquid swallows}}{\text{liquid swallows}} \tag{2}$$

$$\text{Specificity} = \frac{\text{true non liquid events}}{\text{non liquid events}} \tag{3}$$

$$\text{Precision} = \frac{\text{true liquid swallows}}{\text{true liquid swallows} + \text{false liquid swallows}} \tag{4}$$

$$\text{F-Score} = \frac{2(\text{precision} \times \text{sensitivity})}{\text{precision} + \text{sensitivity}} \tag{5}$$

Before proceeding with more sophisticated methods to quantify fluid intake, first, we wanted to verify if the mean sip volume for each subject could be used to predict fluid intake within each subject. The mean and standard deviation of the sips taken by each subject were calculated, and the error was computed as the difference between the mean value as the predicted intake and actual sip volumes.

Secondly, Linear Regression (LR) and a shallow Artificial Neural Network (ANN) were used to perform volume estimation analysis. The performance parameters used for both methods were Root Mean Square Error (RMSE) and mean estimation error as a volume percentage calculated using Equation (6):

$$\text{mean EE } (\%) = \text{mean}\left(\left|\frac{\text{predicted}_\text{i} - \text{actual}_\text{i}}{\text{actual}_\text{i}}\right| \times 100\right) \tag{6}$$

The linear regression model proposed in this study used the recorded sip volumes as response variables and the extracted features as predictor variables. The model specification was set to linear, meaning that the model contained an intercept and a linear term for each predictor, and it used ordinary least squares as a fitting method. Cross-validation was set to 5 k-fold. The ANN model consisted of one layer of 15 neurons with a hyperbolic tangent sigmoid transfer function and a linear transfer function for the output layer, as shown in Figure 2. Data division was performed at random, and the Levenberg–Marquardt algorithm was used for training. As previously described, the optimum feature subset was selected using the forward stepwise selection method.

3. Results

3.1. Classification

The optimal classification performance was given by the KNN classifier using five features, namely: Integrated EMG (IEMG), Sign Slope Change (SSC), Average Amplitude Change (AAC), Area Under the Curve (AUC) and Variance of Central Frequency (VCF). However, results showed that the parameters of LDA classifier, when using five features, were declining compared to the use of four features (IEMG, SSC, AAC and AUC). These results are illustrated in Table 2. ANOVA tests demonstrated that the difference between four-feature and five-feature classification was statistically not significant both for LDA (p = 1) and KNN (p = 0.44). Furthermore, the ANOVA test demonstrated that while for LDA, there is only a significant difference between single-feature and four-feature classifier (p = 0.047), this changes for KNN. Indeed, the results showed that classifiers using one and two features are significantly different from those using three or more features (p < 0.05). Classifiers using three, four, and five features instead were not statistically different from each other (p > 0.05).

Table 2. The table illustrates the resulting performance parameters for four features (IEMG, SSC, AAC and AUC) in the first row and five features (IEMG, SSC, AAC, AUC and VCF) in the second row.

LDA					KNN				
Accuracy	Sensitivity	Specificity	Precision	F Score	Accuracy	Sensitivity	Specificity	Precision	F Score
95.80 ± 4.62	96.02 ± 6.42	95.61 ± 4.69	93.64 ± 7.14	94.71 ± 5.92	94.84 ± 4.32	94.89 ± 6.74	94.72 ± 3.54	92.30 ± 5.52	93.52 ± 5.69
95.51 ± 3.86	94.89 ± 5.46	95.96 ± 4.56	94.18 ± 6.67	94.39 ± 4.86	95.94 ± 2.76	96.02 ± 5.78	95.84 ± 3.62	94.17 ± 4.80	94.93 ± 3.51

3.2. Volume Estimation

The method of using the mean sip as the estimation volume resulted in large estimation errors, as shown in Table 3. When using LR for volume estimation, the best performance was given by a single feature, namely the Absolute value of the Summation of the expth root of the given signal and its Mean (ASM), with an RMSE of 3.90 $\pm$ 1.58 mL and an average estimation error of 24.63 $\pm$ 7.03% of the actual swallowed sip volume, as shown in Table 4. Concerning the ANN, the optimal volume estimation performance was given by SSC and Mean Power Density (MPD) combined with an RMSE of 2.80 $\pm$ 1.22 mL and an average estimation error of 15.43 $\pm$ 8.64% of the actual swallowed volume. Results for the ANN are illustrated in Table 5.

Table 3. The table shows the mean and the standard deviation of the sip volumes ingested by each subject. The last column shows the estimation error when using the mean as the predicted swallowed volume.

Subject	Mean (mL)	SD (mL)	Error (%)
F20	23.83	5.13	15.35
F22	19.42	5.17	22.94
F28	8.73	3.05	29.05
M20	12.19	4.33	34.10
M21	11.40	3.71	30.04
M211	13.67	3.15	17.34
M25	18.72	6.14	28.50
M251	7.14	2.96	40.73
M27	12.73	3.98	28.01
M29	15.13	5.99	32.71
M67	21.33	8.75	43.67
Across All	14.93	5.29	29.31

Table 4. The table shows how the RMSE and the average estimation error change for LR as features are added. As performance did not improve with the addition of the second feature, it was not deemed necessary to proceed with the addition of further features.

Features	RMSE (mL)	Average Estimation Error (%)
ASM	3.90 $\pm$ 1.58	24.63 $\pm$ 7.03
ASM, TM4	3.98 $\pm$ 1.60	25.11 $\pm$ 8.07

Table 5. The table shows how the RMSE and the average estimation error change for the ANN as features are added. Performance deteriorated with the addition of a third feature; thus, it was not deemed necessary to proceed with further feature addition.

Features	RMSE (mL)	Average Estimation Error (%)
SSC	3.84 $\pm$ 2.52	19.35 $\pm$ 11.60
SSC, MPD	2.80 $\pm$ 1.22	15.43 $\pm$ 8.64
SSC, MPD, VAR	3.45 $\pm$ 1.71	16.80 $\pm$ 6.76

4. Discussion

The classification performances obtained in this study seemingly hint at the suitability of using surface EMGs to classify between liquid swallows versus non-liquid swallowing events. By analysing and determining an optimal set of features, classifier performance seemed to improve compared to the results obtained in our previous study [24] when a set of predefined features was used. Furthermore, in our previous study, signals coming from both the digastric and sternohyoid muscles were used, whereas in this study, only the sternohyoid signals were used. The fact that the performance has not deteriorated if classification is performed using data harvested from the left and right sternohyoid muscles alone hints that there is not a necessity to include digastric muscles in future studies. As the latter are submental muscles, the inclusion of surface EMGs in that anatomical region might create discomfort, even if minor, in the movements performed by subjects and even more so in hospitalised patients. Furthermore, digastric muscles are more sensitive to other movements, such as chewing and the swallowing preparation phase. These muscles are recruited to facilitate the ingress of food and fluids in the mouth, thus complicating the extraction of bursts related to fluid swallowing events. In terms of the optimal number of features, as summarised in Table 1, a KNN classifier using five features produces the best performance with an F-score of 94.93. However, statistical analysis demonstrated no significant difference between the performance of four features compared to five features. Therefore, while five features provided the highest accuracy, perhaps the choice of using four features could improve the computational time and cost, especially in continuous online tasks—an assumption that should be validated by potential future studies using larger cohorts.

However, while these results seem promising and demonstrated an improvement on classification performance compared to previous studies [21,24], and they hint at the potential of surface EMGs to differentiate between liquid versus non-liquid swallows, these results need to be validated further. Further validation of these results could be performed by recruiting a larger number of subjects spanning a wide age range (also to verify if age has an influence on performance), as the results presented in this research are generated by a small number of participants; thus, it is necessary to observe if the performance will remain unchanged using a larger pool of participants; by harvesting a larger number of observations compared to the numbers collected in this study; and there is also the need to include in the non-liquid swallowing class more tasks such as coughing, talking, chewing and swallowing food of different viscosities to observe with significant certainty if surface EMGs possess the ability to distinguish all non-liquid swallowing events from liquid swallowing ones.

The evidence gathered in this study and our previous one suggests that surface EMGs also have the potential to estimate fluid intake, thus being an alternative to the use of microphones for fluid estimation purposes [19,20]. Furthermore, if this potential will be corroborated in future studies, this will also confirm the observations of previous studies about the possibility of certain features derived from the neck surface EMGs to differentiate between swallows of different volumes [22,23]. The first method employed in this paper, which used the mean sip volume as an estimator, did not return satisfying results. The estimation error that resulted was larger than that of more sophisticated estimation methods such as using an ANN. This is mainly due to the fact that, as exposed in our previous study, the sip volume is influenced by the shape of the container when this is significantly different in size [24]. Indeed, sips ingested using the straw present a smaller volume compared to the ones consumed from bigger containers such as cups and bottles. Furthermore, factors such as the temperature of the liquid and composition (as an example, carbonated drinks), which were not considered in this study, could influence the sip volume, which might vary from subject to subject, rendering this method quite susceptible to estimation errors. Thus, these factors (fluid temperature and fluid composition such as carbonated drinks), which in real-life scenarios can be commonplace, would render this estimation method ineffective and inaccurate.

Using Linear Regression and ANN improved performance compared with the mean sip estimator with an RMSE of 3.90 ± 1.58 mL and an average estimation error of 24.63 ± 7.03% of the actual swallowed sip volume for the LR. ANN had an RMSE of 2.80 ± 1.22 mL and an average estimation error of 15.43 ± 8.64% for the ANN. As the two means are statistically different ($p < 0.05$), it could be concluded that the best volume estimation performance is given by the ANN. Furthermore, it is worth noting that the ANN developed in this study resulted in a similar performance compared to the two-layer, feed-forward network proposed in our previous study (average RMSE of 2.01 mL), which also used four features instead of two. Hence, this demonstrates that performance can be improved by selecting the appropriate features to combine while also using a network with a lower number of hidden layers, thus reducing the computational costs. We recommend further research to validate the hypothesis presented in this paper with a larger cohort and a more significant number of observations.

However, it is fundamental to reiterate that these findings need to be validated in future studies using a larger cohort that should span across age groups. Indeed, due to the restricted number of participants and limited age range, other factors such as the influence of age on mean sip volume could not be observed. Furthermore, while the classification results shown in this study were promising, there is a need to verify if the performance will remain optimal when including different liquid viscosities and other non-liquid events, such as talking and the ingestion of solid food.

5. Conclusions

The results obtained in this study hint at the potential of surface EMGs not only to differentiate between liquid and non-liquid swallows but also to estimate fluid intake using an optimum set of features. While further research is needed to cater for the limitations presented in this study, our findings could represent a way forward to produce a non-invasive device that could prevent dehydration in older adults and improve the quality of care in healthcare settings.

Author Contributions: Conceptualisation: C.M. and E.N.K.; Machine learning training: C.M.; Data analysis: C.M. and E.N.K.; Critical review of the manuscript: C.M. and E.N.K. All authors have read and agreed to the published version of the manuscript.

Funding: This research was funded by the EPSRC Council UK, grant number EP/R513064/1.

Institutional Review Board Statement: The study was approved by the Research Ethics Committee of King's College London (LRS-18/19-10877) on the 5 July 2019.

Informed Consent Statement: Informed consent was obtained from all subjects involved in the study.

Data Availability Statement: Raw data are available for sharing if requested.

Conflicts of Interest: The authors declare no conflict of interest. The funders had no role in the design of the study; in the collection, analyses, or interpretation of data; in the writing of the manuscript, or in the decision to publish the results.

References

1. United Nations, Department of Economics and Social Affairs. 2019. Available online: https://www.un.org/development/desa/publications/world-population-prospects-2019-highlights.html (accessed on 20 November 2019).
2. Cook, G.; Hodgson, P.; Hope, C.; Thompson, J.; Shaw, L. Hydration practices in residential and nursing care homes for older people. *J. Clin. Nurs.* **2018**, *28*, 1205–1215. [CrossRef] [PubMed]
3. Lea, E.J.; Goldberg, L.R.; Price, A.D.; Tierney, L.T.; McInerney, F. Staff Awareness of Food and Fluid Care Needs for Older People with Dementia in Residential Care: A Qualitative Study. *J. Clin. Nurs.* **2017**, *26*, 5169–5178. [CrossRef] [PubMed]
4. Hooper, L.; Bunn, D.; Jimoh, F.O.; Fairweather-Tait, S.J. Water-loss Dehydration and Ageing. *Mech. Ageing Dev.* **2014**, *136–137*, 50–58. [CrossRef] [PubMed]
5. Botigué, T.; Masot, O.; Miranda, J.; Nuin, C.; Viladrosa, M.; Lavedán, A.; Zwakhalen, S. Prevalence and Risk Factors Associated with Low Fluid Intake in Institutionalized Older Residents. *J. Am. Med. Dir. Assoc.* **2018**, *20*, 317–322. [CrossRef] [PubMed]
6. Archer, S.J.; Barna, A.E.; Holper, R.M.; McNally Forsith, D.; Ellenbecker, S.M.; Smith, L.K.; Clobes, J.E.; Meiers, S.J.; Malone, C.; Lapid, M.I. Associations Among Fluid Intake, Cognitive Function, and Length of Stay in Psychogeriatric Inpatients. *Perspect. Psychiatr. Care* **2015**, *51*, 52–56. [CrossRef] [PubMed]
7. Konings, F.J.C.M.; Mathijssen, J.J.P.; Schellingerhout, J.M.; Kroesbergen, I.H.T.; Goede de, J.; Goor de, I.A.M. Prevention of Dehydration in Independently Living Elderly People at Risk: A Study Protocol of a Randomized Controlled Trial. *Int. J. Prev.* **2015**, *6*, 103. [CrossRef] [PubMed]
8. Lima Ribeiro, S.M.; Morley, J.E. Dehydration is difficult to detect and prevent in nursing homes. *J. Am. Med. Dir. Assoc.* **2015**, *16*, 175–176. [CrossRef] [PubMed]
9. Stookey, J.D.; Allu, P.; Chabas, D.; Pearce, D.; Lang, F. Hypotheses about sub-optimal hydration in the weeks before coronavirus disease (COVID-19) as a risk factor for dying from COVID-19. *Med. Hypotheses* **2020**, *144*, 110237. [CrossRef] [PubMed]

10. Stookey, J.D. Analysis of 2009–2012 Nutrition Health and Examination Survey (NHANES) Data to Estimate the Median Water Intake Associated with Meeting Hydration Criteria for Individuals Aged 12–80 in the US Population. *Nutrients* **2019**, *11*, 657. [CrossRef] [PubMed]
11. Masot, O.; Lavedán, A.; Nuin, C.; Escobar-Bravo, M.A.; Miranda, J.; Botigué, T. Risk Factors Associated with Dehydration in Older People Living in Nursing Homes: Scoping Review. *Int. J. Nurs. Stud.* **2018**, *82*, 90–98. [CrossRef] [PubMed]
12. Thomas, D.R.; Cote, T.R.; Lawhorne, L.; Levenson, S.A.; Rubenstein, L.Z.; Smith, D.A.; Stefanacci, R.G.; Tangalos, E.G.; Morley, J.E.; Dehydration Council. Understanding Clinical Dehydration and Its Treatment. *Rev. JAMDA* **2018**, *9*, 292–301. [CrossRef] [PubMed]
13. Johnson, P.; Hahn, R.G. Signs of Dehydration in Nursing Home Residents. *JAMDA* **2018**, *19*, 1124–1128. [CrossRef] [PubMed]
14. Bennett, J.A. Dehydration: Hazards and Benefits. *Geriatr. Nurs.* **2000**, *21*, 84–88. [CrossRef] [PubMed]
15. Weinberg, A.; Minaker, K.L.; Council on Scientific Affairs; American Medical Association. Dehydration—Evaluation and Management in Older Adults. *JAMA* **1995**, *274*, 1552–1556. [CrossRef] [PubMed]
16. Godfrey, H.; Cloete, J.; Dymond, E.; Long, A. An Exploration of the Hydration Care of Older People: A Qualitative Study. *Int. J. Nurs. Stud.* **2012**, *49*, 1200–1211. [CrossRef] [PubMed]
17. El-Sharkawy, A.M.; Watson, P.; Neal, K.R.; Ljungqvist, O.; Maughan, R.J.; Sahota, O.; Lobo, D.N. Hydration and outcome in older patients admitted to hospital (The HOOP prospective cohort study). *Age Ageing* **2015**, *44*, 943–947. [CrossRef] [PubMed]
18. Xiao, H.; Barber, J.; Campbell, E.S. Economic Burden of Dehydration among Hospitalized Elderly Patients. *Am. J. Health Syst. Pharm.* **2004**, *61*, 2534–2540. [CrossRef] [PubMed]
19. Nakafuji, H.; Imura, M.; Uranishi, Y.; Yoshimoto, S.; Oshiro, O. *Estimation of Amount of Swallowed Water by Analysis of Swallowing Sounds*; Kyoto University Hospital: Kyoto, Japan, 2015; pp. 1–6.
20. Kobayashi, Y.; Mineno, H. Fluid intake recognition for nursing care support by leveraging swallowing sound. In Proceedings of the 2014 IEEE 3rd Global Conference on Consumer Electronics (GCCE), Tokyo, Japan, 7–10 October 2014.
21. Amft, O.; Troster, G. Methods for Detection and Classification of Normal Swallowing from Muscle Activation and Sound. In Proceedings of the Pervasive Health Conference and Workshops, Innsbruck, Austria, 29 November–1 December 2006.
22. Zhu, M.; Yu, B.; Yang, W.; Jiang, Y.; Lu, L.; Huang, Z.; Chen, S.; Li, G. Evaluation of normal swallowing functions by using dynamic high density surface electromyography maps. *BioMed Eng. OnLine* **2017**, *16*, 133. [CrossRef] [PubMed]
23. Miyaoka, Y.; Ashida, I.; Kawakami, S.; Tamaki, Y.; Miyaoka, S. Activity patterns of the suprahyoid muscles during swallowing of different volumes. *J. Oral Rehabil.* **2010**, *37*, 575–582. [CrossRef] [PubMed]
24. Malvuccio, C.; Kamavuako, E.N. Detection of Swallowing Events and Fluid Intake Volume Estimation from Surface Electromyography Signals. In Proceedings of the 2020 IEEE-EMBS Conference on Biomedical Engineering and Sciences (IECBES), Langkawi Island, Malaysia, 1–3 March 2021.
25. Phinyomark, A.; Quaine, F.; Charbonnier, S.; Serviere, C.; Tarpin-Bernard, F.; Laurillau, Y. EMG feature evaluation for improving myoelectric pattern recognition robustness. *Expert Syst. Appl.* **2013**, *40*, 4832–4840. [CrossRef]

sensors

MDPI

Article

Directed Functional Coordination Analysis of Swallowing Muscles in Healthy and Dysphagic Subjects by Surface Electromyography

Yiyao Ye-Lin [1], Gema Prats-Boluda [1,*], Marina Galiano-Botella [1], Sebastian Roldan-Vasco [2], Andres Orozco-Duque [3] and Javier Garcia-Casado [1]

1 Centro de Investigación e Innovación en Bioingeniería, Universitat Politècnica de València, 46022 Valencia, Spain; yiye@ci2b.upv.es (Y.Y.-L.); marinagaliano27@gmail.com (M.G.-B.); jgarciac@ci2b.upv.es (J.G.-C.)

2 Grupo de Investigación en Materiales Avanzados y Energía, Instituto Tecnológico Metropolitano, Medellin 050034, Colombia; sebastianroldan@itm.edu.co

3 Grupo de Investigación e Innovación Biomédica, Instituto Tecnológico Metropolitano, Medellin 050034, Colombia; andresorozco@itm.edu.co

* Correspondence: gprats@ci2b.upv.es

Abstract: Swallowing is a complex sequence of highly regulated and coordinated skeletal and smooth muscle activity. Previous studies have attempted to determine the temporal relationship between the muscles to establish the activation sequence pattern, assessing functional muscle coordination with cross-correlation or coherence, which is seriously impaired by volume conduction. In the present work, we used conditional Granger causality from surface electromyography signals to analyse the directed functional coordination between different swallowing muscles in both healthy and dysphagic subjects ingesting saliva, water, and yoghurt boluses. In healthy individuals, both bilateral and ipsilateral muscles showed higher coupling strength than contralateral muscles. We also found a dominant downward direction in ipsilateral supra and infrahyoid muscles. In dysphagic subjects, we found a significantly higher right-to-left infrahyoid, right ipsilateral infra-to-suprahyoid, and left ipsilateral supra-to-infrahyoid interactions, in addition to significant differences in the left ipsilateral muscles between bolus types. Our results suggest that the functional coordination analysis of swallowing muscles contains relevant information on the swallowing process and possible dysfunctions associated with dysphagia, indicating that it could potentially be used to assess the progress of the disease or the effectiveness of rehabilitation therapies.

Keywords: surface electromyography; Granger causality; functional coordination; dysphagia; swallowing muscle coupling

Citation: Ye-Lin, Y.; Prats-Boluda, G.; Galiano-Botella, M.; Roldan-Vasco, S.; Orozco-Duque, A.; Garcia-Casado, J. Directed Functional Coordination Analysis of Swallowing Muscles in Healthy and Dysphagic Subjects by Surface Electromyography. *Sensors* **2022**, *22*, 4513. https://doi.org/10.3390/s22124513

Academic Editor: Panicos Kyriacou

Received: 17 May 2022
Accepted: 13 June 2022
Published: 15 June 2022

1. Introduction

A complete swallow requires a complex sequence of highly regulated and coordinated oral and pharyngeal events for the safe passage of a bolus into the oesophagus without compromising the airway [1]. Dysphagia is an inability to swallow foods or liquids properly that affects 9.4 million adults every year (1 in 25) in the United States [2]. The prevalence of dysphagia ranges between 12–13% in hospitalized patients, but rises to 30% in the elderly and to 60% in intensive care unit and home-care patients [3–6]. Dysphagia is usually caused by another condition, such as aging, neurological diseases, neuromuscular impairment, head and neck cancers, and gastro-oesophageal reflux disease, among others. Dysphagia can produce malnutrition, dehydration, aspiration, pneumonia, and even death, as well as a reduced quality of life, social isolation, and healthcare-related costs [6–8]. The mean attributable cost of dysphagia is about USD 12,715, 40.36% more than nondysphagic hospitalized patients [9]. Patel estimated that dysphagia was responsible for between USD 4.3 to 7.1 billion in additional hospital costs annually in the US [8].

A videofluoroscopic swallowing study is the reference diagnostic method for assessing oropharyngeal dysphagia [10]. This technique involves patient exposure to ionizing radiation [10], so it is not recommended for patient follow-up when evaluating the effectiveness of rehabilitation, although it does not always identify neuromuscular abnormalities in pharyngeal or laryngeal physiology [10]. An example of the latter could be patients with muscle tension dysphagia who present functional dysphagia but exhibit normal oropharyngeal and oesophageal swallowing function, as evidenced by videofluoroscopic swallow study [11,12]. Surface electromyography (sEMG) has emerged as a simple, nonradioactive, and noninvasive method of measuring the patterns of muscle activity during swallowing and allows clinicians to describe the physiology of swallowing behaviour [13]. sEMG is the extracellular recording of the electrical activity of muscle fibres on the skin surface, which reflects the electrophysiological muscle response to nerve stimulation. Most swallowing-related studies have used sparse sEMG electrodes to sense the activity and have extracted electrophysiological information from face and neck muscles, e.g., from the masseteric, submental (suprahyoid), and infrahyoid regions [14–20]. Unlike classical temporal and spectral parameters from a single sEMG channel, sEMG signal characterization cannot precisely characterize the complete swallowing process.

A complete swallowing process in an adult human is an autonomous motor behaviour requiring not only the coordination of 26 muscles and 5 cranial nerves, but also coordination with mastication and respiration [21]. According to Bernstein's motor control theory and coordination dynamics theory, motor behaviour involves the coupling of different physiologic structures, such as muscles, in task-specific control units known as *synergies* [22]. In most motor control synergies, a common neurologic activation pattern associated with a specific task objective is responsible for providing temporary flexible couplings between muscle systems [22]. As a high-level neuromuscular synergy driven by the skeletal muscles and smooth muscles of the pharynx and oesophagus is required in swallowing to successfully execute the swallowing sequences [23,24], the coordination of different swallowing muscles, also known as *muscle coupling*, exhibits the hallmark characteristics of motor control synergy. This coordination is likely to be altered in dysphagic subjects, and its analysis can, thus, provide new indicators for its early detection or for the quantitative evaluation of rehabilitation therapies.

Several previous works have studied the coordination of swallowing muscles using different approaches. Most have analysed the activation sequences and coordination patterns of the different muscles involved in swallowing using sparse electrode pairs: suprahyoid [25]; oral and laryngeal [26]; and laryngeal, pharyngeal, and submental [16,27]. The analysis of sequence activation has usually been performed by detecting the fiducial point of muscle activation, such as the onset and offset timing, from rectified and integrated sEMGs [15,17,27] to determine the temporal relationship between the muscles and to establish the activation sequence pattern. Using the sEMGs from 15 electrodes in the face and throat, McKeown et al. successfully detected the swallowing pattern by separating laryngeal excursion, tongue movement, and activation of the buccal and masseteric musculature using independent component analysis [28]. Zhu et al. showed the feasibility of obtaining sequential sEMG energy maps from a 2D high-density electrode array on the submental and infrahyoid muscles to analyse spatiotemporal properties during swallowing [1]. They found that the sEMG potential maps constructed from a sliding window mainly reflected the submental and infrahyoid muscles associated with the high intensity on the top and in the centre of the maps, respectively [1]. Dysphagic subjects were shown to present a significantly higher prevalence of inverted muscle activation patterns, i.e., infrahyoid-related was followed by suprahyoid-related activation [15].

Few studies in the literature have been conducted to assess the functional relationship or coordination between swallowing muscles with sEMG. Our previous work assessed the crosstalk and synchronization properties of sEMGs from swallowing muscles based on cross-correlation [14]. We found the correlation of bilateral suprahyoid muscles to be moderate and slightly higher than that of bilateral infrahyoid muscles [14]. Wang et al.

proposed a discoordination index based on cross-correlation coefficients that reflected the differences between the surface EMG patterns of bilateral muscle groups [13]. Lee et al. found that suprahyoid muscle activity showed a significant positive correlation with infrahyoid activity [29]. Covariance was also proposed to examine patterns of functional independence for tongue muscles during speech and swallowing [30]. Steele et al. found that sequential liquid swallowing was associated with increased frequency entrainment through cross-spectral coherence analysis and reduced relative phase variability in tongue–jaw coordination [22]. The EMG–EMG transfer function and coherence function in the first peak frequency were used to assess jaw and neck muscle coordination in rhythmic chewing [31]. The authors found that the coherence and phase in nonchewing, side-neck muscle activities exhibited a significant negative correlation [31].

The correlation coefficient, or lagged correlation, measures the linear relationship in a time domain between two time series [32]. Frequency entrainment [22] and transfer function [31] estimates from cross-spectra or coherence are traditional measures of linear correlations in the frequency domain [32]. These measures are seriously affected by instantaneous interactions or coupling associated with volume conduction, as it is challenging to differentiate them from true interactions [32]. They provide information only on the interchannel interaction strength, not on the directionality of the interaction [32], which is a relevant physiological characteristic of swallowing. Granger causality (G-causality) has been widely used to determine brain functional connectivity to identify regional activations and to characterize functional circuits from functional magnetic resonance imaging, electroencephalography, and magnetoencephalography [33–36]. Based on the hypothesis that causes precede and help to predict effects and that manipulations of the cause change the effects [37], G-causality provides a statistical measurement of functional interaction strength based on the relative prediction improvement to identify linear directional interdependence between multivariate time series [32,36]. This is a data-driven approach that estimates the causal statistical influences without the need for physical intervention that is able to quantify the directional flow of information [33]. The analysis of muscular interactions, including both strength and directionality, can potentially provide a better understanding of the underlying physiology of swallowing and the possible alterations in dysphagic subjects.

This work thus aims to assess functional muscular interactions during swallowing in healthy individuals and the possible changes in dysphagic subjects by analysing the G-causality of their sEMGs. For this, we characterize and compare both the directionality and the coupling strength of multiple swallowing muscles with different bolus consistencies.

2. Materials and Methods

2.1. Data Acquisition

The database was composed of sEMG recordings from 30 healthy volunteers (17 males and 13 females with average ages of 42.2 ± 15.5 and 46.5 ± 17.6 years old, respectively) and 31 subjects with dysphagia (19 males and 12 females with average ages of 42.4 ± 17.1 and 48.7 ± 13.2 years old, respectively). The volunteers signed an informed consent approved by the Ethics Committee of the Instituto Tecnológico Metropolitano (Medellín, Colombia). All the patients included in the study had confirmed diagnoses of functional oropharyngeal dysphagia due to neurogenic causes (multiple sclerosis: 7; amyotrophic lateral sclerosis: 6; cerebral palsy: 4; muscular dystrophy: 4; ischemic stroke: 3; Parkinson's disease: 3; traumatic brain injury: 3; and secondary hydrocephalus (neurocysticercosis: 1)).

To reduce skin-electrode impedance, we first shaved the skin under the electrodes as required and then exfoliated with abrasion gel (Nuprep, Weaver and Company, Aurora, CO, USA) before cleaning the recording surface with isopropyl alcohol. Bipolar sEMG activity was recorded from the following three bilateral muscle groups involved in the oral and pharyngeal swallowing phases using 6 pairs of disposable Ag/AgCl electrodes (3M electrodes Ref. 2228; inter-electrode distance of 25 mm): the left and right masseters (LM and RM), the left and right suprahyoid (LSH and RSH), and the left and right infrahyoid

(LIH and RIH). Figure 1 shows the electrode placement for multichannel sEMG acquisition. A Noraxon Ultium EMG amplifier (Noraxon USA; CMMR > 100 dB, 16 bits A/D converter) was used for signal collection and conditioning. The bipolar sEMG signals were amplified and band-pass filtered within 10 and 500 Hz and were sampled at 2 kHz.

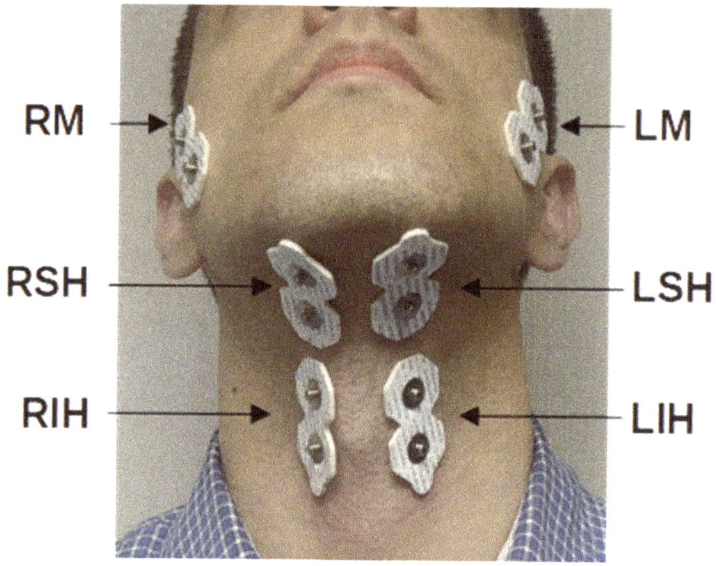

Figure 1. Electrode placement configuration for acquisition of multichannel sEMG signals. RM and LM: right and left masseter, respectively; RSH and LSH: right and left suprahyoid, respectively; RIH and LIH: right and left infrahyoid, respectively.

The recording protocol consisted of three swallowing tasks commonly used in dysphagia screening [38,39]: saliva, 10 mL of water, and 10 mL of liquid yoghurt. A 1.5 oz cup was used to deliver water and yoghurt to the oral cavity. The subjects were seated in an upright position and asked to swallow the bolus as naturally as possible.

2.2. Conditional Granger Causality

Conditional Granger causality (cG-causality) is defined as the degree to which the past of target Y helps to predict source X beyond the degree to which X is already predicted by its own past and the past of the conditioning variable Z [40]. cG-causality can be used to detect real interactions between different muscles, avoiding false causality due to their underlying 'hidden' interactions [40]. Based on vector autoregressive model theory, we considered the full (see Equation (1)) and reduced (Equation (2)) regressions of X, including the conditioning variable Z in both regressions:

$$X_t = \sum_{k=1}^{p} A_{XX,k} \cdot X_{t-k} + \sum_{k=1}^{p} A_{XY,k} \cdot Y_{t-k} + \sum_{k=1}^{p} A_{XZ,k} \cdot Z_{t-k} + \varepsilon_{X,t} \tag{1}$$

$$X_t = \sum_{k=1}^{p} A'_{XX,k} \cdot X_{t-k} + \sum_{k=1}^{p} A'_{XZ,k} \cdot Z_{t-k} + \varepsilon'_{X,t} \tag{2}$$

where p is the vector autoregressive model order, the coefficients $A_{XX,k}$ represent the autodependence of X on its own past, and $A_{XY,k}$ and $A_{XZ,k}$ are the coefficients of the dependence of X on the past of Y and on the past of the conditioning variable Z, respectively. $\varepsilon_{X,t}$ are the model regression residuals with the covariance matrix $\Sigma_{XX} \equiv \text{cov}(\varepsilon_{X,t})$.$A'_{XX,k}$,

$A'_{XZ,k}$ are the corresponding reduced regression coefficients, and $\varepsilon'_{X,t}$ are the residual covariance matrices of the reduced regression $\Sigma'_{XX} \equiv \text{cov}(\varepsilon'_{X,t})$.

cG-causality from Y to X $F_{Y \to X|Z}$ is, thus, defined as the log-likelihood ratio of the determinant of the residual covariance matrix taking into account the joint effects of Z (see Equation (3)). cG-causality attempts to quantify the degree to which the full regression represents a "better" model of the data than the reduced regression.

$$F_{Y \to X|Z} \equiv \frac{|\Sigma'_{XX}|}{|\Sigma_{XX}|} \tag{3}$$

We used pairwise-conditional G-causality [40] of the multichannel EMG data $G_{i,j}(EMG)$, which determines the causality EMG_j–EMG_i (from muscle j to i) conditioning out all the known remaining data (see Equation (4)):

$$G_{i,j}(EMG) \equiv F_{EMG_j \to EMG_i | EMG_{[ij]}} \tag{4}$$

where the subscript [ij] denotes omission of the i_{th} and j_{th} variables in the multivariate EMG data. Details of the algorithm can be found in the original work by Barnett [40].

The optimal order of the vector autoregresssive model to achieve a compromise between the model's precision and complexity was determined with the Akaike information criterion [40].

2.3. Data Analysis

As mentioned above, for a number of reasons we preferred to analyse the interactions between different swallowing muscles with pairwise cG-causality using sliding windows of a fixed length rather than the whole recording or contraction epochs. Firstly, there is no reliable tool for the automatic identification of the onset and offset of the contraction epoch in sEMG, which remains a challenge for the scientific technical community. Secondly, swallowing EMG data may present a nonstationary nature, such as EMGs from other muscles [41], which fails in the assumption of cG-causality on the covariance stationary stochastic process [40]. Both the contraction epoch and the whole EMG recording have a variable duration, which could influence the Granger causal inference analysis [42]. The sliding window length was set to 0.5 s with no overlap based on the timing values of the pharyngeal swallowing phase [38,43]. This selection was experimentally confirmed through comparison with a 1s window, for which the average percentage (see details below) of the cG-causality value did not show consistent trends between healthy and dysphagic subjects.

For each subject, bolus, and muscle pair, we obtained an array of cG-causality values as the results of the sliding window. Since swallowing muscles are expected to present higher interactions during contraction than in basal activity (at rest), we considered the maximum cG-causality value of each interaction pair as their interaction during swallowing-muscle contraction, obtaining maximum interaction matrix—MIM—(see Figure 2), so that one 6×6 MIM was obtained per subject in which each row and column represented one acquisition channel.

We then sought to determine the relevant interaction pairs for each subject by discarding interactions with weak coupling strengths, i.e., below a threshold of 25% of the maximum value of each subject's MIM (MIM_{max}) or the threshold interaction matrix. For each interaction pair, we evaluated the percentage of subjects that exceeded this threshold for each bolus. The average percentage of the three boluses (average percentage matrix) was then used to compare the occurrence of a relevant interaction between different pairs of muscle groups, after which the mean value of the average percentage matrix was computed for the bilateral, ipsilateral, and contralateral muscles of healthy and dysphagic subjects. For the value of the threshold used, we aimed to achieve a trade-off between the sensitivity of detecting relevant interactions by a physiological interpretation and obtaining 'spurious' interactions. True interactions associated with swallowing should be reproducible in almost all healthy subjects, achieving approximately 100% of the subjects that exceeded this threshold. In addition, some of these interactions may be altered in dysphagic subjects. Specifically, in this work, we compared different threshold values, ranging from 10–40%, and obtained similar results for threshold ranges of from 20 to 30%.

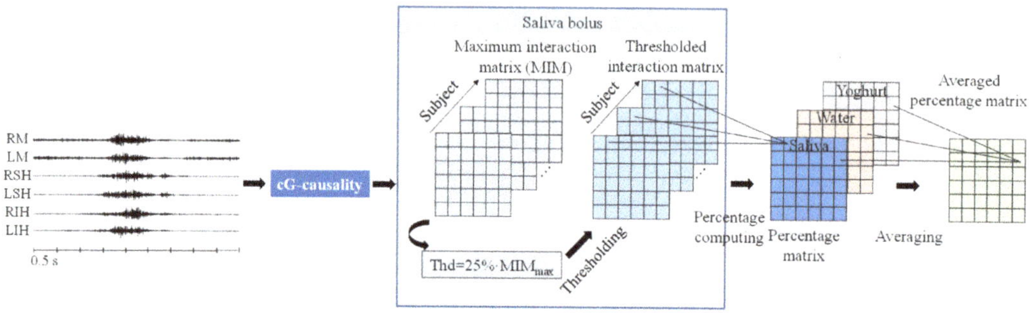

Figure 2. Flowchart of relevant interaction pairs with relatively high coupling strengths.

The cG-causality difference between healthy and dysphagic subjects was assessed using the raw MIM, which represented muscle interaction during swallowing. Firstly, cG-causality values of healthy and dysphagic subjects were compared for each bolus type and muscle pair using the Wilcoxon rank-sum statistical test ($\alpha = 0.05$). Secondly, we analysed the bolus effect on cG-causality for each interaction pair using the Friedman test ($\alpha = 0.05$) for both types of subjects. Finally, we evaluated the interaction asymmetry of each muscle pair (from *i* to *j* muscles vs. from *j* to *i* muscles) using the Wilcoxon signed-rank test ($\alpha = 0.05$) for both subject types.

3. Results

3.1. Assessment of Muscle Interactions

Table 1 shows the average percentage of subjects that exceeded the pre-established threshold of MIM_{max} for the three boluses with each muscle-pair interaction. Bilateral suprahyoid (RSH↔LSH) and infrahyoid (RIH↔LIH) muscles, in general, reacted closely with each other, with high occurrence between patients. The percentage of subjects who exceeded the pre-established threshold was greater than 90%, while the bilateral masseter muscle interaction (RM↔LM) was relevant in a smaller percentage of subjects (70–80%). As expected, regardless of the muscle groups, the bilateral muscles in healthy subjects seemed to activate simultaneously, retrieving a similar average percentage in both directions. Dysphagic subjects obtained a similar average percentage for bilateral muscle interactions. We also found consistent interactions among subjects for ipsilateral suprahyoid-infrahyoid muscles (RSH↔RIH and LSH↔LIH), with 88% of the healthy subjects exceeding the threshold. The ipsilateral couplings between masseter and suprahyoid (RM↔RSH and LM↔LSH) or infrahyoid (RM↔RIH and LM↔LIH) muscles were moderately consistent between healthy subjects, with over 76% of the individuals exceeding the threshold. In dysphagic subjects, the consistency of the average ipsilateral interaction was considerably reduced (87% healthy vs. 78% dysphagia). The greatest differences between healthy and dysphagic subjects occurred in the left ipsilateral interactions: LSH→LM (94% vs. 69%) and LIH→LM (90% vs. 72%). We found that the contralateral interaction was relatively weaker (68% healthy vs. 64% dysphagia). As the bilateral and ipsilateral muscles presented higher consistencies of relevant interactions than contralateral muscles in both healthy and dysphagic subjects, in the succeeding sections we, therefore, focus on the bilateral interactions and ipsilateral masseter–suprahyoid and suprahyoid–infrahyoid interactions due to their sequential activations under physiological conditions.

Table 1. Average percentages of healthy/dysphagic subjects that exceeded the pre-established threshold for each interaction muscle pair. Bilateral, ipsilateral, and contralateral muscles are shaded in grey, green, and blue respectively. The overall average percentages of bilateral, ipsilateral, and contralateral muscle interactions are shown at the bottom of the table. Percentages above 85% are shown in bold.

		From					
		RM	**LM**	**RSH**	**LSH**	**RIH**	**LIH**
To	RM		78% | 79%	**85% | 79%**	66% | 57%	84% | 75%	65% | 53%
	LM	72% | 77%		73% | 63%	**94% | 69%**	59% | 70%	**90% | 72%**
	RSH	83% | 68%	67% | 63%		**100% | 97%**	**88% | 96%**	71% | 58%
	LSH	67% | 62%	76% | 75%	**100% | 91%**		69% | 63%	**92% | 85%**
	RIH	78% | 63%	68% | 65%	**96% | 92%**	74% | 75%		**98% | 95%**
	LIH	57% | 55%	82% | 74%	82% | 78%	**94% | 95%**	**98% | 97%**	
Average		Bilateral	**91% | 89%**	Ipsilateral	**87% | 78%**	Contralateral	68% | 64%

Figure 3 shows the cG-causality for bilateral muscles in healthy and dysphagic subjects for the three boluses. Next, we describe the difference between healthy and dysphagic subjects, the effect of bolus consistency, and the interaction symmetry for each pair of muscles. The results of the statistical analysis are described in Section 3.2.

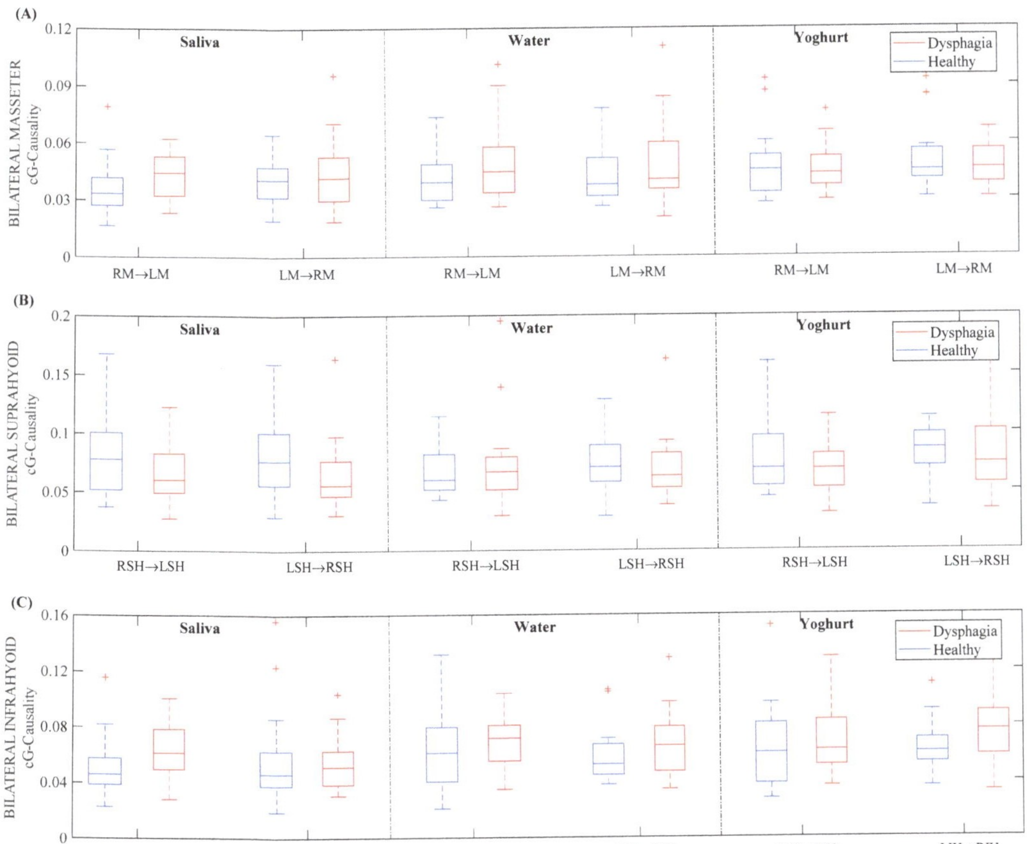

Figure 3. Box-and-whisker plot of the cG-causality values for bilateral muscles in healthy and dysphagic subjects for saliva, water, and yoghurt boluses. (**A**) Masseter, (**B**) suprahyoid, (**C**) infrahyoid.

Bilateral Masseter. Dysphagic subjects showed increased cG-causality in bilateral masseters for saliva and water boluses when compared to healthy subjects with little difference in yoghurt. We found a higher coupling strength of the RM→LM interaction for water and yoghurt than for saliva in the healthy group, but not in dysphagic subjects. The bilateral masseter cG-causality in both directions was similar for both healthy and dysphagic subjects, except for with saliva in the first group.

Bilateral suprahyoid. These muscle interactions tended to decrease more in dysphagic than in healthy subjects. We did not find a clear bolus effect of cG-causality in healthy subjects, while it tended to be higher for water and yoghurt than for saliva in dysphagic subjects. Again, bilateral suprahyoid cG-causality was similar in both directions, except for yoghurt in healthy subjects.

Bilateral infrahyoid. Dysphagic subjects, in general, obtained higher medians of cG-causality than healthy subjects. We found higher medians of cG-causality in the RIH→LIH interaction for the water and yoghurt boluses than for the saliva bolus in healthy subjects. The LIH→RIH interaction showed higher cG-causality for water and yoghurt than for saliva in both subject groups, being more evident in dysphagics. Bilateral infrahyoid muscles showed similar cG-causality in both directions in healthy subjects, while dysphagic subjects had an asymmetric interaction.

Figure 4 gives the cG-causality of ipsilateral muscles in healthy and dysphagic subjects for the three boluses. The results obtained from each muscle pair is again described, differentiating between healthy and dysphagic subjects, bolus consistencies, and the interaction symmetry.

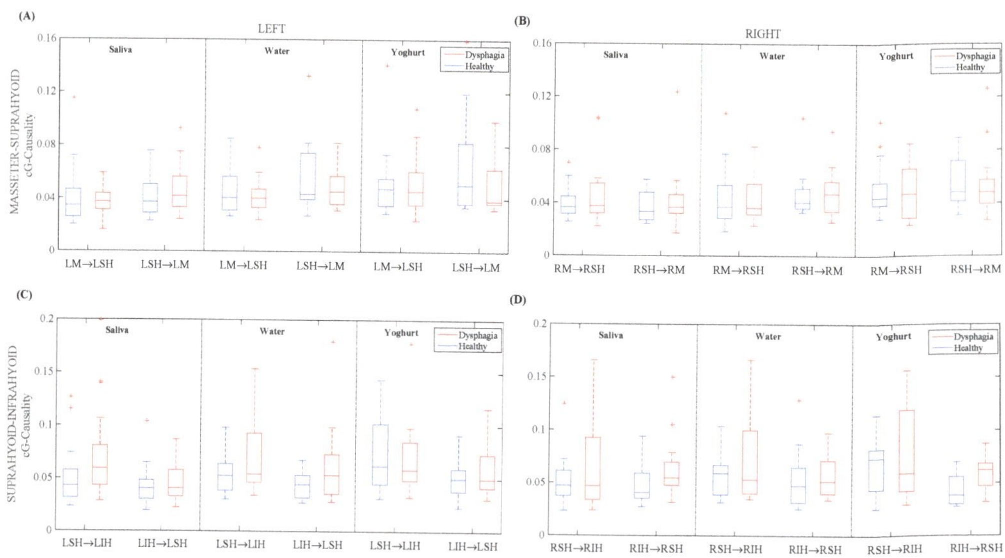

Figure 4. Box-and-whisker plot of the cG-causality for ipsilateral muscle interactions in healthy and dysphagic subjects for saliva, water, and yoghurt boluses. (**A**) LM↔LSH, (**B**) LM↔LSH, (**C**) LSH↔LIH, and (**D**) RSH↔RIH.

Ipsilateral masseter↔ suprahyoid. Dysphagic subjects usually obtained slightly higher than or similar cG-causalities to healthy individuals, with no clear trend. We also found an upward trend in cG-causality from saliva to yoghurt for healthy subjects, while this trend was only apparent in LM→LSH for dysphagic subjects. In healthy subjects, the masseter↔suprahyoid muscle interaction seemed to be symmetric. In general, we found an asymmetric masseter→suprahyoid interaction, except for with the yoghurt bolus on the right side in dysphagic subjects, while the upward direction tended to achieve greater cG-causality.

Ipsilateral suprahyoid↔infrahyoid. Dysphagic subjects generally obtained similar or slightly lower cG-causalities in the downward direction than healthy subjects, while dysphagic subjects tended to obtain higher cG-causality values than healthy subjects in the upward direction, especially in RIH→RSH for both saliva and yoghurt. We found an upward trend in cG-causality from saliva to yoghurt in healthy subjects, except for RIH→RSH, but not in dysphagic subjects. We also found a predominantly downward interaction for both left and right suprahyoid→infrahyoid muscles in healthy subjects, which matched with the direction of swallowing. In dysphagic subjects, this predominance was notably only observed in the LSH→LIH interaction.

3.2. Statistical Analysis

Figures 5–7 illustrate the muscle interactions with statistically significant differences between healthy and dysphagic subjects, between boluses, and between directions, respectively. Statistically significant differences were found between healthy and dysphagic subjects for the RIH→LIH bilateral interaction for saliva (Figure 5A), as well as for the ipsilateral interactions of RIH→RSH and LSH→LIH, also for saliva (Figure 5A), and for RIH→RSH for yoghurt (Figure 5C). Despite weaker contralateral vertical interactions, we found statistically significant differences for RIH→LM in water and yoghurt and for ↔ in yoghurt. The greatest significant differences between healthy and dysphagic subjects were, notably, obtained with yoghurt.

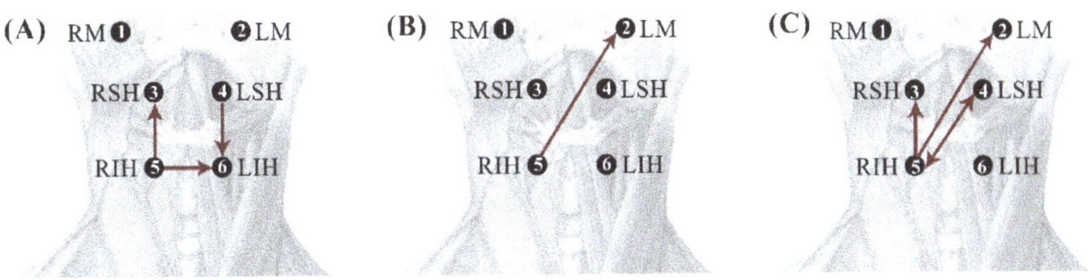

Figure 5. Schematic representations of statistically significant differences in the cG-causality of interaction pairs between healthy and dysphagic subjects for saliva (**A**), water (**B**), and yoghurt (**C**). Significantly higher and lower interactions in dysphagic subjects are shown in red and blue, respectively.

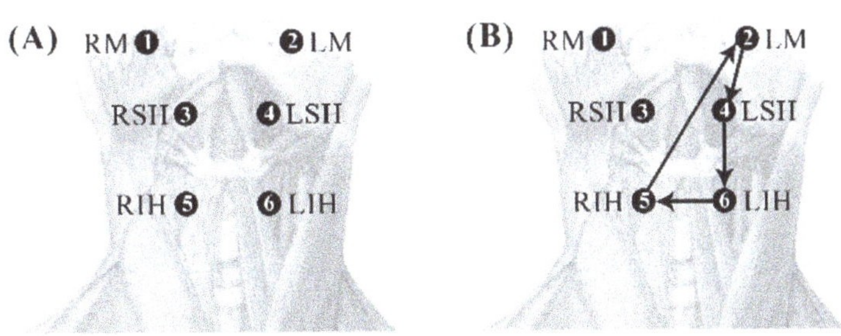

Figure 6. Diagrams with statistically significant differences in the cG-causality of interaction pairs between the different boluses in healthy (**A**) and dysphagic subjects (**B**).

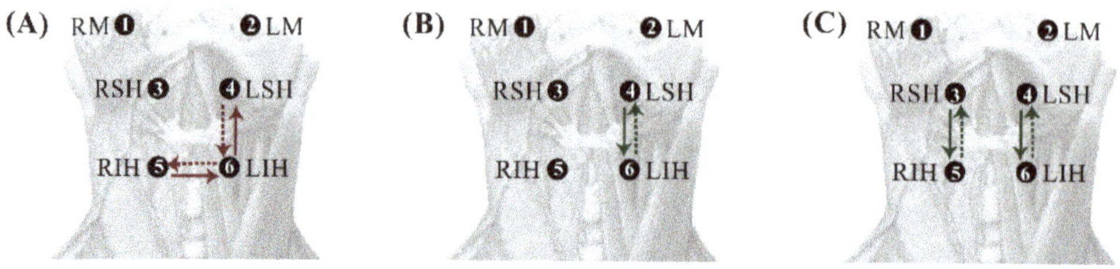

Figure 7. Diagrams with statistically significant differences in the cG-causality of the directionality of interaction pairs (asymmetry) for saliva (**A**), water (**B**), and yoghurt (**C**). Green and red arrows show healthy and dysphagic subjects, respectively. The continuous lines show the dominant direction.

No significant differences were found between the boluses for healthy subjects (Figure 6A). In dysphagic subjects, statistically significant differences between the boluses were found in the bilateral infrahyoid interaction LIH→RIH, the left ipsilateral vertical interactions of LM→LSH and LSH→LIH, and the contralateral vertical interaction of RIH→LM (Figure 6B).

For muscle interaction directionality, we found significant asymmetric interactions with a downward-dominant direction in healthy subjects for the RSH↔RIH interaction for yoghurt and in the LSH↔LIH interactions for water and yoghurt (Figure 7B,C). In dysphagic subjects, the asymmetric muscle interactions only appeared in the bilateral infrahyoid muscle interaction and ipsilateral LSH↔LIH for saliva (Figure 7A).

4. Discussion

4.1. Relevant Muscle Interactions during Swallowing

In this work, we evaluated the functional coordination of three muscle groups involved in swallowing, i.e., masseters, suprahyoid, and infrahyoid. To the best of the authors' knowledge, this is the first exploratory work to study the directed functional coordination of swallowing muscles by means of cG-causality, which is less-influenced by the volume conduction effect than simple correlation or coherence-based approaches [32,44]. Regardless of subject group and bolus type, we found both bilateral and ipsilateral suprahyoid and infrahyoid muscle activities to be strongly coupled, while the interactions of the bilateral masseter, the ipsilateral masseter, and the suprahyoid muscles were weaker. The high degree of supra- and infrahyoid coupling agrees with other authors who found a significant positive correlation between these muscles [29]. We mostly found a high level of bilateral suprahyoid rather than bilateral infrahyoid coupling in healthy subjects, which agrees with our previous study in assessing the functional coordination by cross-correlation [14]. The lower interactions for bilateral masseter muscles may be associated with their being voluntary movements. Only swallowing tasks that involved more supra- and infrahyoid activation than masseteric were assessed, since no chewing was required. In general, we found a high overall occurrence of relevant interactions (high average percentage, see Table 1) of supra- and infrahyoid muscles under physiological conditions, which is consistent with the electrophysiology of swallowing [45]. We also found a predominantly downward direction in ipsilateral supra- and infrahyoid muscles, which matches with the transit of the physiological-descendent bolus. Under physiological conditions, the suprahyoid were activated 95 ms earlier than the infrahyoid muscles [46]. This may justify the relatively higher cG-causality value of suprahyoid-to-infrahyoid interaction (improved predictability of infrahyoid EMG activity with the known EMG activity from the suprahyoid muscle) than the value for infrahyoid-to-suprahyoid interaction. In addition, we found that the overall occurrence of ipsilateral interactions was considerably reduced in dysphagic subjects. The reduced strength of the swallowing interaction coupling could be associated with the alteration of stereotyped motor behaviours and could also be a potential dysphagia biomarker [47].

4.2. Alterations of Muscle Interactions in Dysphagia

We did not find significant differences in functional coordination between the bilateral masseters, ipsilateral masseters, and suprahyoid muscles in either healthy or dysphagic subjects. Our results suggested that the main alterations in dysphagic subjects are mainly found in the supra- and infrahyoid muscles. This finding may be associated with the fact that the primary function of the masseter muscles is to raise the jaw when chewing and to act as mandibular stabilizers during swallowing [48]. In this regard, Zanato et al. showed that swallowing water demanded a greater activation of the suprahyoid than the masseter muscles compared to the values at rest (9.57 μV compared to 3.81 μV and 6.15 μV compared to 3.47 μV, respectively) [49]. Monaco et al. found that masseter muscles showed lower rectified sEMG values than submental groups during spontaneous saliva swallowing, even when activated [50].

Our preliminary results showed a significantly increased interaction in RIH→RSH for both saliva and yoghurt, in RIH→LIH and LSH→LIH for saliva, and in RIH↔LSH for yoghurt in dysphagic subjects, suggesting an altered sequence of the supra- and infrahyoid activation pattern. These findings are consistent with those available in the literature. Koyama et al. found a significantly higher prevalence of inverted muscle activation patterns in dysphagic subjects, with the activation of infrahyoid muscles preceding the suprahyoid muscles [15]. Pre-onset muscle activation is a protective mechanism to prevent neuromuscular degeneration leading to kinematic and functional loss [51] and gives rise to prolonged swallowing times, which has been widely described in dysphagic subjects [52–54]. In fact, the appearance of swallowing with a pre-reflex phase of muscle activation was reported as a compensatory mechanism to adjust for age-related muscle weakness [51]. Koyama et al. found prolonged activation of infrahyoid muscles and shorter activity of suprahyoid muscles in dysphagic subjects, suggesting important changes in the timing of the initiation of swallowing-muscle activity [15]. This phenomenon may be caused by the forceful swallowing secondary to the lack of coordination of the swallowing muscles, which increases muscle activity amplitude [15]. Consequently, the previous activation of the infrahyoid muscle may be the origin of the significantly increased interaction strength in RIH→RSH, producing a loss in the physiological downward-dominant directional RSH→RIH in dysphagic subjects. The symmetry in the RSH↔RIH interaction directionality may, thus, constitute a new dysphagia biomarker.

The different patterns found in RSH→RIH and LSH→LIH in dysphagic subjects may well be associated with the right hemispheric lateralization of the pharyngeal phase. This phenomenon was reported by other authors, who found a reduction in cortical swallowing-related activation in amyotrophic lateral sclerosis patients with progressive dysphagia in comparison to healthy controls, the right sensorimotor cortex being predominant [55]. This right hemispheric lateralization may be associated with the compensatory mechanism to coordinate the pharyngeal phase of swallowing thanks to brain plasticity [55]. The right hemispheric lateralization may also give rise to a delayed activation of LIH with respect to RIH, obtaining a significantly increased RIH→LIH interaction in dysphagic subjects, but not for LIH→RIH. The delayed triggering of the swallowing reflex for voluntarily initiated swallowing has also been observed in both amyotrophic lateral sclerosis patients and in dysphagia for suprabulbar palsy [56,57]. In the latter, when reflex swallowing could be triggered, it was slow and prolonged [57]. Our results suggest that the loss of symmetrical interaction of the bilateral infrahyoid muscles could be another dysphagia biomarker, which agrees with other authors who found that the bilateral muscle discoordination index was significantly greater in dysphagic than in healthy subjects [13]. Krasnodębska et al. also reported that patients with atypical swallowing patterns had significantly greater asymmetry of both the masseter and submental muscles [58].

Previous studies have shown that increasing the bolus consistency in healthy subjects prolonged the duration of oral and pharyngeal swallowing [59–61] as well as discrete and sequential swallowing [62]. Numerous studies have shown that highly viscous liquids significantly increase the duration of the supra- and some infra-hyoid muscle activations [51,59,63,64]. In comparison to swallowing saliva, the highest sEMG amplitude of the supra and infrahyoid muscles was obtained in healthy subjects swallowing 10 mL water and yoghurt [14], suggesting greater muscle recruitment. It was less safe for dysphagic patients to swallow thin liquids rather than thicker ones [65]. In this work, we also found that cG-causality was slightly higher for water and yoghurt than for saliva with no significant difference between them in healthy subjects, who seemed to have a good ability to fine-tune the activation pattern according to the type of bolus ingested, namely swallowing reserve [66], leading to similar interaction strengths among the boluses. The swallowing reserve decline due to neurological and neuromuscular diseases, muscle weakness caused by aging, and positional changes of swallowing-related organs [66]. This could explain the difference in the functional interactions among the boluses observed in dysphagic subjects. Generally, a decline in the swallowing reserve may cause a descent in the positions of the hyoid bone and larynx, a reduced antero-superior movement range for the hyoid bone and larynx elevation, larynx elevation delay, and a delay in the stimulation of the swallowing reflex [66]. The right hemispheric lateralization may also justify the preservation of the right-side muscle swallowing reserve while losing it in the left muscle group [55], which could explain the significant differences among the boluses in the LM→LSH and LSH→LIH interactions in dysphagic subjects.

4.3. Study Limitations

Despite its promising results, this study was not completely exempt from limitations. Firstly, significant differences in the interactions between the healthy and dysphagic subjects, as well as asymmetric interactions, were not consistent for all the boluses. Although the differences in bolus properties could yield inherently different swallowing responses, this finding may also be related to the limited sample size and the high intersubject variability due to population variance and intrasubject variability. The latter could have been affected by diverse biological factors, such as muscle fatigue, the volume of salivary secretions (which may vary according to the volume of liquid swallowed), the time interval between swallows, the number of trials, and the sequence of the food intake [46,67]. In addition, it was reported that dysphagic subjects may show significantly higher intrasubject variability between repetitions [68]. In this work, we only acquired sEMG data for a single swallow of each bolus. Repeated swallowing of these would provide a more robust characterization of the activation pattern of the muscles involved and would reduce intrasubject variability [46]. Future studies with repeated swallowing are still needed to corroborate our preliminary results. In addition, due to the limited sample size, we did not conduct the study by means of dysphagia aetiology to determine the difference in functional coordination between the subjects. Secondly, although cG-causality was originally formulated for linear stationary stochastic processes, sEMG has a nonstationary nature, which was the reason why we carried out the sliding window analysis. Finally, it should be noted that a multimodal analysis using electroencephalography and sEMG would provide a better understanding of the underlying electrophysiological mechanisms involved in swallowing, since the latter requires both voluntary and automatic control involving multiple brain regions [21].

5. Conclusions

In this work, we conducted a preliminary study on the utility of assessing the directed functional coordination between the masseter, supra-, and infrahyoid muscles during swallowing to detect possible alterations in functional coordination. We determined the physiological functional coordination pattern in normal swallowing (bilateral and ipsilateral supra- and infrahyoid-related activity) to be highly coupled. We also found a dominantly downward direction of the ipsilateral supra- and infrahyoid muscles in healthy subjects, which matches with the electrophysiology of swallowing, while the bilateral interactions were symmetric with no significant differences.

The main alterations in dysphagic subjects were found in the supra- and infrahyoid muscles, with no significant differences in the bilateral masseter and ipsilateral masseter↔suprahyoid muscle interactions. Specifically, we found that the right-to-left infrahyoid interaction was significantly higher in dysphagic subjects, suggesting that the loss in symmetry interaction of the bilateral infrahyoid could be potentially used as a dysphagia biomarker. We also found different pattern changes in dysphagic subjects in the left and right supra- and infrahyoid muscle interactions, with the right side being more resistant to a swallowing decline.

The loss in the asymmetric downward direction of the ipsilateral supra- and infrahyoid muscles could also be another dysphagia biomarker. Unlike healthy subjects, dysphagic subjects showed significant differences in the left masseter–suprahyoid, left suprahyoid–infrahyoid, left-to-right infrahyoid, and right infrahyoid–left masseter interactions, depending on the bolus consistency.

Our preliminary results suggested that the functional coordination analysis of swallowing muscles provided relevant information for evaluating motor control synergy and paved the way towards the identification of new, robust biomarkers for the early detection of dysphagia. Our method potentially contributed to developing a noninvasive and objective screening method for the early detection of swallowing dysfunction related to altered functional coordination that is not detectable by a videofluoroscopia swallowing study and could, therefore, be used to quantitatively assess the progress of dysphagia and the effectiveness of rehabilitation therapies.

Author Contributions: Conceptualization, Y.Y.-L., G.P.-B., J.G.-C., S.R.-V. and A.O.-D.; methodology, Y.Y.-L., M.G.-B., G.P.-B., J.G.-C. and A.O.-D.; software, M.G.-B., Y.Y.-L., G.P.-B. and J.G.-C.; validation, M.G.-B., Y.Y.-L., G.P.-B. and J.G.-C.; formal analysis, M.G.-B., Y.Y.-L., G.P.-B., J.G.-C., S.R.-V. and A.O.-D.; investigation, Y.Y.-L., G.P.-B., M.G.-B., J.G.-C., S.R.-V. and A.O.-D.; resources, J.G.-C. and A.O.-D.; data curation, S.R.-V. and A.O.-D.; writing—original draft preparation Y.Y.-L., M.G.-B. and G.P.-B.; writing—review and editing Y.Y.-L., G.P.-B., M.G.-B., J.G.-C., S.R.-V. and A.O.-D.; visualization, Y.Y.-L. and G.P.-B.; supervision, J.G.-C. and A.O.-D.; project administration, J.G.-C. and A.O.-D.; funding acquisition J.G.-C. and A.O.-D. All authors have read and agreed to the published version of the manuscript.

Funding: This work was partially supported by the Spanish Ministry of Economy and Competitiveness, the European Regional Development Fund (MCIU/AEI/FEDER, UE RTI2018-094449-A-I00-AR), and the Regional Joint Call for R+D+i projects (G8+1, Medellín, Colombia; Research Project PE2020-9 (G8 2020-39651)).

Institutional Review Board Statement: The study was conducted in accordance with the Declaration of Helsinki and was approved by the local Ethics Committee of the Instituto Tecnológico Metropolitano (Medellín, Colombia).

Informed Consent Statement: Informed consent was obtained from all subjects involved in the study.

Data Availability Statement: The data are not publicly available due to privacy or ethical restrictions.

Conflicts of Interest: The authors declare no conflict of interest.

References

1. Zhu, M.; Yu, B.; Yang, W.; Jiang, Y.; Lu, L.; Huang, Z.; Chen, S.; Li, G. Evaluation of normal swallowing functions by using dynamic high-density surface electromyography maps. *Biomed. Eng. Online* **2017**, *16*, 1–18. [CrossRef]
2. Bhattacharyya, N. The Prevalence of Dysphagia among Adults in the United States. *Otolaryngol. Neck Surg.* **2014**, *151*, 765–769. [CrossRef] [PubMed]
3. Chen, D.F. Dysphagia in the Hospitalized Patient. *Hosp. Med. Clin.* **2017**, *6*, 38–52. [CrossRef]
4. Daroff, R.B.; Jankovic, J.; Mazziotta, J.C.; Pomeroy, S.L. *Bradley's Neurology in Clinical Practice*, 7th ed.; Elsevier: Amsterdam, The Netherlands, 2016.
5. Hoy, M.; Domer, A.; Plowman, E.K.; Loch, R.; Belafsky, P. Causes of dysphagia in a tertiary-care swallowing center. *Ann. Otol. Rhinol. Laryngol.* **2013**, *122*, 335–338. [CrossRef] [PubMed]
6. Ekberg, O.; Hamdy, S.; Woisard, V.; Wuttge-Hannig, A.; Ortega, P. Social and Psychological Burden of Dysphagia: Its Impact on Diagnosis and Treatment. *Dysphagia* **2002**, *17*, 139–146. [CrossRef]

7. Cichero, J.A.; Altman, K.W. Definition, Prevalence and Burden of Oropharyngeal Dysphagia: A Serious Problem among Older Adults Worldwide and the Impact on Prognosis and Hospital Resources. In *Stepping Stones to Living Well with Dysphagia*; Karger Publishers: Berlin, Germany, 2012; Volume 72, pp. 1–11. [CrossRef]
8. A Patel, D.; Krishnaswami, S.; Steger, E.; Conover, E.; Vaezi, M.F.; Ciucci, M.R.; O Francis, D. Economic and survival burden of dysphagia among inpatients in the United States. *Dis. Esophagus* **2017**, *31*, 1–7. [CrossRef]
9. Attrill, S.; White, S.; Murray, J.; Hammond, S.; Doeltgen, S. Impact of oropharyngeal dysphagia on healthcare cost and length of stay in hospital: A systematic review. *BMC Health Serv. Res.* **2018**, *18*, 1–18. [CrossRef]
10. Vaiman, M. Standardization of surface electromyography utilized to evaluate patients with dysphagia. *Head Face Med.* **2007**, *3*, 26. [CrossRef]
11. Kang, C.H.; Hentz, J.G.; Lott, D.G. Muscle Tension Dysphagia: Symptomology and Theoretical Framework. *Otolaryngol. Head. Neck Surg.* **2016**, *155*, 837–842. [CrossRef]
12. Krasnodębska, P.; Jarzyńska-Bućko, A.; Szkiełkowska, A.; Miaśkiewicz, B.; Skarżyński, H. Diagnosis in Muscle Tension Dysphagia. *Otolaryngol. Polska* **2020**, *74*, 1–5. [CrossRef]
13. Wang, Y.-C.; Chou, W.; Lin, B.-S.; Wang, J.-J. The use of surface electromyography in dysphagia evaluation. *Technol. Health Care* **2017**, *25*, 1025–1028. [CrossRef] [PubMed]
14. Garcia-Casado, J.; Prats-Boluda, G.; Ye-Lin, Y.; Restrepo-Agudelo, S.; Perez-Giraldo, E.; Orozco-Duque, A. Evaluation of Swallowing Related Muscle Activity by Means of Concentric Ring Electrodes. *Sensors* **2020**, *20*, 5267. [CrossRef] [PubMed]
15. Koyama, Y.; Ohmori, N.; Momose, H.; Kondo, E.; Yamada, S.-I.; Kurita, H. Detection of swallowing disorders using a multiple channel surface electromyography sheet: A preliminary study. *J. Dent. Sci.* **2020**, *16*, 160–167. [CrossRef] [PubMed]
16. Mori, H.; Izumi, S.-I.; Kagaya, H.; Oyama, H.; Abe, G.; Yashima, K.; Takagi, T. Evaluation of the myoelectric potential of the infrahyoid muscles as a means of detecting muscle activity of the suprahyoid muscles. *Jpn. J. Compr. Rehabilit. Sci.* **2020**, *11*, 52–58. [CrossRef]
17. Inokuchi, H.; González-Fernández, M.; Matsuo, K.; Brodsky, M.B.; Yoda, M.; Taniguchi, H.; Okazaki, H.; Hiraoka, T.; Palmer, J.B. Electromyography of Swallowing with Fine Wire Intramuscular Electrodes in Healthy Human: Activation Sequence of Selected Hyoid Muscles. *Dysphagia* **2014**, *29*, 713–721. [CrossRef]
18. Restrepo-Agudelo, S.; Roldan-Vasco, S.; Ramirez-Arbelaez, L.; Cadavid-Arboleda, S.; Perez, E.; Orozco-Duque, A. Improving surface EMG burst detection in infrahyoid muscles during swallowing using digital filters and discrete wavelet analysis. *J. Electromyogr. Kinesiol.* **2017**, *35*, 1–8. [CrossRef]
19. Roldan-Vasco, S.; Restrepo-Agudelo, S.; Valencia-Martinez, Y.; Orozco-Duque, A. Automatic detection of oral and pharyngeal phases in swallowing using classification algorithms and multichannel EMG. *J. Electromyogr. Kinesiol.* **2018**, *43*, 193–200. [CrossRef]
20. Sebastian, R.V.; Estefania, P.G.; Andres, O.D. Orozco-Duque. Scalogram-energy based segmentation of surface electromyography signals from swallowing related muscles. *Comput. Methods Programs Biomed.* **2020**, *194*, 105480. [CrossRef]
21. Mosier, K.; Bereznaya, I. Parallel cortical networks for volitional control of swallowing in humans. *Exp. Brain Res.* **2001**, *140*, 280–289. [CrossRef]
22. Steele, C.M.; Van Lieshout, P.H.H.M. The Dynamics of Lingual-Mandibular Coordination During Liquid Swallowing. *Dysphagia* **2007**, *23*, 33–46. [CrossRef]
23. Murakami, K.; Hirano, H.; Watanabe, Y.; Edahiro, A.; Ohara, Y.; Yoshida, H.; Kim, H.; Takagi, D.; Hironaka, S. Relationship between swallowing function and the skeletal muscle mass of older adults requiring long-term care. *Geriatr. Gerontol. Int.* **2015**, *15*, 1185–1192. [CrossRef] [PubMed]
24. Matsuo, K.; Palmer, J.B. Anatomy and Physiology of Feeding and Swallowing: Normal and Abnormal. *Phys. Med. Rehabil. Clin. N. Am.* **2008**, *19*, 691–707. [CrossRef] [PubMed]
25. Spiro, J.; Rendell, J.K.; Gay, T. Activation and Coordination Patterns of the Suprahyoid Muscles During Swallowing. *Laryngoscope* **1994**, *104*, 1376–1382. [CrossRef] [PubMed]
26. Gay, T.; Rendell, J.K.; Spiro, J. Oral and Laryngeal Muscle Coordination During Swallowing. *Laryngoscope* **1994**, *104*, 341–349. [CrossRef] [PubMed]
27. Perlman, A.L.; Palmer, P.M.; McCulloch, T.M.; Vandaele, D.J. Electromyographic activity from human laryngeal, pharyngeal, and submental muscles during swallowing. *J. Appl. Physiol.* **1999**, *86*, 1663–1669. [CrossRef] [PubMed]
28. McKeown, M.J.; Torpey, D.C.; Gehm, W.C. Non-invasive monitoring of functionally distinct muscle activations during swallowing. *Clin. Neurophysiol.* **2002**, *113*, 354–366. [CrossRef]
29. Lee, H.-J.; Kim, K.-D. Analysis of the Correlation between Activity of the Suprahyoid Muscles, Infrahyoid Muscles and the New VFSS Scale in Stroke Patients with Dysphagia. *J. Korean Soc. Phys. Med.* **2018**, *13*, 19–25. [CrossRef]
30. Green, J.R.; Wang, Y.-T. Tongue-surface movement patterns during speech and swallowing. *J. Acoust. Soc. Am.* **2003**, *113*, 2820–2833. [CrossRef]
31. Ishii, T.; Narita, N.; Endo, H. Evaluation of jaw and neck muscle activities while chewing using EMG-EMG transfer function and EMG-EMG coherence function analyses in healthy subjects. *Physiol. Behav.* **2016**, *160*, 35–42. [CrossRef]
32. Bastos, A.M.; Schoffelen, J.-M. A Tutorial Review of Functional Connectivity Analysis Methods and Their Interpretational Pitfalls. *Front. Syst. Neurosci.* **2016**, *9*, 175. [CrossRef]

33. Hnazaee, M.F.; Khachatryan, E.; Chehrazad, S.; Kotarcic, A.; De Letter, M.; Van Hulle, M. Overlapping connectivity patterns during semantic processing of abstract and concrete words revealed with multivariate Granger Causality analysis. *Sci. Rep.* **2020**, *10*, 1–19. [CrossRef]
34. Manomaisaowapak, P.; Nartkulpat, A.; Songsiri, J. Granger Causality Inference in EEG Source Connectivity Analysis: A State-Space Approach. *IEEE Trans. Neural Networks Learn. Syst.* **2021**, 1–11. [CrossRef] [PubMed]
35. Shi, Q.; Chen, H.; Jia, Q.; Yuan, Z.; Wang, J.; Li, Y.; Han, Z.; Mo, D.; Zhang, Y. Altered Granger Causal Connectivity of Resting-State Neural Networks in Patients With Leukoaraiosis-Associated Cognitive Impairment—A Cross-Sectional Study. *Front. Neurol.* **2020**, *11*, 457. [CrossRef] [PubMed]
36. Sitnikova, E.; Dikanev, T.; Smirnov, D.; Bezruchko, B.; van Luijtelaar, G. Granger causality: Cortico-thalamic interdependencies during absence seizures in WAG/Rij rats. *J. Neurosci. Methods* **2008**, *170*, 245–254. [CrossRef]
37. Eichler, M. Causal inference with multiple time series: Principles and problems. *Philos. Trans. R. Soc. Lond. Ser. A Math. Phys. Eng. Sci.* **2013**, *371*, 20110613. [CrossRef]
38. Clavé, P.; DE Kraa, M.; Arreola, V.; Girvent, M.; Farré, R.; Palomera, E.; Serra-Prat, M. The effect of bolus viscosity on swallowing function in neurogenic dysphagia. *Aliment. Pharmacol. Ther.* **2006**, *24*, 1385–1394. [CrossRef]
39. Bozorgi, C.; Holleufer, C.; Wendin, K. Saliva Secretion and Swallowing—The Impact of Different Types of Food and Drink on Subsequent Intake. *Nutrients* **2020**, *12*, 256. [CrossRef]
40. Barnett, L.; Seth, A.K. The MVGC multivariate Granger causality toolbox: A new approach to Granger-causal inference. *J. Neurosci. Methods* **2014**, *223*, 50–68. [CrossRef]
41. Goen, A.; Tiwari, D.C. Review of Surface Electromyogram Signals: Its Analysis and Applications. *Int. J. Electr. Electron. Sci. Eng.* **2013**, *7*, 965–973.
42. Dhamala, M.; Rangarajan, G.; Ding, M. Analyzing information flow in brain networks with nonparametric Granger causality. *NeuroImage* **2008**, *41*, 354–362. [CrossRef]
43. Oommen, E.R.; Kim, Y.; McCullough, G. Stage Transition and Laryngeal Closure in Poststroke Patients with Dysphagia. *Dysphagia* **2010**, *26*, 318–323. [CrossRef] [PubMed]
44. Cohen, D.; Tsuchiya, N. The Effect of Common Signals on Power, Coherence and Granger Causality: Theoretical Review, Simulations, and Empirical Analysis of Fruit Fly LFPs Data. *Front. Syst. Neurosci.* **2018**, *12*, 30. [CrossRef] [PubMed]
45. Alfonsi, E.; Cosentino, G.; Mainardi, L.; Schindler, A.; Fresia, M.; Brighina, F.; Benazzo, M.; Moglia, A.; Alvisi, E.; Fierro, B.; et al. Electrophysiological Investigations of Shape and Reproducibility of Oropharyngeal Swallowing: Interaction with Bolus Volume and Age. *Dysphagia* **2015**, *30*, 540–550. [CrossRef] [PubMed]
46. Park, M.W.; Lee, D.; Gil Seo, H.; Han, T.R.; Lee, J.C.; Kim, H.C.; Oh, B.-M. Reliability of Suprahyoid and Infrahyoid Electromyographic Measurements during Swallowing in Healthy Subjects. *J. Korean Dysphagia Soc.* **2021**, *11*, 128–136. [CrossRef]
47. Cosentino, G.; Alfonsi, E.; Mainardi, L.; Alvisi, E.; Brighina, F.; Valentino, F.; Fierro, B.; Sandrini, G.; Bertino, G.; Berlangieri, M.; et al. The importance of the reproducibility of oropharyngeal swallowing in amyotrophic lateral sclerosis. An electrophysiological study. *Clin. Neurophysiol.* **2017**, *128*, 792–798. [CrossRef] [PubMed]
48. Hiraoka, K. Changes in masseter muscle activity associated with swallowing. *J. Oral Rehabil.* **2004**, *31*, 963–967. [CrossRef]
49. Zanato, L.E.; Chiari, B.M.; Vieira, M.M.; Bommarito, S. Study of the electrical activity of muscles: Masseter, temporal and supra-hyoid during swallowing. *Dent. Oral Craniofacial Res.* **2016**, *3*, 1–4.
50. Monaco, A.; Cattaneo, R.; Spadaro, A.; Giannoni, M. Surface electromyography pattern of human swallowing. *BMC Oral Health* **2008**, *8*, 6. [CrossRef]
51. Ko, J.Y.; Kim, H.; Jang, J.; Lee, J.C.; Ryu, J.S. Electromyographic activation patterns during swallowing in older adults. *Sci. Rep.* **2021**, *11*, 1–10. [CrossRef]
52. Leslie, P.; Carding, P.; Wilson, J. Investigation and management of chronic dysphagia. *BMJ* **2003**, *326*, 433–436. [CrossRef]
53. Karnell, M.P.; Rogus, N.M. Comparison of Clinician Judgments and Measurements of Swallow Response Time. *J. Speech Lang. Hear. Res.* **2005**, *48*, 1269–1279. [CrossRef]
54. Lin, C.-W.; Chang, Y.-C.; Chen, W.-S.; Chang, K.; Chang, H.-Y.; Wang, T.-G. Prolonged Swallowing Time in Dysphagic Parkinsonism Patients With Aspiration Pneumonia. *Arch. Phys. Med. Rehabil.* **2012**, *93*, 2080–2084. [CrossRef] [PubMed]
55. Teismann, I.K.; Warnecke, T.; Suntrup, S.; Steinsträter, O.; Kronenberg, L.; Ringelstein, E.B.; Dengler, R.; Petri, S.; Pantev, C.; Dziewas, R. Cortical Processing of Swallowing in Als Patients with Progressive Dysphagia—A Magnetoencephalographic Study. *PLoS ONE* **2011**, *6*, e19987. [CrossRef]
56. Ertekin, C.; Aydogdu, I.; Yuceyar, N.; Kiylioglu, N.; Tarlacı, S.; Uludag, B. Pathophysiological mechanisms of oropharyngeal dysphagia in amyotrophic lateral sclerosis. *Brain* **2000**, *123*, 125–140. [CrossRef] [PubMed]
57. Ertekin, C.; Aydogdu, I.; Tarlaci, S.; Turman, A.B.; Kiylioglu, N. Mechanisms of dysphagia in suprabulbar palsy with lacunar infarct. *Stroke* **2000**, *31*, 1370–1376. [CrossRef]
58. Krasnodębska, P.; Jarzyńska-Bućko, A.; Szkiełkowska, A.; Bartosik, J. Clinical and Electromyographic Assessment of Swallowing in Individuals with Functional Dysphonia Associated with Dysphagia Due to Muscle Tension or Atypical Swallowing. *Audiol. Res.* **2021**, *11*, 167–178. [CrossRef]
59. Taniguchi, H.; Tsukada, T.; Ootaki, S.; Yamada, Y.; Inoue, M. Correspondence between food consistency and suprahyoid muscle activity, tongue pressure, and bolus transit times during the oropharyngeal phase of swallowing. *J. Appl. Physiol.* **2008**, *105*, 791–799. [CrossRef]

60. Nascimento, W.; A Cassiani, R.; Santos, C.M.; O Dantas, R. Effect of Bolus Volume and Consistency on Swallowing Events Duration in Healthy Subjects. *J. Neurogastroenterol. Motil.* **2015**, *21*, 078–082. [CrossRef]
61. Youmans, S.R.; Stierwalt, J.A.G. Normal swallowing acoustics across age, gender, bolus viscosity, and bolus volume. *Dysphagia* **2011**, *26*, 374–384. [CrossRef]
62. Lee, J.; Sejdić, E.; Steele, C.M.; Chau, T. Effects of liquid stimuli on dual-axis swallowing accelerometry signals in a healthy population. *Biomed. Eng. Online* **2010**, *9*, 7. [CrossRef]
63. Funami, T.; Ishihara, S.; Nakauma, M.; Kohyama, K.; Nishinari, K. Texture design for products using food hydrocolloids. *Food Hydrocoll.* **2012**, *26*, 412–420. [CrossRef]
64. Igarashi, A.; Kawasaki, M.; Nomura, S.-I.; Sakai, Y.; Ueno, M.; Ashida, I.; Miyaoka, Y. Sensory and Motor Responses of Normal Young Adults During Swallowing of Foods with Different Properties and Volumes. *Dysphagia* **2009**, *25*, 198–206. [CrossRef] [PubMed]
65. Sifrim, D.; Vilardell, N.; Clavé, P. Oropharyngeal Dysphagia and Swallowing Dysfunction. *Front. Gastrointest. Res.* **2014**, *33*, 1–13.
66. Suzuki, M.; Sasaki, M.; Kamata, K.; Nakayama, A.; Shibamoto, I.; Tamada, Y. Swallowing Pattern Classification Method Using Multichannel Surface EMG Signals of Suprahyoid and Infrahyoid Muscles. *Adv. Biomed. Eng.* **2020**, *9*, 10–20. [CrossRef]
67. Molfenter, S.M.; Steele, C. Physiological Variability in the Deglutition Literature: Hyoid and Laryngeal Kinematics. *Dysphagia* **2010**, *26*, 67–74. [CrossRef] [PubMed]
68. Archer, S.K.; Garrod, R.; Hart, N.; Miller, S. Dysphagia in Duchenne Muscular Dystrophy Assessed Objectively by Surface Electromyography. *Dysphagia* **2012**, *28*, 188–198. [CrossRef] [PubMed]

sensors

MDPI

Article

Applying Machine Learning to Finger Movements Using Electromyography and Visualization in Opensim

Jose Amezquita-Garcia [1], Miguel Bravo-Zanoguera [1,2,*], Felix F. Gonzalez-Navarro [3], Roberto Lopez-Avitia [1] and M. A. Reyna [3]

[1] Facultad de Ingeniería, Universidad Autónoma de Baja California, Mexicali 21280, Mexico; jose.amezquita@uabc.edu.mx (J.A.-G.); ravitia@uabc.edu.mx (R.L.-A.)
[2] Ingeniería en Mecatrónica, Universidad Politécnica de Baja California, Mexicali 21376, Mexico
[3] Instituto de Ingeniería, Universidad Autónoma de Baja California, Mexicali 21280, Mexico; fernando.gonzalez@uabc.edu.mx (F.F.G.-N.); mreyna@uabc.edu.mx (M.A.R.)
* Correspondence: mbravo@uabc.edu.mx

Abstract: Electromyographic signals have been used with low-degree-of-freedom prostheses, and recently with multifunctional prostheses. Currently, they are also being used as inputs in the human–computer interface that controls interaction through hand gestures. Although there is a gap between academic publications on the control of an upper-limb prosthesis developed in laboratories and its service in the natural environment, there are attempts to achieve easier control using multiple muscle signals. This work contributes to this, using a database and biomechanical simulation software, both open access, to seek simplicity in the classifiers, anticipating their implementation in microcontrollers and their execution in real time. Fifteen predefined finger movements of the hand were identified using classic classifiers such as Bayes, linear and quadratic discriminant analysis. The idealized movements of the database were modeled with Opensim for visualization. Combinations of two preprocessing methods—the forward sequential selection method and the feature normalization method—were evaluated to increase the efficiency of these classifiers. The statistical methods of cross-validation, analysis of variance (ANOVA) and Duncan were used to validate the results. Furthermore, the classifier with the best recognition result was redesigned into a new feature space using the sparse matrix algorithm to improve it, and to determine which features can be eliminated without degrading the classification. The classifiers yielded promising results—the quadratic discriminant being the best, achieving an average recognition rate for each individual considered of 96.16%, and with 78.36% for the total sample group of the eight subjects, in an independent test dataset. The study ends with the visual analysis under Opensim of the classified movements, in which the usefulness of this simulation tool is appreciated by revealing the muscular participation, which can be useful during the design of a multifunctional prosthesis.

Keywords: electromyography; classification model; biomechanical simulation

Citation: Amezquita-Garcia, J.; Bravo-Zanoguera, M.; Gonzalez-Navarro, F.F.; Lopez-Avitia, R.; Reyna, M.A. Applying Machine Learning to Finger Movements Using Electromyography and Visualization in Opensim. *Sensors* **2022**, *22*, 3737. https://doi.org/10.3390/s22103737

Academic Editor: Ernest N. Kamavuako

Received: 22 March 2022
Accepted: 1 May 2022
Published: 14 May 2022

1. Introduction

Research in prosthetic hand control applications most often involves the decoupling of the surface electromyography (sEMG) signal to decipher the natural regulation. Interpretation of the sEMG signals is an active area of research. One objective is to achieve efficient control of prostheses, similar to the natural movements of the body. Within these research publications, one can find studies searching for the ideal place for the electrodes in the muscular region of interest [1–4]; studies focused on the importance of the type, size, shape and material of the electrode to be used [5–9]; studies using data mining to find features in the raw sEMG signals to achieve their decoding [10–12]; and studies evaluating the implementation to detect the target movements [13–15], aiming to implement the processing in systems embedded in microcontrollers.

Prosthetic electronic devices increasingly have more degrees of complexity; however, it is reported that, within these devices, 60% have 1 to 4 degrees of freedom (DoFs), 30% have 5 to 10 DoFs and only 10% have more than 10 DoFs [16]. A DoF within a prosthetic device should be understood as the motion in one direction of the possible movements of a natural joint. Devices and applications evolve into more complex systems; therefore, control systems with a greater number of functionalities are necessary, but they must be intuitive for human control, which is possible with machine learning techniques [15]. A considerable number of investigations work with machine learning for the classification and interpretation of EMG signals [17–20]; however, there is still no model that is applicable in complex systems and outside of controlled environments in laboratories. In our investigations regarding the decoding of the sEMG signal, we found the use of deep learning can be used as a method to improve classic classifiers but demands a large amount of data through many layers of processing [18], the reduction in dimensionality through specific methods [19] allows the selection of a smaller number of features with higher information quality, and the use of time–frequency features [20] allows extracting information hidden in the raw EMG signal and the transfer to new spaces to extract relevant information. Even though diverse research approaches and applications are emerging in the state of the art, such as the mathematically intense genetic algorithms and adaptive neuro-fuzzy systems [21,22] which are powerful methods to solve difficult regression problems, the load of monitoring the progress of each of the DoFs involved in the movement control would make it impractical with these models. The control of upper-limb prosthetic systems requires further research regarding practical utility, and here using traditional methods to classify multiple discrete targets brings an elegant and viable solution.

It is difficult to obtain real-time applications with multifunctional prostheses due to, among other issues, the nature of the sEMG signal and the large number of processing operations required to analyze it. Another drawback regarding progress is the limited availability of standardized EMG databases, but still, there are some databases that help address the complexity of multifunctionality and which have sEMG signal conditioning best practice elements. There are known sites for online multifunction EMG databases that include movements specific to the hand and fingers [23–29]. Some of the sites have more than one dataset with different features [24,25,28,29]. The amounts of electrodes and movements vary, but the movement experiments are captured with EMG surface electrodes, and some of the databases also contain information from other types of sensors.

In a previous work [17], a first attempt was made to create a simple classification model of EMG signals, to be combined with a modified virtual biomechanical model of the wrist and hand in Opensim [30]. In this work, we propose using an online database to decode the coordinated muscle activity obtained from an array of sEMG electrodes in the forearm, to develop a model that classifies 15 hand movements, and to transform the classification sequence into virtual movement with the Opensim environment. In Opensim, the evolution of muscle movements can be virtually observed from the results of the classification model and visually compared with those of the ideal movement.

2. Materials and Methods

2.1. Database of Predefined Finger Movements

Although the relatively poor repeatability of the surface EMG measures is a known issue [31], an exhaustive search was conducted to find databases published online that were open access [23–29], to determine which was the most convenient to use. A database found in an EMG repository was selected [25] since it was one of the most complete with movements that encompassed each finger. This database consists of 15 finger movements, both individual and combined. Data were acquired in line with the standard protocol described in [25,32,33] and summarized next.

2.1.1. Electrode Application Protocol

A ring array of eight sEMG electrodes were equally spaced across the circumference of the right forearm, with an electrode initially placed over the palmaris longus muscle, and the complete electrode set is pictured in Figure 1a,b. According to the electrode position pictures and the given reference descriptions, an illustration that approximates the distribution of the electrodes on the volume of the forearm is shown in Figure 1c. The datasets were recorded using the Bagnoli desktop EMG system (Delsys, Inc., Boston, MA, USA) [34] with the DE-2.1 sensor with 10.0×1.0 mm contact dimensions and 10 mm contact spacing, in differential detection mode, with an overall noise of $\leq$1.2 uV (RMS, R.T.I), and a bandwidth of 20–450 Hz. A 2-slot adhesive skin interface was applied on each of the sensors to firmly stick them to the skin. A conductive adhesive reference electrode (Dermatrode reference electrode) was placed on the wrist of each of the subjects during the experiments. The collected EMG signals were amplified using a total gain of 1000. A 12-bit analog-to-digital converter (National Instruments, BNC-2090) was used to sample the signal at 4000 Hz; the signal data were then acquired using Delsys EMGWorks Acquisition software. The EMG signals were then bandpass filtered between 20 and 450 Hz with a notch filter implemented to remove the 50 Hz line interference.

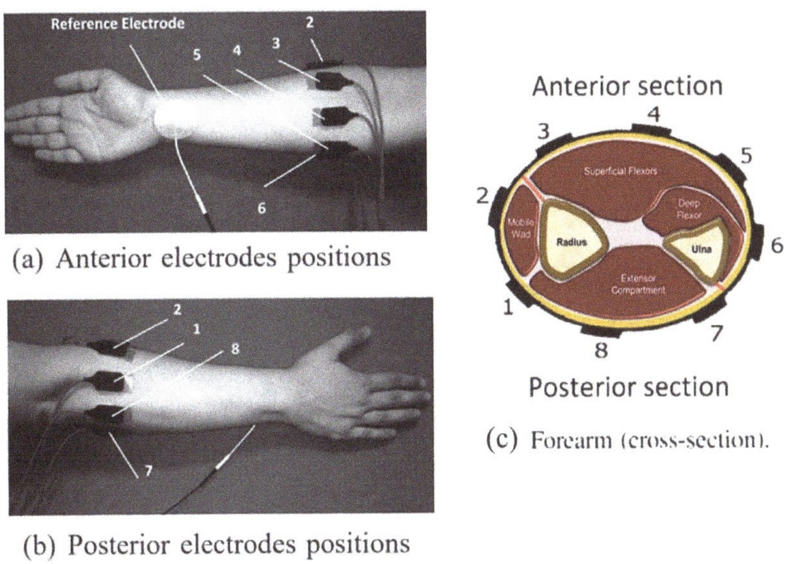

Figure 1. Electrode placement on the right forearm. (**a**) Anterior electrode positions; (**b**) posterior electrode positions; (**c**) muscle zones and electrodes placed on the cross-section of the forearm. (**a**,**b**) are reprinted/adapted with permission from Ref. [32], Copyright 2012, IEEE.

2.1.2. Subjects

Eight normal subjects (six males and two females) aged between 20 and 35 years were recruited to perform the required finger movements, and the execution of each movement was repeated 12 times. They were all from Sydney, with a random sampling; i.e., no specific sampling requirements were reported. The subjects were all normally limbed with no neurological or muscular disorders. All participants provided informed consent prior to participating in the study as was approved by the university research ethics committee and consented to their data being used for research purposes [25,32].

2.1.3. Database Protocol

Subjects were seated on an armchair, with their arms supported and fixed at one position. The 15 movements correspond to flexions, including the flexion of each individual finger and combinations between them: thumb (T_T); index (I_I); middle (M_M); ring (R_R); little finger (L_L); the combinations T_I, T_M, T_R, T_L, I_M, M_R, R_L, I_M_R, and M_R_L; and the closed hand (HC). Table 1 indicates the class label assigned to each of the movement types in the database. The measured region has a signal detection volume that is assumed to lie between the largest number of muscular bodies and the region of the joints of the elbow; the relevance of the use of an array around the circumference of the forearm is because it also applies to trans-radial amputees [33]. In this ring electrode arrangement, the information related to the generated movement is found in the surrounding tissue and is transmitted in all directions by that section of the forearm. Although it is not possible to indicate a specific muscle for each electrode of the ring, all the electrodes pick up signals from all the muscle distribution layers. The information available is that there were electrodes in a given position when the database was captured; repeatability is likely to be acceptable as long as the same experimental protocol conditions are maintained in every test subject.

Table 1. Relationship of movement types of the database and their class labels assigned. Flexion of the fingers involved is indicated.

Class Label	Movements Evaluated	Description
1	HC	Closed hand
2	I_I	Index
3	I_M	Index–middle
4	IMR	Index–middle–ring
5	L_L	Little finger
6	M_M	Middle
7	M_R	Middle–ring
8	MRL	Middle–ring–little
9	R_L	Ring–little
10	R_R	Ring
11	T_I	Thumb–index
12	T_L	Thumb–little
13	T_M	Thumb–middle
14	T_R	Thumb–ring
15	T_T	Thumb

2.2. *Raw Data, Elimination of Outliers, Windowing and Preparation of Feature Matrix*

The database consists of 8 subjects and 15 movements, in which each subject performs 12 repetitions for each movement; that is, the database contains 1440 repetitions of some type of movement. A repetition of a movement is equivalent to collecting, at a rate of 4000 samples per second, the signals of eight sEMGs for 5 s. An sEMG is encoded with 12 bits, but it is read as a floating-point unit. Therefore, a repetition of a movement is equivalent to (4000 samples/s) $\times$ (5 s) = 20,000 data sample packages, including eight electrode signal channels at each sample time.

The database was downloaded from [25], where a zip file can be obtained with eight folders, one folder for each subject. Within each folder are files in CSV format named with the name of the movements. The data were loaded with MATLAB's *csvread* function. For each subject, a matrix was created for each movement, named with the abbreviation of the movement. These matrices comprised eight sEMG columns and 240,000 rows of data, which represent the total of 12 repetitions of a movement per subject. In addition, a column with a class label number was added that identifies the type of movement, according to Table 1, so that at all times it is known which sample corresponds to each subject, type of movement and sEMG electrode number.

A folder was created in MATLAB for each subject in which the 15 matrices of the movements of each subject were stored. The 15 matrices of a subject were concatenated into a new matrix named sub(i). This matrix was 3,600,000 rows by 9 columns, and it was saved for each subject. Then, in a global matrix, the data of the eight subjects were stored, forming a matrix of 28,800,000 rows and 9 columns, all of which are indicated in Figure 2 with circle 1. It should be clarified at this point that the data are ordered from subject 1 to subject 8; that is, the data in the matrix start with movements 1–15 of subject 1, then include movements 1–15 of subject 2, and so on until the movements of subject 8. When in a consecutive row, there is a change in column 9, of movement type, from label class 15 to label class 1, indicating the data edge that exists between two subjects.

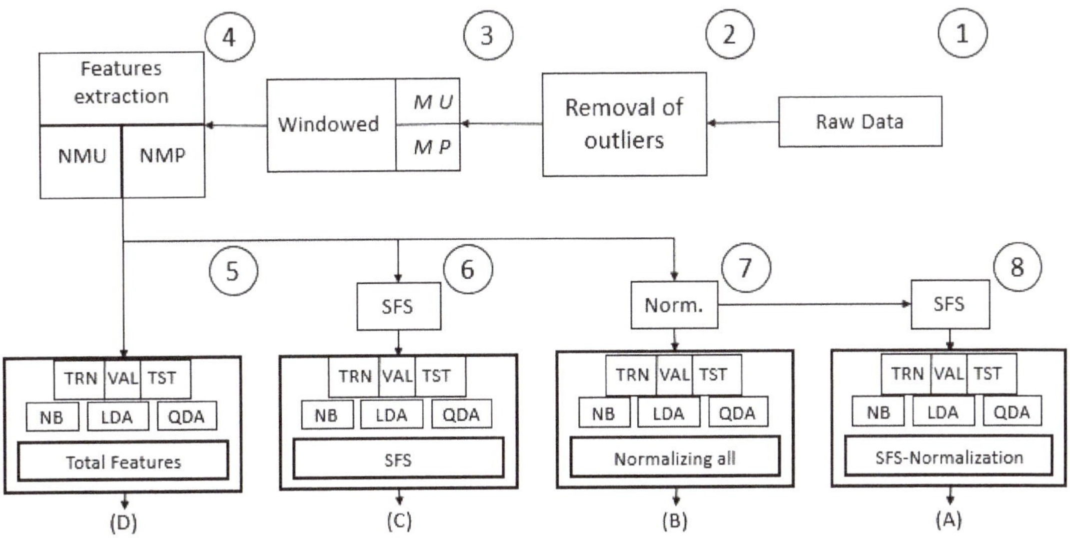

Figure 2. Experimentation conducted to determine the best option for creating a classification model considering recognition percentage and simplicity of the model. MU = single matrix (all eight subjects concatenated), MP = matrix per subject (eight matrices are formed). The number of circles defines the order of processing. The nomenclature (**A–D**) serves for Duncan's significance test treatment identification that compares the output of each experiment.

Due to the nature of the signal and the large number of data, these were preprocessed with cleanup and reduction methods to concentrate the information before creating classifiers. This preprocessing included the steps of removing outliers, grouping clean data into a single global matrix and across multiple matrices (by individual subjects), setting up a data window to extract information, and measuring features per window to characterize the EMG signals.

2.2.1. Method of Elimination of Outliers by a Chi-Square Distribution

According to the literature, an outlier is an observation that deviates from other observations and raises suspicions about being generated by mechanisms external to natural ones. Some statistical methods for the detection of multivariate outliers use the data distribution center and an established distance; in our case, the Mahalanobis distance was used to determine which data are outliers. The outlier elimination method with a chi-square distribution was also used [35]. The global matrix, composed of the eight subjects, was used to find outliers in the entire database. During the outlier elimination, when an irregular reading of an EMG electrode was detected, the entire line with all its eight EMG signal channels was eliminated, when the channel corresponding to that atypical event was included. This method was implemented in MATLAB, delivering a global matrix without outliers; the pseudocode for this process is shown in Algorithm 1. From this updated matrix, a separation into new matrices by individual subjects was carried out manually, and eight matrices with different lengths were obtained given the nature of the elimination of outliers. As a result of this process, the information regarding the data values eliminated and those maintained was known, as the sum of both must coincide with the total data.

Algorithm 1 Pseudocode for Outlier Removal

The data to evaluate are imported.
Statistical parameters (Mean, Covariance) are calculated.
Calculation of the parameters (New_Data, inverse covariance) to obtain the Mahalanobis distance.

- Calculation of New_Data = data minus the mean.
- Inverse covariance is calculated.

Product of New_Data and inverse covariance.
Critical value is obtained according to a chi-square distribution with the function *ncx2inv* with MATLAB.
Data are separated according to the critical value.
Outliers are removed and maintained matrices are created.

2.2.2. Preparation of Matrices and Transformation of Data into Functional Features

With the data no longer containing outliers, we created two datasets: The first was a global matrix updated with the EMG data of all the subjects, which is called MU and has 26,778,435 data packets; 7.02% of data were eliminated for being atypical elements. The second was a dataset including the data of each subject separately, in eight matrices (called MPs), with lengths between a minimum and a maximum of 3,147,328 and 3,540,137 data packets per subject.

Taking a group of 250 rows to form a measurement window, the data were transformed into functional features. Of the 250 rows of data in the window, 12 features were calculated for each EMG channel (see Section 2.3.1), which were concatenated horizontally, resulting in a single vector of 96 electrode features for each window. In addition, the class of the motion type, to which the new vector belongs, was added in column 97. This vector of features was stored in a new matrix, continuing with each window, updating the matrix of features until reaching the end of a movement type. The final window of a specific motion type was calculated with the remaining data, which can range from 1 to 250 data rows.

The arrays have a column with the class label, and depending on the class, the data can be separated into motion subarrays. For any of the two groups of matrices (global or per subject), the MATLAB *find()* function was used to select the movements that belong to classes 1 through 15; *find()* gives us the position in which the rows of the specific movement searched were found. A submatrix of features was created for each movement with subscript j, which went from movement 1 to movement 15. Within this cycle of measurement of features, the first submatrix was taken, that is, movement 1, and once it was finished with the movement type, it started with the next one, and so on until the 15 movements were finished. At this point, 15 submatrices of each movement with their corresponding feature vectors resulted.

Upon having the features' measurements, a rearrangement was performed to obtain the matrix of a particular subject or the one corresponding to the total MU matrix. From having the first subject, the cycle was repeated manually, adding the data of subject 2 and so on until subject 8 was reached. At the end of this process, we have the features in two types of data blocks: a matrix that groups all the subjects and their movements, named NMU, and the eight matrices that group the movements by subjects (called NMPs). These procedures are encapsulated in Figure 3.

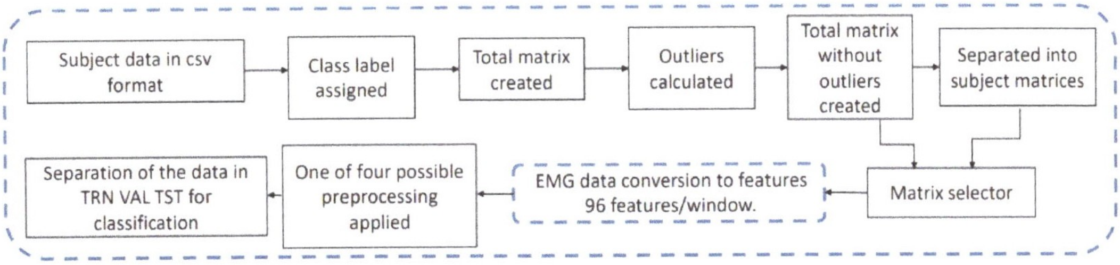

Figure 3. General sequence of data manipulation: from the database files to the preparation of feature matrices, before the experiments being carried out.

Some investigations work only with data from individual subjects, and when modeling the classifiers by subject, very good classification percentages are obtained. Although in other studies the clustered data of a group of subjects are used to generate a single classification model applied to the people in the group, with less promising results, in this study, we compare the classifiers designed with the two types of data groups.

2.2.3. Selected Features from the Literature

The literature consulted revealed that features useful for extracting information from raw EMG signals are divided into those of the time domain and those of the frequency domain, both widely used in classifiers. As one objective of this work is to generate the simplest and quickest processing in terms of execution speed, it was decided to use only features in the time domain. As mentioned in [36], one of the major disadvantages of these types of features is that a stationarity property is assumed, a property that does not coincide with the nature of the EMG signal; however, they have provided good gesture classification results in the past (see [32]).

For the size of the data windows, it is necessary to consider the processing time given to the calculation of the features. In this case, the aim is for the classification to be applied in real time; therefore, it is convenient to work with small data windows. A non-overlapping window was used to process the raw EMG data, with a length of 250 rows of raw data per window. A period of 62.5 ms is required to cover this window length since a row of raw data is captured at the sampling rate of 4 KHz.

Twelve features were proposed to be extracted from the raw EMG data, and their definitions are in [37]: mean absolute value; mean square value; simple square integral; square mean; EMG variance; TM3, TM4 and TM5 time moments; length of wave; zero crossings; myopulse percentage ratio; and curve sign change. The 12 proposed features were measured in each of the eight EMG channels and were concatenated, producing a vector of 96 data (features) for each window per subject. The windowing follows after being an MU or MP matrix, evaluating the 12 features per EMG channel to create new matrices: MU > Windowing > Features > NMU, or MP > Windowing > Features > NMPs. This data transformation process is conceptually visualized in Figures 2 and 3 and was developed to go along with the classical methods in machine learning and pattern recognition to extract information from the raw EMG signal; variations can be observed in the window sizes and in the selected features.

2.3. *Classification Models*

2.3.1. Machine Learning

Classification problems are often divided into two stages: the decision stage and the inference stage. These two stages are visualized in Figure 4. The decision stage is when the model is being trained. The inference stage is used in classification to predict data once the model is already trained. These stages were used extensively in this study. The methodology in Figure 3 was established.

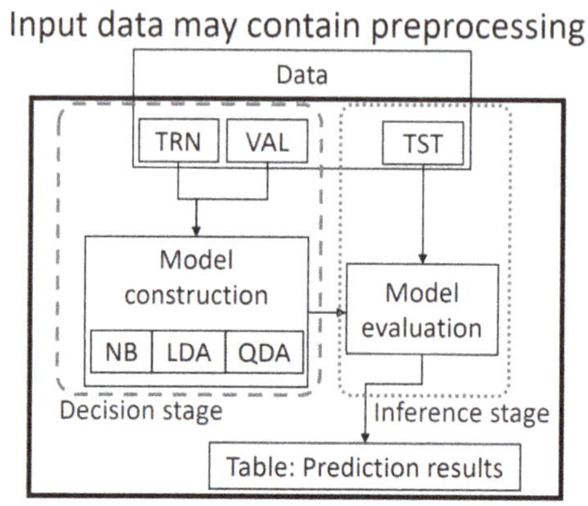

Figure 4. Creation of the classification model. The decision and inference stages of machine learning can be observed in the creation of classification models.

The classifiers considered for their simplicity were naive Bayes, LDA and QDA, all with a Bayesian foundation and probabilistic theory. The importance of these classifiers is that they are very efficient, as well as simple, and they have been used to implement machine learning in microcontrollers [38]. The MATLAB function *fitcdiscr* was used to create LDA and QDA, and the *fitcnb* function was used for naive Bayes.

Four classification schemes were developed, as conceptualized in Figure 2, generating combinations of two preprocessing methods; the forward sequential selection method and the normalized one, were evaluated, in addition to the version without preprocessing. The intention was to investigate the efficiency of these classifiers, comparing the different versions in case there were any statistically significant differences, and to decide the best route for conversion to a new input data space that increases the recognition of the classifier model and/or assists in reducing the number of required features.

2.3.2. Cross-Validation

One of the most used methods in the validation of classification models in machine learning is cross-validation. This enables, within a defined dataset, considering all data as training, validation and test data, and it validates that the results are independent of the data partitions; the calculations were based on [35]. The methodology in Figure 5 establishes the order in the separation of the data to conduct an adequate cross-validation. This process was performed for each matrix of the eight subjects and for the matrix with the data of the eight subjects (NMU). Matrices were created for training, validation and testing. A MATLAB function called *cvpartition* was used to separate the data (partition) for cross-validation. This function was used to separate the original set of data into 10 parts, 9 of which were assigned a TRN-VAL label and 1 of which was assigned a TST label. In this way, we had 10 parts of our original data. Next, the function created 10 datasets by traversing the TST label to each part of the original set (divided into 10). The function was used twice, once to separate the original set of data into 10 parts and create two blocks called TRN-VAL and TST from the original set and once to separate the data from the TRN-VAL block into two TRN sub-blocks and VAL. This process is illustrated in Figure 5: below each cvpartition block are columns of data with 10 sections, which are assigned a name. For the first block, the columns form two blocks: TRN-VAL (with nine parts) and TST (with one part). This method creates 10 datasets with the same total data. In the second cvpartition block are another 10 columns below, which is how one of the previous TRN-VAL datasets of the first block was divided. Then, 1 of those 10 sets was divided into 10 subsets of data composed of 9 parts of TRN and 1 part of VAL in each of the 10 columns.

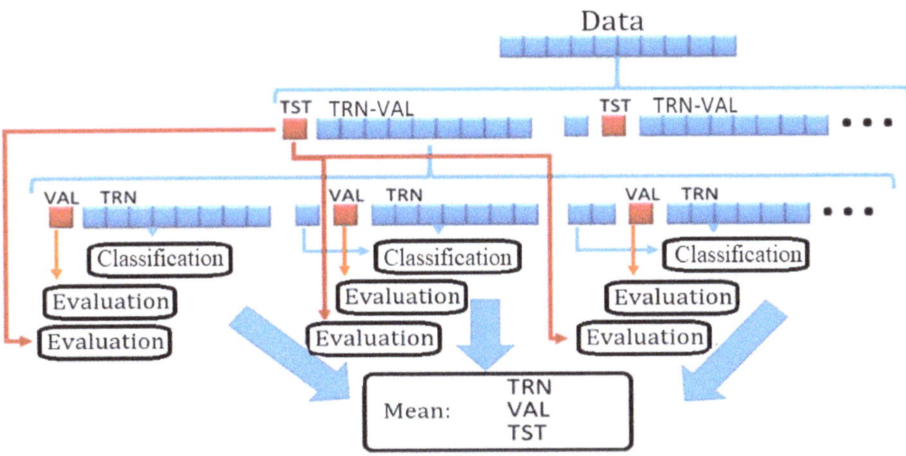

Figure 5. Methodology for cross-validation. The original dataset is split several times to have partitions of data into subsets for model training and evaluation.

At this point, we had 10 datasets divided into TRN-VAL and TST. In the same way, the 10 sets of TRN-VAL were now divided each into 10 new subdatasets in TRN and VAL. That is, we had 100 TRN models validated with 100 sets of VAL. Each result was averaged to calculate a real percentage of training and validation. At the end of the methodology in Figure 5, we obtained 100 classification models in which the first partition of the test (10 in total) of each dataset was tested in 10 subsets of TRN and VAL. Those results were averaged for each classifier separately.

2.4. *Preprocessing Techniques Prior to Classification*

When working with classification models obtained through machine learning, it is desirable to have features that generate maximum differentiation between classes; therefore, it is necessary to use preprocessing of features. This way, the classification models will perform better. Therefore, one task was to find the features that best separate the motion classes in their corresponding spaces and that improve prediction. A feature-selection method was used that reduces the data volume based on the quality of each feature. Another method used was the normalization of the features. Finally, the best possible combination of preprocessing was converted to a new space by the sparse matrix method.

2.4.1. Sequential forward Selection (SFS)

SFS is an iterative method that provides a direct route to determine which features improve the classification results. With 12 features measured for each electrode, 96 features were generated; that is, 96 operations must be executed for each window. A large number of simple operations might work for a real-time procedure; however, the aim was to reduce the number of electrode features since some can impair the performance of the classifier.

Generally, this method analyzes each feature in an orderly manner from 1 to 96 and is an iterative process. Once the method has analyzed each feature individually, it will take the feature that provides the best percentage of recognition; the process will then form groups of two. Once it has determined the best of the individual features, the process will join this feature with each of the 95 remaining ones; this process is detailed in Algorithm 2. Then, the feature that offers the highest percentage to the classifier is selected, and so on until the percentage stagnates.

Algorithm 2 Pseudocode Implemented for forward Sequential Selection

```
SPACE = 1 ... L                              //L = number of total features.
EXIT = false
WHILE NOT EXIT
FOR i =1 TO |SPACE|
TEMP(i) = J(SUBSET V SPACE(i))               //TEMP = Temporal variable.
END
BEST_i = ARGMAX(TEMP)                        //The maximum value of the
                                             //TEMP vector is stored in BEST.
IF TEMP(BEST_i) > BEST_EVAL                  //Comparison between the best stored
SUBSET = [SUBSET, SPACE(BEST_i)]             //evaluation and the best evaluation
BEST_EVAL = TEMP(BEST_i)                     //of the new subset.
SPACE = SPACE-SPACE(BEST_i)
ELSE
OUTPUT = true
END
END
RETURN SUBSET, BEST_EVAL
```

Each classifier considered seeks to separate the classes in its own way, and therefore, it can take the features that offer the best results. All the features' data were taken from each of the subjects and from the NMU matrix, and a classifier was generated for each dataset and for each model (NB, LDA and QDA). That is, the percentage of training was calculated, and the groups of features were saved based on this. By having the groups of features for each classifier considered the best, the data were taken again and separated into TRN, VAL and TST. With 10 × 10 cross-validation, a group of models was generated for each classifier. It is evident how many features remained and what percentage of classification was obtained for TRN, VAL and TST. A more detailed description can be found in [39].

2.4.2. Normalization

If the set of measured features is transformed into a normal dataset (i.e., it has a zero mean and a standard deviation of 1), all the measurements of all the evaluated features are put in the same range of values. Therefore, a feature in a classifier could operate more efficiently. The procedure was conducted in a general way for the three classifier models. The description of this data transformation can be found in [35].

2.4.3. Conversion to Sparse Matrix Space

The conversion of features into new spaces, although it increases the processing load, can provide good results regarding increasing the recognition of the classifier model and/or reducing the number of features. It was decided to use this method because it is an innovative method that has not been widely used and its performance on muscle signals has not been reported. This algorithm is used in the first instance to increase the distances between the evaluated classes and reduce the distances between the two data of the same class. The description of this method is in [40]. The method was carried out only with the preprocessed data that provided the best results; the group of features that increased performance was taken, and these were transformed into a new space, to apply later only the best efficiency classifier model and to calculate the recognition percentage. Figure 6 illustrates the concept of conversion into a new space to improve the classification model.

2.5. Methods of Statistical Validation of the Results

Guided by the statistical tests from similar investigations, the proposed schemes of ANOVA method, multiple comparison test of Duncan and statistical formulations are used below for the comparison of results.

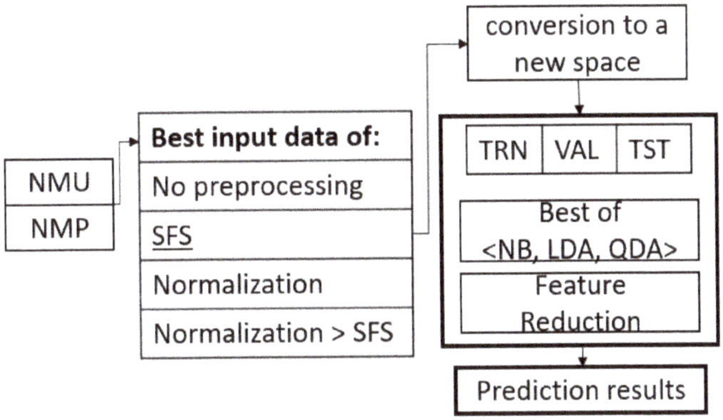

Figure 6. Creation of the classification model after selection of the best preprocessing from the conversion into a new space.

2.5.1. ANOVA Method

The ANOVA method was used to compare and study the effect of the methods applied to our data regarding the means of the recognition percentages obtained from the evaluated classifiers. This analysis is a statistical test to use when comparing the means of two or more groups. The null hypothesis, from which the different types of ANOVA start, is that the means of the groups are statistically equal; that is, the mean is the same in the different groups.

2.5.2. Duncan's Method

This method is a multiple comparison test that enables us to compare the means of the treatments (procedures applied to the data and resulting in a percentage of recognition) after having rejected the null hypothesis.

2.5.3. Statistical Formulations

The sensitivity, specificity, precision and F1 score are statistical parameters that enable us to evaluate the results obtained. These parameters are visualized in Table 2. They can also be obtained via a confusion matrix, in which a class of movement is compared against the other classes of movement, specifically locating each error that occurred when classifying the data. These parameters should be close to 100% for the model to be ideal.

Table 2. Statistical formulations, in which TP = true positive, FP = false positive, TN = true negative and FN = false negative.

Formulation	Operation
A = Sensitivity	TP/(TP + FN)
B = Specificity	TN/(TN + FP)
C = Precision	TP/(TP + FP)
D = F1 score	2TP/(2TP + FP + FN)

2.6. *Opensim Model of the Wrist and Hand Used for the Analysis*

Opensim software is an open access biomechanical simulation program. Among its characteristics, it enables muscle evaluation and movement analysis with visualization files with a MOT extension. In Opensim's repository of models, there is a limited model of the wrist with 10 degrees of freedom, 28 joints and 23 actuators (muscles) with movement in the forearm, wrist, thumb (no flexion) and index [41]. This original model was modified for this study by adding the missing degrees of freedom in the thumb, middle finger, ring finger and little finger to reproduce the total movements that a human hand can perform.

Figure 7 shows how the hand model appears in the graphical interface of the Opensim version 4.0 main screen.

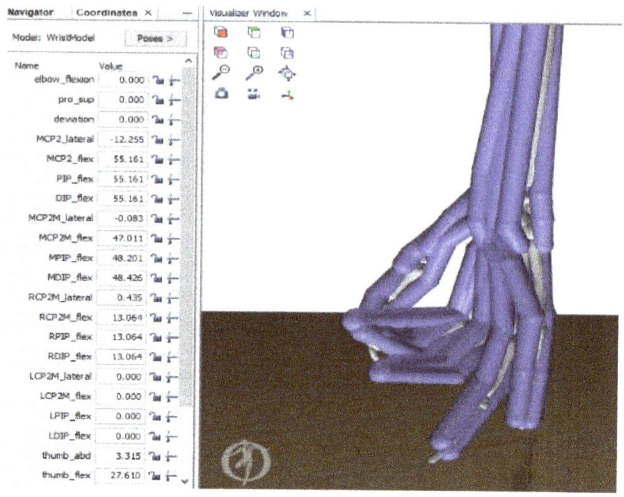

Figure 7. Screenshot of the software application displaying the degrees of freedom of the model used in Opensim.

The completed model was used for the simulation of the hand movements performed in the analyzed database. This model is found in free form in [30] and has 21 degrees of freedom and 36 joints, allowing any possible hand movement to be carried out. Table 3 lists the relationship of the degrees of freedom of the model with the movements considered from the database.

Table 3. Relationship degrees of freedom (DoFs) of the Opensim model of the wrist with the movements of the database.

	Database Finger Movements														
DoF	**HC**	**I**	**IM**	**IMR**	**L**	**M**	**MR**	**MRL**	**RL**	**L**	**TI**	**TL**	**TM**	**TR**	**T**
MCP2_lateral	x	x	x	x							x				
MCP2_flex	x	x	x	x							x				
PIP_flex	x	x	x	x							x				
DIP_flex	x	x	x	x							x				
MCP2M_flex	x		x	x		x	x	x					x		
MPIP_flex	x		x	x		x	x	x					x		
MDIP_flex	x		x	x		x	x	x					x		
RCP2M_lateral	x			x			x	x	x					x	
RCP2M_flex	x			x			x	x	x					x	
RPIP_flex	x			x			x	x	x					x	
RDIP_flex	x			x			x	x	x					x	
LCP2M_lateral	x				x			x	x	x		x			
LCP2M_flex	x				x			x	x	x		x			
LPIP_flex	x				x			x	x	x		x			
LDIP_flex	x				x			x	x	x		x			
thumb_abd	x										x	x	x	x	x
thumb_flex	x										x	x	x	x	x
TCP2M_lateral	x										x	x	x	x	x
TCP2M_flex	x										x	x	x	x	x
TCP2M2_flex	x										x	x	x	x	x

Simulation of Movement in Opensim

The movements were idealized for the simulation of the real movement trajectory; these must be stable and smooth in their execution. We consider that the 15 movements of the database are of this type, and each represents a complete cycle (or repetition) recorded by means of sEMG signals and described by photographs. The information on the database was used, the photographs included, but there were no other spatial measurements provided. With the EMG signal classifiers designed here, the type of movement that an input data window brings is estimated. With the predictions of the type of movement and its window time, a reproduction of the movement can be created virtually through Opensim. Every 62.5 ms (the duration of a data window), an output generated in the classifier demands the next video frame for the virtual model in Opensim. For this reason, advancement vectors were created for each of the movement classes, which contain the degree of rotation of each joint involved capable of reproducing any of the movements in the database.

In principle, we deduce the 15 ideal movements that the database contains. Therefore, the initial and final positions of the movements were taken as a reference based on their description [32]. Considering that each movement lasts for 5 s and that the number of decisions of the classifier in that period is 80 window times, and with the full range of motion of each of the involved joints, their degree of rotation was calculated as the progress of the movement video frame. Thus, the 15 advance vectors were formed for each movement class; Table 4 lists all of them. Then, after obtaining the tag predictions of the classifier and having the table of advance vectors, a motion file can be created for any complete finger movement input, as shown in Algorithm 3. In addition, 15 ideal motion files with the MOT extension can be created by simply accumulating the specific advancement vector itself 80 times, corresponding to the 5 s of a movement repetition. Once the MOT files are obtained, any movements can be viewed using the Opensim platform.

Algorithm 3 Pseudocode for the Creation of a Motion File From Classification. MOV_NUMBER Matrix Corresponds to Advancement Vector Table 4

```
MOV_MATRIX = [zeros]
count = 0
FOR i TO |MAX Window number|
Decision = Model Decision(i)
IF Decision is 1
if count < 18
MOV_MATRIX(i + 1,:) = MOV_MATRIX(i,:) + MOV_NUMBER(1,:)
count = count + 1
else
MOV_MATRIX(i + 1,:)= MOV_MATRIX(i,:) + MOV_NUMBER(2,:)
ELSE
MOV_MATRIX(i + 1,:)= MOV_MATRIX(i,:) + MOV_NUMBER(Decision + 1,:)
END
RETURN MOV_MATRIX
```

It is possible to observe the movements characterized by classification errors. The visualization of a badly classified movement can offer a glimpse of how serious or acceptable that error may be; it is simply a tool to discern a movement classification. The test data can be separated to verify which of the subjects has a worse ranking. This process identifies the worst movement evaluated to appreciate a real reproduction (with complete data) of a subject´s repetition movement in the simulation.

3. Results

3.1. Preprocessing and Processing of the EMG Signal

3.1.1. Elimination of Outliers

The total raw data for each movement were 240,000 lines per subject. However, due to eliminating outliers, these data were reduced, generating a number of data maintained per movement and per subject. The percentage data eliminated by movements ranged from 1.99% (in I_M) to 26.00% (in HC). The percentages of the data eliminated per subject ranged from 1.66% (subject 5) to 12.57% (subject 4). Figure 8 is a graph encompassing the maintained and eliminated values, with information on movement and subject, illustrating what remains of the raw data per movement and per subject.

Table 4. Advancement vectors for each of the movement classes. A vector contains the degrees of rotation for each of the joints considered in the movement. A special case is the hand close motion, which required one of two possible vectors, depending on the time position of the motion. The labels of the joints correspond to the Opensim model [30].

	Joints/DoFs																			
	MCP2_lateral	MCP2_flex	PIP_flex	DIP_flex	MCP2M_lateral	MCP2M_flex	MPIP_flex	MDIP_flex	RCP2M_lateral	RCP2M_flex	RPIP_flex	RDIP_flex	LCP2M_lateral	LCP2M_flex	LPIP_flex	LDIP_flex	thumb_abd	thumb_flex	TCP2M_lateral	TCP2M_flex	TCP2M2_flex
HC1	−0.3225	1.4516	1.4516	1.4516	0	1.4516	1.4516	1.4516	0.0483	1.4516	1.4516	1.4516	0.3225	1.4516	1.4516	1.4516	0	0	0	0	0
HC2	−0.3225	1.4516	1.4516	1.4516	0	1.4516	1.4516	1.4516	0.0483	1.4516	1.4516	1.4516	0.3225	1.4516	1.4516	1.4516	0.14	1.26	−0.4	0.2	0.84
I_I	−0.3225	1.4516	1.4516	1.4516	0	0	0	0	0	0	0	0	0	0	0	0	0	0	0	0	0
I_M	−0.3225	1.4516	1.4516	1.4516	0	1.4516	1.4516	1.4516	0	0	0	0	0	0	0	0	0	0	0	0	0
IMR	−0.3225	1.4516	1.4516	1.4516	0	1.4516	1.4516	1.4516	0.0483	1.4516	1.4516	1.4516	0	0	0	0	0	0	0	0	0
L_L	0	0	0	0	0	0	0	0	0	0	0	0	0.3225	1.4516	1.4516	1.4516	0	0	0	0	0
M_M	0	0	0	0	0	1.4516	1.4516	1.4516	0	0	0	0	0	0	0	0	0	0	0	0	0
M_R	0	0	0	0	0	1.4516	1.4516	1.4516	0.0483	1.4516	1.4516	1.4516	0	0	0	0	0	0	0	0	0
MRL	0	0	0	0	0	1.4516	1.4516	1.4516	0.0483	1.4516	1.4516	1.4516	0.3225	1.4516	1.4516	1.4516	0	0	0	0	0
R_L	0	0	0	0	0	0	0	0	0.0483	1.4516	1.4516	1.4516	0.3225	1.4516	1.4516	1.4516	0	0	0	0	0
R_R	0	0	0	0	0	0	0	0	0.0483	1.4516	1.4516	1.4516	0	0	0	0	0	0	0	0	0
T_I	0	0.3625	0.6[illegible]5	0.75	0	0	0	0	0	0	0	0	0	0	0	0	0	0.7875	−0.25	0.2875	0.1273
T_L	0	0	0	0	0	0	0	0	0	0	0	0	0	0.0585	0.95	0.8812	0.175	1.125	−0.25	0.4125	0.1825
T_M	0	0	0	0	−0.0412	0.28	0.875	0.9875	0	0	0	0	0	0	0	0	0.1875	0.575	−0.25	0.3825	0.325
T_R	0	0	0	0	0	0	0	0	0	0.1691	0.9875	0.875	0	0	0	0	0.1812	0.875	−0.25	0.3678	0.3125
T_T	0	0	0	0	0	0	0	0	0	0	0	0	0	0	0	0	0.14	1.26	−0.4	0.2	0.84

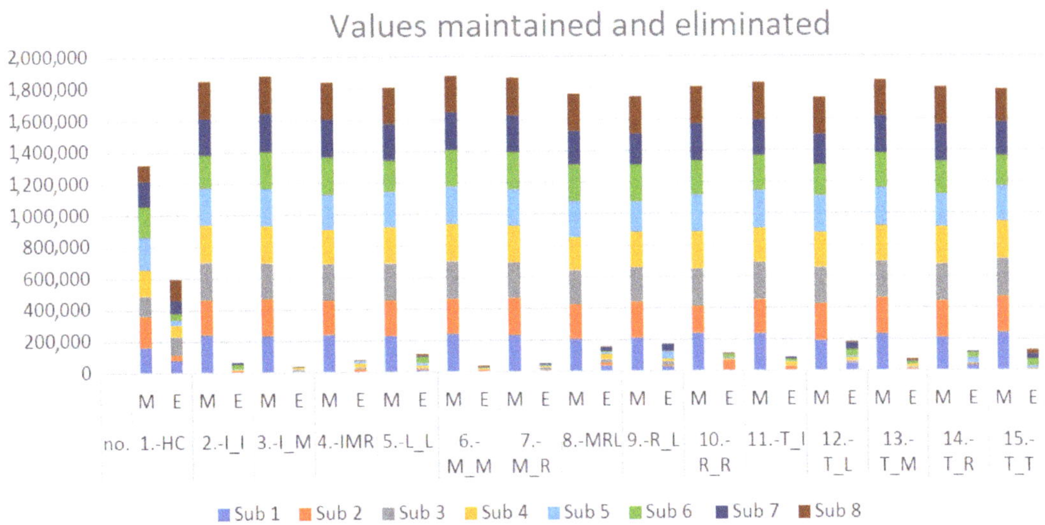

Figure 8. Outliers maintained and eliminated. M = maintained, E = eliminated. The total data for the 8 subjects (sub #, subject number) in the 15 movements are displayed.

3.1.2. Feature Selection

After removing the outliers from the total set of raw data, windows of 250 data rows were created in the resulting matrix, in which each row has the digitized signals of the eight sEMG channels. In each window, 12 features were measured for each sEMG channel, according to the processes in Section 2.2.3, resulting in 96 electrode features per window. Then, the SFS algorithm evaluated the performance of each feature in each of the considered electrodes.

The SFS algorithm was applied in two scenarios, one scenario using the features in their originally measured range and the other scenario using the normalized features, which are the experiments marked 6 and 8, respectively, in Figure 2.

3.1.3. Classification Models

The results regarding the classification models generated the Tables 5–8, which are displayed according to the experimentation set out in Figure 2, carried out to determine which is the best option for creating a classification model considering the recognition percentage. The experimentation was conducted using a 10 $\times$ 10 cross-validation of TRN, VAL and TST for each classifier generated from NB, LDA and QDA. Table 5 lists the recognition percentages when all 96 features were used without preprocessing. Table 6 lists the number of features obtained from the SFS algorithm and the percentage of recognition obtained. In Table 6, with all 96 features, the best classifier created was LDA; however, in Table 6, with the selected features, the best classifier is QDA, which improves significantly and has a lower number of features than those initially considered.

Table 7 contains the results of the classification with the set of 96 features used but normalized. The worst classifier is NB, and the best classifier belongs to QDA. Subsequently, in Table 8, the best normalized features are selected. The worst classifier continues to be NB, and the best QDA, both for individual subjects and for the sample group. Although NB has a smaller number of necessary features, the difference between the recognition percentages is considerable. The total of normalized features for Table 7 is 96, and in Table 8 the total of necessary normalized features is reduced but depends on each classifier with SFS.

Table 5. Experimentation 5: Percentage of recognition using a 10 × 10 cross-validation by dividing the data into training (TRN), validation (VAL) and test (TST). The subject column is divided into two parts for each classifier; 1 ... 8 indicates that an average was performed between the data of the eight subjects; S1:8 indicates that the total data were a single set of the eight individuals.

Experimentation No. 5 without SFS				
	Sub	TRN	VAL	TST
NB	1 ... 8	89.38	89.38	89.38
	S1:8	29.3	29.3	29.3
LDA	1 ... 8	93.22	93.21	93.23
	S1:8	53	52.99	52.96
QDA	1 ... 8	67.5	68.1	68.01
	S1:8	33.43	29.84	29.79

Table 6. Experimentation 6: Percentage of recognition using a 10 × 10 cross-validation by dividing the data into training (TRN), validation (VAL) and test (TST). The subject column is divided into two parts for each classifier; 1 ... 8 indicates that an average was performed between the data of the eight subjects; S1:8 indicates that the total data were a single set of the eight individuals. The N.F column is the number of features selected through SFS.

Experimentation No. 6 with SFS					
	Sub	N.F	TRN	VAL	TST
NB	1 ... 8	21	90.71	90.72	90.72
	S1:8	27	39.15	39.16	39.18
LDA	1 ... 8	57	93.08	93.10	93.11
	S1:8	80	53.18	53.19	53.15
QDA	1 ... 8	30	94.68	94.72	94.73
	S1:8	24	77.71	77.88	77.86

Table 7. Experimentation 7: The format of the table is the same as Table 5, with the total number of features used in the tests; normalization was applied to the data before they were used to create the model.

Experimentation No. 7 without SFS				
	Sub	TRN	VAL	TST
NB	1 ... 8	89.94	89.97	89.96
	S1:8	29.31	29.3	29.30
LDA	1 ... 8	93.6	93.61	93.60
	S1:8	54.79	54.8	54.78
QDA	1 ... 8	94.06	94.06	94.05
	S1:8	51.15	51.04	51.07

Table 8. Experimentation 8: The format of the table is the same as Table 6, with the number of features selected for each subset of tests from SFS; normalization was applied to the data before they were used to create the model.

Experimentation N. 8 Normalization with SFS					
	Sub	N.F	TRN	VAL	TST
NB	**1 ... 8**	21	91.33	91.36	91.34
	S1:8	27	39.15	39.16	39.18
LDA	**1 ... 8**	40	93.10	93.12	93.12
	S1:8	69	54.73	54.74	54.75
QDA	**1 ... 8**	25	96.09	96.14	96.16
	S1:8	49	78.34	78.36	78.36

The effect that the normalization of the data produces is evident in Tables 5 and 7 without SFS, using the 96 features. The normalization effect significantly improves the recognition percentage of the QDA. However, as shown in Tables 6 and 8, when the electrode features are selected, the normalization only produces a slight increase in the recognition percentage of QDA, which had already been improved with the pure selection of features without normalization. From Tables 6 and 8, it is evident how many features-electrodes can be discarded without affecting the percentage of classification obtained for TRN, VAL and TST.

3.2. Statistical Parameters, ANOVA and Duncan Method

The results in Tables 5–8, regarding the performance of the four treatments outlined in Figure 2, indicate not only the good results of QDA, but also that the selection of features and normalization is an important factor in the performance of this type of classifier. An advantageous difference is that SFS reduces the number of features considered.

Among the tests carried out was an inquiry with subsequent statistical validation trials regarding which of the four treatments are significant. Therefore, the eight subjects, who are part of the same population, were taken with the results of their individual QDA classifier. The four data treatments established were the following:

(A) Normalization and SFS (Table 8);
(B) Only normalization (Table 7);
(C) Only with SFS (Table 6);
(D) Without any processing (Table 5).

Table 9 lists the averages of the 10 × 10 cross-validation of the TST data subset of each treatment of each subject. The ANOVA test performed exhibited a significant difference between the means of the results (alpha = 0.05, F = 34.37, p-value < 0.00001), rejecting the null hypothesis.

Table 9. Percentage of recognition for each treatment for each subject with QDA.

	Treatments			
Subjects	**Without Any Processing**	**Only with SFS**	**Only Normalization**	**Normalization and SFS**
1	65.97	96.53	96.68	98.27
2	88.82	95.38	96.37	97.76
3	59.06	91.80	87.57	91.87
4	71.41	92.26	92.32	93.93
5	73.89	97.06	97.48	98.94
6	72.09	95.47	96.92	97.44
7	65.46	95.35	96.06	97.06
8	47.36	94.01	89.04	93.98
Average	68.01	94.73	94.01	96.16

Duncan's method was applied to determine which of the treatments are statistically the same and which are different. As shown in Table 10, Duncan's method revealed that $\mu A = \mu C$, $\mu A = \mu B$, $\mu C = \mu B$, $\mu C \neq \mu D$, $\mu C \neq \mu D$ and $\mu B \neq \mu D$; therefore, treatments A, B and C are statistically the same and D is different. Therefore, we can select any of A, B and C. However, as A provides us with a low number of features and a higher value in the recognition percentage, it was decided to use QDA with normalization and SFS to continue working with these conditions.

3.3. Conversion to a New Sparse Matrix Space

As established, this method was used to reduce further the number of features used, to increase the percentage of model recognition and to generate a simpler model. Figure 9 illustrates the main features versus the recognition curve. The more main features are input, the more the percentage of recognition grows, but it was only possible to reach the highest classification rate as before.

Table 10. Table for results of Duncan's significance testing; the conclusion is that the means of the results obtained from the procedures A, B and C are equal, and D (without any processing) is not equal. Therefore, any of the treatments "normalization and SFS", "only normalization" and "only with SFS" can be chosen.

Population Difference	Sample Difference Compared to Their Rp Range		Decision
$\mu_A - \mu_C$	96.15–94.73 = 1.42	$< 7.1 = R_4$	Not significant
$\mu_A - \mu_B$	96.15–94.05 = 2.10	$< 7.0 = R_3$	Not significant
$\mu_A - \mu_D$	96.15–68 = 28.14	$> 6.6 = R_2$	Significant
$\mu_C - \mu_B$	94.73–94 = 0.67	$< 7.0 = R_3$	Not significant
$\mu_C - \mu_D$	94.73–68 = 26.75	$> 6.6 = R_2$	Significant
$\mu_B - \mu_D$	94–68 = 26.04	$> 6.6 = R_2$	Significant

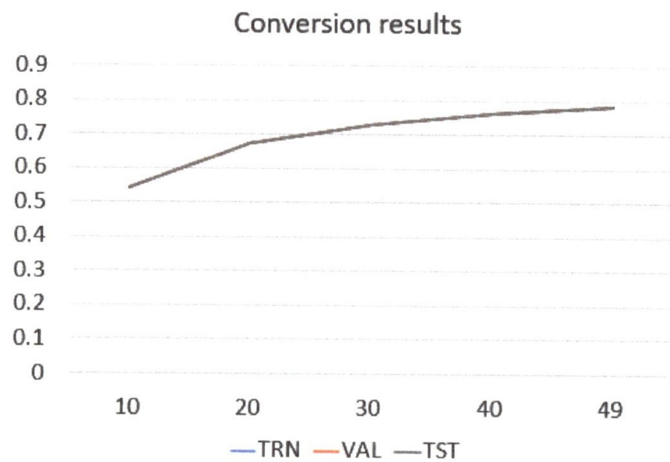

Figure 9. Plot: percentage of classification by selecting a smaller number of features after conversion to sparse matrices.

3.4. *Statistical Formulations for QDA after Space Change*

The statistical parameters considered—sensitivity, specificity, precision and F1 score—for each of the 15 classes were calculated. Table 11 presents these parameters for the average of the eight subjects evaluated individually; data for the group sample of the eight subjects are presented in Table 12. As shown in Table 11, the classes 1 (HC), 8 (MRL), 10 (R_R) and 12 (T_L) are movements that have an ideal classification; Figure 10 illustrates these hand gestures.

3.5. *Evaluation of the Movement in Opensim*

Each new classification of a data window is translated into an advancement vector, which integrates the joints involved in the movement detected, representing a progress proportional to 62.5 ms of a total motion duration of 5 s. Figure 11 visualizes the complete cycle, from the detection of the movement with the classifier to the choice of the corresponding advance vector and the addition of this vector to the movement matrix. Once the movement has finished, the motion file with the MOT extension was used for the subsequent motion visualization in Opensim. With the generated movement files, we performed biomechanical movement analysis in Opensim, verifying the activations and muscular participation within the movements. Some ideal motion files and their videos are in [30].

Table 11. Average of the statistical parameters for the eight subjects evaluated individually.

		QDA		
Class	Sensitivity	Specificity	Precision	F1 Score
1	100	100	100	100
2	88.65	99.09	88.65	88.65
3	96.90	99.74	96.90	96.90
4	98.88	99.83	97.80	98.34
5	98.68	100	100	99.33
6	97.32	99.91	99.09	98.19
7	97.80	99.91	98.88	98.34
8	100	100	100	100
9	97.40	99.42	91.46	94.33
10	100	100	100	100
11	92.52	99.74	97.05	94.73
12	100	100	100	100
13	97.75	99.50	93.54	95.60
14	92.39	99.83	97.70	94.97
15	90.78	99.10	86.25	88.46

Table 12. Statistical parameters for the dataset formed by the group of eight subjects with 49 electrode features.

		QDA		
Class	Sensitivity	Specificity	Precision	F1 Score
1	83.26	99.78	97.11	89.65
2	78.99	98.09	78.77	78.88
3	81.87	96.74	70.18	75.57
4	70.71	98.06	77.05	73.75
5	70.05	98.87	85.00	76.80
6	79.25	96.81	68.87	73.69
7	86.20	94.82	58.36	69.60
8	83.33	98.42	81.63	82.47
9	52.60	98.76	78.73	63.06
10	69.56	98.77	82.90	75.65
11	87.71	98.15	80.85	84.14
12	79.65	99.78	97.25	87.58
13	93.19	97.81	79.06	85.54
14	84.44	98.17	81.02	82.69
15	77.91	98.37	81.18	79.51

Figure 10. Hand gestures with perfect recognition. From left to right, the classes 10 (R_R), 1 (HC), 12 (T_L) and 8 (MRL).

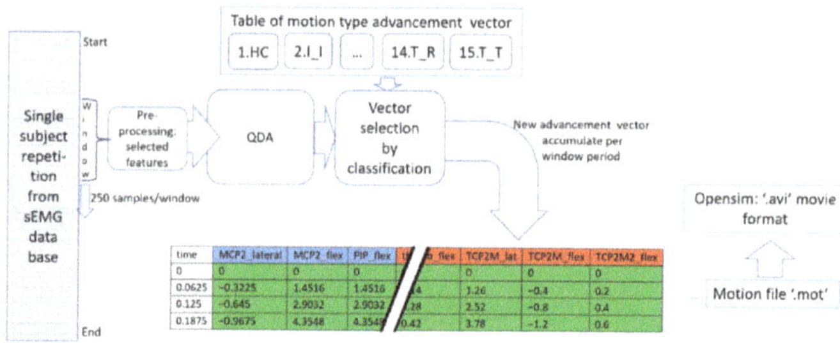

Figure 11. Cycle from classification to generation of the movement file for Opensim.

Evaluating the movement classified with an Opensim visualization is a way of illustrating the performance of the classifier. After analyzing the recognition of each subject and its test data class label (type of movement), it was found that, on average, the worst classified is the RL movement, with subject 6 having the worst recognition percentage for this movement. Figure 12A illustrates the Opensim reproduction of the ideal RL movement for 5 s. Figure 12B visualizes the classification of the complete repetition of the RL movement of subject 6 (repetition 1 was taken). A complete move can take up to a maximum of 80 classification events if there are no outliers removed that reduced the number of windows. In this repetition of RL, only 76 classifications were provided, of which 11 were erroneous for the RM movement (14.5%) and the others were for the IM movement (85.5%). A predominance of the middle finger was observed in the misclassification of movement. In Figure 12B, it is evident how the middle finger progresses most before any other. How serious the misclassification can be depends on the task to be performed and the application.

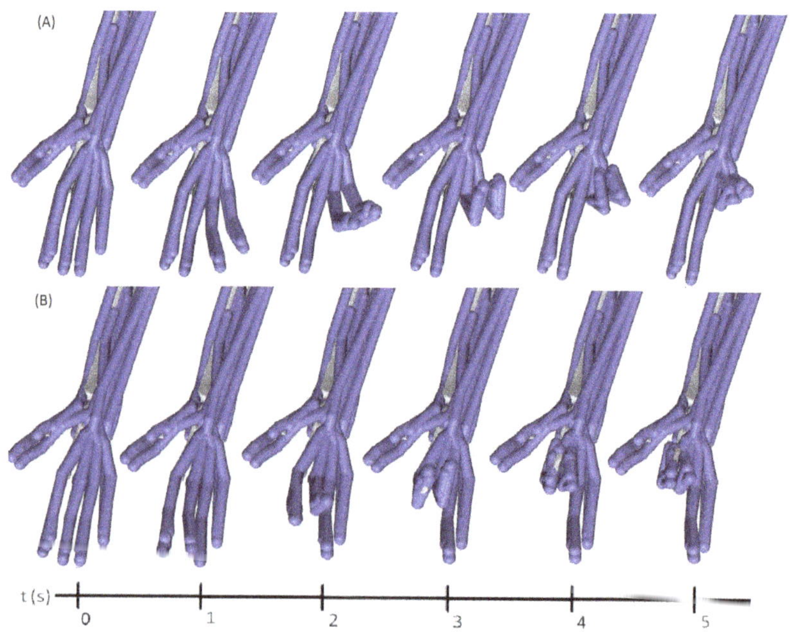

Figure 12. Reproduction of the RL movement for 5 s in Opensim: (**A**) ideal movement; (**B**) total movement of repetition 1 for the movement of subject 6, which was the worst classified.

4. Discussion

The objective of this work was the evaluation of an open access electromyography database for the detection of 15 movements of the fingers of the hand through a simple preprocessing-enhanced classifier to visualize the muscular participation of the predicted movements in the virtual environment of Opensim and generate new insight for prosthetic application. The usefulness of the Opensim musculoskeletal simulation environment in the design of a multifunctional prosthesis was demonstrated, and the evolution of muscle movements was virtually observed from the results of the classification model and visually compared with the ideal movement. Working with multiple specific movements (the flexion of each finger and combinations of flexions) from a database provided a multifunctional complex system outside of laboratory environments, where the control of upper-limb prosthetic systems can be investigated in a practical way. The standard models of classification implemented with a short data window provided a good recognition rate (in accuracy and number of classes) and will be easy to embed into microcontrollers to classify a sequence of movements in real time in the future. A description of specific results follows, along with a discussion of the limitations of the study.

The method for outlier elimination indicates that the closed hand movement was the most affected, as seen in Figure 9. This was the most complex hand gesture because it involved flexing all the fingers, which would have favored the proliferation of their outliers. However, the closed hand flexion is one of the most common movements in daily life, and, at the same time, it was one of the best-classified movements.

After the removal of outliers, without any type of preprocessing, of the four classifiers created in Table 5 (experimentation 5 in Figure 2), the best of the algorithms is LDA for the two groups of data matrices. Table 6 presents the results for when the features were selected in the creation of the classifier. One more column is displayed, in which the number of features selected in each classification algorithm is defined. With a set of features selected from the total set, the classification percentages increase, with the most notable change being in QDA, reaching 77.86% recognition of the global subject matrix and up to 94.73% with the individual matrices of the subjects.

The next set of results is shown in Tables 7 and 8, in which the effects of normalization are evidenced in both cases, either as a single preprocessing or in conjunction with the selection of features. It is clear there is an improvement in all the algorithms, but again the QDA model is especially improved. Selecting the features or making none predominate in the magnitude of their effect helps the classifier. In Table 8, QDA has the best performance for the 15 movements, with 96.15% recognition for subjects individually and 78.36% recognition for the sample group with the eight subjects together.

For the input vector with 96 initial features, these features can be reduced without degrading the recognition; and with selected features, they even improve the recognition of the classifiers. In QDA, 25 electrode features remained, on average, for individual subjects, and 49 remained for the sample group. A curious fact about the electrode features selected and the position of the significant electrodes is that QDA achieved its classification rate using practically only four electrodes. Electrodes 3, 4, 5 and 6 were the ones that behaved the best and correspond to being better positioned to capture the signal from the active flexor muscles, according to Figure 1. This point demonstrates how redundant the EMG electrode ring array setup can become. As revealed by the SFS tests on the NB and QDA classifiers, there is potential to tailor the methodology specifically with a few electrode features to an individual (up to only 17 in a subject) and achieve a good classification percentage.

The results of each procedure performed are supported by the cross-validation included in them and even by an ANOVA test. When the results obtained from the experimental treatments were evaluated with Duncan's multiple comparisons test, the latter three experimental procedures (marked 6, 7 and 8 in Figure 2 and Tables 6–8) were similar; that is, the means of the recognition percentage measurements were not statistically different. Therefore, we could use any of those three treatments; however, for the moment, to continue investigating, the chosen model was the quadratic model with a selection of normalized features.

The statistical formulations of sensitivity, specificity, precision and F1 score of the eight subjects evaluated separately (Table 11) reveal that the movements 1 HC, 8 MRL, 10 R and 12 TL all have a value of 100% recognition; Figure 11 illustrates these hand gestures. This finding indicates that the QDA classifier trained for any subject could distinguish these four gestures without any error. However, a perfect class recognition does not happen when the data of the eight subjects are assembled into a single data group and processed to generate a single model for all subjects (Table 12), as is often the case. In this group, a large specificity is maintained for all classes, and the F1 score indicates a good performance for classes 1 HC and 12 T_L again, confirming them as two robust hand gestures. The class movement 13 T_M might appear to have good performance since it had the greatest sensitivity of the group, but it had a low precision. Regarding the worst class prediction for the subjects evaluated individually (Table 11), the flexion of the index finger (2 I_I) had the lowest sensitivity and specificity, which is in agreement with other research [18], and also the thumb flexion (15 T_T) presents the lowest precision and F1 score. This might correspond to the high levels of fine motor dexterity developed between the thumb and index finger that make them difficult to individually define. As for the whole group classifier (Table 12), the F1 score is above 63%, with most classes being balanced, but class 9 (R_L) shows a loss in sensitivity and class 7 (M_R) shows a loss in precision. In these two movements, which involve the ring finger flexion and the middle finger, misclassification was observed; there were repetitions with large classification errors and mistagging between themselves.

In Figure 9, it is possible to verify that there is no convenience in converting the features into a new space through the sparse matrix method. It simply maintained the recognition percentage already present, so no improvement was seen using the method, and the increased processing load with its use in a microcontroller is not justified.

To simulate the movements in Opensim, after an output result of the classifier, an advance vector was chosen and added to the construction of the simulation; therefore, the class of the movement detected provides each joint of the biomechanical model of the hand with a possible change in the degrees of rotation. At the end of the reading of the 5 s motion, there is a virtual movement matrix representative of the classification. In this way, it was possible to reproduce the classified movements of any of the subjects, whether they were of high hits or poor performance in the classification. Several videos with the movements of all the joints are in [30,42]. With these movement files (MOT format), in Opensim, we performed movement analysis and verified muscle participation within each movement evaluated.

In the videos of a complete movement, the wrong classification of an event does not necessarily affect the total movement generated. For example, a single error of a window period translated into movement can be expressed in the same way as a movement executed at 98%, and perhaps more if the correct movements of some joints are considered within the errors. For example, if a classification indicates flexion of the index–middle–ring fingers, and it is actually a movement of the index finger, this is a partially good classification, as this movement partially helps the correct movement. Even in a normal human hand, usually, when we want to execute a single movement, taking the ring finger as an example, the hand generates some movements in the other fingers.

There are some issues to consider in this study. Although there was no significant difference in the results of the SFS and SFS-normalized tests of the three algorithms, a further discussion is required regarding the final specifications of the system, whether to work with the minimum number of electrodes, whether to have the best recognition or whether it is a matter of real-time implementation. There is a warning related to the findings on the reduction in the numbers of electrodes and features before application; although the database used in this research is of the multifunction type, it only included flexion movements and there was never any extension of a finger. Therefore, further experimentation with all kinds of finger motions would be required to optimize a classifier.

An important argument in the works involving EMG signals is the repeatability in the acquisition of these measurements [43], specifically the difficulty of electrode placement accuracy. Since data mining and machine learning analyze the information as it is acquired, a key point in obtaining more reliable classification models is ensuring the same conditions of the experimental protocol during every measurement. We made sure that the selected database complied with international best practices, although the material in the dataset used did not include information about the repeatability of the EMG measurements, and this is a limitation of the work; at any rate, statistical parameters for the reproducibility of classification were calculated as a measure of performance. Another limitation of the study was that the database did not have motion sensors or video recordings that would allow us to have a direct relationship between EMG signal time and position in space. The model of Opensim was controlled by the classification results of the machine learning algorithms; for this reason, the simulated hand movements might be different from the actual hand movements.

In the future, the validation of the Opensim model using a dataset including the motion data, such as data from a 3D motion capture system or camera, will be important to improve the results of this research and to confirm the virtually created movement with the real movement position data. In addition, another task would be to bring the 15 discriminant functions of the QDA classifier, which are a sum of multiplications, to the world of portability by embedding them in a microcontroller, which allows us to be one step closer to real-time processing.

5. Conclusions

The development presented combined biomechanical simulation with automatic classification of 15 finger movements. An open access database containing the signals from an array of forearm EMG electrodes was used as input, and traditional machine learning and signal preprocessing methods were used for the design of the classifiers. The result turns out to be a nice tool for the practical design of hand prostheses or for human–computer interface control through hand gestures. It allows one to visualize the result of classifying a finger movement to consider its performance.

The algorithms created and their input data preprocessing provided good results in the classification of the finger movements. The QDA algorithm with SFS and data normalization provided the highest recognition rate (96.16%). The experiments suggest it is possible to make a classifier specific to a person, using only 17 electrode features and 4 EMG electrodes. This leads us towards a possible practical implementation and portability of EMG matrix control.

Author Contributions: Conceptualization, J.A.-G. and M.B.-Z.; methodology, J.A.-G. and F.F.G.-N.; software, J.A.-G. and M.B.-Z.; validation, M.B.-Z., F.F.G.-N., R.L.-A. and M.A.R.; formal analysis, J.A.-G. and F.F.G.-N.; investigation, J.A.-G.; resources, M.B.-Z. and F.F.G.-N.; data curation, J.A.-G.; writing—original draft preparation, M.B.-Z.; writing—review and editing, M.B.-Z. and R.L.-A.; visualization, M.B.-Z. and M.A.R. All authors have read and agreed to the published version of the manuscript.

Funding: This research received no external funding.

Institutional Review Board Statement: The study was approved by the university research ethics committee.

Informed Consent Statement: Informed consent was obtained from all subjects involved in the study.

Data Availability Statement: Publicly available datasets were analyzed in this study. This data can be found here: [https://www.rami-khushaba.com/electromyogram-emg-repository.html] (accessed on 21 March 2022), [https://simtk.org/projects/moving-fingers] (accessed on 21 March 2022).

Acknowledgments: We thank Rami N. Kushaba for the availability of his database and his accessibility for questions regarding his work. We thank CONACyT for the scholar fellowship for the author J.A.-G. (CVU 679445).

Conflicts of Interest: The authors declare no conflict of interest.

References

1. Ghapanchizadeh, H.; Ahmad, S.A.; Ishak, A.J. Recommended surface EMG electrode position for wrist extension and flexion. In Proceedings of the ISSBES 2015—IEEE Student Symposium in Biomedical Engineering and Sciences: By the Student for the Student, Kuala Lumpur, Malaysia, 4 November 2015; p. 108112.
2. Paleari, M.; di Girolamo, M.; Celadon, N.; Favetto, A.; Ariano, P. On optimal electrode conguration to estimate hand movements from forearm surface electromyography. In Proceedings of the Annual International Conference of the IEEE Engineering in Medicine and Biology Society, EMBS, Milan, Italy, 25–29 August 2015.
3. Nishihara, K.; Kawai, H.; Gomi, T.; Terajima, M.; Chiba, Y. Investigation of Optimum Electrode Locations by Using an Automatized Surface Electromyography Analysis Technique. *IEEE Trans. Biomed. Eng.* **2008**, *55*, 636–642. [CrossRef] [PubMed]
4. Konishi, Y.; Mizobata, Y.; Yoshida, M. Development of a system for nding best electrode position for myoelectric hand control for derating of upper limb amputee. In Proceedings of the Annual International Conference of the IEEE Engineering in Medicine and Biology—Proceedings, Lyon, France, 22–26 August 2007; Volume 7, p. 56995702.
5. Christov, I.; Raikova, R.; Angelova, S. Separation of electrocardiographic from electromyographic signals using dynamic ltration. *Med. Eng. Phys.* **2018**, *57*, 110. [CrossRef] [PubMed]
6. Green, L.A.; McGuire, J.; Gabriel, D.A. Flexor carpi radialis surface electromyography electrode placement for evoked and voluntary measures. *Muscle Nerve* **2015**, *52*, 818–825. [CrossRef] [PubMed]
7. Rodriguez-Falces, J.; Neyroud, D.; Place, N. Influence of inter-electrode distance, contraction type, and muscle on the relationship between the sEMG power spectrum and contraction force. *Eur. J. Appl. Physiol.* **2014**, *115*, 627–638. [CrossRef]
8. Koh, T.J.; Grabiner, M.D. Evaluation of methods to minimize cross talk in surface electromyography. *J. Biomech.* **1993**, *26* (Suppl. S1), 151–157. [CrossRef]
9. Hary, D.; Bekey, G.A.; Antonelli, D.J. Circuit Models and Simulation Analysis of Electromyographic Signal Sources—I: The Impedance of EMG Electrodes. *IEEE Trans. Biomed. Eng.* **1987**, *BME-34*, 91–97. [CrossRef] [PubMed]
10. Saikia, A.; Kakoty, N.M.; Phukan, N.; Balakrishnan, M.; Sahai, N.; Paul, S.; Bhatia, D. Combination of EMG Features and Stability Index for Finger Movements Recognition. *Procedia Comput. Sci.* **2018**, *133*, 92–98. [CrossRef]
11. Arjunan, S.P.; Kumar, D.K. Decoding subtle forearm exions using fractal features of surface electromyogram from single and multiple sensors. *J. NeuroEng. Rehabil.* **2010**, *7*, 110. [CrossRef]
12. Dennis, T.; He, H.; Todd, K. Study of stability of time-domain features for electromyographic pattern recognition. *J. NeuroEng. Rehabil.* **2010**, *7*, 21.
13. Vonsevych, K.; Goethel, M.F.; Mrozowski, J.; Awrejcewicz, J.; Bezuglyi, M. Fingers Movements Control System Based on Artificial Neural Network Model. *Radioelectron. Commun. Syst.* **2019**, *62*, 23–33. [CrossRef]
14. Shahzad, W.; Ayaz, Y.; Khan, M.J.; Naseer, N.; Khan, M. Enhanced Performance for Multi-Forearm Movement Decoding Using Hybrid IMU–sEMG Interface. *Front. Neurorobotics* **2019**, *13*, 43. [CrossRef]
15. Wang, N.; Lao, K.; Zhang, X. Design and Myoelectric Control of an Anthropomorphic Prosthetic Hand. *J. Bionic Eng.* **2017**, *14*, 47–59. [CrossRef]
16. Gama Melo, E.N.; Aviles Sanchez, O.F.; Amaya Hurtado, D. Anthropomorphic Robotic Hands: A Review. *Ing. Desarro.* **2014**, *32*, 279–313. [CrossRef]
17. Amezquita-Garcia, J.A.; Bravo-Zanoguera, M.E.; González-Navarro, F.F.; Lopez-Avitia, R. Hand Movement Detection from Surface Electromyography Signals by Machine Learning Techniques. In *CLAIB 2019*; IFMBE Proceedings; Díaz, A.G., González, C.C., Eds.; Springer: Cham, Switzerland, 2020; Volume 75, pp. 218–227. [CrossRef]
18. Jia, G.; Lam, H.-K.; Liao, J.; Wang, R. Classification of electromyographic hand gesture signals using machine learning techniques. *Neurocomputing* **2020**, *401*, 236–248. [CrossRef]
19. Rabin, N.; Kahlon, M.; Malayev, S.; Ratnovsky, A. Classification of human hand movements based on EMG signals using nonlinear dimensionality reduction and data fusion techniques. *Expert Syst. Appl.* **2020**, *149*, 113281. [CrossRef]
20. Tuncer, T.; Dogan, S.; Subasi, A. Surface EMG signal classification using ternary pattern and discrete wavelet transform based feature extraction for hand movement recognition. *Biomed. Signal Process. Control* **2020**, *58*, 101872. [CrossRef]
21. Khalaj, O.; Jamshidi, M.B.; Saebnoori, E.; Masek, B.; Stadler, C.; Svoboda, J. Hybrid Machine Learning Techniques and Computational Mechanics: Estimating the Dynamic Behavior of Oxide Precipitation Hardened Steel. *IEEE Access* **2021**, *9*, 156930–156946. [CrossRef]
22. Shafiei, A.; Jamshidi, M.B.; Khani, F.; Talla, J.; Peroutka, Z.; Gantassi, R.; Baz, M.; Cheikhrouhou, O.; Hamam, H. A Hybrid Technique Based on a Genetic Algorithm for Fuzzy Multiobjective Problems in 5G, Internet of Things, and Mobile Edge Computing. *Math. Probl. Eng.* **2021**, *2021*, 1–14. [CrossRef]
23. UCI Machine Learning Repository. Available online: http://archive.ics.uci.edu/ml/datasets/sEMG+for+Basic+Hand+movements (accessed on 21 March 2022).
24. IEEE Lab. Available online: https://www.ufrgs.br/ieelab/IEE_sEMG_db.php (accessed on 21 March 2022).
25. EMG DATASETS REPOSITORY. Available online: https://www.rami-khushaba.com/electromyogram-emg-repository.html (accessed on 21 March 2022).
26. OSFHOME SEEDS Dataset. Available online: https://osf.io/wa3qk/ (accessed on 21 March 2022).

27. Computer Animation & Perception Group. Available online: http://zju-capg.org/research-en.html (accessed on 21 March 2022).
28. The Ninapro Project Database Web Interface. Available online: http://ninaweb.hevs.ch/ (accessed on 21 March 2022).
29. PutEMG: SEMG Gesture and Force Recognition Datasets. Available online: https://biolab.put.poznan.pl/putemg-dataset/ (accessed on 21 March 2022).
30. SimTK Opensim. Modification of Wrist Model to Include All the Movements of the Fingers. Available online: https://simtk.org/projects/moving-fingers (accessed on 21 March 2022).
31. Besomi, M.; Hodges, P.W.; van Dieen, J.; Carson, R.G.; Clancy, E.A.; Disselhorst-Klug, C.; Holobar, A.; Hug, F.; Kiernan, M.C.; Lowery, M.; et al. Consensus for experimental design in electromyography (CEDE) project: Electrode selection matrix. *J. Electromyogr. Kinesiol.* **2019**, *48*, 128–144. [CrossRef]
32. Khushaba, R.N.; Kodagoda, S. Electromyogram (EMG) feature reduction using Mutual Components Analysis for multifunction prosthetic fingers control. In Proceedings of the 12th International Conference on Control Automation Robotics & Vision (ICARCV), Guangzhou, China, 5–7 December 2012; pp. 1534–1539. [CrossRef]
33. Khushaba, R.N.; Takruri, M.; Miro, J.V.; Kodagoda, S. Towards limb position invariant myoelectric pattern recognition using time-dependent spectral features. *Neural Netw.* **2014**, *55*, 42–58. [CrossRef] [PubMed]
34. Delsys Bagnoli, Adquisition Systems. Available online: https://delsys.com/bagnoli/ (accessed on 21 March 2022).
35. Han, J.; Kamber, M.; Pei, J. *Data Mining—Concepts and Techniques*, 3rd ed.; Morgan Kaufmann Publishers: Amsterdam, The Netherlands, 2012.
36. Lei, M.; Wang, Z.; Feng, Z. Detecting nonlinearity of action surface EMG signal. *Phys. Lett. Sect. A Gener. Atom. Solid State Phys.* **2001**, *290*, 297–303. [CrossRef]
37. Phinyomark, A.; Phukpattaranont, P.; Limsakul, C. Feature reduction and selection for EMG signal classification. *Expert Syst. Appl.* **2012**, *6*, 7420–7431. [CrossRef]
38. Garcia-Breijo, E.; Atkinson, J.K.; Gil-Sanchez, L.; Masot, R.; Ibañez, J.; Garrigues, J.; Glanc, M.; Laguarda-Miro, N.; Olguin, C. A comparison study of pattern recognition algorithms implemented on a microcontroller for use in an electronic tongue for monitoring drinking waters. *Sens. Actuators A Phys.* **2011**, *172*, 570–582. [CrossRef]
39. Liu, H.; Motoda, H. *Feature Selection for Knowledge Discovery and Data Mining*; Kluwer Academic: Norwell, MA, USA, 1998; p. 46. [CrossRef]
40. Lisboa, P.; Ellis, I.; Green, A.; Ambrogi, F.; Dias, M. Cluster-based visualisation with scatter matrices. *Pattern Recognit. Lett.* **2008**, *29*, 1814–1823. [CrossRef]
41. SimTK Opensim. Wrist Model. Available online: https://simtk.org/projects/wrist-model (accessed on 21 March 2022).
42. Amezquita-Garcia, J.A.; Bravo-Zanoguera, M.E.; Avitia, R.L.; Reyna, M.A.; Cuevas-González, D. OpenSim Visualization of the Classification of Finger Movements Based on Electromyography Signal as the Single-Input Variable during Predefined Movements. In Proceedings of the 8th International Electronic Conference on Sensors and Applications, Online, 1–15 November 2021.
43. Merletti, R.; Farina, D. (Eds.) *Surface Electromyography: Physiology, Engineering and Applications*; IEEE Press: Piscataway, NJ, USA; John Wiley & Sons: Hoboken, NJ, USA, 2016.

sensors

MDPI

Article

A New Labeling Approach for Proportional Electromyographic Control

Annette Hagengruber [1,2,*], Ulrike Leipscher [1], Bjoern M. Eskofier [2] and Jörn Vogel [1]

[1] German Aerospace Center (DLR), Institute of Robotics and Mechatronics, 82234 Weßling, Germany; ulrike.Leipscher@dlr.de (U.L.); joern.vogel@dlr.de (J.V.)

[2] Machine Learning and Data Analytics Lab, Department Artificial Intelligence in Biomedical Engineering, Friedrich-Alexander-University Erlangen-Nürnberg (FAU), 91052 Erlangen, Germany; bjoern.eskofier@fau.de

* Correspondence: annette.hagengruber@dlr.de

Abstract: Different control strategies are available for human machine interfaces based on electromyography (EMG) to map voluntary muscle signals to control signals of a remote controlled device. Complex systems such as robots or multi-fingered hands require a natural commanding, which can be realized with proportional and simultaneous control schemes. Machine learning approaches and methods based on regression are often used to realize the desired functionality. Training procedures often include the tracking of visual stimuli on a screen or additional sensors, such as cameras or force sensors, to create labels for decoder calibration. In certain scenarios, where ground truth, such as additional sensor data, can not be measured, e.g., with people suffering from physical disabilities, these methods come with the challenge of generating appropriate labels. We introduce a new approach that uses the EMG-feature stream recorded during a simple training procedure to generate continuous labels. The method avoids synchronization mismatches in the labels and has no need for additional sensor data. Furthermore, we investigated the influence of the transient phase of the muscle contraction when using the new labeling approach. For this purpose, we performed a user study involving 10 subjects performing online 2D goal-reaching and tracking tasks on a screen. In total, five different labeling methods were tested, including three variations of the new approach as well as methods based on binary labels, which served as a baseline. Results of the evaluation showed that the introduced labeling approach in combination with the transient phase leads to a proportional command that is more accurate than using only binary labels. In summary, this work presents a new labeling approach for proportional EMG control without the need of a complex training procedure or additional sensors.

Keywords: electromyography; human machine interface; robotcontrol; EMG-control schemes

Citation: Hagengruber, A.; Leipscher, U.; Eskofier, B.M.; Vogel, J. A New Labeling Approach for Proportional Electromyographic Control. *Sensors* **2022**, *22*, 1368. https://doi.org/10.3390/s22041368

Academic Editors: Ernest N. Kamavuako

Received: 28 November 2021
Accepted: 31 January 2022
Published: 10 February 2022

1. Introduction

Human machine interfaces based on electromyography (EMG) are a technology used in many different applications. Besides the control of prosthesis, which is the most common application of EMG control, it is nowadays also used in rehabilitation [1], robot control [2], in computer gaming interaction [3], or for teleoperation in space applications [4].

Depending on the application, different control strategies are realized to use the voluntary muscle activity as an input signal for an external device. In commercially available prosthesis conventional control techniques, such as threshold-based methods [5,6], or linear classification [7] are still widely used. However, usually just a few degrees of freedom (DoFs) are sequentially controlled by these techniques. More complex devices, such as a robot or multi-fingered prosthesis, require a more natural and versatile control scheme. Therefore, proportional and simultaneous control strategies are becoming more widely adapted.

While proportional control allows the user to continuously change the control output by varying the control input, i.e., the EMG signal, simultaneous control further enables the user to command multiple available motor functions or DoFs of the system at the same time [8]. One way to realize proportional control with classification is given by Simon et al. [9], where the authors present a two-step method. First, a classifier is trained to differentiate between the classes. Second, the mean absolute value (MAV) of all EMG channels per class is applied to calculate a continuous control output. However, this method only allows to control one motor function or class at a time. To overcome the problem of non-simultaneous control, regression can be used instead of classification. Regression allows for a continuous output for multiple DoF and thereby enables simultaneous control. Hahne et al. [10] could further show that regression leads to an improvement in performance and also allows for a better user correction of control commands, when comparing regression to classification in an online test. Various machine learning methods have been applied to realize regression methods with EMG. For example, artificial neural networks (ANN) are commonly used [11–13], as well as support vector regression [14], or more recently convolutional neural networks [15]. To realize proportional control, the machine learning methods are also often combined with continuous data gathered by additional sensors. One example for this is given by Castellini et al. [16], where EMG signals have been combined with force measurements to control a dexterous multi-fingered hand.

Independent of the mapping method, proportional control methods come with typical problems of myocontrol, such as sensor shift, signal drift, or muscle fatigue [17,18]. On top of that, these methods exhibit the additional challenge of acquiring a suitable training data set with correct labels for model building. Within the last few years, different research groups investigated a variety of methods to provide suitable training data sets for proportional control. A quite common way to generate the labels is to ask the user to track a visual stimulus on a screen. The stimulus can be illustrated, for example, by a moving cursor or the motion of an animated or video-recorded hand shown on a computer screen such as in [19–21]. EMG data are recorded while the user is following the stimulus on the screen and the label is calculated based on the state of the visual stimulus. An alternative way to generate the labels is to use additional tracking devices such as a camera [13,22] or data glove [23]. Here, the EMG signal is fitted to the motion gathered by the additional sensors during the training procedure. In terms of amputees, the contra-lateral hand/arm can be tracked with the sensors while EMG signals are recorded from the arm used with the prosthesis. This method is called mirroring. EMG signals are also often interpreted as force. Therefore, additional force sensors can be used to generate the training data set. Users exert, for instance, finger force to a force sensor during the training procedure [24].

Additional sensors, such as force sensors, measure the motor output directly, which can be considered as the ground truth of the EMG decoding, which makes it the optimal signal to generate labels for the mapping. However, in terms of people with disabilities, additional sensors are usually not an option. People with motor impairments often have a limited proprioception of muscular activity and in case of amputees, finger force can not even be measured [24]. In addition, mirroring can cause wrong data, as it is hard for the subjects to provide exact mirror movements. In Hahne et al. [25], the performance of hand movements dropped when using labels generated with the co-lateral hand in comparison to that of the ipsy-lateral hand.

Generating labels without additional sensors, e.g., using a visual stimulus, comes with the challenge to synchronize the EMG signals to the data used for the labels. When generating EMG activation during the training procedure, the user may lag behind the visualization on the screen. Poor synchronization of the data can lead to mismatches between EMG data and labels, which may result in unreliable or simply wrong control commands during usage. This can lead to unintended movements of the remote device and therefore to frustrated users [26].

An additional challenge for the training data set used in proportional EMG control is to gather the required variations in EMG signals in correlation with the desired continuous control output. A relevant influencing factor for a robust EMG-based control lies in the in- or exclusion of the transient phase. Fougner et al. [8] state that the training data needs to be as realistic as possible, including continuous movements to achieve good proportional control. The transient phase of an EMG signal is defined as part of muscular activity in which the signal rises from rest to contraction level. It includes the burst of the sudden muscular activity in which not all motor-units (MUs) are activated yet. Compared to this, the steady-state phase is defined as the phase of a constantly maintained muscle contraction [27]. Although Englehart et al. [28] showed in an online study that the usage of steady-state data leads to more accurate performance than using transient data, the transient phase can give information for continuous labeling. Kanitz et al. [21], for example state that the onset of a muscle contraction gives predictive information about the upcoming class. According to Raghu et al. [29], the inclusion of transient data is not trivial, as the segmentation, synchronization of additional sensors, as well as labeling is challenging with this dynamic data. As a result to all these challenges, the transient phases of the signal are often omitted, while the steady-state phase is regularly used for training, as, for example, shown in [30].

Nevertheless, literature presents various possibilities to generate continuous labels with varying EMG data. One option is to record and label EMG signals in a graded representation. Therefore, the user provides muscular activity with different levels of activation, e.g., at a low, medium, and high level, depending on the maximum voluntary contraction (MVC) during the training procedure [31]. Another option is to continuously ramp the motion during training from no contraction to a defined contraction level [32,33]. In Hahne et al. [10], the subject had to increase intensity up to 80% of MVC while a cursor on the screen moved along the axis in order represent the required intensity. Zia et al. [34] asked the subjects to provide muscular activity with contraction and relaxing periods of 4 s. In Jiang et al. [35], the forces, which had to be produced during the training procedure, were visualized on a screen. Subjects were asked to ramp muscle contractions to a medium force level which was recorded with a force-torque sensor. The force levels were then used as labels. Gailey at al. [36] also used force sensors to measure finger force during the training procedure. Phases of in- and decreasing forces allow one to measure different force levels corresponding to the EMG data stream.

In this work, we focus on an EMG-based interface designed for robot control. The main application is for people suffering from muscular atrophy. The interface maps available residual muscular activity to a velocity-based output by using Gaussian process regression. As presented above, generating continuous labels for proportional control is often associated with the effort to fuse data from multiple sources. In terms of people with disabilities, this is often not an option. Furthermore, complex training procedures are often used to generate the labels, which can cause unreliable control commands.

The interface we are using provides a proportional output and thus comes with the given challenges. In this work, we address these challenges and introduce a new labeling approach to generate continuous labels for proportional control in an easy and direct way. The introduced method directly uses the EMG feature stream of the training data and creates continuous labels without the use of any additional sensors. As the label is calculated directly based on the feature stream of the EMG signal, delays are reduced to a minimum. Furthermore, a simple training procedure leads to easy application for users. We compare the method to a standard binary labeling method and investigate the effect of the different labeling strategies on task performance when operating a continuous and simultaneous EMG-based interface. Additionally, the influence of in- and excluding the transient phase of the EMG signal is analyzed.

To validate this approach experimentally, a user study with 10 subjects was conducted using the velocity-based EMG interface. The task performance was analyzed with the help of a 2D aiming and tracking task on a screen.

To summarize, the contribution of this work is to introduce a new labeling approach, which generates continuous labels for proportional control in an easy and direct way. Furthermore, the work includes a validation of this new labeling approach during task performance and the effect compared to a binary labeling approach.

2. Materials and Methods

2.1. The EMG-Based Interface

The used interface is interpreting the muscular activity of an operator to control a remote device, e.g., a robot or cursor on a screen. Muscular activity is measured using surface EMG sensors at different locations on the operator's upper and lower arm. Based on these EMG signals, the interface generates a continuous and velocity-based control signal in 2D or 3D. The main use case of this interface is for people with severe muscular atrophy. It provides people with the possibility to control a robot in 3D, when the usage of a joystick is not an option anymore [37]. An assistive device like the robotic wheelchair EDAN can present such a system, which is commanded via EMG signals by people suffering from severe muscular atrophy (c.f. Figure 1 on the right) [38]. Depending on the user, the sensors are either placed on the prominent muscle bellies, or in case of users with muscular atrophy on spots along the arm where muscles can be still voluntarily activated. In this study, a 2D control input was generated to perform tasks on a screen. The schematic overview of the interface is given in Figure 1.

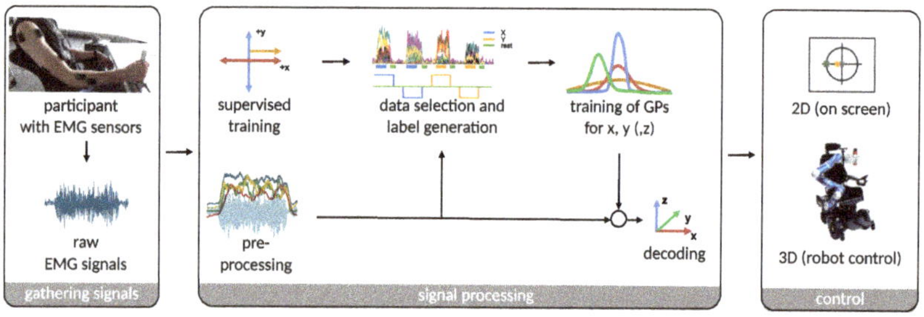

Figure 1. A schematic overview of the EMG-based interface used in this study. **Gathering signals:** A participant is shown with the EMG sensors, holding a handle to generate isometric muscle contractions. Raw EMG signals are wirelessly transferred for further processing. **Signal processing:** Four time-domain features are calculated from the raw EMG signals. A supervised training procedure allows to generate the labels for the prediction. The labeled data is mapped to a velocity-based control command by the use of Gaussian process regression. **Control:** The interface realizes a proportional and simultaneous control output. 2D as well as 3D applications are feasible, e.g., for a commanding an assistive robot. In this study, only 2D control was realized to perform tasks on a screen.

2.2. Experimental Setup

In this work, eight wireless electromyography Trigno® sensors from the company Delsys were used to record muscular activity. A medical grade double-sided tape allows for an easy attachment of the sensors to the surface of the skin. To record hand and wrist activity, electrodes were placed close to the muscles M. flexor digitorum superficialis, M. flexor carpi radialis, M. extensor carpi radialis, and M. extensor digitorum, respectively. Two sensors were attached to the upper arm, i.e., the M. biceps brachii and to the M. triceps brachii, and two to the M. deltoid (anterior and posterior). For each participant, the sEMG electrodes were placed on the same physiological spots along the dominant arm. Figure 1 on the left shows a subject wearing the eight EMG sensors. The raw biosignals were amplified and wirelessly transferred to the Delsys Trigno® base station. An analog-to-digital converter of the company Beckhoff digitized the ±5 V analog signal from the base station into a 12-bit signal at a rate of 1 kHz. A linux real-time computer received the data via EtherCAT, where the signal was further processed with 1 kHz. The time domain (TD) feature set was used for preprocessing, which was originally proposed by Hudgins et al. [39] to classify myoelectric patterns for the control of a multifunction prosthesis. This feature set includes: waveform-length, slope sign-change, zero-crossing-rate, and sEMG-amplitude. All features are calculated on each EMG channel with a sliding window of 150 samples.

Gaussian process (GP) regression was used to map EMG data to the directional control command ($\pm x$, $\pm y$). Here, the *pyGP* library [40] was utilized, which is based on the implementation of [41]. Participants were asked to exert forces and torques against a rigid handle, which was placed in front of them. Thus, they were able to generate reproducible muscle contraction in an isometric fashion.

2.3. The Training Procedure

A training procedure was conducted with each participant at the beginning of an experimental session, in order to acquire data to be used for calibration of the GP. Therefore, participants hold on to a handle with their dominant hand (the same side on which the electrodes are placed) in front of them with an angled elbow. First, the rest signal of the arm was recorded while grasping the handle in a comfortable position without any specific muscle contraction. EMG signals were recorded during this rest state to determine and remove the signal's DC offset. Furthermore, this rest data allowed to define an individual activity threshold to distinguish between rest and voluntarily activated muscles. Based on this activity threshold, a supervised training procedure was performed in order to generate a training data set for the GP.

In total, four different directions are decoded: left and right ($\pm x$), up and down ($\pm y$). To do so, visual cues on a screen guided the participants through the training procedure. The screen displayed a coordinate system illustrating these four directions. One direction at a time was highlighted and participants were asked to exert forces and torques against the handle to be associated with motion along this highlighted direction. A marker moving along the axis and into the direction of interest served as a progress bar, to visualize the amount of collected active samples. Samples were counted as active as soon as the activity threshold was exceeded. Only these active samples were considered as potential training data for the direction. Thus, the activity signal allows to track the users state of participation during the training and the effect of the reaction time is eliminated.

Subjects were asked to provide muscular activity at a level they are comfortable with and stay within the steady-state phase for at least 3 s. They were not asked to ramp the EMG data or reach a special activation level. During data acquisition, raw EMG signals and features were recorded in combination with the indicated direction of motion and the information of an exceeded activity threshold. For each direction, the visual cue was displayed until 3000 active samples have been collected. However, participants had no feedback about their activation level or which muscles were activated during muscle contraction. Once 3000 active samples had been recorded for the requested direction, subjects had to return to a resting state (i.e., stay below the activity threshold). Once remaining in rest for 1000 consecutive data samples, the next direction of motion would be indicated to the subject, until data had been collected for all four directions.

This data acquisition procedure was repeated four times, while only the last three repetitions were used to build the training data sets used for mapping. Gaussian process regression was used to decode continuous velocity commands for each DoF, i.e., one GP for $\pm x$, and one for $\pm y$, respectively. Further details about the decoder pipeline are given in Vogel et al. [42].

2.4. Labeling Approach

The goal is to create continuous labels based on the simple data acquisition procedure described above and without additional sensors. The preprocessed EMG signals, gathered during the training procedure, serve as the basis for the continuous labels. The onset of activation was determined by the activity threshold measured during the rest state and can be used as the onset of the labels. Therefore, unwanted delays due to the participants reaction time can be avoided. However, an intrinsic delay of 47 ms is introduced by the data acquisition system. Furthermore, the feature extraction, which uses a 150 ms sliding window results in additional delay. However, these delays are present not exclusively during training but they also occur during usage of the interface.

In previous work [42], an offline analysis was conducted to reveal the features with the most information content used for the decoding. The results indicated that all four TD features (waveform-length, slope sign-change, zero-crossing-rate, and amplitude) are involved in the prediction. Hence, all four features were used to maintain the influence of the individual features. Furthermore, the data of all eight sensors were considered for the labels, since each directional command is a composition of different sensors. In total, a data stream of 32 features (4 features and 8 electrodes) was processed per time step. All of the following calculation steps have been done separately for each direction.

In a first step, all eight signal values x of the eight electrodes e per TD feature f were summed up for each time step i, in order to maintain the influence of each electrode proportionally. This is shown in Equation (1) with $n_e = 8$ for any direction dir. Thus, the feature stream of each direction, was reduced from 32 to 4 features:

$$x_{f,dir}[i] = \sum_{e=1}^{n_e} x_{e,f,dir}[i]. \tag{1}$$

Further on, each feature stream is normalized for itself between 0 and 1. Since each TD feature has a different unit, this step provides equal weight to the features. Equation (2) illustrates this step, where $max(x_{f,dir})$ presents the maximum value of each feature f and x_{min} is chosen to be the mean of the recorded rest states of all training sequences:

$$x'_{f,dir}[i] = (x_{f,dir}[i] - x_{min})/(max(x_{f,dir}) - x_{min}). \tag{2}$$

Finally, the four features ($n_f = 4$) are summed up at each time step i and again normalized to be between 0 and 1, in order to get a maximum label of 1 for each direction (c.f. Equation (3)):

$$label_{dir}[i] = normalize(\sum_{f=1}^{n_f} x'_{f,dir}[i]). \tag{3}$$

Based on these equations, different labeling strategies can be realized. In our previous work [2,42], we used a binary labeling method, which is considered as the baseline for evaluation. In this binary labeling method, samples of the training set in which the activity threshold was exceeded are labeled as ± 1 for the respective direction. Accordingly, "inactive" samples as well as samples of a non-active direction were labeled as 0. It has to be noted that even when the labels for the mapping are binary, the output of the decoder still allows for continuous signals, when a regression method is used.

Binary labels ($label_{dir}[i] \in \{-1, 0, 1\}$) are simple to realize but in contrast to continuous labels ($label_{dir}[i] \in \{x \in \mathbb{Q} \mid -1 \leq x \leq 1\}$), they can not represent the continuous change of the EMG signal.

Additionally, the effect of in- and excluding the transient phase of the EMG signal was investigated. Therefore, data either included the transient phase or relied purely on the steady-state phase of the EMG signal. The first 1000 samples were discarded in case of steady-state labeling. In total, five labeling strategies were evaluated. All labels were derived from the EMG data of the same three training data sets per subject. From the total of 3000 available samples per direction and repetition, each method took 2000 samples into account. The sample selection was dependent on the particular method. The label strategies were realized as followed:

- **(A)** Binary label, including the rising transient phase of the EMG signal in the activ label;
- **(B)** Continuous label, including the rising transient phase of the EMG signal in the active label;
- **(C)** Binary label, using only the steady state of the EMG signal (excluding transient phases);
- **(D)** Continuous label, using only the steady state of the EMG signal (excluding transient phases);
- **(E)** Continuous label, including both transient phases (rising and falling) of the EMG signal.

We limited the evaluation to these 5 combinations of labels in order to keep the duration of one experimental session below 90 min. A visualization of the different labeling strategies can be found in Figure 2.

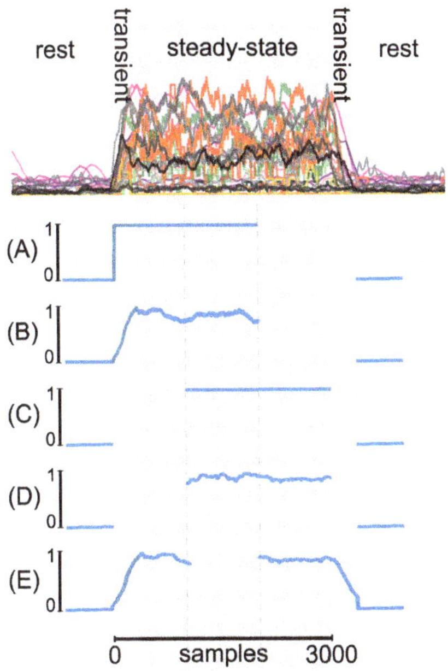

Figure 2. Example for the generated labels. (**A**) Binary label (0 = not active, 1 = active), including the rising transient phase of the EMG signal in the active label. (**B**) Continuous label, including the rising transient phase of the EMG signal in the active label. (**C**) Binary label, using only the steady-state signal (excluding transient phases). (**D**) Continuous label, only using the steady-state part of the signal. (**E**) Continuous label, including the rising and falling transient phases of the EMG signal.

2.5. Experimental Procedure

An online experiment was conducted to evaluate performance in a 2D task on a computer screen. The decoded EMG signals of the participants were used to control the velocity of a cursor on the screen. Subjects were able to move the cursor in 2 DoF, i.e., up, down and left, right. As one individual predictive model was used for each DoF, simultaneous movements (e.g., diagonal movements) of the cursor were possible. Using this interface, subjects performed two different tasks during the experiments: An aiming task (AT) and a tracking task (TT). The application GUI for both tasks contains crosshairs with an x- and y-axis. During the AT, a target circle was visible on the screen (c.f. Figure 3, left side). It was placed 400 px from the starting point along one of the main axes. The starting point of each trial was set to the middle of the crosshairs. Subjects were asked to move the cursor as quickly and accurately as possible into the target. The cursor had to stay within the target circle for at least 500 ms to finish the task successfully. After each trial, the cursor was set back to the starting position, where it had to steadily remain for 2 s before the next trial would start. A countdown visualized the 2 s duration before a new trial was initiated. A test sequence involved all four possible directions of a target, presented in random order. One experiment included five test sequences, which leads to a total of 20 trials per AT experiment.

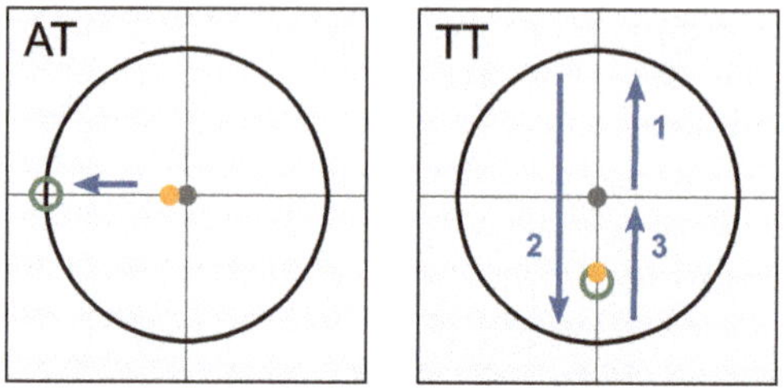

Figure 3. Description of the aiming task (AT) on the left and the tracking task (TT) on the right. The yellow dot displays the cursor controlled by the participants. The gray dot displays the starting point in the middle of the coordinate axes. The green circle is the respective target. During the AT, a target located along the cardinal axis had to be reached as quickly and accurately as possible. During the TT, participants had to track the target which moved with constant velocity along the horizontal or vertical DoF.

During the TT, the target moved with a constant speed either horizontal or vertical along the axis. The target started in the middle of the coordinate system and traveled in total 1600 px along one DoF ($\pm x$ or $\pm y$), including two turns (c.f. Figure 3, right side). Subjects were asked to follow the target and stay within the target diameter as accurately as possible. One test sequence included both DoFs: horizontally and vertically. Five test sequences completed one TT experiment.

Each AT and TT was conducted with all calculated models based on the different labeling strategies. In total, each subject performed five rounds of experiments of AT and TT. The model order was randomized based on a Latin square design over all subjects. The subjects could familiarize to the new control output prior to each experiment round. This included control of the cursor freely on the screen for 30 s, followed by one test sequence, which was not considered in the evaluation.

2.6. Subjects

Ten right-handed subjects (1 female, 9 males, age range 21–28 years) took part in this study. None of them reported known neurological diseases or other physical impairments. All of the subjects had prior experience with the EMG-based interface. In particular, they had used the interface in combination with labeling method A as this was used in previous applications. All subjects gave written consent to the procedure, which was explained to them orally and in a written form. The study was conducted according to the guidelines of the Declaration of Helsinki, and approved by the Ethics Committee of the Technical University of Munich, School of Medicine (approval number: 6/14S).

2.7. Performance Measurement and Data Analysis

For analyzing the data of the AT following performance measures were used:

- Success rate;
- Completion time, gross, and fine motion time;
- Average and maximum speed during gross motion;
- Path efficiency.

A trial is counted to be successfully finished when the cursor is placed within the target circle for more than 500 ms. If a task could not be finished within 10 s, the trial is considered to be failed. The path efficiency describes the ratio of the shortest distance to the target to the traveled distance of the cursor. The gross motion time is defined as the time from the beginning of the task until the cursor touches the target circle for the first time. The fine motion time starts as soon as the cursor touches the target circle for the first time (end of gross motion) and lasted until the task is finished successfully. Accordingly, the completion time is the sum of gross and fine motion time. The evaluation of the TT was done using the average distance from the cursor to the center of the target. Furthermore, the average travel speed was analyzed.

A statistical analysis was performed on the performance measures gross motion time, fine motion time, completion time, path efficiency, as well as on the performance measures of the TT. Thereby, the following hypothesis should be proofed:

Hypothesis 1 (H_1). *The different labeling methods do not influence task performance when using the EMG-based interface in a 2D task.*

The mean value of each subject per labeling approach and performance measure was used for this analysis to obtain a more expressive result. Since the data was not normally distributed, a Kruskal–Wallis (KW) test was used for statistical analysis. A pairwise Wilcoxon test with Bonferroni correction was used as a post hoc test. The effect size r was estimated through the equation given by Rosenthal, 1991 [43]. One sample of the TT in method D showed a value more than twice as the mean, which was not explainable. This value was identified as an outlier and eliminated for further analysis. In addition, a questionnaire was made after each experiment to gather information on the subjects' confidence in controlling the cursor.

3. Results

In total, 200 trials were performed per labeling method across all subjects during the AT. The highest success rate was reached by the labeling method B with 98.5%, and E with 99% (c.f. Table 1). Both methods are based on the new labeling approach and include a transient phase. Methods C and D show success rates of about 93%, while the baseline method A has a success rate of 88.5%. Most failures occurred during the fine motion time independent of the method (A: 19/23, B: 3/3, C: 12/13, D: 12/14, E: 1/2). Gross motion failures occurred rather rarely. Figure 4 illustrates the failed trials, successful trials with overshoots, and trials which could be finished directly (without overshooting). It can be observed that subjects were able to finish the AT in 92% of the cases without any overshoots using method B. Subjects were able to finish the AT without overshoots in 88.5% of the trials with method E, while method A shows 56.5% of the trials without overshoots.

Table 1. Success rates and failures over all subjects during the AT.

		A	B	C	D	E
Complete trial	success rate	88.5%	98.5%	93.5%	93%	99%
	failure	23	3	13	14	2
Fine motion	failure	19	3	12	12	1
Gross motion	failure	4	0	1	2	1

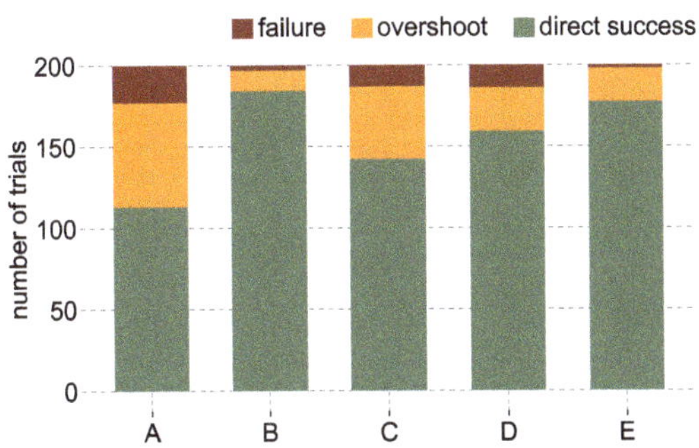

Figure 4. Overshoots over the whole experiment. For each labeling method A–E, the bar plot illustrates how often the trials were finished directly by moving the cursor straight into the target circle (green), a target was overshot before finishing the trial (orange), or the trials failed (red).

Figure 5 illustrates the completion time, gross motion, and fine motion time for successfully finished trials of all subjects. For illustration, the mean values per labeling method and subject are used, analogous to the statistical analysis. The gross motion time shows stable results over the different methods. A significant effect could not be identified. In a direct comparison between all methods using continuous labels (B, D, E) to the methods based on binary labels (A and C), no significant effect of the gross motion time could be found either. The results of this comparison can be seen in Figure 5 on the right side.

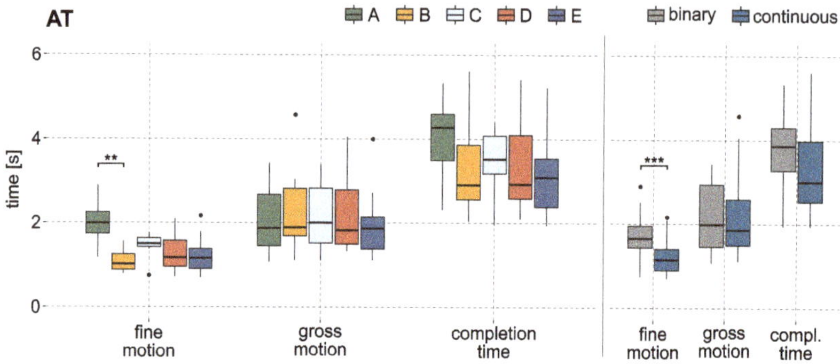

Figure 5. Results of completion time, fine motion, and gross motion time over all subjects during AT, based on the mean values of each subject per labeling method. (**Left**): boxplots of the needed time sorted by labeling method. The statistical analysis identified a significant effect in fine motion time between method A and B. (**Right**): boxplots of needed time sorted by methods based on binary and continuous labels. Binary includes results performed with model A and C; continuous includes results from B, D, and E. The KW test could identify a significant effect for the fine motion time. '•' indicates outliers; '*' indicates statistical significance ('*' $p < 0.05$, '**' $p < 0.01$, '***' $p < 0.001$).

In contrast to the gross motion time, the fine motion time shows stronger variation between the methods. A significant effect can be reported by the KW test (p-value < 0.01). The post hoc test showed that it took significantly longer to finish the fine motion part when using the binary model A compared to model B (p-value < 0.01, $r = 0.66$), which is based on continuous labels. A direct comparison of methods using binary and continuous labels shows a significant effect as well (p-value < 0.001, $r = 0.80$).

The analysis could identify a significant effect of the path efficiency during gross motion. The KW test identified methods based on binary labels with an average path efficiency of less than 79% to be less efficient (p-value < 0.05, $r = 0.52$) compared to the methods based on continuous labels (path efficiency of B,D,E > 83%). These results can be found in Table 2.

Table 2. Path efficiency and results of the questionnaire. The table show the path efficiency during the gross motion section of the AT over all subjects. The questionnaire asked the subjects how well they could control the cursor during each experiment. They could rate from 0 to 20, where 0 represented bad control of the cursor and 20 represented a perfect control.

		A	B	C	D	E
path efficiency (PE)	mean in %	76.7	83.1	78.9	83.5	84.0
	±sd	17.7	15.0	17.9	14.2	14.4
questionnaire: How good was the control of the cursor?	mean (0–20)	13.6	16.3	15.3	15.4	15.9
	±sd	3.2	3.0	1.9	2.3	2.2

When comparing the cursor speeds during the AT, differences can be observed for the used methods (c.f. Figure 6 on the left side). The maximum cursor speed during gross motion was reached by method A followed by method C, both with an average maximum speed of more than 400 px/s. The cursor would reach a velocity of approximately 600 px/s, if the output of the predictive model is 1. The KW test identified a significant difference in maximum speed. The baseline method A shows an effect when compared to the methods based on continuous labels (B, D, E). A faster maximum cursor speed can also be reported (p-value < 0.001, $r = 0.93$) in a direct comparison between methods using binary or continuous labels. The results are visualized in Figure 6 on the right side.

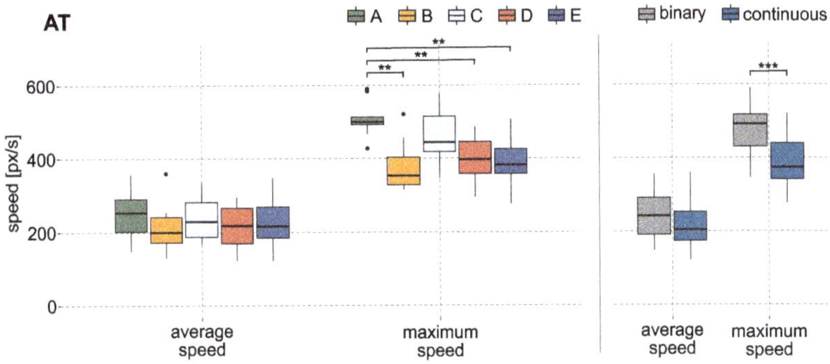

Figure 6. Results of average and maximum speed over all subjects during AT, based on the mean values of each subject per labeling method. (**Left**): boxplot of the average speed and maximum speed in px/s during the gross motion section of the AT. A significant difference was identified for maximum speed (A-(B, D, E) by the KW test. (**Right**): boxplot for the average and maximum speed during the gross motion for methods based on binary and continuously labeled data. A significant difference was identified for maximum speed. '•' indicates outliers; '*' indicates statistical significance ('*' $p < 0.05$, '**' $p < 0.01$, '***' $p < 0.001$).

During the TT, no significant effect on average distance to the target could be identified. The left side of Figure 7 illustrates these results of the TT over all subjects. The statistical analysis obtained a significant difference in average speed during the TT (p-value < 0.01, $r = 0.66$). Hence, labeling method A showed a higher average speed compared to method B, which is based on continuous labels. Results can be seen in Figure 7 on the right side. While the average speed of method A lies at about 130 px/s, methods using continuous labels show mainly values lower than 120 px/s. The target moved with a constant speed of 100 px/s.

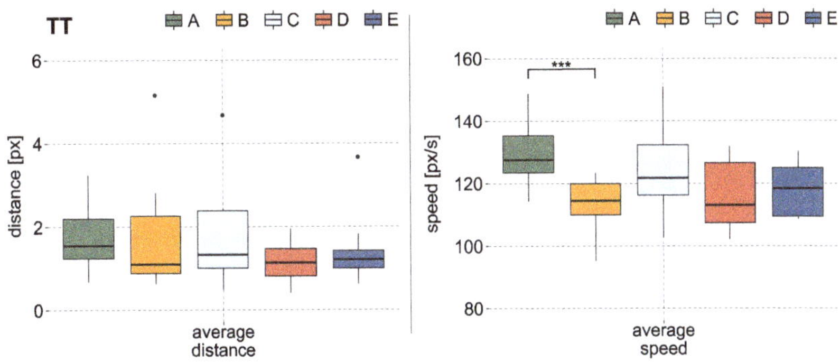

Figure 7. Results of the TT over all subjects, based on the mean values of each subject per labeling method. (**Left**): boxplot of the average distance from cursor to target. The total distance traveled per trial was 1600 px. There was no significant effect between the methods. (**Right**): boxplot of the average travel speed of the cursor. The target moved with a constant velocity of 100 px/s. A significant effect was identified between method A and B by the KW test and the post hoc test. '•' indicates outliers; '*' indicates statistical significance ('*' $p < 0.05$, '**' $p < 0.01$, '***' $p < 0.001$).

Subjects had to rate their ability to control the cursor on a scale from 0 (very bad control) to 20 (very good control) after performing each labeling method. Method B reached on average the best rating with 16.3 ± 3 points, while method A was rated worst with 13.6 ± 3.2 points. The results of the questionnaire can be found in Table 2.

4. Discussion

In this work, we introduced a new labeling approach to generate continuous labels for proportional electromyographic control. The continuous labels originated from the preprocessed EMG data, recorded during a simple data acquisition procedure. No additional sensors were used to generate the labels. Since the labels were created directly from the EMG feature stream, they reflect the intensity of the muscular activity provided by the participant. By using the EMG data without additional sensors, the challenge of synchronizing external values to the proportional EMG signals as described in Raghu et al. [29] can be avoided. Mismatches that may occur from tracking a visual stimulus during the training procedure e.g., through an incorrect activity level or a delayed muscle contraction can be excluded with this method. The simple training procedure specifies just the category of the label (in this case, the direction) as well as the time to gather enough data. No ramp or MVC levels had to be reached. The predefined activity threshold enables to coincide the onset of the muscular activity with the onset of the labels, including the transient phase. Thus, no additional time delay occurs between the EMG signals and the labels during the training. The time delay given by the system, to transfer and process the data, is the only delay present. However, this delay is inherent to the system and is present during training but also during the usage of the interface.

The realized EMG-based interface achieved success rates between 88.5% and 99% during the aiming task. In the literature it has been shown that aiming tasks are viable to test the performance of EMG-based interfaces [44,45]. In Scheme et al. [44] e.g., two proportional control schemes where compared using a Fitts' Law test, which presents a special type of an aiming task. Proportional control was added here as a post-processing step to the classification. Since the experimental design differs from our approach (target distances, target and cursor sizes, and simultaneous control), the results are not directly comparable, however a good estimate was given for task performance with EMG-based interfaces during aiming tasks. In Scheme et al. [44], a success rate of up to 96% could be reached. Kamavuako et al. [45] used as well a Fitts' Law test to measure performance and reached success rates of about 91% with a surface EMG approach and 96% in a combined approach of surface and intramuscular EMG. The success rate of our study (between 88.5% and 99%) lies in a similar range as the results achieved in the mentioned publications. This reveals that our approach provides, generally, a useful control output.

A success rate of up to 99% (B: 98.5%, E: 99%) could be achieved by methods based on the new introduced labeling approach. It is noteworthy that the two continuous labeling methods that include the transient phase (B and E) performed best. Exclusion of transient phases in method D led to a drop in success rate to 93%. Here, the result is comparable to method C (93.5%), a method based on binary labels which also excludes the transient phase. Figure 8 illustrates the histogram of the labels generated by the new labeling approach. The effect of in- or exclusion of the transient phase on the labels can be observed. Methods, including the transient phase, contains more variability in the labels. As an example, in method B and E, 11 % and 21 % of the labels have a value below 0.5. These are considerably more as in method D (1.5 %). We assume that the label-variability, given by the transient phase, improves the control accuracy. This statement is supported by the fact that failures in method D occurred mainly during fine motion (c.f. Table 1), in which a precise commanding is needed to reach the target. Including the transient phase allows for more variability in EMG control while maintaining the simple training procedure. The short transient phases that are required to reach the steady state signal already provide enough variability in the EMG signal to improve the level of proportional control, compared to using steady state data only.

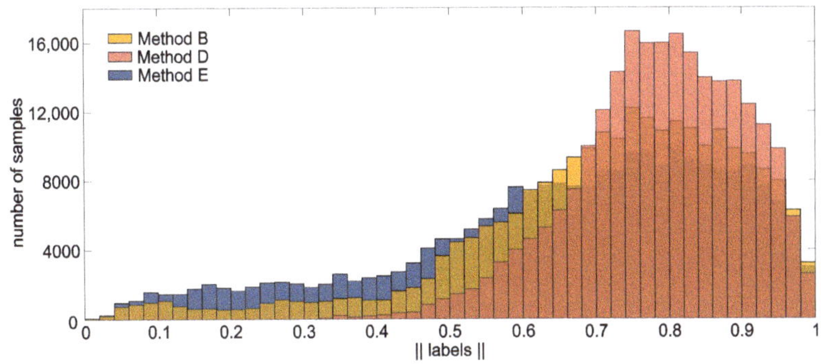

Figure 8. Histogram of the continuous labels. Shown is a histogram of the generated labels of the methods using continuous labels B, D, and E. The data includes the labels between 0 and 1 of all subjects and directions (0 is excluded).

Compared to method B, method E includes two transient phases: the rising as well as the falling flank of EMG data. The success rate of both models shows similar results. Further, a significant effect between these models could not be identified. This indicates that considering the second flank has little effect on task performance. The experience from our prior studies with people suffering severe muscular atrophy showed that it is often a problem to relax the muscles abruptly after voluntary muscle contraction. This might influence the falling EMG signal during training. Therefore, we hypothesize that in our method, no advantage is given by including the falling flank in the training data.

When comparing methods based on continuous labels with methods based on binary labels, differences in success rate, fine motion time, and maximum cursor speed can be observed. Trials performed with a predictive model based on continuous labels show a higher success rate as methods based on binary labels. The baseline method A showed, with 88.5% success rate, the lowest value for all methods. On closer inspection, the effect can be attributed to the fine control of the cursor. While the gross motion time is almost equal over all methods, the outcome of the fine motion time varies. This means that the cursor could be moved correctly in the direction of the target and reach it, however the cursor could not always be moved precisely into the target circle. The statistical analysis confirmed that the fine motion could be controlled more precisely with the methods based on continuous labels. We assume that better fine motion indicates better proportional control as the ability for slow and precise commands is required to place the cursor inside the goal area.

The analysis of the cursor speed showed that a binary labeling strategy leads to faster maximum cursor commands during the gross motion. The active EMG data was labeled with 1 during the data acquisition when a binary label is applied. The generated labels of the newly-introduced approach (c.f. Figure 8) are spread between 0 and 1. This leads to a slower control output, which in turn leads to the ability to control the output more precisely. Although, the maximum cursor velocity during the AT is faster for methods based on binary labels, the average speed did not increase and the gross motion time did not decrease for these models. However, the path efficiency showed that the traveled path was more efficient in methods based on continuous labels (PE for binary labels < 79%; PE for continuous labels > 83%). The average PE of methods based on continuous labels correspond to those found in the literature. A PE of up to 82% was reported in [44] during a Fitts' Law test. Kamavuako [45] reported a PE of up to 87% in the Fitts' Law test, which was performed by nine subjects. Both realized the EMG control by the use of a proportional classifier. In Ameri et al. [14], a support vector machine was used to achieve simultaneous and proportional control. Here, a PE of up to 85% could be reported during a 2D aiming task for one subject with transradial limb deficiency and about 76% for 10 healthy subjects during a 3D task.

A detailed inspection shows that especially method A causes differences to the methods with continuous labels. Besides the lower success rate, method A also produces an effect in fine motion time and maximum cursor velocity. Method C, which presents a method that excludes the transient phase, does not cause these effects. This reveals that the inclusion of the transient phase to the binary label further increases the agility in the prediction. Thus, model A appears most agile at the cost of less accuracy.

These findings can also be observed for the tracking task. Higher average speed was achieved when method A was active. A statistical effect was found between A and B. Method B shows the lowest average speed, which is also closest to that of the moving target. Nevertheless, the results of the average distance to the target indicates that an equivalent performance could be reached by all methods. Apparently the selected speed of the target of 100 px/s can be tracked equally well with all five labeling methods. We suggest further tests with varying velocities of the target to investigate more detailed effects regarding task performance during tracking.

With the results shown above, the Hypothesis 1 (H_1) can be partially rejected, since the task performance of fine motion and cursor speed are influenced by the different labeling methods, especially when distinguishing between continuous and binary labeling methods.

The evaluation of the different labeling strategies indicates that a continuous label leads to more precision during fine motion control. The control of prostheses or assistive robots often requires precise proportional control in order to master tasks of daily living. Therefore, our method can potentially help to improve the control of assistive devices. The results of the questionnaire indicate that the participants preferred a precise control as given by method B over a reactive control as given by method A for the given task. This is expected, as the performance is worse using method A, with participants rating this method lower. The assessment of the participants supports the results of the quantitative measures, such as the success rate and fine motion.

A high frustration level due to a lack of control is one of the reasons to reject an assistive device [26]. An increased confidence about the control may therefore increase the acceptance of the assistive device. However, the findings about the sensitive control given through a binary label and the inclusion of the transient phase, as given in method A, are of interest to systems using shared control. Assistive devices with an integrated shared control algorithm support the user during complex tasks of daily living. Operators command the device roughly along a task. Precise movements and fine motions are then supported by the system [46], which makes the task easier for the user. Here, the advantage of more reactive control may outweigh that of precise control because deficits in precision can be compensated by shared control. A detailed analysis of the labels during tasks of daily living and with and without shared control would be necessary to confirm that assumption.

Prior research shows that the presented EMG-based interface can be used in 3D applications with a robotic system [38]. Baseline method A was used as labeling method to perform tasks of daily living with an assistive robot. The interface allows a proportional and simultaneous control of the robotic device in 3D. Although, in this investigation the labeling methods were evaluated in a 2D task on a screen, we assume that the results may be valid for proportional and simultaneous control in 3D.

As sensor shift and drift are still problems in EMG-based control, techniques of online adaptations are useful to update the predictive model during the application if control commands are not fitting anymore [47]. The introduced approach provides continuous labels that can be created during the usage of the interface. It is neither dependent on a defined training procedure nor on additional sensors to generate the continuous labels. Thus, the training data can be gathered during the application and interrupting the application is avoidable. This makes the introduced approach ideal to update the predictive model online with continuous labels.

5. Conclusions

This work presents a new labeling approach for continuous labels used in proportional EMG control. Labels are directly extracted from EMG features calculated during data acquisition. The method allows for a simple training procedure and still covers the variability of the EMG signal by including the transient phase of muscle contraction during training. No additional sensors were needed and the method ensures a minimum time delay between the EMG training data and the generated labels. Furthermore, mismatches due to wrong contraction levels are avoided. Thus, this labeling method presents an ideal method for people with physical disabilities, especially if additional sensors are not an option. However, this work does not provide a comparison between the introduced approach and other labeling strategies, such as tracking a visual stimulus or additional sensors. At this point we can also not make any statement on the control with prosthesis or other EMG applications. Further studies are mandatory to verify if the introduced method can be used or even improve e.g., prosthesis control.

We investigated the effect of the introduced labeling approach on task performance when using an EMG-based interface. To quantify the effect, participants performed tasks with five different mappings based on the same training data, however with different labeling methods. The five different methods included continuous labels, a baseline method with binary labels, as well as variations thereof with in- and excluding the transient phases of the EMG signal. In total, three variations of the new labeling approach (called: B, D, and E) were compared with two variations of a binary labeling strategy (called: A and C). The evaluation showed that the methods based on continuous labels and including transient and steady-state phases led to high success rates of 98.5% with B and 99% with method E, compared to methods based on binary labels, which showed success rates of 88.5% with A and 93.5% with method C. Differences between the methods were particularly evident during fine motion. A significant difference could be identified between methods based on binary and continuous labels, as well as between method A and B. This indicates improved fine motion capabilities with methods using continuous labels and a higher command accuracy in comparison with methods using binary labels.

To conclude, the introduced continuous labels in combination with the transient phase (corresponding to labeling method B) presents an efficient way to map muscular activity to a proportional control input with good command accuracy. The baseline method, which uses binary labels, proved to be more reactive and less accurate. In applications where no precise control is required, as e.g., in systems providing shared control, these strategies may be advantageous. In future, the introduced approach must be deployed with a robotic system in 3D to confirm the achieved results with the intended system of use. Furthermore, the approach must be evaluated with actual users suffering from severe muscular atrophy, where this labeling method is of particular interest, as additional sensors are not an option for generating appropriate labels for this group of users.

Author Contributions: Conceptualization, A.H. and J.V.; methodology, A.H. and J.V.; software, U.L.; validation, A.H., J.V. and B.M.E.; data analysis A.H. and J.V.; writing—original draft preparation, A.H.; writing—review and editing, A.H., J.V. and B.M.E.; visualization, A.H.; supervision, J.V., B.M.E. All authors have read and agreed to the published version of the manuscript.

Funding: This research received no external funding.

Institutional Review Board Statement: The study was conducted according to the guidelines of the Declaration of Helsinki, and approved by the Ethics Committee of the Technical University of Munich, School of Medicine (approval number: 6/14S).

Informed Consent Statement: Informed consent was obtained from all subjects involved in the study.

Acknowledgments: The authors would like to thank Markus Nowak as well as Andrea Gigli for their valuable input to this manuscript as well as all subjects that took part in this study. B.M. Eskofier gratefully acknowledges the support of the German Research Foundation (DFG) within the framework of the Heisenberg professorship programme (grant number ES 434/8-1).

Conflicts of Interest: The authors declare no conflict of interest.

Abbreviations

The following abbreviations are used in this manuscript:

EMG	Electromyography
DoF	Degree of Freedom
GP	Gaussian Process
AT	Aiming Task
TT	Tracking Task
TD	Time Domain
PE	Path Efficiency
KW	Kruskal Wallis
GUI	Graphical User Interface

References

1. Mulas, M.; Folgheraiter, M.; Gini, G. An EMG-controlled exoskeleton for hand rehabilitation. In Proceedings of the 9th International Conference on Rehabilitation Robotics, 2005. ICORR 2005, Chicago, IL, USA, 28 June–1 July 2005; IEEE: Piscataway, NJ, USA, 2005; pp. 371–374.
2. Hagengruber, A.; Vogel, J. Functional tasks performed by people with severe muscular atrophy using an semg controlled robotic manipulator. In Proceedings of the 2018 40th Annual International Conference of the IEEE Engineering in Medicine and Biology Society (EMBC), Honolulu, HI, USA, 18–21 July 2018; IEEE: Piscataway, NJ, USA, 2018; pp. 1713–1718.
3. Nacke, L.E.; Kalyn, M.; Lough, C.; Mandryk, R.L. Biofeedback game design: Using direct and indirect physiological control to enhance game interaction. In Proceedings of the SIGCHI Conference on Human Factors in Computing Systems, Vancouver, BC, Canada, 7–12 May 2011; pp. 103–112.
4. Hagengruber, A.; Leipscher, U.; Eskofier, B.M.; Vogel, J. Electromyography for Teleoperated Tasks in Weightlessness. *IEEE Trans. Hum.-Mach. Syst.* **2021**, *51*, 130–140. [CrossRef]
5. Samuel, O.W.; Asogbon, M.G.; Geng, Y.; Al-Timemy, A.H.; Pirbhulal, S.; Ji, N.; Chen, S.; Fang, P.; Li, G. Intelligent EMG pattern recognition control method for upper-limb multifunctional prostheses: Advances, current challenges, and future prospects. *IEEE Access* **2019**, *7*, 10150–10165. [CrossRef]
6. Muzumdar, A. *Powered Upper Limb Prostheses: Control, Implementation and Clinical Application; 11 Tables*; Springer Science & Business Media: Berlin/Heidelberg, Germany, 2004.
7. Parajuli, N.; Sreenivasan, N.; Bifulco, P.; Cesarelli, M.; Savino, S.; Niola, V.; Esposito, D.; Hamilton, T.J.; Naik, G.R.; Gunawardana, U.; et al. Real-time EMG based pattern recognition control for hand prostheses: A review on existing methods, challenges and future implementation. *Sensors* **2019**, *19*, 4596. [CrossRef]
8. Fougner, A.; Stavdahl, Ø.; Kyberd, P.J.; Losier, Y.G.; Parker, P.A. Control of upper limb prostheses: Terminology and proportional myoelectric control—A review. *IEEE Trans. Neural Syst. Rehabil. Eng.* **2012**, *20*, 663–677. [CrossRef] [PubMed]
9. Simon, A.M.; Stern, K.; Hargrove, L.J. A comparison of proportional control methods for pattern recognition control. In Proceedings of the 2011 Annual International Conference of the IEEE Engineering in Medicine and Biology Society, Boston, MA, USA, 30 August–3 September 2011; IEEE: Piscataway, NJ, USA, 2011; pp. 3354–3357.
10. Hahne, J.M.; Markovic, M.; Farina, D. User adaptation in myoelectric man-machine interfaces. *Sci. Rep.* **2017**, *7*, 4437. [CrossRef]
11. Fougner, A.L. Proportional myoelectric control of a multifunction upper-limb prosthesis. Master's Thesis, Institutt for Teknisk Kybernetikk, Trondheim, Norway, 2007.
12. Nielsen, J.L.; Holmgaard, S.; Jiang, N.; Englehart, K.B.; Farina, D.; Parker, P.A. Simultaneous and proportional force estimation for multifunction myoelectric prostheses using mirrored bilateral training. *IEEE Trans. Biomed. Eng.* **2010**, *58*, 681–688. [CrossRef]
13. Muceli, S.; Farina, D. Simultaneous and proportional estimation of hand kinematics from EMG during mirrored movements at multiple degrees-of-freedom. *IEEE Trans. Neural Syst. Rehabil. Eng.* **2011**, *20*, 371–378. [CrossRef]
14. Ameri, A.; Kamavuako, E.N.; Scheme, E.J.; Englehart, K.B.; Parker, P.A. Support vector regression for improved real-time, simultaneous myoelectric control. *IEEE Trans. Neural Syst. Rehabil. Eng.* **2014**, *22*, 1198–1209. [CrossRef]
15. Ameri, A.; Akhaee, M.A.; Scheme, E.; Englehart, K. Regression convolutional neural network for improved simultaneous EMG control. *J. Neural Eng.* **2019**, *16*, 036015. [CrossRef]
16. Castellini, C.; Van Der Smagt, P. Surface EMG in advanced hand prosthetics. *Biol. Cybern.* **2009**, *100*, 35–47. [CrossRef]

17. Vidovic, M.M.C.; Hwang, H.J.; Amsüss, S.; Hahne, J.M.; Farina, D.; Müller, K.R. Improving the robustness of myoelectric pattern recognition for upper limb prostheses by covariate shift adaptation. *IEEE Trans. Neural Syst. Rehabil. Eng.* **2015**, *24*, 961–970. [CrossRef]
18. Kyranou, I.; Vijayakumar, S.; Erden, M.S. Causes of performance degradation in non-invasive electromyographic pattern recognition in upper limb prostheses. *Front. Neurorobot.* **2018**, *12*, 58. [CrossRef] [PubMed]
19. Ameri, A.; Kamavuako, E.N.; Scheme, E.J.; Englehart, K.B.; Parker, P.A. Real-time, simultaneous myoelectric control using visual target-based training paradigm. *Biomed. Signal Process. Control* **2014**, *13*, 8–14. [CrossRef]
20. Nowak, M.; Aretz, B.; Castellini, C. Wrist and grasp myocontrol: Online validation in a goal-reaching task. In Proceedings of the 2016 25th IEEE International Symposium on Robot and Human Interactive Communication (RO-MAN), New York, NY, USA, 26–31 August 2016; IEEE: Piscataway, NJ, USA, 2016; pp. 132–137.
21. Kanitz, G.; Cipriani, C.; Edin, B.B. Classification of transient myoelectric signals for the control of multi-grasp hand prostheses. *IEEE Trans. Neural Syst. Rehabil. Eng.* **2018**, *26*, 1756–1764. [CrossRef] [PubMed]
22. Ceolini, E.; Taverni, G.; Khacef, L.; Payvand, M.; Donati, E. Sensor fusion using EMG and vision for hand gesture classification in mobile applications. In Proceedings of the 2019 IEEE Biomedical Circuits and Systems Conference (BioCAS), Nara, Japan, 17–19 October 2019; IEEE: Piscataway, NJ, USA, 2019; pp. 1–4.
23. Krasoulis, A.; Vijayakumar, S.; Nazarpour, K. Effect of user practice on prosthetic finger control with an intuitive myoelectric decoder. *Front. Neurosci.* **2019**, *13*, 891. [CrossRef] [PubMed]
24. Gijsberts, A.; Bohra, R.; Sierra González, D.; Werner, A.; Nowak, M.; Caputo, B.; Roa, M.A.; Castellini, C. Stable myoelectric control of a hand prosthesis using non-linear incremental learning. *Front. Neurorobot.* **2014**, *8*, 8. [CrossRef]
25. Hahne, J.M.; Biessmann, F.; Jiang, N.; Rehbaum, H.; Farina, D.; Meinecke, F.C.; Müller, K.R.; Parra, L.C. Linear and nonlinear regression techniques for simultaneous and proportional myoelectric control. *IEEE Trans. Neural Syst. Rehabil. Eng.* **2014**, *22*, 269–279. [CrossRef]
26. Simon, A.M.; Hargrove, L.J.; Lock, B.A.; Kuiken, T.A. A decision-based velocity ramp for minimizing the effect of misclassifications during real-time pattern recognition control. *IEEE Trans. Biomed. Eng.* **2011**, *58*, 2360–2368. [CrossRef]
27. Oskoei, M.A.; Hu, H. Myoelectric control systems—A survey. *Biomed. Signal Process. Control* **2007**, *2*, 275–294. [CrossRef]
28. Englehart, K.; Hudgin, B.; Parker, P.A. A wavelet-based continuous classification scheme for multifunction myoelectric control. *IEEE Trans. Biomed. Eng.* **2001**, *48*, 302–311. [CrossRef]
29. Raghu, S.T.P.; MacIsaac, D.; Scheme, E. Analyzing the impact of class transitions on the design of pattern recognition-based myoelectric control schemes. *Biomed. Signal Process. Control* **2022**, *71*, 103134. [CrossRef]
30. Côté-Allard, U.; Fall, C.L.; Drouin, A.; Campeau-Lecours, A.; Gosselin, C.; Glette, K.; Laviolette, F.; Gosselin, B. Deep learning for electromyographic hand gesture signal classification using transfer learning. *IEEE Trans. Neural Syst. Rehabil. Eng.* **2019**, *27*, 760–771. [CrossRef] [PubMed]
31. Scheme, E.; Englehart, K. Training strategies for mitigating the effect of proportional control on classification in pattern recognition based myoelectric control. *J. Prosthet. Orthot. JPO* **2013**, *25*, 76. [CrossRef] [PubMed]
32. Smith, L.H.; Kuiken, T.A.; Hargrove, L.J. Evaluation of linear regression simultaneous myoelectric control using intramuscular EMG. *IEEE Trans. Biomed. Eng.* **2015**, *63*, 737–746. [CrossRef] [PubMed]
33. Hahne, J.M.; Schweisfurth, M.A.; Koppe, M.; Farina, D. Simultaneous control of multiple functions of bionic hand prostheses: Performance and robustness in end users. *Sci. Robot.* **2018**, *3*, eaat3630. [CrossRef] [PubMed]
34. Zia ur Rehman, M.; Waris, A.; Gilani, S.O.; Jochumsen, M.; Niazi, I.K.; Jamil, M.; Farina, D.; Kamavuako, E.N. Multiday EMG-based classification of hand motions with deep learning techniques. *Sensors* **2018**, *18*, 2497. [CrossRef] [PubMed]
35. Jiang, N.; Englehart, K.B.; Parker, P.A. Extracting simultaneous and proportional neural control information for multiple-DOF prostheses from the surface electromyographic signal. *IEEE Trans. Biomed. Eng.* **2008**, *56*, 1070–1080. [CrossRef]
36. Gailey, A.; Artemiadis, P.; Santello, M. Proof of concept of an online EMG-based decoding of hand postures and individual digit forces for prosthetic hand control. *Front. Neurol.* **2017**, *8*, 7. [CrossRef]
37. Vogel, J.; Bayer, J.; Van Der Smagt, P. Continuous robot control using surface electromyography of atrophic muscles. In Proceedings of the 2013 IEEE/RSJ International Conference on Intelligent Robots and Systems, Tokyo, Japan, 3–7 November 2013; IEEE: Piscataway, NJ, USA, 2013; pp. 845–850.
38. Vogel, J.; Hagengruber, A.; Iskandar, M.; Quere, G.; Leipscher, U.; Bustamante, S.; Dietrich, A.; Höppner, H.; Leidner, D.; Albu-Schäffer, A. EDAN: An EMG-controlled Daily Assistant to Help People With Physical Disabilities. In Proceedings of the 2020 IEEE/RSJ International Conference on Intelligent Robots and Systems (IROS), Las Vegas, NV, USA, 25–29 October 2020; IEEE: Piscataway, NJ, USA, 2020; pp. 4183–4190.
39. Hudgins, B.; Parker, P.; Scott, R.N. A new strategy for multifunction myoelectric control. *IEEE Trans. Biomed. Eng.* **1993**, *40*, 82–94. [CrossRef]
40. Neumann, M.; Huang, S.; Marthaler, D.E.; Kersting, K. pyGPs: A Python library for Gaussian process regression and classification. *J. Mach. Learn. Res.* **2015**, *16*, 2611–2616.
41. Rasmussen, C.E.; Nickisch, H. Gaussian processes for machine learning (GPML) toolbox. *J. Mach. Learn. Res.* **2010**, *11*, 3011–3015.
42. Vogel, J.; Hagengruber, A. An sEMG-based interface to give people with severe muscular atrophy control over assistive devices. In Proceedings of the 2018 40th Annual International Conference of the IEEE Engineering in Medicine and Biology Society (EMBC), Honolulu, HI, USA, 18–21 July 2018; IEEE: Piscataway, NJ, USA, 2018; pp. 2136–2141.

43. Field, A.; Miles, J.; Field, Z. *Discovering Statistics Using R*; Sage Publications: Thousand Oaks, CA, USA, 2012; pp. 822–824.
44. Scheme, E.J.; Englehart, K.B. Validation of a selective ensemble-based classification scheme for myoelectric control using a three-dimensional Fitts' law test. *IEEE Trans. Neural Syst. Rehabil. Eng.* **2012**, *21*, 616–623. [CrossRef] [PubMed]
45. Kamavuako, E.N.; Scheme, E.J.; Englehart, K.B. On the usability of intramuscular EMG for prosthetic control: A Fitts' Law approach. *J. Electromyogr. Kinesiol.* **2014**, *24*, 770–777. [CrossRef] [PubMed]
46. Quere, G.; Hagengruber, A.; Iskandar, M.; Bustamante, S.; Leidner, D.; Stulp, F.; Vogel, J. Shared control templates for assistive robotics. In Proceedings of the 2020 IEEE International Conference on Robotics and Automation (ICRA), Paris, France, 31 May–31 August 2020; IEEE: Piscataway, NJ, USA, 2020; pp. 1956–1962.
47. Schiel, F.; Hagengruber, A.; Vogel, J.; Triebel, R. Incremental learning of EMG-based control commands using Gaussian processes. In Proceedings of the Conference on Robot Learning 2020, Virtual Conference, 16–18 November 2020.

sensors

MDPI

Article

sEMG-Based Hand Posture Recognition Considering Electrode Shift, Feature Vectors, and Posture Groups

Jongman Kim, Bummo Koo, Yejin Nam and Youngho Kim *

Department of Biomedical Engineering, Yonsei University, Wonju 26493, Korea; jmkim0127@ybrl.yonsei.ac.kr (J.K.); bmk726@ybrl.yonsei.ac.kr (B.K.); namyj_1007@naver.com (Y.N.)
* Correspondence: younghokim@yonsei.ac.kr; Tel.: +82-33-760-2859

Abstract: Surface electromyography (sEMG)-based gesture recognition systems provide the intuitive and accurate recognition of various gestures in human-computer interaction. In this study, an sEMG-based hand posture recognition algorithm was developed, considering three main problems: electrode shift, feature vectors, and posture groups. The sEMG signal was measured using an armband sensor with the electrode shift. An artificial neural network classifier was trained using 21 feature vectors for seven different posture groups. The inter-session and inter-feature Pearson correlation coefficients (PCCs) were calculated. The results indicate that the classification performance improved with the number of training sessions of the electrode shift. The number of sessions necessary for efficient training was four, and the feature vectors with a high inter-session PCC ($r > 0.7$) exhibited high classification accuracy. Similarities between postures in a posture group decreased the classification accuracy. Our results indicate that the classification accuracy could be improved with the addition of more electrode shift training sessions and that the PCC is useful for selecting the feature vector. Furthermore, hand posture selection was as important as feature vector selection. These findings will help in optimizing the sEMG-based pattern recognition algorithm more easily and quickly.

Keywords: surface electromyography; pattern recognition; artificial neural network; electrode shift; hand posture; feature vector; human-computer interaction; armband sensor

Citation: Kim, J.; Koo, B.; Nam, Y.; Kim, Y. sEMG-Based Hand Posture Recognition Considering Electrode Shift, Feature Vectors, and Posture Groups. *Sensors* **2021**, *21*, 7681. https://doi.org/10.3390/s21227681

Academic Editor: Ernest N. Kamavuako

Received: 19 October 2021
Accepted: 15 November 2021
Published: 18 November 2021

Publisher's Note: MDPI stays neutral with regard to jurisdictional claims in published maps and institutional affiliations.

1. Introduction

Gestures, involving the physical movements of the hands, face, or body, is a form of communication used to convey meaningful information or interact with the environment [1]. Among the various gestures, those typically applied in machine learning algorithms as the interface of human-computer interaction (HCI) are hand gestures. This is because they constitute the most natural and efficient movements in daily life [2]. As an HCI interface, a hand gesture recognition system has three advantages [3]. The first advantage is the ease of hygiene management through a contactless interface. This contactless interface helps in maintaining hygienic conditions for the user by preventing contamination due to contact. Therefore, hand gesture recognition systems are useful in clinical applications such as healthcare systems. The second advantage is that a hand gesture recognition system can be applied as an alternative to overcome physical disabilities. It is easy to apply this system to an assistive device, such as a home-care system or an IoT system controller, for the disabled or elderly who have difficulty in moving. Furthermore, the need for a gesture-based HCI interface is increasing, owing to the increasing number of people who can only communicate through hand gestures (e.g., sign language for the deaf). The third advantage is that considerable data and commands can be easily managed by the intuitive movements.

Studies on hand gesture recognition have predominantly used one of the two following technologies: computer vision or wearable sensors. Computer vision-based systems use one or more cameras to recognize hand gestures. Shin et al. developed a six-hand gesture recognition system using a low-priced USB color camera and entropy analysis; they achieved a classification accuracy of 97.2% [4]. Stergiopoulou and Papamarkos were developed a neural network-based hand gesture recognition system using one camera and thirty-one classified hand gestures with an accuracy of 90.0% [5]. However, these systems encountered the problem of environmental factors (e.g., shadow, lighting, background, and camera position) affecting the classification performance. They exemplify the difficulty involved in optimizing computer vision-based recognition systems under environmental factors, varying with respect to time and place. Furthermore, the camera, which has a high resolution for accurate gesture recognition, is expensive and has low portability, and the output data file is very large.

Wearable sensor-based systems use non-invasive sensors on the user's skin. The bio-signal or the motion of the user is detected and measured by the non-invasive sensors, and the measured data are used to recognize the gestures. A data glove is predominantly used in studies on wearable sensor-based hand gesture recognition, and is useful for measuring the posture and gesture of the hand accurately. Nam et al. classified ten hand gestures with an accuracy of 80.0% using the VLP data glove and hidden Markov model (HMM) [6]. Additionally, Yin et al. developed a hand gesture recognition system that recognizes nine hand gestures, with an accuracy of 99.8%, using the data glove and neural network algorithm [7]. However, it is difficult to use the data glove-based gesture recognition system in daily life, owing to the high price of the glove and consequent contamination from sweat and oil after long-term use. Furthermore, the data glove restricts the natural hand gestures of the user, and the glove design causes discomfort because of repetitive donning and doffing in daily life.

Recently, several studies have applied sEMG sensors to overcome problems encountered by hand gesture recognition systems. sEMG is a non-invasive method for measuring the fine bio-signal of muscle activation, and the sEMG signal contains extensive information about the activity of neurons from the spinal cord to the muscle fibers. Therefore, sEMG is widely used in clinic and rehabilitation and bio-signal-based control systems for HCI. Kim et al. classified four-hand gestures from one sEMG sensor and achieved a classification accuracy of approximately 94.0% using a combination of the K-nearest neighbor (KNN) and Bayes classifier [8]. Shi et al. developed a four-hand gesture recognition algorithm and achieved an accuracy of 94.0% using two sEMG sensors and a KNN classifier [9]. These previous studies suggested using each muscle belly for the positions of the sEMG sensors, to avoid crosstalk between sEMG signals; however, this decreases the practicality of using the gesture recognition system because finding the muscle belly is difficult for non-expert users in daily life.

A wearable sensor, which has the design of an armband or a wristband, has been suggested to increase the practicality for non-expert users in daily life. Jiang et al. developed a wristband-type sEMG sensor that includes four channel sEMG modules and one inertial measurement unit (IMU). They classified eight hand gestures with an accuracy of 92.6% using a linear discriminant analysis classifier [10]. Abreu et al. classified twenty static hand gestures with an accuracy of 98.6% using a support vector machine (SVM) classifier and a commercial sensor called the Myo Armband (Thalmic Labs, Kitchener, Canada) [11]. However, the two main problems—electrode shift and feature vector selection—remain unsolved.

Electrode shift is a common issue that arises during donning and doffing of a sensor in daily life. Many previous studies fixed the positions of both the sEMG and wearable sensors to avoid misclassifications that result from electrode shift. Feature vector selection is an important process in the development of a pattern recognition algorithm [12,13]. Previous studies analyzed the classification performance of each feature vector and applied the feature vector that delivered the best performance in the hand gesture recognition algorithm. However, various factors, such as the sensor performance, limb position, and electrode shift, easily affect the classification performance of the feature vector. Therefore, feature vector selection, based on classification performance, is inefficient in the development of the pattern recognition algorithm. In previous studies on efficient pattern recognition, a principal component analysis (PCA) and a genetic algorithm (GA), both of which reduce the dimension of the feature vector and minimize the data complexity, were suggested for feature vector selection [14,15]; however, few studies have been conducted on feature vector selection that consider the electrode shift.

Many previous studies developed hand gesture recognition systems, but the number of target gestures are limited because of the limitations of the classification performances of the algorithms and the efficiencies of the systems. Therefore, target gestures were selected with reference to previous studies. For the HCI interface, Wahid et al. developed a classification algorithm for the target gestures of a fist, wave in, and wave out [16], and Zhang et al. developed an algorithm for the target gestures of a double tap, wave in, wave out, fingers spread, and fist [17]. Castiblanco et al. selected twelve hand wrist gestures to develop a gesture recognition algorithm for stroke rehabilitation [18]. Kim et al. selected thirty-eight Korean finger languages in a finger language recognition study [19], and seven hand gestures were selected in a myoelectric hand prosthesis control study [20]. Andrad et al. developed a hand gesture recognition algorithm for the target gestures of a cylindrical grasp, tip pinch, and hook (snap, palmar pinch, spherical grasp, and lateral pinch) [21]; furthermore, they reported that the classification accuracy decreased with similarity in the same-gesture group (precision grasp: tip, palmar and lateral; power grasp: cylindrical, hook, and spherical). Therefore, these previous studies were performed to improve the classification performance with a selected target gesture. However, selected gestures could be replaced with other gestures which had a similar form or function. Therefore, the analysis about the gesture selection was important to improve the gesture recognition algorithm, but few studies have been conducted on classification performance in accordance with the gesture selection type.

In this study, a hand posture recognition algorithm was developed, taking into consideration the electrode shift, feature vectors, and posture types. A custom armband-type multi-channel sEMG sensor was used to measure the sEMG signal on the forearm. Experiments were conducted with electrode shifts and the measured sEMG signal was used to calculate twenty-one feature vectors. Twelve hand postures were selected after a literature review, and seven posture groups were formed considering the function and muscle activation of each hand posture. The classification accuracy and inter-session and inter-feature PCCs were analyzed for the PCC-based feature vector selection.

2. Materials and Methods

2.1. Participants

Ten healthy right-handed adults (7 males, 3 females, 24.1 $\pm$ 0.7 years) without neurological disorders were recruited to participate in this study. All participants were fully informed of any of the risks associated with the experiments, and they gave their written consent to participate in this study. The experimental procedure was approved by the Yonsei University Mirae Institutional Review Board (1041849-201704-BM-018-01).

2.2. Equipment

Figure 1 shows the armband-type sEMG sensor and Baseline hand dynamometers (Fabrication Enterprises, Inc., White Plains, NY, USA) used in this experiment. The custom armband-type eight-channel sEMG sensor was used to measure the sEMG signal with a sampling frequency of 500 Hz [19]. Each participant wore the armband sensor on their right forearm, and the main module of the armband sensor was placed on the belly area of the anterior part of the forearm during wrist flexion (around the flexor carpi radialis). The hand dynamometers were used to perform each hand posture with fixed grasp force.

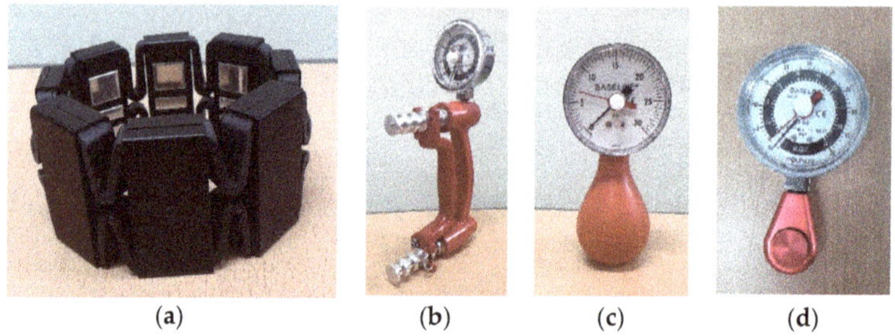

Figure 1. Armband-type multi-channel sEMG sensor (**a**) and hand dynamometers (**b–d**).

2.3. Experimental Procedures

Twelve hand postures were selected after a literature review [10,21–30] (Figure 2); the function of each of these hand postures is presented in Table 1.

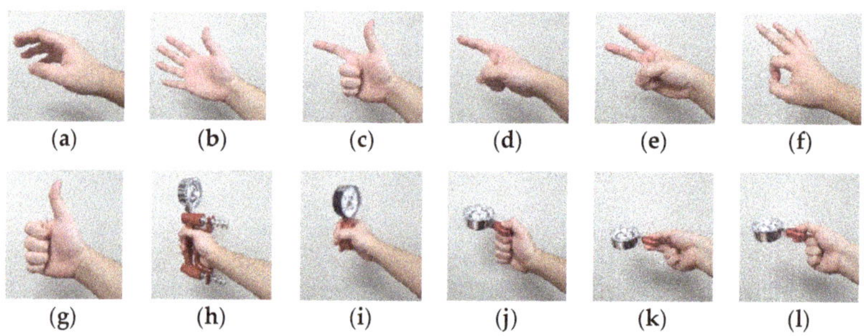

Figure 2. Twelve hand postures selected for use in this study: (**a**) rest, (**b**) spread, (**c**) scissor-sign, (**d**) finger pointing, (**e**) V-sign, (**f**) O.K.-sign, (**g**) thumb-up (hook), (**h**) cylindrical grasp, (**i**) spherical grasp, (**j**) lateral pinch, (**k**) palmar pinch, (**l**) tip pinch.

In this study, seven hand posture groups were constructed, and the classification performances of each posture group were analyzed to find the most efficient hand posture group for the development of a hand posture recognition algorithm. The posture groups were determined using the function and activated muscles of each hand posture. Group 1 was composed of the most frequently used and important hand postures identified in previous studies. Group 2 and Group 5 consisted of the postures in Group 1 and the finger-pointing or scissor-sign postures, respectively, which are postures used to point to objects. Group 3 and Group 4 consisted of the postures in Group 1 and postures such as a tip pinch and spherical grasp, which have the same function as those included in Group 1. Seven different hand postures were selected to be part of Group 6, considering the overlap of the functions and activated muscles on each hand posture. Finally, Group 7 included all of the listed hand postures.

Table 1. Hand posture functions and groups.

Function	Hand Postures		Posture Groups	
Holding objects	Cylindrical grasp [21,27,28] (for cylindrical objects)	Spherical grasp [21,28](for spherical objects)	Group 1	Rest, cylindrical grasp, palmar pinch, hook (thumb-up), lateral pinch, spread
Holding small/thin/flat objects	Palmar pinch [21,28] (using thumb, index, and middle fingers for palm facing the objects)	Tip pinch [21,28] (using thumb and index fingers for palm facing the objects)	Group 2	Group 1 + finger pointing
	Lateral pinch [21,27] (using a thumb pad and the radial side of the index finger)		Group 3	Group 1 + tip pinch
			Group 4	Group 1 + spherical grasp
Releasing objects	-	Spread [10,24,27,29,30]	Group 5	Group 1 + scissor-sign
Supporting loads	Thumb-up (hook) [10,21,23,28–30]			
Expression of emotion		V-sign [10,25,29,30] O.K.-sign [10,22,29,30]	Group 6	Rest, cylindrical grasp, palmar pinch, V-sign, finger pointing, thumb-up (hook), lateral pinch
No activation		Rest		
Pointing objects	Finger pointing [10,23,27] (using index finger, only)	Scissor-sign [23,26] (using thumb and index finger)	Group 7	All hand postures (12)

Hand postures with gray background: the most frequently used and important hand postures in previous studies.

In an sEMG-based gesture recognition algorithm, force and muscle fatigue were classed as critical factors to increase the variability of the sEMG signal. Therefore, all hand postures were performed with 20% maximum voluntary contraction (MVC) to avoid muscle fatigue and the confounding factor with the grasp force [20,31]. Hydraulic-hand and pneumatic-hand dynamometers, and hydraulic pinch gauge (Fabrication Enterprises, Inc., White Plains, NY, USA) were used for the grasp and pinch postures, respectively. All participants practiced maintaining 20% MVC in all hand postures except the rest posture. Postures without the use of the hand dynamometers were performed with the same force as displayed in the practice. The participants were seated on chairs and performed each hand posture for 5 s in a random order, and the experimental session was repeated 10 times. All participants took sufficient rests, removing the sensor between experimental sessions, and the sensor was worn again before the next session.

2.4. Feature Vector Extraction

The sEMG signal was filtered using the fourth bandpass filter with a bandwidth of 15–250 Hz, and the filtered sEMG signal was used to calculate the feature vectors. In this study, the feature vectors were selected in the time domain corresponding to real-time application [32]. Twenty-one time-domain feature vectors and their corresponding formulas are presented in Table 2. The hand posture recognition algorithm was applied to feature vectors of a single type. The feature vectors were calculated with a window size of 250 ms and a window shift of 10 ms, as recommended in a previous study [33]. The feature vectors of AR and CC were calculated using various orders of 1 to 10.

Table 2. Time-domain feature vectors.

N: window size, i: data sample, EMG_i: sEMG signal, w_i: white noise error term; p: function order			
$RMS = \sqrt{\frac{1}{N}\sum_{i=1}^{N} EMG_i^2}$	$WL = \sum_{i=1}^{N-1} \lvert EMG_{i+1} - EMG_i \rvert$	$MAV = \frac{1}{N}\sum_{i=1}^{N} \lvert EMG_i \rvert$	$MAVSLP_i = MAV_{i+1} - MAV_i$
$MAV1\ \&\ MAV2 = \frac{1}{N}\sum_{i=1}^{N} w_i \lvert EMG_i \rvert$		$ZC = \sum_{i=1}^{N-1} [f(x_i \times x_{i+1}) \cap \lvert x_i - x_{i+1} \rvert \geq threshold]$	
$MAV1: w_i = \begin{cases} 1, \ if\ 0.25N \leq i \leq 0.75N \\ 0.5, \quad otherwise \end{cases}$			
$MAV2: w_i = \begin{cases} 1, \ if\ 0.25N \leq i \leq 0.75N \\ \frac{4i}{N}, \ elseif\ i < 0.25N \\ \frac{4(i-N)}{N}, \ otherwise \end{cases}$		$WAMP = \sum_{i=1}^{N-1} [f(\lvert x_i - x_{i+1} \rvert)]$ $SSC = \sum_{i=2}^{N-1} [f[(x_i - x_{i-1}) \times (x_i - x_{i+1})]]$	$f(x) = \begin{cases} 1, \ if\ x \geq threshold \\ 0, \quad otherwise \end{cases}$
$IEMG = \sum_{i=1}^{N} \lvert EMG_i \rvert$	$VAR = \frac{1}{N-1}\sum_{i=1}^{N} EMG_i^2$	$SSI = \sum_{i=1}^{N} EMG_i^2$	$DASDV = \sqrt{\frac{1}{N-1}\sum_{i=1}^{N-1} (x_{i+1} - x_i)^2}$
$TM3 = \left\lvert \frac{1}{N}\sum_{i=1}^{N} EMG_i^3 \right\rvert$	$LOG = e^{\frac{1}{N}\sum_{i=1}^{N} log(\lvert EMG_i \rvert)}$	$MYOP = \frac{1}{N}\sum_{i=1}^{N} [f(x_i)]$	$AAC = \frac{1}{N}\sum_{i=1}^{N-1} \lvert x_{i+1} - x_i \rvert$
$TM4 = \frac{1}{N}\sum_{i=1}^{N} EMG_i^4$	$AR = a_p; Auto-regressive\ mode: \ x_i = \sum_{p=1}^{P} a_p x_{i-p} + w_i$		
$TM5 = \left\lvert \frac{1}{N}\sum_{i=1}^{N} EMG_i^5 \right\rvert$	$CC = c_p; c_1 = -a_1; c_p = -a_p - \sum_{l=1}^{p-1}\left(1 - \frac{l}{p}\right) a_p c_{p-l}; 1 \leq l \leq p$		

The threshold-based feature vectors, such as ZC, MYOP, WAMP, and SSC, were calculated using the predefined threshold values. In previous studies, the threshold value was selected from 50 μV to 100 mV considering the gain of the sEMG sensor and background noise. The optimization of the threshold value is crucial because a considerably low threshold value leads to the transmission of unwanted information from the background noise, and a considerably high threshold value misses important information for pattern recognition. However, finding the optimized threshold value requires significant amounts of time and high costs for the gesture recognition system. Therefore, many previous studies used threshold values from other studies, rather than optimizing the threshold values in their systems; moreover, few studies have been conducted to find the most suitable threshold value for each feature vector. Kamavuako et al. suggested Equation (1) for defining various threshold values using a factor R and the RMS value from the sEMG signal at rest [34]. The equation proved useful in reducing the time and cost for the optimization of the threshold values in the gesture recognition algorithm:

$$Threshold\ value = R \times RMS_{sEMG^{at\ rest}}, \tag{1}$$

Consequently, the equation was used in this study to optimize the threshold value for each feature vector, and a factor R was applied in increments of 0.5, from 0.0 to 10.0. In addition, the threshold values used in previous studies [13,20,35–43] were applied in the hand posture recognition algorithm.

2.5. Classifier

An artificial neural network (ANN) is a machine learning algorithm that was developed by simulating a biological neural network in the brain of a human or animal. The ANN was constructed with the input layer, hidden layer, and output layer, using artificial neurons, known as the node, which then classified the input signals through learning processes such as the backpropagation algorithm. The joint function between the input/hidden/output layers was easily estimated, and the classification was performed quickly.

In the input and the output layers, the number of nodes were determined by the feature vector and the target class. However, the hidden layer was able to change the number of nodes. The optimization of the hidden layer's nodes was important because insufficient nodes caused underfitting, while excessive nodes caused overfitting [44,45]. In this study, 8 nodes in the input layer (1 sEMG feature × 8 channels) and 6–12 nodes in the output layer (the number of hand posture) were determined in ANN, and there were 17 nodes in the hidden layer according to a previous study [46]. Both the training and testing of the ANN classifier were performed through 10-fold cross validation using the MATLAB Deep Learning Toolbox (Mathworks, Inc., Natick, MA, USA).

2.6. Performance Evaluation

In this study, twenty-one time-domain feature vectors and two feature combinations (Hudgins' set: MAV, WL, ZC, and SSC [38] and Du's set: IEMG, VAR, WL, ZC, SSC, and WAMP [47]) from previous studies were applied to the ANN classifier. The classification accuracy was analyzed based on the number of training sessions and considering the electrode shift, and the PCC was calculated to analyze the linear relationship between the sessions or the feature vectors. Many previous studies suggested various methods to select the feature vectors for the improvement of the classification performance [10,12,14,15]. Correlation-based feature vector selection was primarily used to evaluate the way each feature vector is able to distinguish a gesture or verify the similarity of the information of feature vectors. In this study, the inter-session PCC was used to analyze the correlation between the sessions with the electrode shift, and the similarity of the feature vectors was analyzed by inter-feature PCC. A factor r represented the linear relationships as follows: weak linear relationship: $0 < r \leq 0.3$; moderate linear relationship: $0.3 < r \leq 0.7$; and strong linear relationship: $0.7 < r \leq 1.0$.

A statistical analysis was performed using IBM SPSS Statistics (IBM, Corp., Armonk, NY, USA), and the results of the evaluation were determined as nonparametric. The Kruskal–Wallis H test and pairwise comparison were performed to examine the differences in classification accuracy in accordance with the threshold values, feature vector orders, and the number of training sessions. The statistical significance was set at $p < 0.05$ for all tests.

3. Results

3.1. Classification Accuracy Based on Threshold Values and Feature Vector Orders

In this study, various threshold values obtained using Equation (1), and from previous studies, were applied to calculate the threshold-based feature vectors. The RMS value of the sEMG signal at rest was 3.3 mV, and the threshold values were defined in the range of 0.0–33.0 mV following accordance with Equation (1), with a step size of 1.65 mV. The range of the threshold values from Equation (1) included all threshold values from previous studies.

Figure 3 shows the classification accuracy and inter-session PCC of Group 1 in accordance with the threshold values (the results for the other groups are shown in Figures S1–S6). All threshold-based feature vectors showed improved classification accuracies corresponding with an increasing number of training sessions. The classification accuracy and inter-session PCC were dramatically improved by increasing the threshold value, and then maintained at the constant level. However, excessive threshold values were found to cause a degradation of the classification accuracy and inter-session PCC. The best threshold values depended on the feature vector.

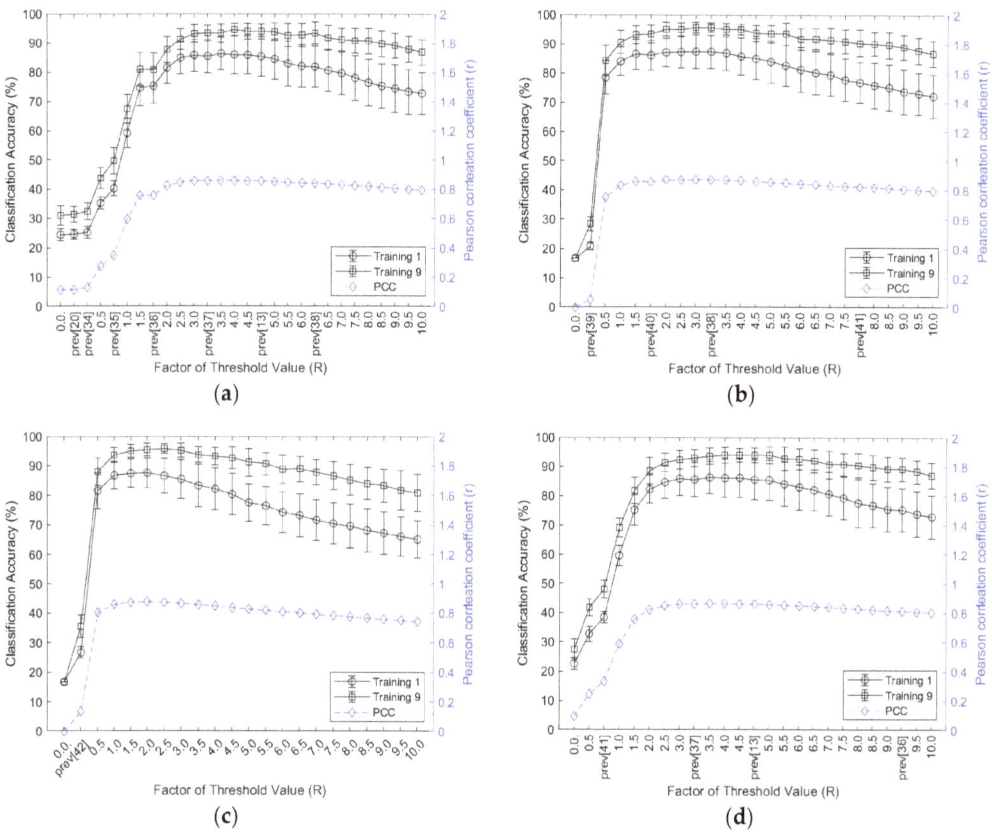

Figure 3. Classification accuracy and inter-session PCC of Group 1 based on threshold values: (**a**) ZC; (**b**) WAMP; (**c**) MYOP; (**d**) SSC; prev: threshold value from the previous study.

The threshold value of 13.2 mV exhibited the best classification accuracy in ZC with nine training sessions (Group 1: 94.5 ± 2.1%, Group 2: 91.2 ± 2.3%, Group 3: 85.7 ± 4.3%, Group 4: 86.6 ± 4.6%, Group 5: 89.4 ± 2.9%, Group 6: 93.1 ± 3.8%, Group 7: 76.7 ± 4.7%). The best threshold value of WAMP with nine training sessions was 9.9 mV (Group 1: 95.6 ± 1.8%, Group 2: 92.7 ± 1.6%, Group 3: 86.1 ± 4.4%, Group 4: 89.5 ± 3.7%, Group 5: 90.8 ± 3.2%, Group 6: 94.5 ± 3.5%, Group 7: 79.3 ± 4.9%). In MYOP and SSC, with nine training sessions, the best threshold values were 6.6 mV (Group 1: 95.4 ± 2.2%, Group 2: 92.5 ± 2.2%, Group 3: 86.7 ± 4.7%, Group 4: 89.8 ± 3.8%, Group 5: 90.7 ± 3.0%,

Group 6: 94.7 ± 3.1%, Group 7: 79.1 ± 4.7%) and 13.2 mV (Group 1: 93.9 ± 2.8%, Group 2: 91.1 ± 2.4%, Group 3: 84.6 ± 4.6%, Group 4: 86.9 ± 5.0%, Group 5: 88.9 ± 2.4%, Group 6: 92.9 ± 3.7%, Group 7: 76.6 ± 5.1%), respectively. Furthermore, the appropriate ranges of the threshold values were 8.25–23.1 mV, 6.6–18.15 mV, 3.3–14.85 mV, and 8.25–23.1 mV in the feature vectors of ZC, WAMP, MYOP, and SSC, respectively. These ranges of threshold values were similar for the posture group in each feature vector. The classification accuracy and inter-session PCC were high ($r > 0.8$) for the feature vectors with appropriate ranges of threshold values.

The classification accuracy and inter-session PCC of Group 1, for the feature vectors of AR and CC, are shown in Figure 4 (results of the other groups are presented in Figures S7–S12). The second order exhibited the best classification accuracies for both AR (Group 1: 53.6 ± 7.2%; Group 2: 49.1 ± 6.0%; Group 3: 47.6 ± 6.6%; Group 4: 49.1 ± 7.0%; Group 5: 48.9 ± 6.2%; Group 6: 52.2 ± 8.2%; Group 7: 35.9 ± 5.8%) and CC (Group 1: 53.7 ± 7.2%; Group 2: 49.5 ± 6.6%; Group 3: 46.8 ± 6.3%; Group 4: 49.6 ± 6.5%; Group 5: 49.2 ± 6.9%; Group 6: 52.2 ± 8.6%; Group 7: 36.5 ± 5.9%); however, there was no significant difference between the orders of AR and CC.

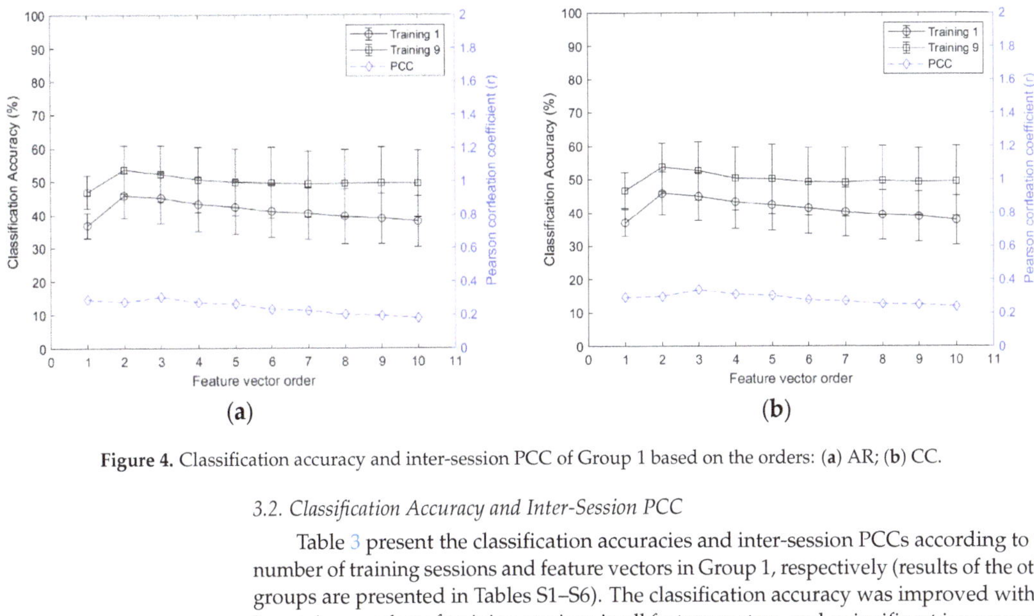

(a) (b)

Figure 4. Classification accuracy and inter-session PCC of Group 1 based on the orders: (**a**) AR; (**b**) CC.

3.2. Classification Accuracy and Inter-Session PCC

Table 3 present the classification accuracies and inter-session PCCs according to the number of training sessions and feature vectors in Group 1, respectively (results of the other groups are presented in Tables S1–S6). The classification accuracy was improved with an increasing number of training sessions in all feature vectors, and a significant improvement was observed for four or more training sessions. Although the classification accuracies for five to nine training sessions (more than four) were higher than those for four training sessions, no significant difference was found. Across the four training sessions, feature vectors with high inter-session PCCs ($r > 0.7$; strong linear relationship) exhibited higher classification accuracies (Group 1: >90.0%, Group 2: >88.0%, Group 3: >81.0%, Group 4: >85.0%, Group 5: >86.0%, Group 6: >90.0%, and Group 7: >70.0%) than those of the feature vectors with low inter-session PCCs ($r < 0.7$). The feature vectors, which were used in the feature combination from previous studies, exhibited high inter-session PCCs ($r > 0.7$).

Table 3. Classification accuracy and inter-session PCC of the feature vectors in Group 1.

Feature Vector	Classification Accuracy (%): Mean (Standard Deviation)									PCC (*r*)
	TRN1	**TRN2**	**TRN3**	**TRN4**	**TRN5**	**TRN6**	**TRN7**	**TRN8**	**TRN9**	
IEMG	84.6 (5.7)	90.1 (3.1)	92.3 (2.2)	93.2 (2.0)	93.8 (1.7)	94.4 (1.6)	94.9 (1.3)	95.4 (1.4)	95.4 (1.2)	0.837
MAV	84.9 (5.8)	90.2 (3.0)	92.5 (1.9)	93.2 (1.8)	93.8 (1.7)	94.3 (1.7)	94.7 (1.5)	95.5 (1.3)	95.4 (1.4)	0.837
MAV1	85.2 (5.5)	90.1 (3.0)	92.2 (2.1)	93.1 (1.8)	93.5 (1.6)	94.3 (1.5)	94.5 (1.3)	94.9 (1.5)	95.4 (1.4)	0.835
MAV2	84.4 (5.3)	89.1 (2.9)	90.4 (2.9)	91.3 (2.6)	91.7 (2.7)	92.2 (2.5)	92.5 (2.5)	93.0 (2.3)	93.2 (2.3)	0.808
SSI	79.6 (6.5)	86.0 (3.8)	89.3 (2.5)	90.5 (1.8)	91.9 (1.3)	93.0 (1.3)	93.3 (1.1)	94.1 (1.3)	94.4 (1.1)	0.735
VAR	79.5 (6.7)	86.0 (3.3)	89.3 (2.1)	90.6 (1.8)	92.2 (1.4)	93.2 (1.3)	93.3 (1.2)	94.0 (1.0)	94.3 (1.2)	0.735
TM3	65.9 (6.1)	73.9 (4.7)	77.7 (4.3)	79.9 (3.6)	81.5 (3.8)	82.4 (3.8)	83.1 (3.7)	83.4 (3.5)	83.6 (4.2)	0.392
TM4	69.6 (7.7)	77.8 (5.5)	82.1 (4.1)	84.1 (3.7)	85.8 (3.5)	86.6 (3.1)	87.6 (3.1)	88.2 (3.4)	88.9 (3.1)	0.428
TM5	61.3 (6.6)	68.7 (5.9)	71.3 (6.2)	71.9 (5.8)	72.4 (6.0)	72.5 (6.3)	72.5 (6.1)	72.1 (6.7)	72.4 (6.9)	0.248
RMS	84.9 (5.9)	90.0 (3.3)	92.4 (2.5)	93.2 (1.8)	93.7 (1.7)	94.3 (1.7)	94.7 (1.4)	95.1 (1.3)	95.5 (1.3)	0.829
LOG	85.8 (5.3)	90.1 (3.1)	92.1 (2.6)	93.0 (2.4)	93.1 (2.4)	93.8 (2.3)	93.8 (2.4)	94.2 (2.3)	94.3 (2.2)	0.859
WL	84.5 (5.9)	90.3 (3.0)	92.4 (2.1)	93.1 (1.9)	93.7 (1.7)	93.9 (1.8)	94.4 (1.6)	94.8 (1.6)	95.0 (2.1)	0.832
AAC	84.6 (5.7)	90.0 (3.3)	92.4 (2.1)	93.2 (1.8)	93.6 (1.9)	94.0 (1.8)	94.5 (1.9)	94.7 (1.8)	95.1 (2.0)	0.832
DASDV	84.5 (5.9)	89.8 (3.7)	92.0 (2.2)	93.2 (1.9)	93.6 (1.9)	94.2 (1.9)	94.3 (1.8)	94.9 (2.1)	95.0 (1.6)	0.824
MAVSLP	39.2 (3.8)	45.5 (4.2)	47.9 (4.3)	49.1 (4.2)	49.8 (4.1)	50.2 (4.2)	50.5 (4.2)	50.9 (4.1)	51.3 (4.0)	0.005
ZC	85.9 (6.7)	90.0 (3.8)	92.0 (3.1)	92.9 (2.6)	93.2 (2.6)	93.6 (2.3)	93.8 (2.3)	94.2 (2.4)	94.5 (2.1)	0.858
WAMP	87.3 (5.8)	91.3 (3.8)	92.9 (2.9)	93.6 (2.4)	94.1 (2.3)	94.3 (2.3)	94.7 (2.0)	94.7 (2.3)	95.6 (1.8)	0.873
MYOP	87.5 (5.0)	91.5 (3.2)	93.1 (2.7)	93.9 (2.4)	94.4 (2.1)	94.5 (2.3)	94.9 (2.4)	95.2 (2.1)	95.4 (2.2)	0.876
SSC	86.0 (6.5)	90.1 (3.9)	91.8 (3.1)	92.4 (2.8)	93.1 (2.6)	93.1 (2.5)	93.6 (2.5)	94.0 (2.6)	93.9 (2.8)	0.862
AR	45.8 (6.6)	48.1 (7.8)	48.3 (8.2)	48.4 (8.0)	49.4 (7.7)	50.6 (7.4)	52.2 (7.3)	53.0 (6.6)	53.6 (7.2)	0.278
CC	45.8 (6.5)	48.2 (7.6)	48.1 (8.2)	48.6 (8.4)	49.7 (7.8)	50.9 (7.7)	52.1 (7.1)	52.5 (7.3)	53.7 (7.2)	0.299

3.3. Classification Accuracy and Inter-Feature PCC

The classification accuracies of all of the feature vectors and their combinations in Group 1 are presented in Figure 5 (the results for the other groups are presented in Figures S13–S18). The classification accuracies of all feature vectors, including the combinations of feature vectors, were improved by increasing the number of training sessions. However, certain feature vectors, such as MAVSLP, AR and CC, demonstrated lower classification accuracy (51.3%~53.7%) and inter-session PCC ($r < 0.3$) than other feature vectors, even though nine training sessions were applied. On the contrary, the feature vectors with a high inter-session PCC ($r > 0.7$) showed a classification accuracy of higher than 90.0%, which was statistically similar with the classification accuracy of the feature vector combinations. Furthermore, Figure 6 and Figures S19–S24 showed that strong linear relationship in the inter-feature PCC appeared between the feature vectors with a high inter-session PCC ($r > 0.7$), while a weak linear relationship in the inter-feature PCC appeared between the feature vectors with a high inter-session PCC ($r > 0.7$) and the feature vectors with a low inter-session PCC ($r < 0.3$). These results reveal that information was similar between the feature vectors that had a high inter-session PCC and high classification accuracy.

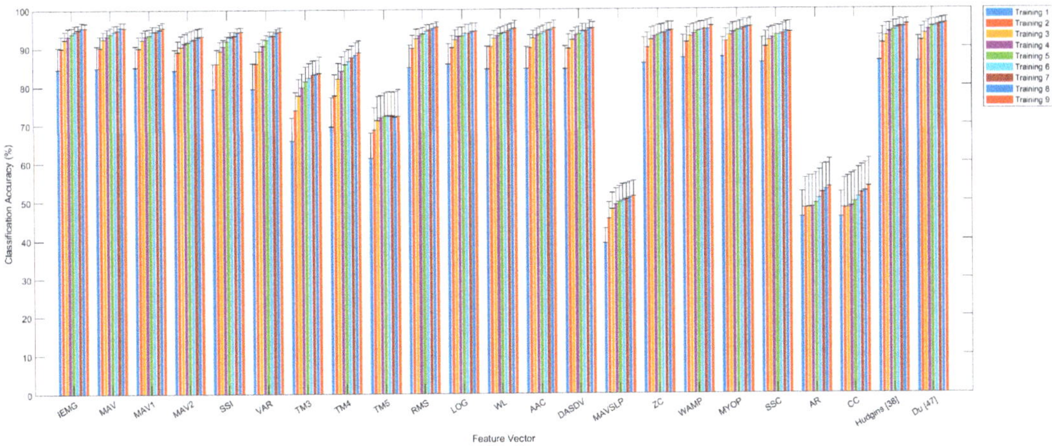

Figure 5. Classification accuracies of feature vectors and feature combinations according to training in Group 1.

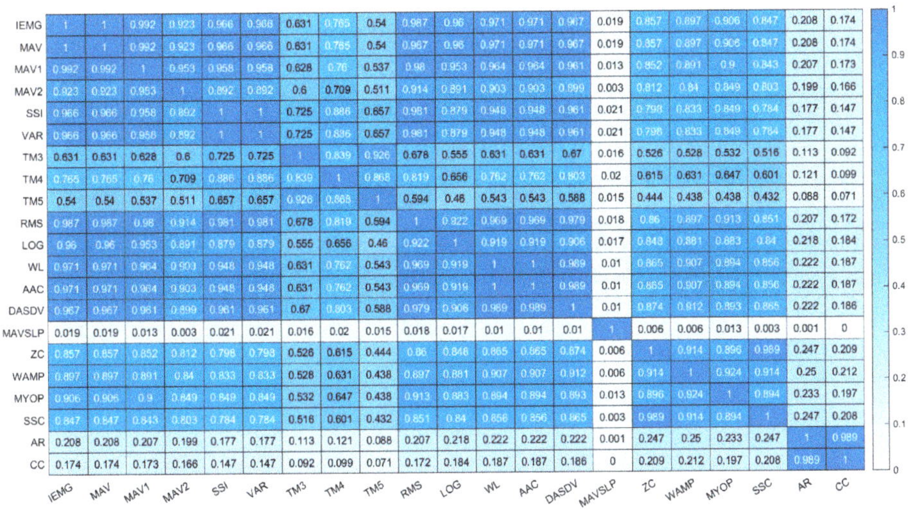

Figure 6. Confusion matrix of inter-feature PCC in Group 1.

3.4. Classification Accuracies According to Hand Posture Groups

Figure 7 shows the confusion matrices of the Group 1 and Group 7 in MAV, which is a well-known feature vector (all results are presented in Figures S25–S31). In Group 1, the hand gestures of rest and cylindrical grasp showed the high classification accuracy, while the worst classification accuracy appeared in palmar pinch. The misclassification between palmar pinch and lateral pinch was also minimized from 13.0% to 4.2% by increasing the number of training sessions. The change in the misclassification in relation to the number of training sessions in Group 2 were from 12.6% to 4.0% and 14.4% to 3.7% for 'palmar pinch vs. lateral pinch' and 'finger pointing vs. rest', respectively (Figure S26). However, the misclassifications of 'palmar pinch vs. tip pinch vs. lateral pinch', 'cylindrical grasp vs. spherical grasp', and 'scissor sign vs. thumb up (hook)' were still high despite the increasing number of training sessions in Groups 3–5 (Figures S27–S29). Group 6 (Figure S30) exhibited a good classification performance independent of the number of training sessions. Group 7 revealed all types of misclassifications. As shown in the confusion matrices, the hand postures of a similar form or function showed more misclassifications, as they were more difficult to differentiate from one other. However, most misclassifications were significantly reduced given an increasing number of training sessions. The same results were obtained in the other feature vectors.

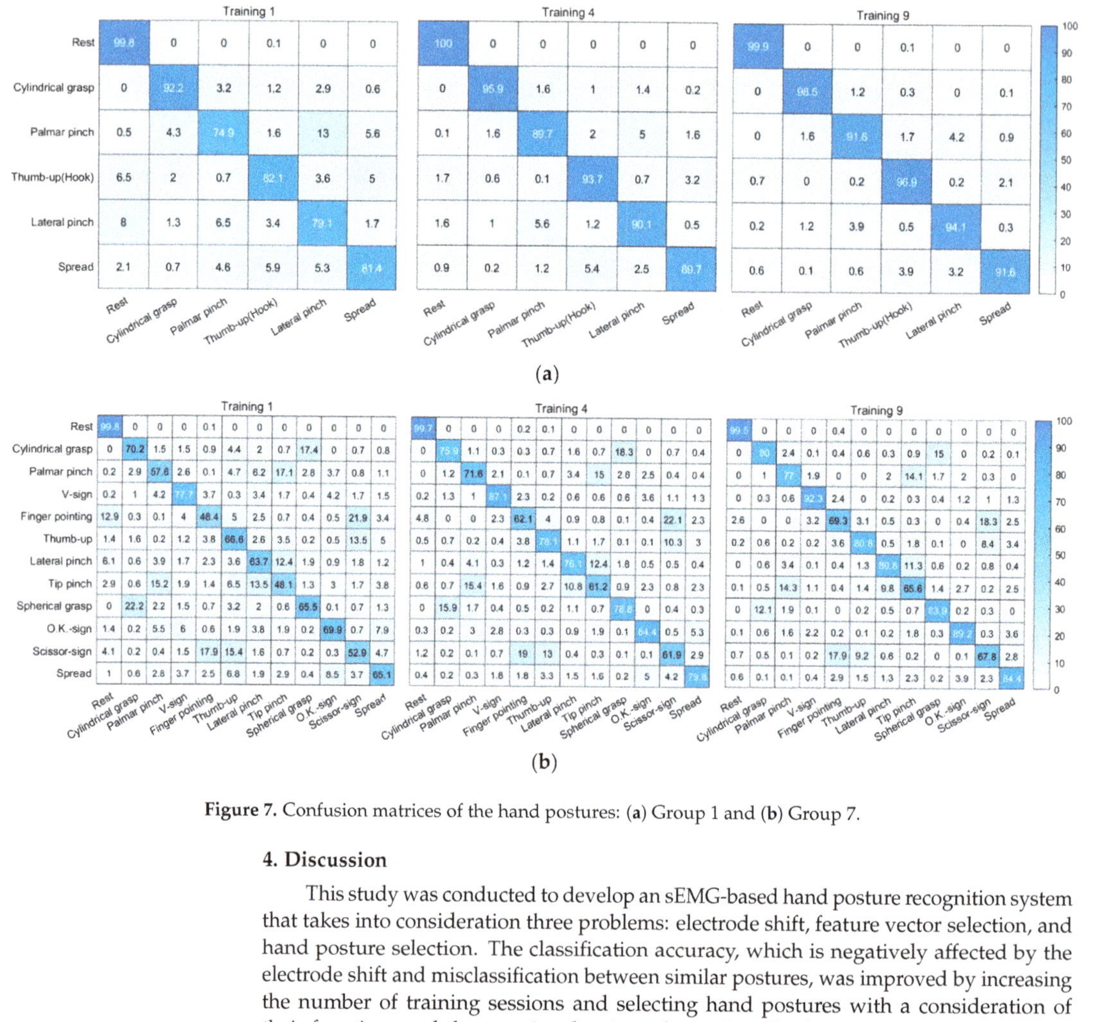

Figure 7. Confusion matrices of the hand postures: (**a**) Group 1 and (**b**) Group 7.

4. Discussion

This study was conducted to develop an sEMG-based hand posture recognition system that takes into consideration three problems: electrode shift, feature vector selection, and hand posture selection. The classification accuracy, which is negatively affected by the electrode shift and misclassification between similar postures, was improved by increasing the number of training sessions and selecting hand postures with a consideration of their functions and the associated activated muscles. Furthermore, an efficient feature vector optimization method was developed by analyzing the relationship between the classification and inter-session PCC. These findings provide a method for developing an sEMG-based hand posture recognition system displaying a high practicality.

Electrode positional changes are common in sEMG-based gesture recognition systems in daily life; however, many previous studies reported that the positions of the sEMG sensors were fixed to avoid the occurrence of misclassification resultant of the electrode shift. Lu et al. used seven sEMG sensors with fixed positions on the muscle belly to control a robotic hand [48]. The measured sEMG signal was applied to a Gaussian Naive Bayes classifier and an SVM, and six hand gestures were classified with an accuracy of 84.1%. sEMG-based gesture recognition algorithms with fixed sEMG sensor positions are appropriate for robotic prosthesis, which uses a socket for preventing the electrode shift;

however, it is not appropriate in the HCI interface for non-expert users. Phinyomark et al. emphasized the importance of the solution for the confounding factors, such as the electrode shift, to improve its re-usability and sustainability for real-world application with long-term use [49]. These confounding factors were addressed by a big dataset that was measured over a long-term experiment using the electrode shift.

In this study, an armband sensor was designed to measure the EMG signal pattern, which is related to the hand gesture of the user, even if the individual sensor is not placed on the specific muscles. Furthermore, the position of the main module was only suggested for non-expert users, and the experiment was repeated 10 times with the donning and doffing of the armband sensor to induce the electrode shift. Relatively low classification accuracies appeared in all posture groups with one training session (in MAV, Group 1: 84.9 ± 5.8%, Group 2: 80.9 ± 6.6%, Group 3: 75.5 ± 7.9%, Group 4: 79.2 ± 7.8%, Group 5: 79.5 ± 6.4% Group 6: 83.6 ± 8.6%, and Group 7: 65.5 ± 10.2%) owing to a lack of training on the electrode shift. The highest classification accuracies appeared after nine training sessions (in MAV, Group 1: 95.4 ± 1.4%, Group 2: 93.6 ± 2.2%, Group 3: 87.6 ± 3.6%, Group 4: 91.4 ± 3.6%, Group 5: 91.3 ± 3.7%, Group 6: 95.5 ± 2.0%, and Group 7: 80.9 ± 5.2%, respectively). Hence, in all feature vectors and posture groups, the classification accuracy improved with an increasing number of training sessions for the electrode shift, and a statistically significant improvement was observed when at least three training sessions were conducted. The classification accuracy improved when more than four training sessions were conducted, but the difference was not as significant. These results indicate that the problem of classification accuracy degradation could be resolved by increasing the number of training sessions on the electrode shift; the number of training sessions for efficiency was four.

Oskoei et al. reported that feeding a myoelectric signal presented as a time sequence directly to a classifier is impractical because of the large number of inputs and the randomness of the signal [50]. Therefore, many previous studies used feature vectors to compress the data and to normalize the pattern. Most of the previous studies selected the feature vectors according to the results of the classification performance evaluation [9,51,52]; however, accuracy-based feature vectors selection requires significant time and high costs because the pattern recognition algorithms are affected by various confounding factors. Data processing techniques, such as PCA and GA, were suggested to improve the classification performance of the feature vectors. Kakoty et al. reported that the classification accuracy improved by more than 8.0% through the compressed information from the PCA-based dimension reduction [12], which was useful for reducing the complexity of data or to reinforce the important information for pattern recognition, but the process of selecting the feature vector still remains unsolved. Oskoei et al. proposed a method for selecting feature vectors using entropy and GA [32,50], and Phinyomark et al. reported an RES-index-based feature vector selection, based on Euclidean distance and standard deviation [12]. These methods of selecting feature vectors are efficient because the classification performance of each feature vector can be evaluated before the development of the classifiers; however, few studies have been conducted on feature vector selection considering the electrode shift.

Twenty-one feature vectors were selected in the time domain, considering the real-time application in this study. These feature vectors were applied to the ANN-based classifier as a single type, and the classification performance of each feature vector was analyzed with the inter-session and inter-feature PCCs. The results indicate that the feature vectors with a strong linear relationship in inter-session PCC ($r > 0.7$) had a higher classification accuracy than that of the feature vectors with low inter-session PCC ($r < 0.7$), and these results were obtained for all training conditions and posture groups. Furthermore, in the threshold-based feature vectors of ZC, WAMP, MYOP, and SSC, the appropriate ranges of threshold values were found from the high inter-session PCC ($r > 0.8$). These results indicate that the inter-session PCC was well-correlated with the classification accuracy of each feature vector, and the feature vectors and the threshold values could be selected easily and efficiently by analyzing the inter-session PCC.

There were two differences in the feature vectors between this study and previous studies. The first difference was in the classification accuracies of AR and CC, and the second was in the effects of the feature combinations. AR and CC, which are calculated by the auto-regressive model, are well known as feature vectors with excellent classification performances [53]. However, AR and CC exhibited the lowest classification accuracies in this study, owing to the lack of data to calculate the feature vectors. In this study, the amount of data available for calculating the feature vectors was less than for previous studies because the armband sensor had a sampling rate of 500 Hz, whereas previous studies used sEMG sensors with a sampling rate of at least 1000 Hz. Phinyomark et al. evaluated the gesture recognition algorithm using various sEMG datasets and reported that the classification performances of AR and CC degraded owing to a lack of data points [54]. The problem resulting from the lack of data could be solved by increasing the sampling rate, but this solution will lead to an increase in the computational load and power consumption in the HCI interface and wearable device that is used over a long period in daily life. Increasing the window size used to calculate the feature vectors is another solution for the lack of data, but this solution is not thought to be appropriate because the increased window size introduces a delay in the feature vector calculation process. Therefore, AR and CC are inappropriate in the HCI interface and wearable device, as they set high values for efficiency and real-time classification.

Feature combinations were applied to improve the classification performance of pattern recognition algorithms in previous studies, with Hudgins' set [38] and Du's set [47] being the main combinations used. Hudgins' set comprises MAV, WL, ZC, and SSC whereas Du's set comprises IEMG, VAR, WL, ZC, SSC, and WAMP. Among the feature vectors in these combinations, those of ZC, WAMP, and SSC are well known to possess the frequency information, although these are included in the time domain [53] and are useful for gathering various information in real-time applications. Phinyomark et al. compared the classification performances of the single feature vectors and feature combinations (Hudgins' set and Du's set) and reported that the latter is superior [53]. However, they also reported that the difference in classification accuracy between individual feature vectors with high inter-session PCCs ($r > 0.7$) and the feature combinations is not significant. These results were obtained using the sampling rate of the sensor and noise from the crosstalk via the electrode shift. Many previous studies that reported improved classification performances with feature combinations used sEMG sensors with a high sampling rate (approximately 1000 Hz) and fixed positions on specific muscles. These experimental methods had the advantages of avoiding noise from crosstalk and gathering enough frequency information with the calculation of ZC, WAMP, and SSC; however, in this study, the frequency information was not sufficient in the ZC, WAMP, and SSC, owing to the low sampling frequency (500 Hz) and noise from the crosstalk. Furthermore, the inter-feature PCC between feature vectors with a high inter-session PCC ($r > 0.7$) was higher than 0.8 (strong linear relationship). These results indicate that these feature vectors (even the ZC, WAMP, and SSC) had the same information. Therefore, the classification performance of the feature combinations degraded. These problems, such as the degradation of the classification performance by the crosstalk, could be solved by fixing the position of the electrode, but finding the position for each specific muscle is difficult for non-expert users in daily life. Additional sensors, such as the IMU, will help provide diverse information on the feature combinations to improve the classification performance.

For the development of a gesture recognition algorithm, many studies have applied various groups of gestures and postures based on their research aim. However, the classification accuracy degraded with similar gestures in a gesture group because the function alone of each gesture was considered. Andrade et al. selected six gestures—cylindrical grasp, tip pinch, hook (snap), palmar pinch, spherical grasp, and lateral pinch—to develop an sEMG-based gesture recognition algorithm [21]. Their results indicated that the misclassification for similar gestures that are activated by the same muscle (precision grasp: tip pinch, palmar pinch, and lateral pinch; power group: cylindrical grasp, hook, and spherical grasp) was high for the gesture recognition algorithm.

In this study, twelve hand postures—which are frequently used and are important in hand posture recognition algorithms—were selected after a literature review. Seven different posture groups were formed according to the functions and activated muscles of hand postures to analyze the effects of similar postures. In Andrade's study, the rate of misclassification between similar postures was high, and this misclassification was not addressed in various feature vectors with an increase in the number of training sessions. The classification accuracies were in the following order: Group 1 (six hand postures) > Group 6 (seven hand postures considering the function and the muscle activation) > Group 2 (Group 1 + finger pointing) > Group 4 (Group 1 + similar posture: spherical grasp) ≥ Group 5 (Group 1 + similar posture: scissor-sign) > Group 3 (Group 1 + similar posture: tip pinch) > Group 7 (twelve hand postures). For Groups 2 to 6, the number of postures was 7, but the classification accuracies differed with the addition of postures (similar or not). More specifically, Group 3 exhibited the lowest classification accuracy because of the tip pinch posture, which was similar to both the palmar pinch and lateral pinch. Groups 2 and 6 were similar in that the postures were selected with consideration to the activated muscles. Group 2 had the spread posture and Group 6 had the V-sign posture. Group 6 had a higher classification accuracy than Group 2, but the difference was not significant. Furthermore, the spread posture, which opens the hand and supports loads, is more useful than the V-sign posture, which expresses emotions. Hence, Group 2 is more efficient in the HCI interface. These results indicate that the development of the gesture recognition algorithm could be made more efficient by the selection of gestures, giving consideration to both functions and the activated muscles.

This study has three limitations. The first is related to pattern information. All EMG feature vectors were selected in the time domain considering real-time application. In addition, because the armband sensor had a lower sampling frequency than that of the sEMG sensor in the previous study, the feature vectors information had to be similar. The second limitation is that the hand posture recognition algorithm was optimized for each subject, but normalization did not occur for optimization in all subjects. The gesture recognition algorithm without the normalization required more time and a higher cost because the classifier had to be trained for each user. The last limitation is that the gestures of this study were static postures only, and dynamic gestures were not considered.

5. Conclusions

This paper presented an sEMG-based hand posture recognition algorithm using an armband sensor, considering the following three problems: electrode shift, feature vector selection, and postures selection. This study showed that the electrode shift could degrade the classification performance of the pattern recognition algorithm, and this problem could be solved by increasing the number of training sessions on the electrode shift. Additionally, the inter-session PCC was verified as a means for selecting feature vectors because it exhibits a strong relationship with feature vectors and threshold values, with a high classification accuracy. Furthermore, information on each feature vector was compared with the inter-feature PCC, and the results of this analysis confirm that an additional sensor, such as an IMU, is required to provide diverse information for improving the hand posture recognition algorithm. In addition, selecting the target postures with consideration to the functions and activated muscles was as important as selecting feature vectors with a high classification accuracy for the development of an efficient posture recognition algorithm. These findings will be helpful in assisting the development process of sEMG-based gesture recognition algorithms more efficiently. In future works, the IMU sensor and the normalized algorithm will be applied in the pattern recognition system to provide diverse information and to reduce the training time and associated costs. Furthermore, the reaction speed and the practicality of the pattern recognition algorithm will be improved via the recognition of dynamic gestures to expand the application range.

Supplementary Materials: The following are available online at https://www.mdpi.com/article/10.3390/s21227681/s1, Figure S1: Classification accuracy and inter-session PCC of Group 2 by threshold values; Figure S2. Classification accuracy and inter-session PCC of Group 3 by threshold values; Figure S3. Classification accuracy and inter-session PCC of Group 4 by threshold values; Figure S4. Classification accuracy and inter-session PCC of Group 5 by threshold values; Figure S5. Classification accuracy and inter-session PCC of Group 6 by threshold values; Figure S6. Classification accuracy and inter-session PCC of Group 7 by threshold values; Figure S7. Classification accuracy and inter-session PCC of Group 2 by feature vector order; Figure S8. Classification accuracy and inter-session PCC of Group 3 by feature vector order; Figure S9. Classification accuracy and inter-session PCC of Group 4 by feature vector order; Figure S10. Classification accuracy and inter-session PCC of Group 5 by feature vector order; Figure S11. Classification accuracy and inter-session PCC of Group 6 by feature vector order; Figure S12. Classification accuracy and inter-session PCC of Group 7 by feature vector order; Figure S13. Classification accuracy of feature vectors and combinations by number of training sessions in Group 2; Figure S14. Classification accuracy of feature vectors and combinations by number of training sessions in Group 3; Figure S15. Classification accuracy of feature vectors and combinations by number of training sessions in Group 4; Figure S16. Classification accuracy of feature vectors and combinations by number of training sessions in Group 5; Figure S17. Classification accuracy of feature vectors and combinations by number of training sessions in Group 6; Figure S18. Classification accuracy of feature vectors and combinations by number of training sessions in Group 7; Figure S19. Confusion matrix of inter-feature PCC in Group 2; Figure S20. Confusion matrix of inter-feature PCC in Group 3; Figure S21. Confusion matrix of inter-feature PCC in Group 4; Figure S22. Confusion matrix of inter-feature PCC in Group 5; Figure S23. Confusion matrix of inter-feature PCC in Group 6; Figure S24. Confusion matrix of inter-feature PCC in Group 7; Figure S25. Confusion matrix of the hand postures of Group 1; Figure S26. Confusion matrix of the hand postures of Group 2; Figure S27. Confusion matrix of the hand postures of Group 3; Figure S28. Confusion matrix of the hand postures of Group 4; Figure S29. Confusion matrix of the hand postures of Group 5; Figure S30. Confusion matrix of the hand postures of Group 6; Figure S31. Confusion matrix of the hand postures of Group 7; Table S1. Classification accuracy and inter-session PCC of the feature vectors in Group 2; Table S2. Classification accuracy and inter-session PCC of the feature vectors in Group 3; Table S3. Classification accuracy and inter-session PCC of the feature vectors in Group 4; Table S4. Classification accuracy and inter-session PCC of the feature vectors in Group 5; Table S5. Classification accuracy and inter-session PCC of the feature vectors in Group 6; Table S6. Classification accuracy and inter-session PCC of the feature vectors in Group 7.

Author Contributions: Conceptualization, J.K. and Y.K.; Data curation, J.K., B.K. and Y.N.; Formal analysis, J.K.; Funding acquisition, Y.K.; Investigation, J.K.; Methodology, J.K., B.K. and Y.N.; Project administration, J.K. and Y.K.; Resources, Y.K.; Software, J.K. and B.K.; Supervision, Y.K.; Validation, J.K.; Visualization, J.K.; Writing—original draft, J.K. and Y.K.; Writing—review & editing, J.K. and Y.K. All authors have read and agreed to the published version of the manuscript.

Funding: This research was funded by The Bio & Medical Technology Development Program, grant number NRF-2017M3A9E2063270, through the National Research Foundation of Korea (NRF) funded by the Ministry of Science and ICT and the Institute of Information & Communications Technology Planning & Evaluation (IITP) grant funded by the Korea government (MSIT), grant number 2021-0-01980.

Institutional Review Board Statement: The study was conducted according to the guidelines of the Declaration of Helsinki, and approved by the Institutional Review Board (protocol code: 1041849-201704-BM-018-01; approval date: 14 April 2017) of Yonsei University Mirae Campus.

Informed Consent Statement: Informed consent was obtained from all subjects involved in the study.

Data Availability Statement: The data presented in this study are available on request from the corresponding author. The data are not publicly available owing to continuing study by the authors.

Conflicts of Interest: The authors declare no conflict of interest. The funders had no role in the design of the study; in the collection, analyses, or interpretation of data; in the writing of the manuscript, or in the decision to publish the results.

References

1. Mitra, S.; Acharya, T. Gesture recognition: A survey. *IEEE Trans. Syst. Man. Cybern. Part C Appl. Rev.* **2007**, *37*, 311–324. [CrossRef]
2. Chakraborty, B.K.; Sarma, D.; Bhuyan, M.K.; MacDorman, K.F. Review of constraints on vision-based gesture recognition for human–computer interaction. *IET Comput. Vis.* **2018**, *12*, 3–15. [CrossRef]
3. Wachs, J.P.; Kölsch, M.; Stern, H.; Edan, Y. Vision-based hand-gesture applications. *Commun. ACM* **2011**, *54*, 60–71. [CrossRef]
4. Shin, J.H.; Lee, J.S.; Kil, S.K.; Shen, D.F.; Ryu, J.G.; Lee, E.H.; Hong, S.H. Hand region extraction and gesture recognition using entropy analysis. *Int. J. Comput. Sci. Netw. Secur.* **2006**, *6*, 216–222.
5. Stergiopoulou, E.; Papamarkos, N. Hand gesture recognition using a neural network shape fitting technique. *Eng. Appl. Artif. Intell.* **2009**, *22*, 1141–1158. [CrossRef]
6. Nam, Y.; Wohn, K. Recognition of space-time hand-gestures using hidden Markov model. In Proceedings of the ACM Symposium on Virtual Reality Software and Technology, Hong Kong, China, 1–4 July 1996; pp. 51–58.
7. Yin, S.; Yang, J.; Qu, Y.; Liu, W.; Guo, Y.; Liu, H.; Wei, D. Research on gesture recognition technology of data glove based on joint algorithm. In Proceedings of the International Conference on Mechanical, Electronic, Control and Automation Engineering, Qingdao, China, 30–31 March 2018; pp. 1–10.
8. Kim, J.; Mastnik, S.; André, E. EMG-based hand gesture recognition for real-time biosignal interfacing. In Proceedings of the 13th International Conference on Intelligent User Interfaces, Gran Canaria, Spain, 13–16 January 2008; pp. 30–39.
9. Shi, W.T.; Lyu, Z.J.; Tang, S.T.; Chia, T.L.; Yang, C.Y. A bionic hand controlled by hand gesture recognition based on surface EMG signals: A preliminary study. *Biocybern. Biomed. Eng.* **2018**, *38*, 126–135. [CrossRef]
10. Jiang, S.; Lv, B.; Guo, W.; Zhang, C.; Wang, H.; Sheng, X.; Shull, P.B. Feasibility of wrist-worn, real-time hand, and surface gesture recognition via sEMG and IMU sensing. *IEEE Trans. Ind. Inf.* **2017**, *14*, 3376–3385. [CrossRef]
11. Abreu, J.G.; Teixeira, J.M.; Figueiredo, L.S.; Teichrieb, V. Evaluating sign language recognition using the myo armband. In Proceedings of the 2016 XVIII Symposium on Virtual and Augmented Reality, Gramado, Brazil, 21–24 June 2016; pp. 64–70.
12. Phinyomark, A.; Hirunviriya, S.; Limsakul, C.; Phukpattaranont, P. Evaluation of EMG feature extraction for hand movement recognition based on Euclidean distance and standard deviation. In Proceedings of the ECTI-CON2010: The 2010 ECTI International Conference on Electrical Engineering/Electronics, Computer, Telecommunications and Information Technology, Chiang Mai, Thailand, 19–21 May 2010; pp. 856–860.
13. Tkach, D.; Huang, H.; Kuiken, T.A. Study of stability of time-domain features for electromyographic pattern recognition. *J. NeuroEng. Rehabil.* **2010**, *7*, 1–13. [CrossRef] [PubMed]
14. Kakoty, N.M.; Hazarika, S.M.; Gan, J.Q. EMG feature set selection through linear relationship for grasp recognition. *J. Med. Biol. Eng.* **2016**, *36*, 883–890. [CrossRef]
15. Oskoei, M.A.; Hu, H. GA-based feature subset selection for myoelectric classification. In Proceedings of the 2006 IEEE International Conference on Robotics and Biomimetics, Kunming, China, 17–20 December 2006; pp. 1465–1470.
16. Wahid, M.F.; Tafreshi, R.; Al-Sowaidi, M.; Langari, R. Subject-independent hand gesture recognition using normalization and machine learning algorithms. *J. Comput. Sci.* **2018**, *27*, 69–76. [CrossRef]
17. Zhang, Z.; Yang, K.; Qian, J.; Zhang, L. Real-time surface EMG pattern recognition for hand gestures based on an artificial neural network. *Sensors* **2019**, *19*, 3170. [CrossRef] [PubMed]
18. Castiblanco, J.C.; Ortmann, S.; Mondragon, I.F.; Alvarado-Rojas, C.; Jöbges, M.; Colorado, J.D. Myoelectric pattern recognition of hand motions for stroke rehabilitation. *Biomed. Signal Process. Control* **2020**, *57*, 101737. [CrossRef]
19. Kim, S.; Kim, J.; Ahn, S.; Kim, Y. Finger language recognition based on ensemble artificial neural network learning using armband EMG sensors. *Technol. Health Care* **2018**, *26*, 249–258. [CrossRef]
20. Kim, S.; Kim, J.; Koo, B.; Kim, T.; Jung, H.; Park, S.; Kim, Y. Development of an armband EMG module and a pattern recognition algorithm for the 5-finger myoelectric hand prosthesis. *Int. J. Precis. Eng. Manuf.* **2019**, *20*, 1997–2006. [CrossRef]
21. De Andrade, F.H.C.; Pereira, F.G.; Resende, C.Z.; Cavalieri, D.C. Improving sEMG-based hand gesture recognition using maximal overlap discrete wavelet transform and an autoencoder neural network. In Proceedings of the XXVI Brazilian Congress on Biomedical Engineering, Armação dos Búzios, Brazil, 21–25 October 2019; pp. 271–279.
22. Suarez, J.; Murphy, R.R. Hand gesture recognition with depth images: A review. In Proceedings of the 2012 IEEE RO-MAN: The 21st IEEE International Symposium on Robot and Human Interactive Communication, Paris, France, 9–13 September 2012; pp. 411–417.
23. Murthy, G.R.S.; Jadon, R.S. Hand gesture recognition using neural networks. In Proceedings of the 2010 IEEE 2nd International Advance Computing Conference (IACC), Patiala, India, 19–20 February 2010; pp. 134–138.
24. Li, W.J.; Hsieh, C.Y.; Lin, L.F.; Chu, W.C. Hand gesture recognition for post-stroke rehabilitation using leap motion. In Proceedings of the 2017 International Conference on Applied System Innovation (ICASI), Sapporo, Japan, 13–17 May 2017; pp. 386–388.
25. Chonbodeechalermroong, A.; Chalidabhongse, T.H. Dynamic contour matching for hand gesture recognition from monocular image. In Proceedings of the 2015 12th International Joint Conference on Computer Science and Software Engineering (JCSSE), Hatyai, Thailand, 22–24 July 2015; pp. 47–51.
26. Ren, Z.; Meng, J.; Yuan, J.; Zhang, Z. Robust hand gesture recognition with Kinect sensor. In Proceedings of the 19th ACM International Conference on Multimedia, New York, NY, USA, 1–28 December 2011; pp. 759–760.

27. Sayin, F.S.; Ozen, S.; Baspinar, U. Hand gesture recognition by using sEMG signals for human machine interaction applications. In Proceedings of the 2018 Signal Processing: Algorithms, Architectures, Arrangements, and Applications (SPA), Poznan, Poland, 19–21 September 2018; pp. 27–30.
28. Yang, Y.; Fermuller, C.; Li, Y.; Aloimonos, Y. Grasp type revisited: A modern perspective on a classical feature for vision. In Proceedings of the IEEE Conference on Computer Vision and Pattern Recognition, Boston, MA, USA, 7–12 June 2015; pp. 400–408.
29. Plouffe, G.; Cretu, A.M. Static and dynamic hand gesture recognition in depth data using dynamic time warping. *IEEE Trans. Instrum. Meas.* **2015**, *65*, 305–316. [CrossRef]
30. Apostol, B.; Mihalache, C.R.; Manta, V. Using spin images for hand gesture recognition in 3D point clouds. In Proceedings of the 2014 18th International Conference on System Theory, Control and Computing (ICSTCC), Sinaia, Romania, 17–19 October 2014; pp. 544–549.
31. Frey Law, L.A.; Avin, K.G. Endurance time is joint-specific: A modelling and meta-analysis investigation. *Ergonomics* **2010**, *53*, 109–129. [CrossRef]
32. Oskoei, M.A.; Hu, H. Support vector machine-based classification scheme for myoelectric control applied to upper limb. *IEEE Trans. Biomed. Eng.* **2008**, *55*, 1956–1965. [CrossRef] [PubMed]
33. Englehart, K.; Hudgin, B.; Parker, P.A. A wavelet-based continuous classification scheme for multifunction myoelectric control. *IEEE Trans. Biomed. Eng.* **2001**, *48*, 302–311. [CrossRef]
34. Kamavuako, E.N.; Scheme, E.J.; Englehart, K.B. Determination of optimum threshold values for EMG time domain features; a multi-dataset investigation. *J. Neural Eng.* **2016**, *13*, 046011. [CrossRef] [PubMed]
35. Amrani, M.Z.E.A.; Daoudi, A.; Achour, N.; Tair, M. Artificial neural networks based myoelectric control system for automatic assistance in hand rehabilitation. In Proceedings of the 2017 26th IEEE International Symposium on Robot and Human Interactive Communication (RO-MAN), Lisbon, Portugal, 28 August–1 September 2017; pp. 968–973.
36. Phinyomark, A.; Quaine, F.; Laurillau, Y. The relationship between anthropometric variables and features of electromyography signal for human–computer interface. In *Applications, Challenges, and Advancements in Electromyography Signal Processing*; Ganesh, R.N., Ed.; IGI Global: Hershey, PA, USA, 2014; pp. 321–353.
37. Phinyomark, A.; Limsakul, C.; Phukpattaranont, P. A novel feature extraction for robust EMG pattern recognition. *J. Comput.* **2009**, *1*, 71–80.
38. Hudgins, B.; Parker, P.; Scott, R.N. A new strategy for multifunction myoelectric control. *IEEE Trans. Biomed. Eng.* **1993**, *40*, 82–94. [CrossRef] [PubMed]
39. Phinyomark, A.; Limsakul, C.; Phukpattaranont, P. EMG feature extraction for tolerance of white Gaussian noise. In Proceedings of the International Workshop and Symposium Science Technology, Nong Khai, Thailand, 2 December 2008; pp. 178–183.
40. Boostani, R.; Moradi, M.H. Evaluation of the forearm EMG signal features for the control of a prosthetic hand. *Physiol. Meas.* **2003**, *24*, 309–319. [CrossRef] [PubMed]
41. Phinyomark, A.; Limsakul, C.; Phukpattaranont, P. EMG feature extraction for tolerance of 50 Hz interference. In Proceedings of the PSU-UNS International Conference on Engineering Technologies, ICET, Novi Sad, Serbia, 28–30 April 2009; pp. 289–293.
42. Phinyomark, A.; Hirunviriya, S.; Nuidod, A.; Phukpattaranont, P.; Limsakul, C. Evaluation of EMG feature extraction for movement control of upper limb prostheses based on class separation index. In Proceedings of the 5th Kuala Lumpur International Conference on Biomedical Engineering 2011, Kuala Lumpur, Malaysia, 20–23 June 2011; pp. 750–754.
43. Fougner, A.L. Proportional Myoelectric Control of a Multifunction Upper-Limb Prosthesis. Master's Thesis, Norwegian University of Science and Technology, Trondheim, Norway, June 2007.
44. Cybenko, G. Approximation by superpositions of a sigmoidal function. *Mat. Control. Signals Syst.* **1989**, *2*, 303–314. [CrossRef]
45. Hornik, K. Approximation capabilities of multilayer feedforward networks. *Neural Netw.* **1991**, *4*, 251–257. [CrossRef]
46. Heaton, J. The Number of Hidden Layers. Available online: https://www.heatonresearch.com/2017/06/01/hidden-layers.html (accessed on 31 August 2021).
47. Du, Y.C.; Lin, C.H.; Shyu, L.Y.; Chen, T. Portable hand motion classifier for multi-channel surface electromyography recognition using grey relational analysis. *Expert Syst. Appl.* **2010**, *37*, 4283–4291. [CrossRef]
48. Lu, Z.; Tong, K.Y.; Zhang, X.; Li, S.; Zhou, P. Myoelectric pattern recognition for controlling a robotic hand: A feasibility study in stroke. *IEEE Trans. Biomed. Eng.* **2018**, *66*, 365–372. [CrossRef] [PubMed]
49. Phinyomark, A.; Campbell, E.; Scheme, E. Surface electromyography (EMG) signal processing, classification, and practical considerations. In *Biomedical Signal Processing*; Springer: Singapore, 2020; pp. 3–29. [CrossRef]
50. Oskoei, M.A.; Hu, H. Myoelectric control systems—A survey. *Biomed. Signal. Process. Control* **2007**, *2*, 275–294. [CrossRef]
51. Abbaspour, S.; Lindén, M.; Gholamhosseini, H.; Naber, A.; Ortiz-Catalan, M. Evaluation of surface EMG-based recognition algorithms for decoding hand movements. *Med. Biol. Eng. Comput.* **2020**, *58*, 83–100. [CrossRef]
52. Phinyomark, A.; Quaine, F.; Charbonnier, S.; Serviere, C.; Tarpin-Bernard, F.; Laurillau, Y. EMG feature evaluation for improving myoelectric pattern recognition robustness. *Expert Syst. Appl.* **2013**, *40*, 4832–4840. [CrossRef]

53. Phinyomark, A.; Phukpattaranont, P.; Limsakul, C. Feature reduction and selection for EMG signal classification. *Expert Syst. Appl.* **2012**, *39*, 7420–7431. [CrossRef]
54. Phinyomark, A.; Khushaba, R.N.; Scheme, E. Feature extraction and selection for myoelectric control based on wearable EMG sensors. *Sensors* **2018**, *18*, 1615. [CrossRef] [PubMed]

sensors

MDPI

Article

Affordable Embroidered EMG Electrodes for Myoelectric Control of Prostheses: A Pilot Study

Ernest N. Kamavuako [1,2,*], Mitchell Brown [1], Xinqi Bao [1], Ines Chihi [3,4], Samuel Pitou [1] and Matthew Howard [1]

1 Department of Engineering, King's College London, London WC2R 2LS, UK; mitchell.brown@kcl.ac.uk (M.B.); xinqi.bao@kcl.ac.uk (X.B.); samuel.pitou@kcl.ac.uk (S.P.); matthew.j.howard@kcl.ac.uk (M.H.)
2 Faculté de Médecine, Université de Kindu, Kindu, DR, Congo
3 National Engineering School of Bizerta, Carthage University, Tunis 2070, Tunisia; ines.chihi@uni.lu
4 Department of Engineering (DOE), The Faculty of Science, Technology and Medicine (FSTM), University of Luxembourg, 4365 Luxembourg, Luxembourg
* Correspondence: ernest.kamavuako@kcl.ac.uk; Tel.: +44-207-848-8666

Abstract: Commercial myoelectric prostheses are costly to purchase and maintain, making their provision challenging for developing countries. Recent research indicates that embroidered EMG electrodes may provide a more affordable alternative to the sensors used in current prostheses. This pilot study investigates the usability of such electrodes for myoelectric control by comparing online and offline performance against conventional gel electrodes. Offline performance is evaluated through the classification of nine different hand and wrist gestures. Online performance is assessed with a crossover two-degree-of-freedom real-time experiment using Fitts' Law. Two performance metrics (Throughput and Completion Rate) are used to quantify usability. The mean classification accuracy of the nine gestures is approximately 98% for subject-specific models trained on both gel and embroidered electrode offline data from individual subjects, and 97% and 96% for general models trained on gel and embroidered offline data, respectively, from all subjects. Throughput (0.3 bits/s) and completion rate (95–97%) are similar in the online test. Results indicate that embroidered electrodes can achieve similar performance to gel electrodes paving the way for low-cost myoelectric prostheses.

Keywords: myoelectric prostheses; embroidered EMG electrodes; pilot study; online and offline performance; conventional gel electrodes

Citation: Kamavuako, E.N.; Brown, M.; Bao, X.; Chihi, I.; Pitou, S.; Howard, M. Affordable Embroidered EMG Electrodes for Myoelectric Control of Prostheses: A Pilot Study. *Sensors* **2021**, *21*, 5245. https://doi.org/10.3390/s21155245

Academic Editor: Roberto Pilloton

Received: 17 June 2021
Accepted: 31 July 2021
Published: 3 August 2021

1. Introduction

Upper extremity loss is a highly disabling family of injuries that ranges from partial hand loss to loss of an entire arm. It can dramatically reduce a person's quality of life by impairing their ability to interact with their environment creating an economic and social burden. The total number of people with upper-limb loss is often difficult to quantify because many countries do not keep track of the incidence of amputation. However, it is estimated that over half a million people were living with some degree of upper limb loss in the United States in 2005. This figure is projected to double by 2050 [1]. Upper limb loss is estimated to be even more prevalent in the developing world, with most of the world's amputees and disabled living in low- and middle-income countries [2].

Prostheses, or artificial limbs, can replace lost functionality and improve the quality of life in people who have suffered upper limb loss. Historically, artificial upper limbs have been either cosmetic devices that restore the natural appearance of the lost limb or body-powered prosthetics that offer a limited restoration of functionality. Over the last century, however, various active artificial upper limbs have entered the literature and, increasingly, the commercial market. Most commercial functional prostheses, such as the Michelangelo hand (Otto Bock, Duderstadt, Germany) and the LUKE arm (Mobius Bionics, New Hampshire USA), are expensive devices for wealthy individuals in developed countries. However, a growing number of projects across the world aimed at harnessing the emerging technologies of the 4th industrial revolution, such as 3D printing, to produce affordable active prostheses. The Hero Arm (Open Bionics, Bristol, UK) and the Touch hand (Touch Prosthetics, Cape Town, South Africa) are examples of affordable active prostheses aimed at high-income and low- to middle-income nations, respectively.

An active prosthetic may be further classified as either an electric or myoelectric device. The former is controlled using external buttons and joysticks, while the latter is controlled using electromyographic (EMG) signals [3]. In myoelectric control, time and frequency domain features are extracted from the user via surface electrodes and are then fed into a control algorithm. The algorithm, in turn, converts them into motor commands fed into the prosthesis, thereby allowing the user to control the prosthetic and replace some of their lost functionality [4].

The development of a more affordable surface EMG electrode is one way that the cost of producing and maintaining myoelectric-controlled artificial limbs can be reduced and is an area that has received less attention than other prosthetic components. The two most common electrode types used in surface EMG are solid metal dry electrodes, which are widely used in prosthetics [5], and Ag-AgCl gel' wet' electrodes, which are the standard in clinical measurements [6].

Gel electrodes have lower impedance than other surface electrodes due to the large contact area provided by the electrolyte gel, resulting in minimal impedance and noise. The adhesive coating on the electrodes also produces a stable skin–electrode contact, which minimises motion artefacts [6]. Despite their high signal quality, gel electrodes are ill-suited for the heavy use required in powered prosthetics. The electrolyte gel dries out over time, causing the signal to decay and eventually fail, after which the electrodes or gel must be disposed of and replaced. Furthermore, the adhesive coating used in the electrodes can irritate the skin during use.

The solid metal dry electrodes used in current prostheses are typically made from stainless steel or titanium with a flat or domed shape. They are usually active with a preamplifier and other circuitry built directly into the electrode. These electrodes solve some of the problems posed by gel electrodes: they can be used for prolonged periods without signal failure, are reusable, and do not require adhesive coatings. However, they have shortcomings of their own. The metal plates can cause skin irritation through friction, and they are more susceptible to motion artefacts than gel-based designs. They are also more expensive to manufacture than gel and other electrode designs.

E-textiles may provide an alternative to the commercial electrode designs currently on the market. Textile electrodes can be constructed by embedding conductive fibres into a textile substrate using traditional fabrication techniques, such as embroidery and weaving [7]. Several textile electrodes for EMG have been presented in the literature [8–10], and their potential performance in myoelectric control has been demonstrated in wrist and hand gesture EMG classification experiments [11,12]. Because they are soft and breathable and do not use any abrasive chemicals, these textile electrodes do not irritate the user's skin. Their flexibility enables the electrode to conform to the shape of the arm and ensures that solid skin contact is maintained throughout the measurement, thereby minimising motion artefacts. Embroidered electrodes can be easily integrated into commercial prosthesis sockets [13] or fixed locations on smart garments or sleeves, making the process of placing the electrodes simple and appropriate for use by amputees. Like other dry electrodes, the textile designs are also reusable and can be worn for long periods without signal failure. The use of silver-coated thread has also been shown to have an antibacterial effect that helps to keep electrodes hygienic [14].

This paper evaluates the usability of affordable embroidered textile electrodes in real-time myoelectric control against a gel electrode standard. The significant advantage of this textile electrode design is its simple construction, which could potentially enable it to be handmade [15]. The ability to produce the electrodes by hand, coupled with the advantages inherent to textile electrodes, such as reusability, could make the design as much as 40% cheaper than gel alternatives [16] and ultimately facilitate the construction of more affordable myoelectric prostheses.

2. Materials and Methods

2.1. Embroidered Electrode Design

The electrodes were developed in the Centre for Robotics Research (CORE) at King's College London in a series of investigations [15–19] and consist of silver-coated thread (Electro-Fashion Conductive Thread, Kitronik, 40 Ωm^{-1}) embroidered into linen fabric with a Vilene cut away stabiliser. The embroidered pattern used is a 20 mm diameter circle with a cross-hatched fill pattern. The hatching has a 2 mm separation, and two iterations of embroidery are performed. An example of an electrode pair can be seen in Figure 1. The design parameters are selected to minimise resistance through the electrode in accordance with [18]. The electrodes are implemented in bipolar pairs with an inter-electrode distance (IED) of 25 mm. A standard snap fastener (Hemline, 13 mm, brass rust-proof fastener) is sewn onto each electrode and used to connect the electrode to an external amplifier. The electrodes are manufactured using a Pfaff Creative 3.0 (Pfaff, Kaiserslautern Germany) programmable sewing machine designed in the companion 6D Embroidery Software application.

Figure 1. Example of embroidered electrodes used in this study. The outer side shows snap connectors (**top**), and the inner side shows conductive thread (**bottom**).

2.2. Subjects

Three subjects volunteered for the experiments reported here, two female and one male. The participants were healthy, normally limbed, and ranged in age from 21 to 24. Due to the pandemic, we are not allowed to recruit more subjects, and thus this is a pilot study. We are planning to run a follow-up study soon with proper statistical analysis. All provided written informed consent before testing, and all tests had King's College Research Ethics Committee approval (Approval number: LRS-16/17-4213).

2.3. Experimental Setup

Two sets of surface electrodes are used for the experiments, one set of gel electrodes and one set of embroidered textile electrodes. When applying each set, subjects are seated, and four bipolar electrode pairs are placed on the forearm. Two pairs are placed over the extensor muscle group—the extensor carpi ulnaris and extensor digitorum—and two over the flexor group—the flexor carpi ulnaris and flexor carpi radialis. For experimental convenience, the textile electrodes are secured using kinesiology tape and covered using a tubular bandage. Alcohol wipes are used to clean the target areas before the application of the electrodes to ensure a clean connection following SENIAM recommendations [6]. For both sets, an additional reference gel electrode is attached at the elbow. The arrangement can be seen in Figure 2.

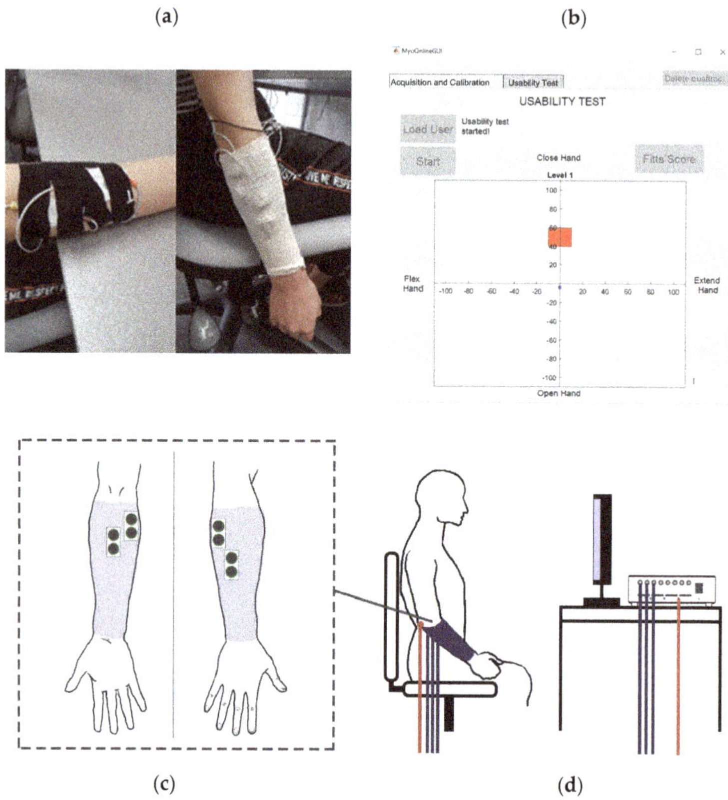

Figure 2. Experimental setup. (**a**) The electrodes are first placed with kinesiology tape and then secured with a bandage. (**b**) Usability test with the blue cursor and red target. (**c**) Positions of electrodes on the anterior and posterior sides of the right forearm. (**d**) Subject arrangement.

Each electrode is connected via snap connectors to a Quattracento amplifier (OT Bioelettronica, Torino, Italy), and the EMG signals are amplified with a gain of 1500, band-pass filtered (bandwidth 10–500 Hz), sampled at 2048 Hz, and A/D converted on 16 bits.

2.4. Experimental Protocol

The protocol consists of an offline test and an online usability test, both of which are conducted using a custom application implemented in Matlab 2019b (Mathworks, Natick, MA, USA).

For the offline test, the patients are taken through a series of nine gestures: closed hand, open hand, wrist extension, wrist flexion, chuck grip, index pointing, supination, pronation, and rest. These can be seen in Figure 3. Each gesture is performed four times, each time for six seconds following a three-second count down, which allows the subject to prepare. There is a six-second rest period between each gesture to prevent muscle fatigue.

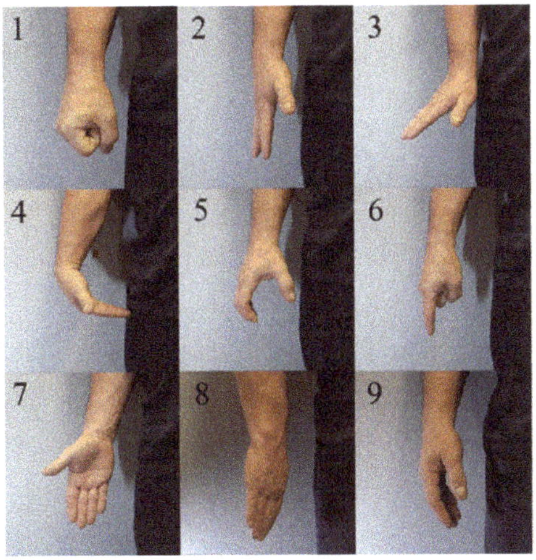

Figure 3. Gestures used in offline experiments. (1) closed hand, (2) open hand, (3) wrist extension, (4) wrist flexion, (5) chuck grip, (6) index point, (7) supination (8) pronation, and (9) rest.

For the usability test, only the first four gestures, and rest, are used. The subjects are again asked to perform each motion for six seconds following a three-second countdown with six seconds' rest between gestures, but now each gesture is performed three times. The recorded measurements are used to train an artificial neural network (ANN) for gestures classification (see Section 2.5). The subjects are then asked to complete a game in which they move a cursor to a series of square target areas on a two-dimensional grid (see Figure 2b). Both grid axes range from −110 to +110 units, and each combination of axis and direction is associated with one of four gestures, while rest causes the cursor to remain in its previous position on the grid. For each target, the cursor begins at the origin. The subject can move the cursor a fixed step size in one of the four orthogonal directions by performing the associated gesture. The subject was asked to hold the position within the target for one second (dwell time) for the trial to be considered successful. If the subject was unable to reach the target within a 20 s time limit, the online trial was considered unsuccessful, and the participant moved on to the next target with the cursor back at the origin.

There are 24 targets in a complete game, and six configurations for the width and distance from the origin of the targets, each of which is used four times during a game and has a particular index of difficulty. Each configuration is associated with four target numbers between 1 and 24, and, before the start of the game, MatLab's randperm() function is used to randomise an integer array from 1 to 24 using pseudo uniform random number generation, which then determines the order in which targets will appear. Table 1 shows the width and distance combinations for each configuration.

Table 1. Target configurations with associated difficulties and targets numbers.

Target Configuration	Distance from Origin	Target Width	Index of Difficulty	Associated Target Number
1	50	5	3.46	1, 7, 13, 19
2	50	10	2.59	2, 8, 14, 20
3	50	20	1.81	3, 9, 15, 21
4	100	5	4.39	4, 10, 16, 22
5	100	10	3.46	5, 11, 17, 23
6	100	20	2.59	6, 12, 18, 24

The step size for each gesture classification is 110/28. The numerator is the total length of the axis in each direction, so each step is a fixed fraction of the length. At the same time, the denominator was chosen through trial and error. Subjects complete the game three times.

The order of the electrode type (gel first or textile first) was controlled through counterbalancing. The first subject was randomly assigned to one order using a coin flip, and the order was reversed with each subsequent subject. In practice, because there are only three subjects, two subjects started in one condition, and one in the other.

2.5. Signal Processing

All offline data are processed using the same procedure. The first and last second of each EMG recording in a dataset is removed to account for subjects beginning and ending each gesture, leaving a four-second sample of each gesture. The sample for each gesture is pre-processed by subtracting the mean value in each channel was s. The sample is then segmented with a 200 ms window, which is passed over the sample such that each segment has a 50 ms overlap with the preceding segment. Six features are extracted from each segment: zero crossings, root mean square, mean absolute value, waveform length, slope sign change, and Willison amplitude. The feature matrix dimensionality reduction using principal component analysis (PCA) with a 0.95 variance threshold typically reduced the number of features to four.

The offline classification performance is evaluated using 10-fold cross-validation. For each subject, the raw EMG data for each electrode type is pooled, and pre-processing, feature extraction, and dimensionality reduction is performed. The prepared data are then divided into training (80%), validation (10%), and testing sets (10%), with the test set being the kth fold of the cross-validation. Individual subject models are trained using a feedforward ANN with a single 18-node hidden layer with tansig transfer function and nine-node softmax output layer. Levenberg–Marquardt error backpropagation is used for training with validation-based early stopping for regularisation. This is repeated for each fold, and the mean and standard deviation of the test fold accuracies is taken. The individual subject data is then pooled together into all gel and all fabric datasets, and the above cross-validation procedure is repeated to evaluate a general model for each electrode type.

For the usability test, a similar processing procedure is used as in the offline test to train an ANN for each subject. Online pattern recognition is achieved by recording 200 ms samples and feeding each sample to the trained algorithm for real-time classification. A Fitts' Law approach is used to assess real-time performance in each round of the online game, and the completion rate and throughput are calculated as performance metrics. The completion rate is defined as the percentage of targets successfully reached out of the total number of targets. Throughput (TP) is the amount of information transmitted by the user via the EMG electrodes. This can be expressed mathematically as the average ratio of the index of difficulty (I) of each target and the completion time (C), which is the time required to reach each target:

$$T = \frac{I}{C}. \tag{1}$$

The I for the usability test is defined according to the distance of the target centres from the origin (D) and the width of the target area (W) and is derived from Shannon's extension of Fitts' law in accordance with [20]:

$$I = log_2\left(\frac{D}{W} + 1\right) \tag{2}$$

The values of I for the targets used in the experiment can be found in Table 1.

3. Results

Offline models were trained and evaluated using the individual subject datasets for each electrode type and on a pooled dataset containing all electrode types. Table 2 shows the offline classification accuracy of models trained on individual subject data and their means and standard deviations. The mean accuracy across the three subject-specific models is 98.4% for the gel electrodes and 97.9% for the embroidered electrodes overall nine gestures. The classification accuracy for the general model trained on all gel electrode data is 97.3 ± 1.2%, and the accuracy for the model trained on all embroidered electrode data is 95.9 ± 1.2%. These can be seen in the respective confusion matrices (Figures 4 and 5). The percentages in the main (red and green) square are total classifications made across all classes, so they sum to 100. Therefore, the numbers on the diagonal are the number of correct classifications for each class as a percentage of total classifications made. There are nine classes, so they would each be ~11.1% if the model classified each class perfectly. The accuracy in practice is close to this for both models, reflecting good accuracy.

Table 2. Percentage classification accuracies of models trained on individual subject data for the recognition of nine gestures. Results are mean ± standard deviation over ten folds.

Subject	Electrode Type	
	Gel	Fabric
1	99.55 ± 0.29	99.23 ± 0.36
2	99.51 ± 0.58	95.63 ± 1.61
3	96.05 ± 1.06	98.74 ± 0.58
Mean	98.37 ± 2.01	97.87 ± 1.95

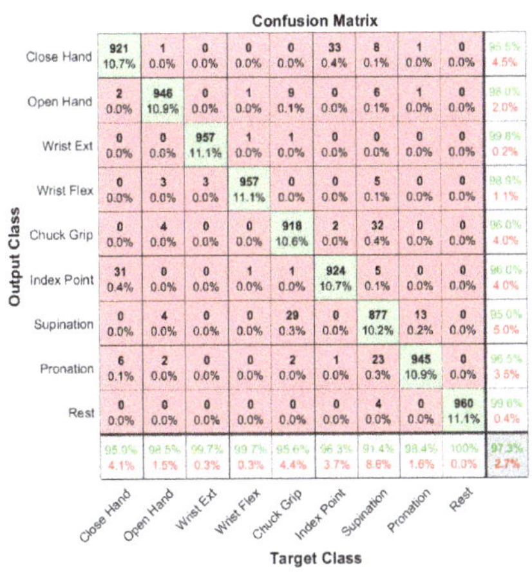

Figure 4. Confusion matrix for 10-fold cross-validation of a model trained on combined gel data from all subjects. The bottom row is the recall, the right column is the precision, and the bottom right is the overall accuracy.

The results from the usability test, found in Table 3, show similar performance between electrode types, with a throughput of 0.30 bit/s for both gel and embroidered electrodes. There is a slight variation in throughput across individual subjects for both electrode types. Completion rates are slightly higher for the textile electrodes across the three subjects, with an average of 97.2 ± 3.0% for the embroidered electrodes and 95.4 ± 5.7% for the gel electrodes. Examples of high-pass filtered EMG recordings and power spectra from both types of electrodes are shown in Figure 6. The spectra reveal similar frequency characteristics in both the textile and gel electrodes. Both have peak frequencies around

100 Hz and trail to zero above 500 Hz (note that no 50 Hz power line noise filtering is applied to the signals).

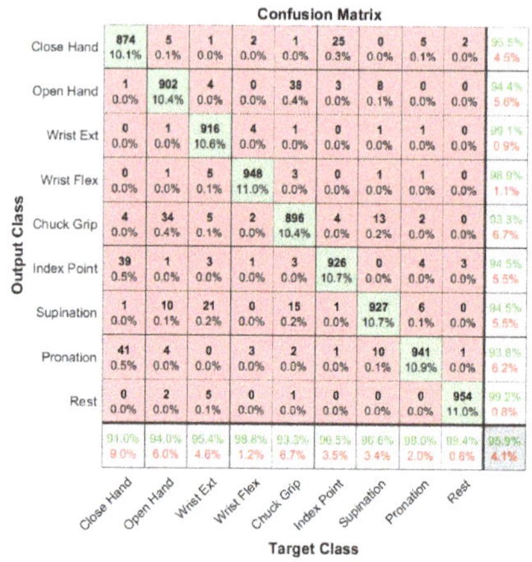

Figure 5. Confusion matrix for 10-fold cross-validation of a model trained on combined fabric data from three subjects. The bottom row is the recall, the right column is the precision, and the bottom right is the overall accuracy.

Table 3. The mean and standard deviation of throughput (TP) for each experimental subject and overall during the usability test.

Subject	Electrode Type	
	Gel	Fabric
1	0.31 ± 0.08	0.29 ± 0.08
2	0.29 ± 0.09	0.30 ± 0.09
3	0.29 ± 0.08	0.30 ± 0.09
All	0.30 ± 0.01	0.30 ± 0.01

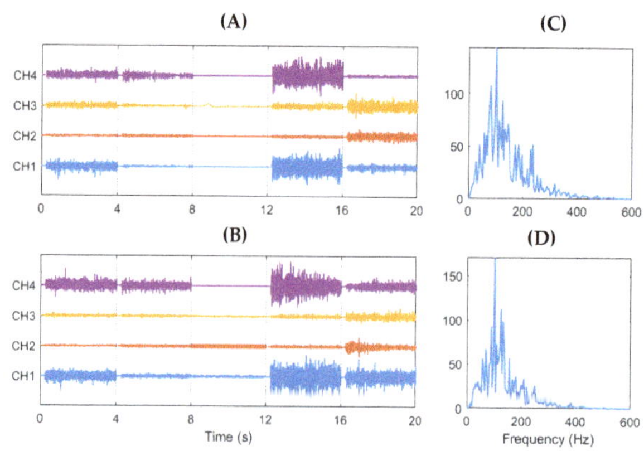

Figure 6. Example of recorded high-pass filtered signals from 4th gel and embroidered offline sessions of subject 1. EMG from gel electrodes (**A**) and embroidered electrodes (**B**). Power spectra of gel (**C**) and embroidered (**D**) signals. Samples of each gesture are four seconds in length, and gestures used, in order, are open hand, closed hand, relax, wrist extension, and wrist flexion. *Y*-axis are of arbitrary units.

4. Discussion

This study compares the performance of embroidered textile EMG electrodes with disposable gel EMG electrodes in gesture recognition and real-time usability tasks with a view towards use in affordable upper limb prostheses. The results demonstrate that embroidered electrodes can achieve comparable performance to gel electrodes both online and offline. The general classification models of nine hand and wrist gestures trained on data from all three subjects achieved an average accuracy of approximately 96% for the embroidered and 97% for the gel electrodes. In contrast, the individual subject-specific models achieved an average of 98% accuracy for both electrode types, although the accuracy for the embroidered electrodes is slightly lower in both cases. The lower accuracy for the models trained on all data was expected as they are more generalised than the individual models. The accuracy would be expected to improve with a larger sample of subjects, as the models would become less dependent on subject variation and noise. The individual models trained using embroidered electrodes also showed slightly higher variability than the gel electrodes (Table 2), which may be due to a lower signal-to-noise ratio (SNR) in the embroidered type. The usability of the embroidered and gel electrodes was also shown to be similar as measured by throughput and completion rate in the Fitts' law-based reaching task. The difference in completion rate between electrode types may be a consequence of the limited number of participants.

Although the results of this investigation are promising, future work is required to determine the suitability of embroidered electrodes to myoelectric control as there were several limitations to this pilot study. Examples of limitations include the small number of subjects, the use of subjects with intact limbs only, and lack of gender balance (two female subjects and one male), making it difficult to generalise the findings to real-world situations. Each subject was also only tested in a single session, so the study does not account for the variation of EMG signal over time due to natural biological fluctuations and the potential degradation of the fabric electrodes.

Another limiting factor was that the electrodes had to be secured in place using tape and tubing to ensure good skin contact and prevent motion artefacts, which made the applied pressure on the electrodes difficult to control and may have produced differences in signal quality between electrodes and subjects. This method of securing the electrodes also differs significantly from how the electrodes would be implemented in an actual prosthesis, making it difficult to generalise the results to practical myoelectric control. Pressure sensors combined with adjustable bands could have been used to ensure consistent pressure on the electrodes while maintaining good contact. Using such a setup, the sensitivity of the fabric electrodes to motion artefacts could be investigated, allowing an optimal level of pressure to be found that balances signal quality with user comfort.

The experiment is also limited by the electrode design because the 20 mm diameter of the electrodes restricted the number that could be applied to the forearm and made it difficult to secure them in place comfortably. Two-dimensional high-density EMG electrode arrays have shown promise for myoelectric control in recent years, including superior positional shift robustness and classification accuracy [21]. Such an arrangement has been proven viable with textile electrodes [11]. Implementing the embroidered electrodes in a high-density array form may therefore have improved performance.

Future work could also be conducted to analyse the potential for practical use in commercial myoelectric control. Although the classification algorithm used in the experiment is typical of those used in current sequential commercial prosthetics, testing with a simultaneous and proportional control algorithm would have better-assessed serviceability in future prosthetics. The assessment of applicability in myoelectric control could also have been improved by testing on amputee subjects rather than able-bodied subjects and by using an active prosthetic in an actual real-time reaching test rather than using a reaching simulation. It could also be improved by testing with the electrodes embedded into a fabric socket, garment, or sleeve, similar to how it would need to be implemented in a commercial device.

This would allow many of the limiting factors to be investigated simultaneously, including the ease of placement for users, the susceptibility to motion artefacts, and user comfort.

Another necessary area of investigation would be analysis of the durability of the fabric electrodes during long-term use, such as would be expected of electrodes in prosthetics, and should include measuring the evolution of skin-electrode impedance and SNR over time. The effects of environmental factors, such as temperature and moisture, on signal quality also require investigation.

Despite these limitations, the results are in line with previous studies showing that using embroidered electrodes can achieve similar performance with gel sensors in upper limb movement classification for control of myoelectric prostheses [11,16,21,22]. Besides EMG, the embroidered electrodes also show promise for use in electroencephalography (EEG) [23] and electrocardiography (ECG) [24,25]. For instance, a recent study using the same type of embroidered electrodes for ECG measurement [26] showed that embroidered electrodes can capture high quality ECG signals, albeit with less stability than gel electrodes due to issues with skin contact. Some studies, such as [27], have investigated the possibility of using textile EMG electrodes in elasticated fabric bands to ensure skin contact and improve wearability in prosthetics, which is another potential development for embroidered electrodes.

Author Contributions: Conceptualization: All; data collection: X.B., M.B., E.N.K., S.P.; Data analysis: M.B., E.N.K.; critical review of the manuscript: All. All authors have read and agreed to the published version of the manuscript.

Funding: This research received no external funding.

Institutional Review Board Statement: The study was conducted according to the guidelines of the Declaration of Helsinki, and approved by the Institutional Review Board (or Ethics Committee) of King's College London (LRS-16/17-4213 approved 27 March 2017).

Informed Consent Statement: Informed consent was obtained from all subjects involved in the study.

Data Availability Statement: Data sharing is available per request.

Conflicts of Interest: The authors declare no conflict of interest.

References

1. Ziegler-Graham, K.; MacKenzie, E.J.; Ephraim, P.L.; Travison, T.G.; Brookmeyer, R. Estimating the Prevalence of Limb Loss in the United States: 2005 to 2050. *Arch. Phys. Med. Rehabil.* **2008**, *89*, 422–429. [CrossRef] [PubMed]
2. Phillips, B.; Zingalis, G.; Ritter, S.; Mehta, K. A Review of Current Upper-Limb Prostheses for Resource Constrained Settings. In Proceedings of the 5th Annual IEEE Global Humanitarian Technology Conference (GHTC 2015), Seattle, WA, USA, 8–11 October 2015; pp. 52–58.
3. Cordella, F.; Ciancio, A.L.; Sacchetti, R.; Davalli, A.; Cutti, A.G.; Guglielmelli, E.; Zollo, L. Literature review on needs of upper limb prosthesis users. *Front. Neurosci.* **2016**, *10*, 209. [CrossRef] [PubMed]
4. Roche, A.D.; Rehbaum, H.; Farina, D.; Aszmann, O.C. Prosthetic Myoelectric Control Strategies: A Clinical Perspective. *Curr. Surg. Rep.* **2014**, *2*, 3. [CrossRef]
5. Chadwell, A.; Kenney, L.; Thies, S.; Galpin, A.; Head, J. The Reality of Myoelectric Prostheses: Understanding What Makes These Devices Difficult for Some Users to Control. *Front. Neurorobot.* **2016**, *10*, 7. [CrossRef] [PubMed]
6. Day, B.S. *Important Factors in Surface EMG Measurement*; Bortec. Biomed. Ltd.: Calgary, AB, Canada, 2002; pp. 1–17.
7. Gonçalves, C.; Ferreira da Silva, A.; Gomes, J.; Simoes, R. Wearable E-Textile Technologies: A Review on Sensors, Actuators and Control Elements. *Inventions* **2018**, *3*, 14. [CrossRef]
8. Lintu, N.; Holopainen, J.; Hänninen, O. *Usability of Textile-Integrated Electrodes for EMG Measurements*; University of Kuopio, Department of Physiology, Laboratory of Clothing Physiology: Kuopio, Finland, 2005; pp. 64–69.
9. Finni, T.; Hu, M.; Kettunen, P.; Vilavno, T.; Cheng, S. Measurement of EMG activity with textile electrodes embedded into clothing. *Physiol. Meas.* **2007**, *28*, 1405–1419. [CrossRef] [PubMed]
10. Oliveira, C.C.; Machado da Silva, J.; Trindade, I.G.; Martins, F. Characterization of the electrode-skin impedance of textile electrodes. In Proceedings of the 29th Conference on Design of Circuits and Integrated Systems (DCIS 2014), Madrid, Spain, 26–28 November 2014; pp. 1–6.
11. Farina, D.; Lorrain, T.; Negro, F.; Jiang, N. High-density EMG E-textile systems for the control of active prostheses. In Proceedings of the 32nd Annual International Conference of the IEEE Engineering in Medicine and Biology Society (EMBC 2010), Buenos Aires, Argentina, 31 August–4 September 2010; pp. 3591–3593.

12. Zhang, H.; Tian, L.; Zhang, L.; Li, G. Using textile electrode EMG for prosthetic movement identification in transradial amputees. In Proceedings of the 2013 IEEE International Conference on Body Sensor Networks (BSN 2013), Boston, MA, USA, 6–9 May 2013; pp. 1–5.
13. Lee, S.; Jamil, B.; Kim, S.; Choi, Y. Fabric Vest Socket with Embroidered Electrodes for Control of Myoelectric Prosthesis. *Sensors* **2020**, *20*, 1196. [CrossRef] [PubMed]
14. Spadaro, J.A.; Berger, T.J.; Barranco, S.D.; Chapin, S.E.; Becker, R.O. Antibacterial Effects of Silver Electrodes with Weak Direct Current. *Antimicrob. Agents Chemother.* **1974**, *6*, 637–642. [CrossRef] [PubMed]
15. Pitou, S.; Michael, B.; Thompson, K.; Howard, M. Hand-made embroidered electromyography: Towards a solution for low-income countries. *Sensors* **2020**, *20*, 3347. [CrossRef] [PubMed]
16. Pitou, S.; Michael, B.; Howard, M.; Wu, F.; Shafti, A.; Stopforth, R. Embroidered Electrodes for Control of Affordable Myoelectric Prostheses. In Proceedings of the 2018 IEEE International Conference on Robotics and Automation (ICRA 2018), Brisbane, QLD, Australia, 21–25 May 2018; pp. 1812–1817.
17. Manero, R.B.R.; Shafti, A.; Michael, B.; Grewal, J.; Fernández, J.; Althoefer, K.; Howard, M. Wearable embroidered muscle activity sensing device for the human upper leg. In Proceedings of the 38th Annual International Conference of the IEEE Engineering in Medicine and Biology Society (EMBC 2016), Orlando, FL, USA, 16–20 October; pp. 6062–6065.
18. Shafti, A.; Manero, R.B.R.; Borg, A.M.; Althoefer, K.; Howard, M. Designing embroidered electrodes for wearable surface electromyography. In Proceedings of the 2016 IEEE International Conference on Robotics and Automation (ICRA 2016), Qingdao, China, 3–7 December 2016; pp. 172–177.
19. Mangezi, A.; Rosendo, A.; Howard, M.; Stopforth, R. Embroidered archimedean spiral electrodes for contactless prosthetic control. In Proceedings of the 15th Annual International Conference on Rehabilitation Robotics (ICORR 2017), London, UK, 17–20 July 2017; pp. 1343–1348.
20. Williams, M.R.; Kirsch, R.F. Evaluation of head orientation and neck muscle EMG signals as command inputs to a human–computer interface for individuals with high tetraplegia. *IEEE Trans. Neural Syst. Rehabil. Eng.* **2008**, *16*, 485–496. [CrossRef] [PubMed]
21. Hakonen, M.; Piitulainen, H.; Visala, A. Current state of digital signal processing in myoelectric interfaces and related applications. *Biomed. Signal Process. Control* **2015**, *18*, 334–359. [CrossRef]
22. Li, G.; Geng, Y.; Tao, D.; Zhou, P. Performance of Electromyography Recorded Using Textile Electrodes in Classifying Arm Movements. In Proceedings of the 33rd Annual International Conference of the IEEE Engineering in Medicine and Biology Society (EMBS 2011), Boston, MA, USA, 30 August–3 September 2011; pp. 4243–4246.
23. Löfhede, J.; Seoane, F.; Thordstein, M. Textile Electrodes for EEG Recording—A Pilot Study. *Sensors* **2012**, *12*, 16907–16919. [CrossRef]
24. Pola, T.; Vanhala, J. Textile electrodes in ECG measurement. In Proceedings of the 3rd International Conference on Intelligent Sensors, Sensor Networks and Information (ISSNIP 2007), Melbourne, VIC, Australia, 3–6 December 2007; pp. 635–639.
25. Park, S.; Noh, Y.; Park, S.; Yoon, H. An improved algorithm for respiration signal extraction from electrocardiogram measured by conductive textile electrodes using instantaneous frequency estimation. *Med. Biol. Eng. Comput.* **2008**, *46*, 147–158. [CrossRef] [PubMed]
26. Bao, X.; Howard, M.; Niazi, I.K.; Kamavuako, E.N. Comparison between Embroidered and Gel Electrodes on ECG-Derived Respiration Rate. In Proceedings of the 42nd Annual International Conference of the IEEE Engineering in Medicine and Biology Society (EMBS 2020), Montreal, QC, Canada, 20–24 July 2020; pp. 2622–2625.
27. Lee, S.; Kim, M.O.; Kang, T.; Park, J.; Choi, Y. Knit band sensor for myoelectric control of surface EMG-based prosthetic hand. *IEEE Sens. J.* **2018**, *18*, 8578–8586. [CrossRef]

sensors

Article

Decoding of Ankle Joint Movements in Stroke Patients Using Surface Electromyography

Afaq Noor [1], Asim Waris [1,*], Syed Omer Gilani [1], Amer Sohail Kashif [1], Mads Jochumsen [2], Javaid Iqbal [1] and Imran Khan Niazi [2,3,4,*]

1 Department of Biomedical Engineering & Sciences, School of Mechanical & Manufacturing Engineering, National University of Sciences and Technology (NUST), Islamabad 44000, Pakistan; afaqahmed.pg@smme.edu.pk (A.N.); omer@smme.nust.edu.pk (S.O.G.); amer.kashif@smme.nust.edu.pk (A.S.K.); j.iqbal@ceme.nust.edu.pk (J.I.)
2 Department of Health Science and Technology, Aalborg University, 9220 Aalborg Øst, Denmark; mj@hst.aau.dk
3 Center for Chiropractic Research, New Zealand College of Chiropractic, Auckland 1060, New Zealand
4 Health and Rehabilitation Research Institute, AUT University, Auckland 1010, New Zealand
* Correspondence: asim.waris@smme.nust.edu.pk (A.W.); imran.niazi@nzchiro.co.nz (I.K.N.)

Abstract: Stroke is a cerebrovascular disease (CVD), which results in hemiplegia, paralysis, or death. Conventionally, a stroke patient requires prolonged sessions with physical therapists for the recovery of motor function. Various home-based rehabilitative devices are also available for upper limbs and require minimal or no assistance from a physiotherapist. However, there is no clinically proven device available for functional recovery of a lower limb. In this study, we explored the potential use of surface electromyography (sEMG) as a controlling mechanism for the development of a home-based lower limb rehabilitative device for stroke patients. In this experiment, three channels of sEMG were used to record data from 11 stroke patients while performing ankle joint movements. The movements were then decoded from the sEMG data and their correlation with the level of motor impairment was investigated. The impairment level was quantified using the Fugl-Meyer Assessment (FMA) scale. During the analysis, Hudgins time-domain features were extracted and classified using linear discriminant analysis (LDA) and artificial neural network (ANN). On average, 63.86% ± 4.3% and 67.1% ± 7.9% of the movements were accurately classified in an offline analysis by LDA and ANN, respectively. We found that in both classifiers, some motions outperformed others ($p < 0.001$ for LDA and $p = 0.014$ for ANN). The Spearman correlation (ρ) was calculated between the FMA scores and classification accuracies. The results indicate that there is a moderately positive correlation (ρ = 0.75 for LDA and ρ = 0.55 for ANN) between the two of them. The findings of this study suggest that a home-based EMG system can be developed to provide customized therapy for the improvement of functional lower limb motion in stroke patients.

Keywords: stroke rehabilitation; surface electromyography (sEMG); pattern recognition (PR); ankle joint movements; home-based physical therapy; lower limb functional recovery

Citation: Noor, A.; Waris, A.; Gilani, S.O.; Kashif, A.S.; Jochumsen, M.; Iqbal, J.; Niazi, I.K. Decoding of Ankle Joint Movements in Stroke Patients Using Surface Electromyography. *Sensors* **2021**, *21*, 1575. https://doi.org/10.3390/s21051575

Academic Editor: Toshiyo Tamura

Received: 27 January 2021
Accepted: 19 February 2021
Published: 24 February 2021

Publisher's Note: MDPI stays neutral with regard to jurisdictional claims in published maps and institutional affiliations.

1. Introduction

Stroke remains one of the leading causes of social isolation, disability, and death [1]. In children, the incidence of stroke is rare [2] and it has been estimated that in both men and women, the risk of stroke increases with age [3] while women have more stroke events than men [4].

As the average age of population is increasing across the world due to multiple reasons such as advances in medical technology, health care system, and provision of cheap and readily available medicines, it is expected that the number of stroke patients will rise [5,6]. Consequently, more patients will need physical rehabilitation in the future and governments will require induction of an increased number of healthcare professionals than usual to provide physical rehabilitation to these individuals. It is also more likely that the economic burden of stroke will increase and pose challenges to those health systems with limited resources [7].

A stroke survivor faces long-term effects after the acute phase of stroke. These effects include the development of impairment, limitations of activities (disability), and reduced participation (handicap) [8]. Although stroke results in a variety of physical and cognitive abnormalities, the most widely recognized is motor impairment, which affects 80% of stroke patients [9]. Commonly, stroke results in loss of movement control of one side of the body, impacting locomotion.

In neurological disorders including stroke, the restoration of physical function heavily depends on the onset, the type of injury, and the paradigm being followed for motor function recovery [10]. For the rehabilitation of upper limb motor function in stroke patients, constraint-induced movement therapy (CIMT), robotics, brain–computer interfaces (BCIs), electromyographic biofeedback, and mental practice (MP) combined with motor imagery have shown improvements in motor function [11–14]. Additionally, repetitive task training, high intensity physiotherapy/physical therapy (PT) and PT in combination with MP [15] have resulted in improved functional outcomes for lower limb mobility [16].

Assessing the outcomes of PT over time is very important in evaluating the functional performance of patients as well as the administered intervention. There are many scales used to access motor performance after stroke, but the most commonly used scale for assessing motor impairments in clinical practice is the Fugl-Meyer Assessment (FMA) scale [17].

The use of the upper limb is more frequent in performing activities of daily life (ADLs) and the upper limb has been targeted vastly in the areas of physical rehabilitation. Currently, there are various commercially available and widely used rehabilitative systems for upper limb rehabilitation after stroke, such as exoskeletons [18], rehabilitation robots [19], gaming devices [20], and virtual reality (VR) based systems [21]. Many of these rehabilitative devices are electromyography (EMG)-based. However, there is no device commercially available and clinically proven for stroke patients with lower limb motor impairments. Some of the main challenges hindering the commercial availability of the many proposed lower limb rehabilitative devices include design limitations and do not account for the physical requirements of stroke patients [22]. Additionally, more research focus on upper limbs is also an important factor that has resulted in researchers' comparatively less technical inclination toward the development of lower limb rehabilitative devices.

The first step toward the development of an EMG-controlled and home-based lower limb motor rehabilitation device is to investigate the movements of the lower limb in stroke patients. In the normal functionality of the lower limbs, movements that occur at the ankle joint complex have major significance in gait and balance. The available literature on the decoding of ankle joint motions from the movement intention of a user using surface electromyography (sEMG) in healthy subjects as well as in stroke patients is limited. However, Al-Quraishi et al. [23] successfully decoded ankle joint movements in healthy subjects while investigating the impact of different feature extraction and dimensionality reduction techniques on classification accuracies using autoregressive (AR) features and the following classifiers: K-nearest neighbor (k-NN), multilayer perceptron (MLP), and linear discriminant analysis (LDA). Their findings suggested that k-NN along with fuzzy neighborhood preserving analysis with QR (FNPA-QR) decomposition, as a dimensionality reduction technique, provides superior results with an average accuracy of 96.20% $\pm$ 4.1%. In another study exploring the biomechanical strategies used by healthy individuals during walking over uneven terrain, Gregory et al. [24] utilized time domain (TD) features (second-order AR coefficients, integrated EMG (IEMG), variance (VAR), waveform length (WL), moving average, and root mean square (RMS)) to predict user intent of performing ankle joint motions using LDA and the classification tree (CART) from sEMG signals. They reported the highest classification accuracy of 77.2% using LDA. Furthermore, Waris et al. [25] evaluated six different classifiers (LDA, ANN, K-NN, SVM, TREE, and naïve Bayes) in a multiday evaluation to identify the most suitable algorithm for sEMG classification of hand motions. In their study, ANN performed better of all classifiers.

The purpose of this study was to investigate the potential use of sEMG for a home-based ankle joint rehabilitative device using PR approaches and to evaluate the performance of two classifiers (ANN and LDA). In this study, the intent of performing different ankle joint movements was decoded from the recorded sEMG of stroke patients and the relationship between motor impairment and functional movements was explored. Previously, it has been reported that the classification accuracy for upper limbs is affected by impairment level in stroke patients [26,27]. Additionally, patients were observed by the data collection team throughout the experimental protocol and notes were taken. A total of four movements take place at the ankle joint complex: dorsiflexion and plantar flexion in the sagittal plane, and eversion and inversion in the frontal plane [27].

2. Materials and Methods

2.1. Participants

Fourteen stroke patients participated in this study from Railway General Hospital in Rawalpindi, Pakistan. All the participants were male. Only those patients who had a Mini-Mental State Examination (MMSE) score of greater than 24 were included in this study [28], which means they did not suffer from any severe cognitive disorder and were able to understand the given instructions. Based on the mentioned criteria, three patients were dropped (see the patient demographics in Table 1). An informed consent was provided by all patients prior to their participation in this study. This study was carried out according to the rules of the Declaration of Helsinki of 1975 and the experimental protocol for this study was approved by the local ethical committee (Riphah /RCRS /REC /00651). The Fugl-Meyer Assessment was carried out by a registered Physical Therapist to provide the scores for motor impairment. The motor section of the Fugl-Meyer Assessment scale consists of 100 points in total, which are used for the assessment of motor function in the upper extremity (66 points) and lower extremity (34 points) [29].

Table 1. Demographics of the patients who participated in this study.

Patient No.	Age	Sex	Months Since Injury	Affected Side	Injury	Fugl-Meyer Score of Lower Limb (E-F)
1	59	Male	15	Right	Ischemic	30
2	61	Male	7	Right	Ischemic	29
3	50	Male	35	Right	Ischemic	32
4	56	Male	8	Left	Ischemic	26
5	49	Male	20	Right	Ischemic	25
6	58	Male	12	Right	Ischemic	22
7	62	Male	34	Left	Ischemic	29
8	48	Male	8	Left	Hemorrhagic	31
9	57	Male	60	Right	Hemorrhagic	29
10	60	Male	20	Left	Ischemic	20
11	40	Male	40	Left	Hemorrhagic	34

2.2. Recordings: Surface EMG

Six surface EMG electrodes (Ambu Neuroline 720 surface electrodes, REF 72000-S/25, Baltropbakken 13, DK-2750, Ballerup, Denmark) were placed on the paretic leg on Tibialis anterior, gastrocnemius, and peroneus longus [23]. Two electrodes were placed in bipolar configuration 2 cm apart on the belly of each muscle to acquire signal from a given muscle. A moist wristband was used as a reference for signals. The sampling frequency was 2048 Hz, and the signals were amplified with a gain of 10,000 (EMG-USB2+, OT-Bioelettronica Metropolitan City of Turin, Italy).

2.3. Experimental Setup

Initially, the sEMG electrodes were placed on the affected leg and the quality of the signals were examined. The experimental protocol was explained in detail to each participant and a visual cue (photo of each motion class) was used to guide the subjects during the experiment (Figure 1). Furthermore, each subject was first asked to perform ankle joint movements with both limbs to familiarize them with the experimental protocol. Each subject performed one recording session while seated in a comfortable chair. A digital trigger was transmitted to the amplifier both at the beginning and at the end of the recording session so that the EMG recordings and the visual cue were synchronized. The following motion classes were included: plantar flexion, dorsiflexion, eversion, inversion, and rest. The time duration of performing a single instance of motion was six seconds (each patient was told to sustain the muscle contraction for six seconds) and every movement was repeated ten times. A rest time of six seconds was included between each repetition. The order of the motion classes was randomized and after completion of one motion class, the next motion class began.

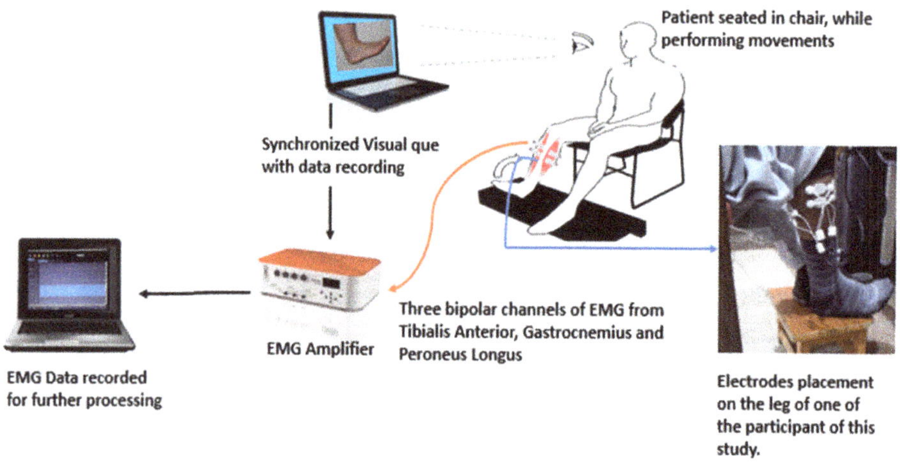

Figure 1. An illustration of the electromyography (EMG) data recording setup while the patient was performing movements.

3. Data Analysis

3.1. Pre-Processing and Feature Extraction

A bandpass filter with cut-off frequencies of 20 and 500 Hz and a Notch filter from 49–51 Hz were applied to the EMG signals using a second-order Butterworth filter with no phase shift to remove noise. Every six second repetition of each motion class was extracted and the first and last seconds were excluded from the analysis. As a result, 4 s epochs for every repetition of each motion class were obtained. After pre-processing, the following five (TD) features were extracted: ZC, WL, MAV, SSC [29] and Wilsons amplitude (WAMP) [30,31]. A 280 ms overlapping data window with a step size of 20% was used for feature extraction [32]. Figure 2 shows an example of a rectified and filtered EMG signal for every motion class and the associated channels (muscles) used for recording the EMG.

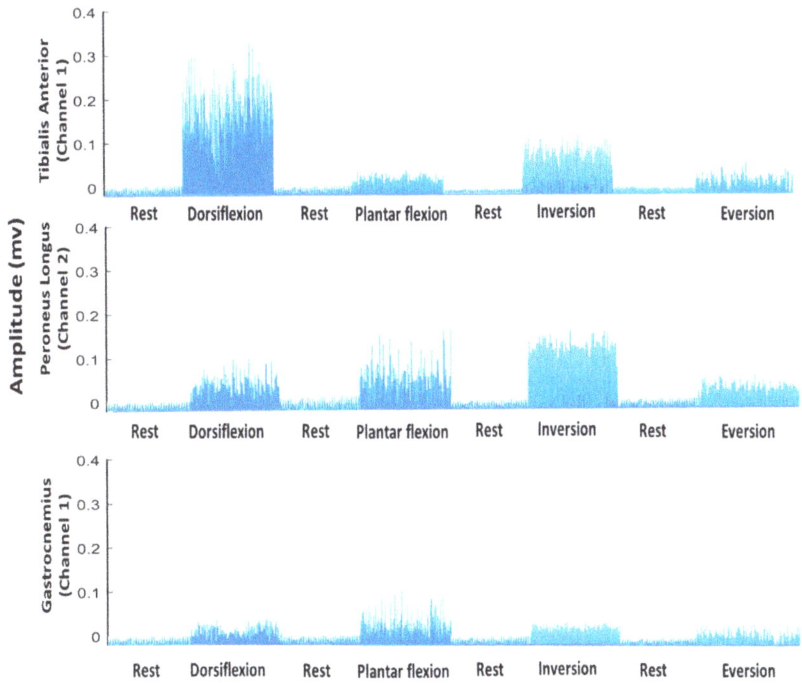

Figure 2. Rectified and bandpass-filtered EMG signal (only for understanding and visualization) of all motion classes and all channels separately for subject 3.

3.2. Classification

LDA and ANN (1 hidden layer and 15 neurons) were used to classify the different lower limb motions. The classification problem consisted of 5 classes. In this analysis, a four-fold cross-validation was used and the data were randomly divided into four folds for both the classifiers. Furthermore, one fold was used for testing while the remaining three folds were used to train the classifier. The overall average across the four folds is reported. Within the session, calibration was used to perform the classification, i.e., the training and testing were performed on the same session's data. All the extracted features were included in the analysis. MATLAB R2020a (MathWorks®, Natick, MA, USA) was used to perform data processing and classification. All the analyses were performed on a laptop with a 64-bit operating system, a core i5 processor, and 8 GB RAM.

3.3. Statistics

One-way repeated measures of analysis of variance (ANOVA) were used for separate inferential statistical analysis of the classification results of ANN and LDA to investigate the performance variance in the different motion classes. A probability value (p-Value of < 0.05) was used to infer the significance of all statistical analysis. Significant results were followed by a post-hoc test with Bonferroni correction for multiple comparisons. Lastly, the Spearman correlation coefficient (ρ) and coefficient of determination (R^2) were calculated between the lower limb Fugl-Meyer score and the average class accuracies of ANN and LDA for individual motions, as well as for all motions combined. All statistical analyses were performed using IBM® SPSS®, Chicago, IL, USA).

4. Results

The results of this study related to motions classification and the association between level of motor impairment and its effect on performing ankle joint movements are summarized in Table 2 and in Figures 2–8. As studies have reported differences in EMG signals on the basis of sex, age, and anthropometric variables; therefore, the results of this study are not representative of the whole population [33].

Table 2. Values of the Spearman correlation (ρ) and coefficient of determination (R^2) analysis between classification accuracies and movements (individual movement and average of all movements).

Motions	LDA			ANN		
	ρ	(*p*-Value)	R^2	ρ	(*p*-Value)	R^2
All Movements	0.75	<0.001	0.71	0.55	0.07	0.27
Dorsiflexion	0.26	0.42	0.05	0.21	0.53	0.00787
Eversion	0.64	0.03	0.34	0.28	0.39	0.05
Inversion	0.22	0.49	0.05	0.43	0.18	0.2
Plantar flexion	0.51	0.1	0.25	0.54	0.08	0.17
Rest	−0.15	0.64	0.03	0.12	0.7	0.02

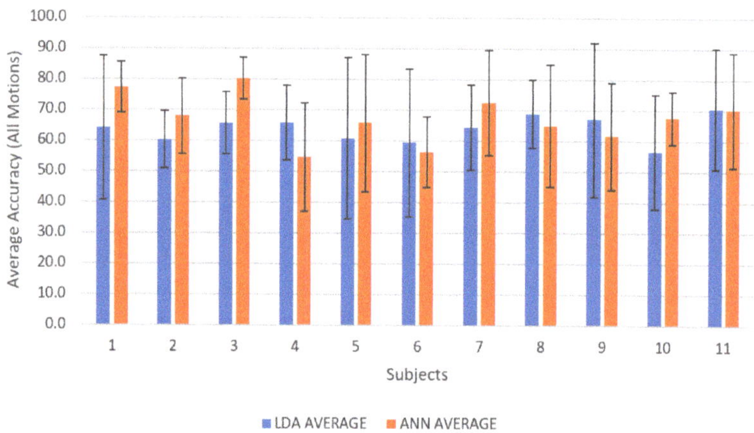

Figure 3. A column graph representing the average classification accuracies for all subjects using linear discriminant analysis (LDA) and artificial neural network (ANN) reported in the form of mean ± standard deviation.

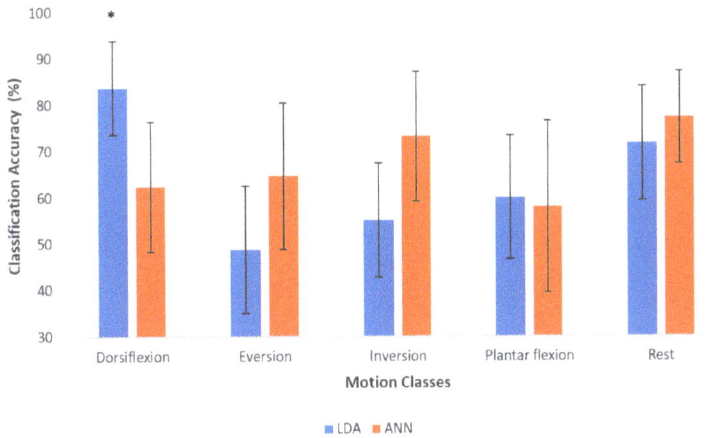

Figure 4. The average classification accuracy for all motion classes across all participants. The results are reported as mean ± standard deviation and asterisk denotes significant differences between motion class for each classifier. The asterisk denotes statistically significant difference.

	Dorsiflexion	Eversion	Inversion	Plantarflexion	Rest
Dorsiflexion	**83.6%**	1.9%	4.6%	8.3%	1.4%
Eversion	4.5%	**48.7%**	17.2%	9.3%	20.0%
Inversion	5.1%	15.2%	**55.1%**	14.7%	9.7%
Plantarflexion	10.0%	8.9%	16.1%	**60.0%**	4.8%
Rest	0.2%	15.8%	7.7%	4.3%	**71.6%**

Figure 5. Confusion matrix based on the average classification accuracies of all subjects for LDA with overall average accuracy of 63.86% ± 4.3%. The Highlighted boxes represents the correct percentage of predictions made by the classifier.

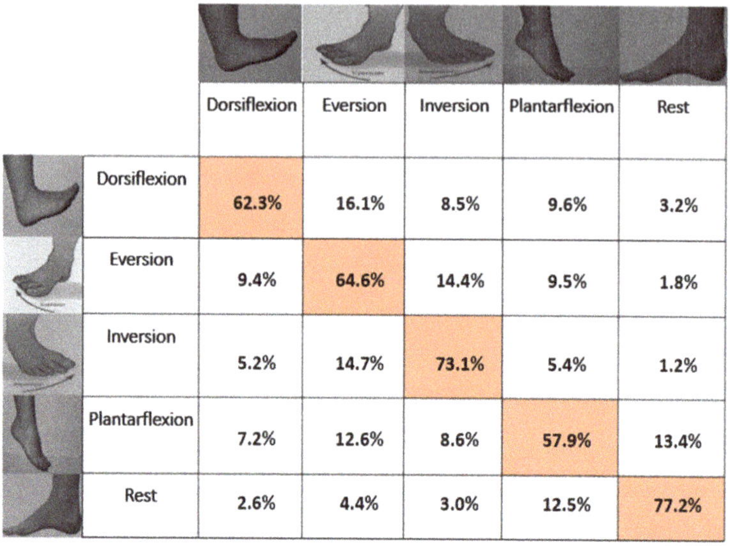

	Dorsiflexion	Eversion	Inversion	Plantarflexion	Rest
Dorsiflexion	**62.3%**	**16.1%**	**8.5%**	**9.6%**	**3.2%**
Eversion	**9.4%**	**64.6%**	**14.4%**	**9.5%**	**1.8%**
Inversion	**5.2%**	**14.7%**	**73.1%**	**5.4%**	**1.2%**
Plantarflexion	**7.2%**	**12.6%**	**8.6%**	**57.9%**	**13.4%**
Rest	**2.6%**	**4.4%**	**3.0%**	**12.5%**	**77.2%**

Figure 6. Confusion matrix based on the average classification accuracies of all subjects for ANN with overall average accuracy of 67.1% ± 7.9%. The Highlighted boxes represents the correct percentage of predictions made by the classifier.

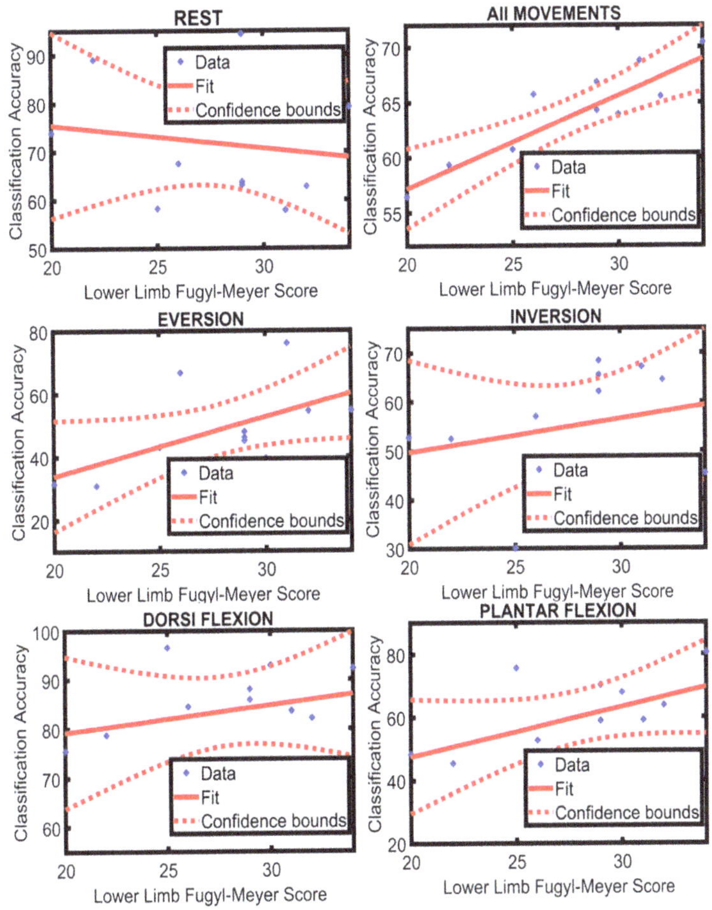

Figure 7. A regression line fitted to the participants' data of Limb Fugl-Meyer score and their classification accuracies (LDA) of individual movement and all movements combined.

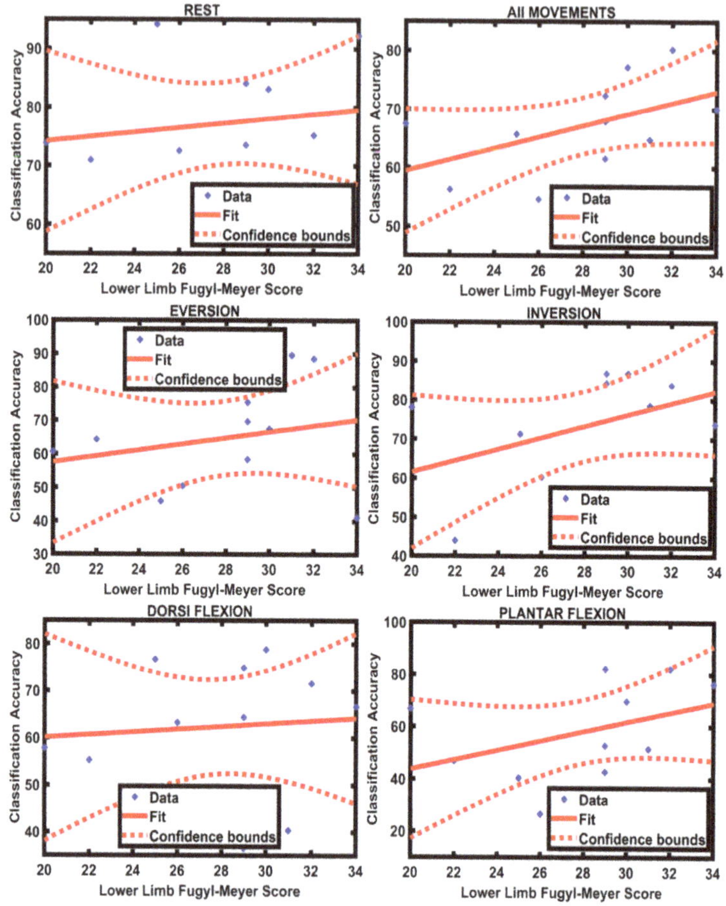

Figure 8. A regression line fitted to the participants' data of Limb Fugl-Meyer score and their classification accuracies (ANN) of individual movements and all movements combined.

4.1. Classification Results

On average across all participants, 63.85% ± 4.2% and 67.09% ± 7.8% of the movements were accurately classified in the offline analysis using LDA and ANN, respectively. One-way repeated measures of analysis of variance (ANOVA) revealed a significant difference between motion classes for both LDA ((F (4, 40) = 12.48; $p < 0.001$; partial $\eta^2 = 0.55$) and ANN ((F (4, 40) = 3.57; $p < 0.001$; partial $\eta^2 = 0.26$). For LDA, the post-hoc test with Bonferroni correction for multiple comparisons revealed that dorsiflexion was significantly different from eversion, inversion, and plantar flexion. Additionally, using LDA, dorsiflexion was the easiest to discriminate (83.6%). The accuracies of the other motion classes were in the range between 48% and 72%. The accuracies of dorsiflexion and the rest class were higher than all other motion classes, while the accuracies of plantar flexion and inversion were higher than that of eversion. The confusion matrix in Figure 5 shows that rest and eversion were mostly confused by LDA. In the classification using ANN, rest was classified most accurately with an accuracy of 77.1%, and plantar flexion had the lowest accuracy of 57.9%. For ANN, the post-hoc test with Bonferroni correction for multiple comparisons revealed that dorsiflexion was significantly different from rest.

On average, ANN outperformed LDA (see Figure 3 showing subject-wise results), and similarly in motion-wise results, the mean classification accuracy of ANN was better than LDA; however, LDA identified dorsiflexion and plantar flexion better than ANN. In addition, there was a significant difference between LDA and ANN when identifying dorsiflexion (F (9, 90) = 7.07; $p < 0.001$; partial $\eta^2 = 0.414$). Figure 4 reports the average classification accuracy for all the motion classes across all participants in the form of mean $\pm$ standard deviation. The confusion matrices (mean across the movements and across participants) are also shown in Figures 5 and 6 for LDA and ANN, respectively.

4.2. Relationship between Functional Score and Classification Accuracy

The Spearman correlation was calculated between the lower limb FMA score (section E-F of FMA scale) and the class accuracy across all movements as well as each movement individually for ANN and LDA separately. Furthermore, the coefficient of determination (R^2) was also calculated between the lower limb FMA score and the classification accuracy of all movements and each movement separately for both the classifiers. There was a positive, moderate association between the functional scores and the classification accuracies of the motion classes; the results for the correlation analysis and coefficient of determination are presented in Table 2. Additionally, a regression line was fitted to the values of the lower limb FMA score and the average classification accuracy of LDA and ANN for all movements and each movement individually (Figures 7 and 8). It can be observed from Table 2 that the value of R^2 for all movements is greater in LDA ($R^2 = 0.71$) compared to ANN ($R^2 = 0.27$). This suggests that LDA was better at identifying a possible trend between the classification accuracies and the lower limb FMA scores.

4.3. Patients' Feedback

For the participating stroke patients, data recording via sEMG was a new experience. At the end of every recording session, each patient was asked to share their experience with the computer-guided training, and 9 among the 11 participants responded positively. Table 3 provides the questions that were posed to each subject at the end of the recording session along with their answers. Their positive response was foreseeable because of the interactive nature of the experimental protocol.

Table 3. Set of questions that were posed to each subject at the end of recording session and each subject's response.

Q.	Questions (Total Participants 11)	Yes	No
1	Have you ever participated in a scientific study like this one?	0	11
2	Was it convenient for you to follow the series of images and perform the exercise?	11	0
3	Do you want to participate in another session?	10	1
4	Do you feel in control while doing the exercise on your own without the help of a Physical therapist?	10	1
5	Did you feel fatigued?	1	10
6	Are you in favor of a rehabilitative device that will provide physical therapy in your own environment?	10	1
7	Did you feel pain at any time during the experimental protocol?	0	11
8	Did you feel relaxed during the experiment?	10	1

5. Discussion

The aim of this study was to investigate the potential use of sEMG as a controlling mechanism for a home-based lower limb rehabilitative device for use in post-stroke PT. In this study, we were able to decode five different motion classes: dorsiflexion, eversion, inversion, plantar flexion, and rest (also included as a motion class) in stroke patients with an average accuracy of 63.86% $\pm$ 4.3% and 67.1% $\pm$ 7.9% using LDA and ANN, respectively. We found that in both classifiers, some motions outperformed others ($p < 0.001$ for LDA and $p = 0.014$ for ANN). The results revealed that ANN performed better than LDA, which is in agreement with a recent study aimed at characterizing distinct motions in healthy subjects and amputees using sEMG and intramuscular electromyography (IEMG) signals [34].

From earlier sources, it was known that ankle joint movements can be decoded from the recorded sEMG data of healthy subjects [23]. Furthermore, it was reported that the intent of performing lower limb movement (as decoded from sEMG signals) can be used to control a rehabilitative device [35]. We were able to successfully decode ankle joint movements from recorded sEMG data of stroke patients. The findings of the current study are in accordance with previous studies indicating that EMG activity of attempted movements can be decoded from stroke patients with motor impairments [26,36–40]. However, the number of sEMG channels used in the current setup was less than in other studies [40,41]. Hudgins (TD) features [30,31,42–44] were used in this study, but adding other features derived from autoregressive coefficients or wavelets and increasing the number of sEMG channels will probably improve the classification accuracy

In the current analysis, we found a positive and moderate association between the extent of motor impairment (lower limb FMA score) and average classification accuracies for both the classifiers, indicating that as the classification accuracy increases (in case of both LDA and ANN) moderately, there is small increase in the lower limb FMA score. This is in accordance with a previous study, which reported that the classification accuracies decrease as a function of the severity of stroke [39]. It is more likely that a perfect association was not found because of the small sample size (limited number of stroke patients) and the possibility that the lower limb FMA score is not a very sensitive measure for quantifying the level of impairment in these movements specifically, since the FMA score also includes reflexes, and muscle synergies, knee movements, etc.

During our investigation, we asked patients to perform various lower limb movements in order to familiarize themselves with the experimental protocol. We observed that patients felt relaxed when they were asked to perform the same movement using both limbs simultaneously. This can be attributed to the interaction between the paralyzed and the normal cerebral cortex, which results in additional stimulation [45]. It is also known that exercising both the normal and the affected side in stroke patients is more effective in upper limb function recovery compared to exercising the affected side alone [46]. New studies involving lower limbs can improve current knowledge on this factor. It was also observed that patients seemingly performed movements well when the recording session was closer to ending, which means that some level of dexterity can be achieved in performing these motions within only one session of training, but this needs to be further explored in a specially designed experiment.

For lower limb rehabilitation after stroke, manual PT is mostly used in clinical practice [47,48]. However, neuro-rehabilitation with the help of robotic devices presents a bright future [49–51]. These devices facilitate personalized treatment, offer reliable assessments, and ease PT from rigorous manual therapy [52–54]. Based on mechanical design and actuation, there are two categories of ankle rehabilitation device: end-effector-based robots and robotic exoskeletons/orthoses [55]. Furthermore, many lower limb rehabilitation devices have been developed in the past two decades, but most of these mechatronic systems use a combination of force sensors and feedback signals from position encoders. Besides, sEMG has been successfully used as a controlling mechanism for many upper limb rehabilitation devices but rarely investigated for lower limb rehabilitative devices [56].

Despite the successful use of sEMG for upper limb rehabilitative devices, its potential has not been studied extensively for lower limb PT devices.

The possibility of using sEMG as a controlling mechanism for a home-based rehabilitative device was demonstrated in this pilot study. In the current investigation, the trial arrangement was moderately cumbersome. However, the use of compact sEMG technologies such as the Myo-armband can make the use of sEMG much easier for patients. Furthermore, it was shown that comparable accuracies can be achieved using such setups [57]. New studies should be carried out involving a large sample size and smaller setups to validate these findings in the future.

Presently, because stroke patients are at great risk to coronavirus disease (COVID-19) [58], their rehabilitation regimen was severely affected in the months following the pandemic. The possibility of unavoidable circumstances like COVID-19 also advocates for shifting the focus of stroke-related research toward home-based PT devices so that in the future, stroke patients can continue their PT even if physical therapists are not available or the therapy is not feasible. Many systems have been introduced for the rehabilitation of the ankle joint, but the majority of these systems cannot be used in lower limb rehabilitation after stroke [22]. A dire need exists to address the challenges encountered in the development of EMG-based lower limb rehabilitation devices and the promotion of home-based PT for stroke patients.

6. Conclusions

It is possible to decode various motion classes of the ankle joint in stroke patients using a PR-based technique (LDA and ANN) requiring a smaller experimental setup and offering a high level of classification performance, which is a key factor for the efficacy of home-based rehabilitative devices. However, the availability of the number of stroke patients was limited and we propose that more studies are needed to validate these findings with a bigger sample size of stroke patients. Furthermore, from the current study, we concluded that research involving experts from different areas and an interdisciplinary approach is more fruitful and viable.

Author Contributions: Conceptualization, A.N., A.W.; M.J., A.S.K., J.I. and I.K.N.; methodology, A.N., A.W., S.O.G., M.J.; A.S.K., J.I. and I.K.N.; data collection, A.N., A.W., A.S.K. and S.O.G.; data curation, A.N. and A.W.; formal analyisis, A.N. and A.W.; validation, A.W., A.S.K., M.J., J.I. and I.K.N.; Resources, J.I. and I.K.N.; writing—original draft preparation, A.N. and A.W.; writing—review and editing, A.N., A.W., S.O.G., A.S.K., M.J., J.I. and I.K.N.; supervision, J.I. and I.K.N. All authors have read and agreed to the published version of the manuscript.

Funding: The current study received no funding.

Institutional Review Board Statement: This study was conducted according to the guidelines of the Declaration of Helsinki and approved by the Institutional Review Board (Riphah /RCRS /REC /00651).

Informed Consent Statement: Informed consent was obtained from all subjects involved in this study.

Data Availability Statement: The data for this study has been archived at the Department of Biomedical Engineering, School of Mechanical and Manufacturing Engineering (SMME), National University of Sciences and Technology (NUST), Islamabad, Pakistan. The data is private property of NUST. The data is available on request as per the University Policy.

Acknowledgments: We would like to acknowledge two master students from the department of biomedical engineering and sciences, SMME, NUST, Islamabad, Ahmad Saadullah Khan and Nadir Abbas for their help during the data collection, and Maham (Doctor of Physical Therapy) from Islamic International Medical College (IIMC), Rawalpindi, Pakistan.

Conflicts of Interest: The authors declare no conflict of interest.

References

1. Vos, T.; Allen, C.; Arora, M.; Barber, R.M.; Brown, A.; Carter, A.; Casey, D.C.; Charlson, F.J.; Chen, A.Z.; Coggeshall, M.; et al. Global, regional, and national incidence, prevalence, and years lived with disability for 310 diseases and injuries, 1990–2015: A systematic analysis for the Global Burden of Disease Study 2015. *Lancet* **2016**, *388*, 1545–1602. [CrossRef]
2. Tsze, D.S.; Valente, J.H. Pediatric Stroke: A Review. *Emerg. Med. Int.* **2011**, *2011*, 1–10. [CrossRef]
3. Lui, S.K.; Nguyen, M.H. Elderly Stroke Rehabilitation: Overcoming the Complications and Its Associated Challenges. *Curr. Gerontol. Geriatr. Res.* **2018**, *2018*. [CrossRef] [PubMed]
4. Reeves, M.J.; Bushnell, C.D.; Howard, G.; Gargano, J.W.; Duncan, P.W.; Lynch, G.; Khatiwoda, A.; Lisabeth, L. Sex differences in stroke: Epidemiology, clinical presentation, medical care, and outcomes. *Lancet Neurol.* **2008**, *7*, 915–926. [CrossRef]
5. Lackland, D.T.; Roccella, E.J.; Deutsch, A.F.; Fornage, M.; George, M.G.; Howard, G.; Kissela, B.M.; Kittner, S.J.; Lichtman, J.H.; Lisabeth, L.D.; et al. Factors influencing the decline in stroke mortality a statement from the american heart association/american stroke association. *Stroke* **2014**, *45*, 315–353. [CrossRef]
6. Brewer, L.; Horgan, F.; Hickey, A.; Williams, D. Stroke rehabilitation: Recent advances and future therapies. *QJM* **2013**, *106*, 11–25. [CrossRef] [PubMed]
7. Rajsic, S.; Gothe, H.; Borba, H.H.; Sroczynski, G.; Vujicic, J.; Toell, T.; Siebert, U. Economic burden of stroke: A systematic review on post-stroke care. *Eur. J. Health Econ.* **2019**, *20*, 107–134. [CrossRef] [PubMed]
8. Langhorne, P.; Bernhardt, J.; Kwakkel, G. Stroke rehabilitation. *Lancet* **2011**, *377*, 1693–1702. [CrossRef]
9. Lawrence, E.S.; Coshall, C.; Dundas, R.; Stewart, J.; Rudd, A.G.; Howard, R.; Wolfe, C.D.A. Estimates of the prevalence of acute stroke impairments and disability in a multiethnic population. *Stroke* **2001**, *32*, 1279–1284. [CrossRef]
10. Maciejasz, P.; Eschweiler, J.; Gerlach-Hahn, K.; Jansen-Troy, A.; Leonhardt, S. A survey on robotic devices for upper limb rehabilitation. *J. Neuroeng. Rehabil.* **2014**, *11*, 1–29. [CrossRef]
11. Cervera, M.A.; Soekadar, S.R.; Ushiba, J.; del Millán, J.R.; Liu, M.; Birbaumer, N.; Garipelli, G. Brain-computer interfaces for post-stroke motor rehabilitation: A meta-analysis. *Ann. Clin. Transl. Neurol.* **2018**, *5*, 651–663. [CrossRef] [PubMed]
12. Ramos-Murguialday, A.; Broetz, D.; Rea, M.; Läer, L.; Yilmaz, Ö.; Brasil, F.L.; Liberati, G.; Curado, M.R.; Garcia-Cossio, E.; Vyziotis, A.; et al. Brain-machine interface in chronic stroke rehabilitation: A controlled study. *Ann. Neurol.* **2013**, *74*, 100–108. [CrossRef] [PubMed]
13. Biasiucci, A.; Leeb, R.; Iturrate, I.; Perdikis, S.; Al-Khodairy, A.; Corbet, T.; Schnider, A.; Schmidlin, T.; Zhang, H.; Bassolino, M.; et al. Brain-actuated functional electrical stimulation elicits lasting arm motor recovery after stroke. *Nat. Commun.* **2018**, *9*, 1–13. [CrossRef] [PubMed]
14. Pichiorri, F.; Morone, G.; Petti, M.; Toppi, J.; Pisotta, I.; Molinari, M.; Paolucci, S.; Inghilleri, M.; Astolfi, L.; Cincotti, F.; et al. Brain-computer interface boosts motor imagery practice during stroke recovery. *Ann. Neurol.* **2015**, *77*, 851–865. [CrossRef]
15. García Carrasco, D.; Aboitiz Cantalapiedra, J. Effectiveness of motor imagery or mental practice in functional recovery after stroke: A systematic review. *Neurología (Engl. Ed.)* **2016**, *31*, 43–52. [CrossRef]
16. Langhorne, P.; Coupar, F.; Pollock, A. Motor recovery after stroke: A systematic review. *Lancet Neurol.* **2009**, *8*, 741–754. [CrossRef]
17. Lyden, P.D.; Hantson, L. Assessment scales for the evaluation of stroke patients. *J. Stroke Cerebrovasc. Dis.* **1998**, *7*, 113–127. [CrossRef]
18. Islam, M.R.; Spiewak, C.; Rahman, M.H.; Fareh, R. A Brief Review on Robotic Exoskeletons for Upper Extremity Rehabilitation to Find the Gap between Research Porotype and Commercial Type. *Adv. Robot. Autom.* **2017**, *06*. [CrossRef]
19. Aggogeri, F.; Mikolajczyk, T.; O'Kane, J. Robotics for rehabilitation of hand movement in stroke survivors. *Adv. Mech. Eng.* **2019**, *11*, 1–14. [CrossRef]
20. Joo, L.Y.; Yin, T.S.; Xu, D.; Thia, E.; Chia, P.F.; Kuah, C.W.K.; He, K.K. A feasibility study using interactive commercial off-the-shelf computer gaming in upper limb rehabilitation in patients after stroke. *J. Rehabil. Med.* **2010**, *42*, 437–441. [CrossRef]
21. Ikbali Afsar, S.; Mirzayev, I.; Umit Yemisci, O.; Cosar Saracgil, S.N. Virtual Reality in Upper Extremity Rehabilitation of Stroke Patients: A Randomized Controlled Trial. *J. Stroke Cerebrovasc. Dis.* **2018**, *27*, 3473–3478. [CrossRef]
22. Alvarez-Perez, M.G.; Garcia-Murillo, M.A.; Cervantes-Sánchez, J.J. Robot-assisted ankle rehabilitation: A review. *Disabil. Rehabil. Assist. Technol.* **2020**, *15*, 394–408. [CrossRef]
23. Al-Quraishi, M.S.; Ishak, A.J.; Ahmad, S.A.; Hasan, M.K. Impact of feature extraction techniques on classification accuracy for EMG based ankle joint movements. In Proceedings of the 2015 10th Asian Control Conference, Kota Kinabalu, Malaysia, 31 May–3 June 2015. [CrossRef]
24. Gregory, U.; Ren, L. Intent Prediction of Multi-axial Ankle Motion Using Limited EMG Signals. *Front. Bioeng. Biotechnol.* **2019**, *7*, 1–18. [CrossRef]
25. Waris, A.; Niazi, I.K.; Jamil, M.; Englehart, K.; Jensen, W.; Kamavuako, E.N. Multiday Evaluation of Techniques for EMG-Based Classification of Hand Motions. *IEEE J. Biomed. Health Inform.* **2019**, *23*, 1526–1534. [CrossRef] [PubMed]
26. Ramos-Murguialday, A.; García-Cossio, E.; Walter, A.; Cho, W.; Broetz, D.; Bogdan, M.; Cohen, L.G.; Birbaumer, N. Decoding upper limb residual muscle activity in severe chronic stroke. *Ann. Clin. Transl. Neurol.* **2015**, *2*, 1–11. [CrossRef] [PubMed]
27. Lee, S.W.; Wilson, K.M.; Lock, B.A.; Kamper, D.G. Subject-Specific Myoelectric Pattern Classification of Functional Hand Movements for Stroke Survivors. *IEEE Trans. Neural Syst. Rehabil. Eng.* **2011**, *19*, 558–566. [CrossRef] [PubMed]
28. Zwipp, H.; Randt, T. Ankle joint biomechanics. *Foot Ankle Surg.* **1994**, *1*, 21–27. [CrossRef]

29. Folstein, M.F.; Folstein, S.E.; McHugh, P.R. "Mini-mental state". A practical method for grading the cognitive state of patients for the clinician. *J. Psychiatr. Res.* **1975**, *12*, 189–198. [CrossRef]
30. Jochumsen, M.; Niazi, I.K.; Zia, M.; Amjad, I.; Shafique, M.; Gilani, S.O.; Waris, A. Decoding Attempted Hand Movements in Stroke Patients Using Surface Electromyography. *Sensors* **2020**, *20*, 6763. [CrossRef]
31. Abbaspour, S.; Lindén, M.; Gholamhosseini, H.; Naber, A.; Ortiz-Catalan, M. Evaluation of surface EMG-based recognition algorithms for decoding hand movements. *Med. Biol. Eng. Comput.* **2020**, *58*, 83–100. [CrossRef]
32. Spiewak, C. A Comprehensive Study on EMG Feature Extraction and Classifiers. *Open Access J. Biomed. Eng. Biosci.* **2018**, *1*, 17–26. [CrossRef]
33. Ashraf, H.; Waris, A.; Jamil, M.; Gilani, S.O.; Niazi, I.K.; Kamavuako, E.N.; Gilani, S.H.N. Determination of Optimum Segmentation Schemes for Pattern Recognition-Based Myoelectric Control: A Multi-Dataset Investigation. *IEEE Access* **2020**, *8*, 90862–90877. [CrossRef]
34. Phinyomark, A.; Campbell, E.; Scheme, E. Surface Electromyography (EMG) Signal Processing, Classification, and Practical Considerations. In *Biomedical Signal Processing: Advances in Theory, Algorithms and Applications*; Naik, G., Ed.; Springer: Singapore, 2020; pp. 3–29. ISBN 978-981-13-9097-5.
35. Saeed, B.; Zia-ur-Rehman, M.; Gilani, S.O.; Amin, F.; Waris, A.; Jamil, M.; Shafique, M. Leveraging ANN and LDA Classifiers for Characterizing Different Hand Movements Using EMG Signals. *Arab. J. Sci. Eng.* **2020**. [CrossRef]
36. Meng, W.; Liu, Q.; Zhou, Z.; Ai, Q. Active interaction control applied to a lower limb rehabilitation robot by using EMG recognition and impedance model. *Ind. Robot. An. Int. J.* **2014**, *41*, 465–479. [CrossRef]
37. Kopke, J.V.; Hargrove, L.J.; Ellis, M.D. Applying LDA-based pattern recognition to predict isometric shoulder and elbow torque generation in individuals with chronic stroke with moderate to severe motor impairment. *J. Neuroeng. Rehabil.* **2019**, *16*, 1–11. [CrossRef]
38. Balasubramanian, S.; Garcia-Cossio, E.; Birbaumer, N.; Burdet, E.; Ramos-Murguialday, A. Is EMG a Viable Alternative to BCI for Detecting Movement Intention in Severe Stroke? *IEEE Trans. Biomed. Eng.* **2018**, *65*, 2790–2797. [CrossRef]
39. Lu, Z.; Tong, K.; Zhang, X.; Li, S.; Zhou, P. Myoelectric Pattern Recognition for Controlling a Robotic Hand: A Feasibility Study in Stroke. *IEEE Trans. Biomed. Eng.* **2019**, *66*, 365–372. [CrossRef] [PubMed]
40. Zhang, X.; Zhou, P. High-Density Myoelectric Pattern Recognition Toward Improved Stroke Rehabilitation. *IEEE Trans. Biomed. Eng.* **2012**, *59*, 1649–1657. [CrossRef] [PubMed]
41. Zhang, X.; Zhou, P. Myoelectric pattern identification of stroke survivors using multivariate empirical mode decomposition. *J. Healthc. Eng.* **2014**, *5*, 261–274. [CrossRef]
42. Xi, X.; Tang, M.; Miran, S.M.; Luo, Z. Evaluation of feature extraction and recognition for activity monitoring and fall detection based on wearable sEMG sensors. *Sensors* **2017**, *17*, 1229. [CrossRef] [PubMed]
43. Turgunov, A.; Zohirov, K.; Ganiyev, A.; Sharopova, B. Defining the Features of EMG Signals on the Forearm of the Hand Using SVM, RF, k-NN Classification Algorithms. In Proceedings of the 2020 Information Communication Technologies Conference (ICTC), Nanjing, China, 29–31 May 2020; pp. 260–264. [CrossRef]
44. Nazmi, N.; Rahman, M.A.A.; Yamamoto, S.I.; Ahmad, S.A.; Zamzuri, H.; Mazlan, S.A. A review of classification techniques of EMG signals during isotonic and isometric contractions. *Sensors* **2016**, *16*, 1304. [CrossRef] [PubMed]
45. Lim, K.B.; Lee, H.J.; Yoo, J.; Yun, H.J.; Hwang, H.J. Efficacy of mirror therapy containing functional tasks in poststroke patients. *Ann. Rehabil. Med.* **2016**, *40*, 629–636. [CrossRef] [PubMed]
46. Cauraugh, J.H.; Summers, J.J. Neural plasticity and bilateral movements: A rehabilitation approach for chronic stroke. *Prog. Neurobiol.* **2005**, *75*, 309–320. [CrossRef]
47. Park, D.; Lee, J.-H.; Kang, T.-W.; Cynn, H.-S. Four-week training involving ankle mobilization with movement versus static muscle stretching in patients with chronic stroke: A randomized controlled trial. *Top. Stroke Rehabil.* **2019**, *26*, 81–86. [CrossRef]
48. Ardestani, M.M.; Kinnaird, C.R.; Henderson, C.E.; Hornby, T.G. Compensation or Recovery? Altered Kinetics and Neuromuscular Synergies Following High-Intensity Stepping Training Poststroke. *Neurorehabil. Neural Repair* **2019**, *33*, 47–58. [CrossRef]
49. Michmizos, K.P.; Rossi, S.; Castelli, E.; Cappa, P.; Krebs, H.I. Robot-Aided Neurorehabilitation: A Pediatric Robot for Ankle Rehabilitation. *IEEE Trans. Neural Syst. Rehabil. Eng.* **2015**, *23*, 1056–1067. [CrossRef]
50. Fazekas, G.; Tavaszi, I. The future role of robots in neuro-rehabilitation. *Expert Rev. Neurother.* **2019**, *19*, 471–473. [CrossRef] [PubMed]
51. Deng, W.; Papavasileiou, I.; Qiao, Z.; Zhang, W.; Lam, K.Y.; Han, S. Advances in Automation Technologies for Lower Extremity Neurorehabilitation: A Review and Future Challenges. *IEEE Rev. Biomed. Eng.* **2018**, *11*, 289–305. [CrossRef]
52. Young, A.J.; Ferris, D.P. State of the art and future directions for lower limb robotic exoskeletons. *IEEE Trans. Neural Syst. Rehabil. Eng.* **2017**, *25*, 171–182. [CrossRef] [PubMed]
53. Hussain, S.; Jamwal, P.K.; Ghayesh, M.H. State-of-the-art robotic devices for ankle rehabilitation: Mechanism and control review. *Proc. Inst. Mech. Eng. Part H J. Eng. Med.* **2017**, *231*, 1224–1234. [CrossRef]
54. Jamwal, P.K.; Hussain, S.; Xie, S.Q. Review on design and control aspects of ankle rehabilitation robots. *Disabil. Rehabil. Assist. Technol.* **2015**, *10*, 93–101. [CrossRef] [PubMed]
55. Hussain, S.; Jamwal, P.K.; Vliet, P.V.; Brown, N.A.T. Robot Assisted Ankle Neuro-Rehabilitation: State of the art and Future Challenges. *Expert Rev. Neurother.* **2020**, 1–11. [CrossRef]
56. Gull, M.A.; Bai, S.; Bak, T. A review on design of upper limb exoskeletons. *Robotics* **2020**, *9*, 1–35. [CrossRef]

57. Pizzolato, S.; Tagliapietra, L.; Cognolato, M.; Reggiani, M.; Müller, H.; Atzori, M. Comparison of six electromyography acquisition setups on hand movement classification tasks. *PLoS ONE* **2017**, *12*, 1–17. [CrossRef]
58. Markus, H.S.; Brainin, M. COVID-19 and stroke—A global World Stroke Organization perspective. *Int. J. Stroke* **2020**, *15*, 361–364. [CrossRef] [PubMed]

sensors MDPI

Article

Ablation Analysis to Select Wearable Sensors for Classifying Standing, Walking, and Running †

Sarah Gonzalez [1,*], Paul Stegall [1], Harvey Edwards [2], Leia Stirling [3] and Ho Chit Siu [2]

1 Department of Aeronautics and Astronautics, Massachusetts Institute of Technology, 77 Massachusetts Avenue, Cambridge, MA 02139, USA; stegall@mit.edu
2 Lincoln Laboratory, Massachusetts Institute of Technology, 244 Wood Street, Lexington, MA 02421-6426, USA; harvey.edwards@ll.mit.edu (H.E.); hochit.siu@ll.mit.edu (H.C.S.)
3 Department of Industrial and Operations Engineering, Robotics Institute, University of Michigan, 1205 Beal Avenue, Ann Arbor, MI 48109, USA; leias@umich.edu
* Correspondence: sarahmg@mit.edu
† Distribution Statement A. Approved for public release. Distribution is unlimited. This material is based upon work supported by the United States Air Force under Air Force Contract No. FA8702-15-D-0001. Any opinions, findings, conclusions or recommendations expressed in this material are those of the author(s) and do not necessarily reflect the views of the United States Air Force.

Abstract: The field of human activity recognition (HAR) often utilizes wearable sensors and machine learning techniques in order to identify the actions of the subject. This paper considers the activity recognition of walking and running while using a support vector machine (SVM) that was trained on principal components derived from wearable sensor data. An ablation analysis is performed in order to select the subset of sensors that yield the highest classification accuracy. The paper also compares principal components across trials to inform the similarity of the trials. Five subjects were instructed to perform standing, walking, running, and sprinting on a self-paced treadmill, and the data were recorded while using surface electromyography sensors (sEMGs), inertial measurement units (IMUs), and force plates. When all of the sensors were included, the SVM had over 90% classification accuracy using only the first three principal components of the data with the classes of stand, walk, and run/sprint (combined run and sprint class). It was found that sensors that were placed only on the lower leg produce higher accuracies than sensors placed on the upper leg. There was a small decrease in accuracy when the force plates are ablated, but the difference may not be operationally relevant. Using only accelerometers without sEMGs was shown to decrease the accuracy of the SVM.

Keywords: human activity recognition; surface electromyography; inertial measurement units; feature selection; wearable sensors

Citation: Gonzalez, S.; Stegall, P.; Edwards, H.; Stirling, L.; Siu, H.C. Ablation Analysis to Select Wearable Sensors for Classifying Standing, Walking, and Running. *Sensors* **2021**, *21*, 194. https://doi.org/10.3390/s21010194

Received: 11 November 2020
Accepted: 19 December 2020
Published: 30 December 2020

1. Introduction

Human Activity Recognition (HAR) aims to classify motions with the goal of characterizing the behaviors. HAR has previously been performed while using both wearable and external sensors [1]. External sensors, such as cameras, photometric sensors, and motion capture systems, are common tools in activity monitoring, but they have their drawbacks. Cameras and photometric sensors are commonly used to represent "ground truth" labeling [2], but they are limited by occlusions and shadows from other objects in the environment. Additionally, activities present in a variety of ways when viewed from different angles, so picking the proper view for observation becomes an issue for two-dimensional (2D) systems [3]. Marker-based motion capture systems, where only the three-dimensional (3D) marker coordinates are recorded, have been shown to be more reliable at activity detection than cameras alone, but the system requires a setup that can be quite costly and it can also be affected by occlusions and extraneous reflections [3].

Wearable sensors, such as surface electromyography sensors (sEMGs) and inertial measurement units (IMUs), have also been explored for HAR [4,5]. Walking and running are not only biomechanically different, but they also present differences in EMG activity over a range of fixed gait speeds [6]. The different EMG profiles make EMG useful for observing activity differences on various subjects with self-selected natural gait speeds, where one person's running speed might be another person's walking speed. Although, sEMGs are not without their drawbacks. If the primary muscle used for a specific activity has a weakened signal like in cases of patients going through rehabilitation, the sEMG would need to be placed on another, related muscle [7], potentially decreasing the accuracy of the activity classification in situations where the sEMG needs precise placement. If the muscle of interest is in a region with multiple muscles, then sEMG signals might need to be processed in order to separate out confounding signals [7]. Another common wearable sensor is the accelerometer. Although useful on their own, these widely available sensors increase HAR accuracy in combination with other sensors, such as sEMGs [5,8]. Every sensor has its benefits and drawbacks, but it has been shown that sensors can be used in tandem with HAR, and this fusion leads to increased accuracies in recognition of defined activities [9]. Examples of these fusions are using EMGs and IMUs (which contain accelerometers) for HAR for flexion and extension motions of the trunk [10], as well as IMUs and pressure sensors to recognize sitting, standing, walking, and running [11].

Data from sensors can be used in order to classify activities while using supervised machine learning techniques, such as Support Vector Machines (SVMs). The accuracy of an SVM is affected by the features selected and activities defined, as the methods are trained while using these data. Supervised models are trained while using a set of features and known activity labels. In machine learning, a feature is a characteristic of the data collected. Creating a definition of activities and their specific characteristics is one of the largest challenges of HAR [2]. Previous efforts across a range of activities have had accuracies that range from 70 to 95% [5,12,13]. These studies differ in the sensors and features used and the activities detected, leading to the different measured accuracies. It is unclear which features or sensor signals drive accuracy and further effort is needed in order to assess sensitivity and specificity for selected features.

The data and time needed for training will increase with the dimensionality of the system, which can occur by adding additional sensors and features [2]. Dimensionality reduction techniques, such as singular value decomposition (SVD) and principal component analysis (PCA), can be used in order to eliminate noise and reduce the need for large sets of training data [14,15]. When PCA is used for HAR in literature, it is typically used only for feature selection [1,8].

This paper seeks to understand which sensors drive accuracy, sensitivity, and specificity in human activity classification. The categories of "standing", "walking", "running", and "sprinting" were chosen because they are common activities that humans perform. Furthermore, data ablation is used in order to show that sensors on the lower leg produce higher classification accuracies than sensors on the upper leg. Through the ablation analysis, this paper provides the novel contribution of determining which sensors and sensor placements contribute to the classification accuracy and an interpretation of these findings. The subjects performed these motions using a self-paced treadmill, which enables users to select their own walk and run speeds, creating a natural variability within and across subjects. In order to use data that are transformed by PCA to train an SVM, it must be established that these classifications are appropriate for use between subjects and trials. PCA weightings are vectors in a high dimensional space; therefore, to determine that these weightings are similar and can be applied across subjects and trials, the angles between the weighting vectors were analyzed. The effect of the sensor set on classification accuracy will be presented. First, the paper will discuss the methods that were used to conduct the study. Subsequently, the results of the principal component similarity test and accuracies of the SVM training will be presented and discussed. This research informs how wearable sensors can be selected for classifying standing, walking, and running.

2. Materials and Methods

2.1. Experimental Setup

The experiment was performed at the MIT Lincoln Lab Sensorimotor Technology Realization in Immersive Virtual Environments (STRIVE) Center in the Computer Assisted Rehabilitation Environment (CAREN) (Motek Medical, Amsterdam, The Netherlands). The CAREN is a 24-foot diameter virtual-reality dome with a six-degree of freedom platform. The platform had a self-paced, split belt treadmill that was equipped with two force plates, one under each foot. Though these force plates were not wearable sensors, they could reasonably be replaced with force sensitive shoe insoles. For data collection, each subject was outfitted with wireless Trigno sEMG sensors with onboard accelerometers (Delsys Inc., Natick, MA, USA) that were placed precisely on certain muscles (Figure 1). Additionally, each subject was outfitted with motion capture markers, so the Vicon system (Vicon, Oxford, UK) in the CAREN could record the subject's motion. The marker set used was a modified Plug-in-Gait marker set, where the marker that was set on the upper body was reduced, as the focus of the study is on the lower body. These markers were used for the self-paced treadmill control and they were not used for activity classification.

The experiment had five subjects with a mean weight of 69.6 ± 19.9 kg and a mean age of 24.5 ± 4.3 years, all of whom were physically capable of completing all tasks. Subjects performed six trials on a treadmill, where they were given commands of "stand", "walk", "run", and "sprint", via on-screen text. New commands were issued every ten seconds, and each trial lasted 150 s. The order of the commands was switched for each trial in order to capture a roughly even distribution of transitions between different states (Figure 2). The MIT Committee on the Use of Humans as Experimental Subjects approved the procedure, and all of the subjects gave voluntary, informed, written consent to participate.

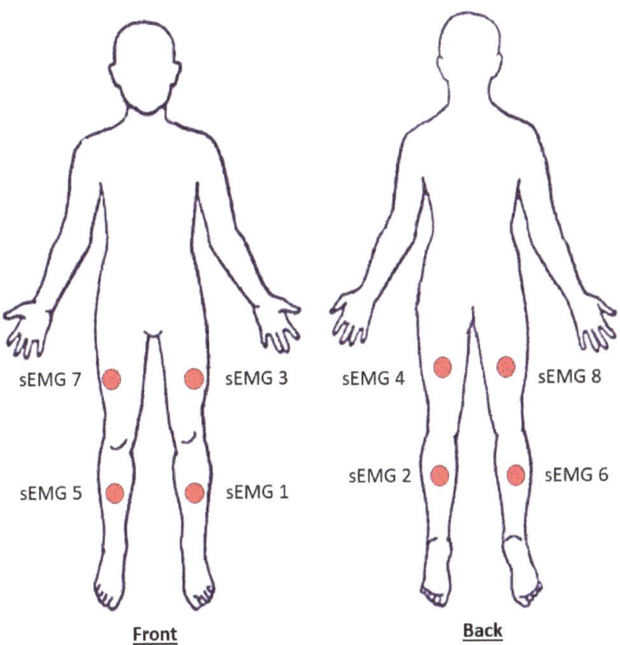

sEMG 1	Left Anterior Tibialis
sEMG 2	Left Medial Gastrocnemius
sEMG 3	Left Vastus Medialis
sEMG 4	Left Hamstring
sEMG 5	Right Anterior Tibialis
sEMG 6	Right Medial Gastrocnemius
sEMG 7	Right Vastus Medialis
sEMG 8	Right Hamstring

Figure 1. Trigno sensor locations on the subjects. These sensors include both a surface electromyography sensor (sEMG) and an accelerometer

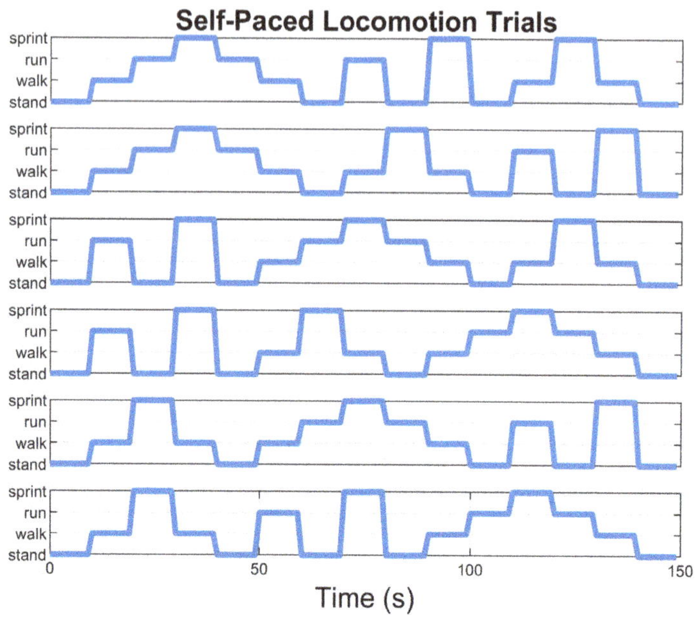

(a) Experiment Commands

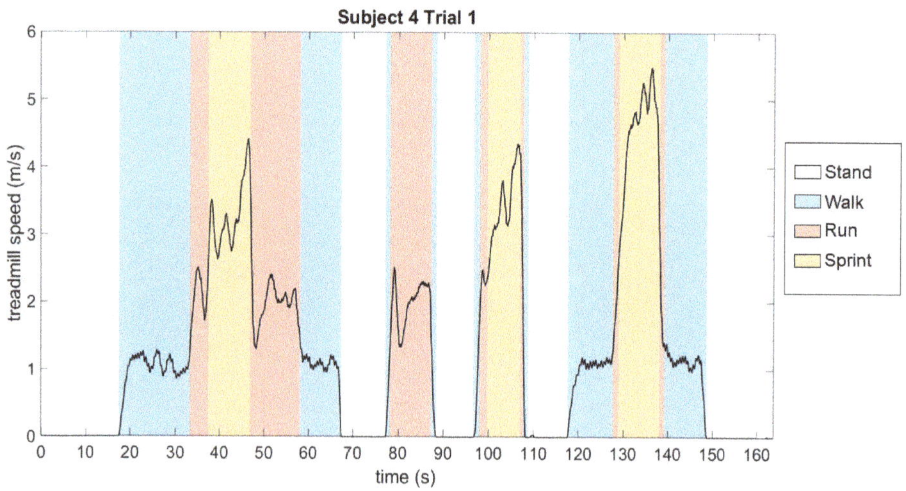

(b) Treadmill Speed

Figure 2. (**a**) The distribution of the commands each subject was given across time. There are six plots, one for each of the six trials. (**b**) Example of the self-paced treadmill speed for the first trial Subject 4 completed, where the commands are reflected in the speed at which the subject was moving. Note that while commands were given for the same duration, the activities are not all performed for the same duration, justifying the decision to label data points by speed instead of commanded activity.

2.2. Feature Selection

All of the features were extracted while using a sliding window of 0.5 s with 0.30 s overlap, similar to previous work [16,17]. The accelerometer data from all eight sensors were used as features by taking the vector norm of the accelerometer data in each window (ACC). The mean vertical force that was applied to the individual force plates (FP) as well as the standard deviation of the force applied to the force plates were also used as features. The median frequency (MF) of each sEMG sensor was used as a feature. It has been found that the frequency domain is useful in assessing muscle fatigue, which is important, because, as the trials progress, the subject tires, so it is important to ensure that our classification works for all levels of fatigue [18]. The sEMG signal samples in each window were binned by assessing the mean and standard deviation (SD) within that window in order to generate amplitude bin features (H1, H2, H3). Three bins were used for 0 to 1 SD from the mean (H1), 1 to 2 SD from the mean (H2), and >2 SD from the mean (H3). This method was chosen, because histogram features work well in high dimensional space, leading to higher robustness in recognizing activities [19].

In total, there were 44 features used: two force plate means (FPM), two force plate standard deviations (FPSD), eight sets of accelerometer data (ACC), eight sets of median sEMG frequencies (MF), and eight sets of three sEMG bins (H1, H2, H3). All of the features were z-score normalized in order to ensure that all data are on the same scale.

2.3. PCA Analysis

PCA is a dimensionality reduction technique that creates new axes for a set of data that maximizes the variance that occurs along the axes. From an $m \times n$ set of data, where m is the number of observations and n is the number of features, PCA creates n orthogonal principal components, where component 1 is a vector in the feature space where the data have the most variance and each successive component contains decreasing variance, so component n is a vector where the data show the least variance. These components are defined by a $n \times n$ set of weights, known as coefficients, for each feature. The coefficients indicate how much each feature is correlated to that particular principal component. Each feature has a corresponding score, where the scores are representations of the data in the space of the PCA coefficients, such that $T = XW$, where T is the original data set, the rows of X are the coefficients of each score, and the columns of W are the scores of each component [20].

The comparison of PCA weightings between subjects was performed by determining the angle between the orthogonal vector sets. The angle between every combination of trials without repeating combinations and the median of every angle was found. An angle between the PC coefficients can be calculated, because these weightings are the basis of an orthogonal set of vectors. The angle between components can be found while using the dot product of coefficients of the trials being compared. Smaller angles between principal components indicates a greater similarity in the features. Similarity in a matrix comparing two vectors would mean low values on the diagonal. Because PCA weightings indicate variance in either direction of the corresponding vector in feature space, we can take the supplementary angle when differences in vector angles are more than 90 degrees in order to account for the axis being identical, whether it is srepresented as positive or negative. In this study, the supplementary angle was taken for comparisons between like-components. In this context, "like-components" refer to the same principal components that originate from different trials and/or subjects (e.g., principal component 1 from trial 1 and principal component 1 from trial 2). Before the supplementary angle was taken, there was a bi-modal distribution that was centered around zero. To more accurately reflect how far from 0° these angles were, the like-components were reported in a range from 0° to 90°. All of the comparisons were done between the first three principal components of each dataset, as they describe approximately 85% of variance of each dataset. These comparisons were performed across all trials and subjects.

2.4. Support Vector Machines

A SVM was used in order to classify the PCA data into categories of motion. SVM finds separating hyperplanes, such that data points with different labels are maximally separated. These data can be "classified" by using the side of the hyperplane on which they are located [21]. A SVM was chosen from other supervised learning techniques, because visual inspection and initial observations of the data supported the use of a hyperplane to separate the data into categories. A Gaussian SVM, which uses a Gaussian kernel, was used in order to separate the data into the categories of stand, walk, run, and sprint. Because there was a natural delay between the subject reading the command and then performing it, we were unable to classify actions based on the time that the command was initially displayed. Instead, actions were classified manually (Table 1). For training, speed thresholds were manually set based on the speed that the user was maintaining at each point in time, as determined by an inspection of the treadmill speed across time. This labelling also included the transition regions. For example, if the subject was commanded to transition from standing to running, then the portion of data where the subject was moving with their determined walking speeds (Table 1) was labelled as walking (Figure 2). Subsequently, the class labels that were used for training were filtered with a sixth order Butterworth filter with a cutoff frequency of 50 Hz to smooth changes in speed due to high frequency noise.

Table 1. Thresholds of Speeds Used for the Labelling of Each Subject's Activities.

	Speed Threshold (m/s)			
	Stand	**Walk**	**Run**	**Sprint**
Subject 1	0	(0, 1.3]	(1.3, 2.4]	(2.4, ∞]
Subject 2	0	(0, 1.6]	(1.6, 3.3]	(3.3, ∞]
Subject 3	0	(0, 1.6]	(1.6, 2.8]	(2.8, ∞]
Subject 4	0	(0, 1.4]	(1.4, 2.5]	(2.5, ∞]
Subject 5	0	(0, 1.2]	(1.2, 2.7]	(2.7, ∞]

2.5. Accuracy

The accuracy of the SVM was found while using six-fold cross validation, where whole trials were reserved one at a time for testing. The accuracies for each subject were only calculated within each subject. During the cross-validation, PCA was performed on the data that were reserved for training. The data reserved for testing were transformed into the PCA space of the training data while using the coefficients found; no separate PCA was performed on the testing data. The accuracy was defined as the percentage of data points that were correctly classified by the SVM. Because the main difference between running and sprinting is speed rather than a biomechanical difference, such as presence or lack of double support as in walk vs. run, the accuracy was also computed when prediction confusion between running and sprinting was allowed by grouping running and sprinting labels together. Additionally, the sensitivity and specificity of each set of sensors are reported. Sensitivity is the proportion of correctly-classified data points with respect to the overall data points in that class, and specificity is the the proportion of correctly-classified negative data points versus the number of data points that are truly negative. Positive data is a correct activity classification, while the negative data is an incorrect classification. All of the machine learning and mathematical calculations were completed using MATLAB.

2.6. Ablation

In order to determine which sensors were driving the SVM classification of activities, features that were derived from certain sensors were removed from the total set in a process that is known as ablation [22] (Figure 3). The ablation process serves two purposes: it will help to determine which sensors and sensor placements lead to a higher classification accuracy, and it will lead to a greater understanding regarding which elements of the sensor data gathered drives the classification. The PCA and SVM procedure were performed on the case selected. Comparisons of mean accuracy between cases were evaluated while using Cohen's *d* effect size in order to consider the differences.

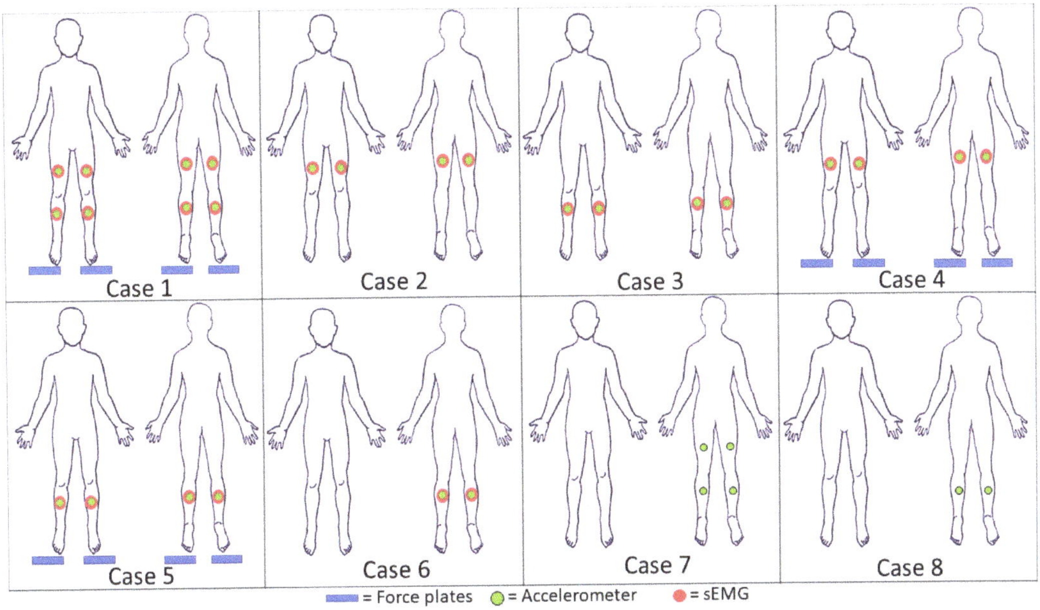

Figure 3. Locations of sensors during ablation. Case 1 includes all sensors. Case 2 includes sensors on the upper leg. Case 3 includes sensors on the lower leg. Case 4 includes sensors on the upper leg and the force plate. Case 5 includes sensors on the lower leg and the force plate. Case 6 includes sensors on the gastrocnemius. Case 7 includes one accelerometer on each segment of the leg. Case 8 includes one accelerometer per shank.

3. Results and Discussion

3.1. PCA Comparisons

Before discussing the comparisons between principal components, we must first understand what these principal components represent. Principal Component 1 was observed to be correlated with treadmill speed, with a median correlation across subjects and trials of $\rho = 0.96$ with an inter-quartile range of $\rho = 0.02$ (Figure 4). The correlations could be positive or negative, depending on the directionality of the principal components of each trial, so the absolute value of each trial's correlation was taken. Additionally, the type of locomotion was also observed to show distinct clusters across Principal Component 1 (Figure 4). Principal Component 2 of Subjects 2–5 appears to be correlated with the difference between the left and right force plate, which can be used to infer mediolateral positioning. The median correlation value was $\rho = 0.83$ with an inter-quartile range of $\rho = 0.16$ (Figure 5). The difference in force plates readings was observed to show a distinct gradient across the Principal Component 2 axis (Figure 5). Subject 1 did not appear to have a correlation with the force plate data and PC 2, which was likely due to cross-plate strikes, as discussed in more detail in Section 3.4. There was no clear relationship between Principal Component 3 and the collected data observed.

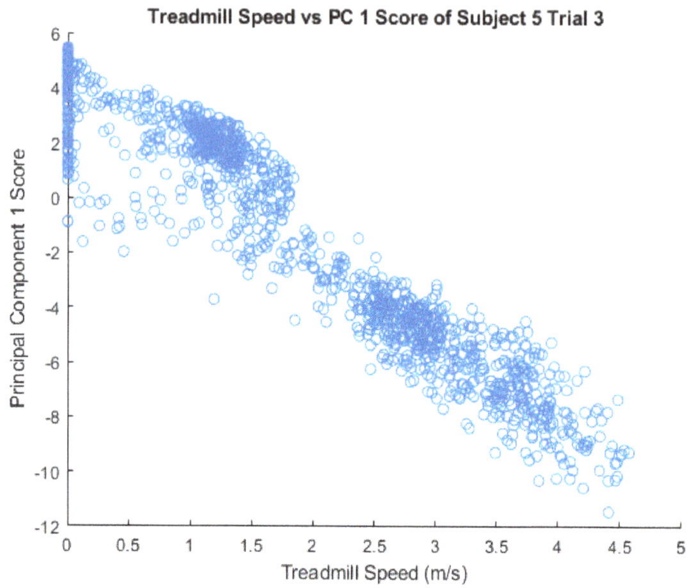

(**a**) Principal Component 1 and Treadmill Speed

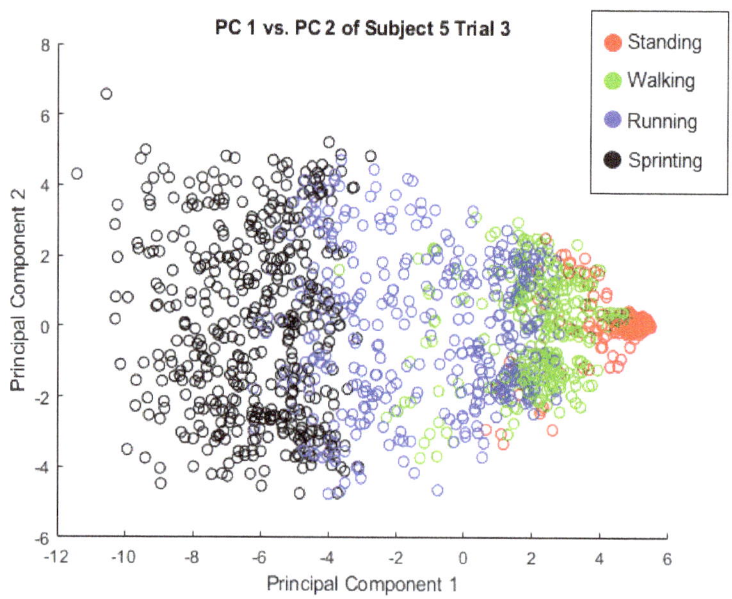

(**b**) Striation of Locomotion Modes on Principal Component Plot

Figure 4. (**a**) A plot of the treadmill speed versus the scores of Principal Component 1. A high level of correlation between these two parameters can be seen with the strong negative slope. This correlation can be seen over all trials, but Subject 5 Trial 3 was selected as a representative case. (**b**) A plot of principal component scores where each data point is colored according to the mode of locomotion the subject was performing. A clear pattern emerges on the Principal Component 1 axis.

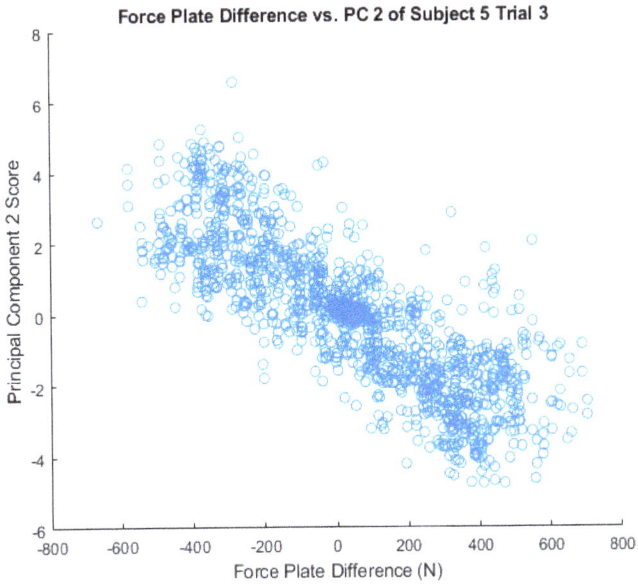

(**a**) Experiment Commands

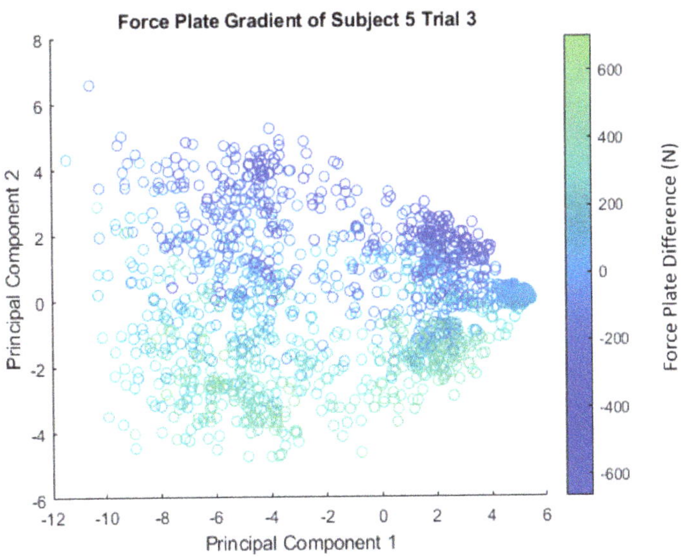

(**b**) Treadmill Speed

Figure 5. (**a**) A plot of the difference in force plate readings versus the scores of Principal Component 2. A high level of correlation between the two can be seen with the strong negative slope. This correlation can be seen over all trials, but Subject 5 Trial 3 was selected as a representative case. (**b**) Principal component scores plotted with a gradient representing the difference in force plate readings. When the subject places all of their weight on a single leg (single support), the difference between force plates is at the maximum absolute value. The gradient can clearly be seen along the axis of the second principal component, which indicated that this principal component is driven by the variation of right and left leg motion.

Table 2 presents the median angle between the first three principal components of each trial across subjects and they are individually plotted in Appendix A.

Table 2. Median Angle Between Components Across Trial and Subject Combinations.

	Component 1	Component 2	Component 3
Component 1	21.05°	90.65°	89.97°
Component 2	91.15°	51.05°	87.42°
Component 3	90.58°	90.01°	59.71°

The comparisons of both Components 1 (the first row and the first column) show the smallest angle, 21.05°, is between the like-components, while the other angles are nearly orthogonal. If two sets of principal components were exactly the same, then the diagonal would show 0 degree differences, and all off-diagonal entries would show 90 degree differences. These separation angles support that all first components are similar. The comparisons of Components 2 (the second row and the second column) and of Components 3 (the third row and the third column) have the smallest angle between the like-components as opposed to non-like components. Although the comparisons between the second and third components did not have ideal separations, since the like-components are the smallest, we consider the components that are similar enough to be used for classification, because the like-components are closer to each other than the non-like components. This assumption is further investigated by determining the classification accuracy.

The spread observed in comparing Component 2 across trials (Figure A5) may be from the alternation of left and right legs during locomotion, as supported by the correlation of Component 2 and the force plate readings (Figure 5).

3.2. Classification Accuracy

The median accuracy across all classes for all sensors (Case 1) was 86.4%, with a median absolute deviation (MAD) of 2.9%. Without allowing for run/sprint confusion, all of the effect sizes for all comparisons between cases were over 0.8 and considered to be large. Consequently, the discussion will only focus on the results when run and sprint confusion is allowed (Table 3). With this confusion, the median accuracy for all sensors was 91.6% with a MAD of 2.3% (Figure 6). A self-paced treadmill allows for subjects to perform locomotion close to their natural gait, but it leads to increased variability in speeds when compared to other studies that use fix-paced treadmills [23]. Despite the increased variability in speeds, a high classification accuracy was obtained with the full suite of sensors and the first three principal components.

The Cohen's d effect sizes across all comparisons that are listed in Table 3 were all large when directly using the predicted labels from the SVM. Only when run/sprint confusion was allowed did the effect sizes range across small, medium, and large, meaning that our reduced sensor sets do not contain enough information to accurately distinguish between running and sprinting. Our subjects were not expert sprinters, so the run and sprint were also similar to each other for this reason.

The results show that a higher level of accuracy is achieved when sensors are on the lower leg (Cases 3 and 5) than when sensors are placed on the upper leg (Cases 2 and 4). Case 3 has a higher accuracy than Case 2 (large effect size), which demonstrates that there is a significant impact on placing sensors on upper versus lower legs. Removing sensors on the lower leg and maintaining sensors on the upper leg, with the force plates (Case 4) as well as without the force plates (Case 2), led to losses in accuracy (large effect size). This loss in accuracy might be because not enough information about the activity is provided from the hamstring and vastus medialis, as they have similar EMG profiles during walking and running [24]. The effect size when both of these cases are compared to Case 1 is large, and Figure 6 shows that the median classification accuracy is lower than in Case 1.

Table 3. Cohen's *d* effect sizes.

Comparison	Cohen's *d* Value- Allowed Run/Sprint Confusion	Effect Size
Case 1 vs. Case 2	1.12	Large
Case 1 vs. Case 3	0.65	Medium
Case 1 vs. Case 4	0.82	Large
Case 1 vs. Case 5	0.29	Small
Case 1 vs. Case 6	0.74	Large
Case 1 vs. Case 7	0.53	Medium
Case 1 vs. Case 8	0.91	Large
Case 7 vs. Case 8	1.66	Large
Case 2 vs. Case 3	−0.85	Large
Case 2 vs. Case 4	−1.04	Large
Case 3 vs. Case 5	−0.67	Medium
Case 3 vs. Case 6	0.65	Medium
Case 6 vs. Case 8	0.52	Medium

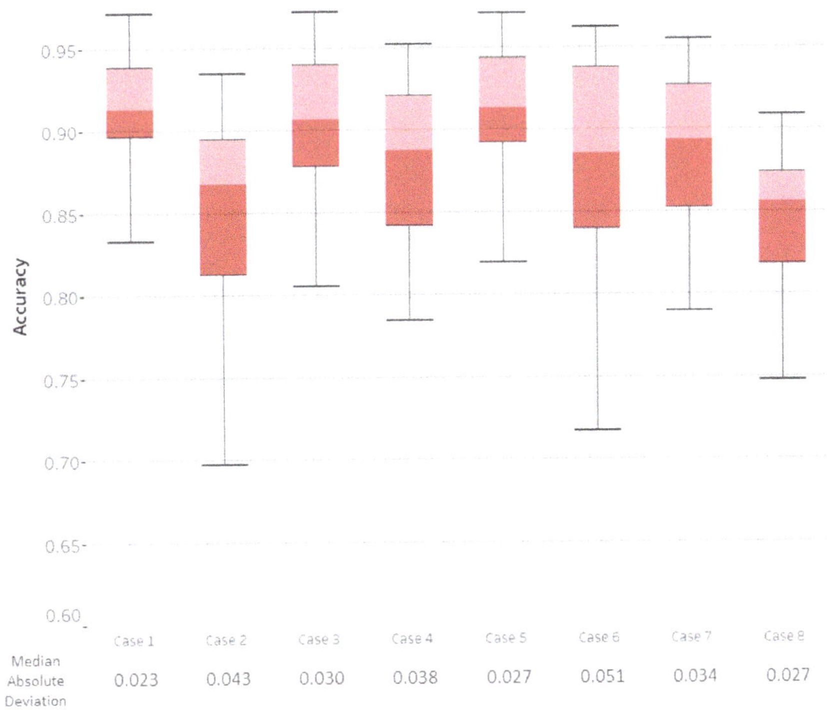

Figure 6. A box plot of the classification accuracy for all eight cases across all subjects with allowed run/sprint confusion. The line on the box represents the median, while the whiskers reach 1.5 times the interquartile range. The cases here refer to the cases that are defined in Figure 3.

The number of sensors that are placed on the lower leg matters in accurately classifying activities. When the sensors are only on the gastrocnemius muscle (Case 6), there is a decrease in accuracy when compared to Case 1 (large effect size) and a decrease in accuracy when compared to Case 3 (medium effect size). This difference in effect size means that only having two sensors on the lower leg might be insufficient for classification between stand–walk–run. During standing, walking, and running, the gastrocnemius and tibialis anterior co-contract in order to provide stability to the ankle. The removal of the tibialis sEMG sensors eliminates the ability for co-contraction to be captured within the algorithm. These results support the method having a higher accuracy when co-contraction is captured. Additionally, the lower leg is farther away from the center of rotation (the hips), so the differences in gait are magnified in the lower-leg sensors. There is a greater difference in how much the sensors on the lower leg swing during walking and running than the sensors on the upper leg.

Removing the force plates results in changes to accuracy when compared to cases with the force plate. The removal of the force plate led to reductions in accuracy as observed by comparing the sensors on the lower leg, Case 3 vs. Case 5 (medium effect), as well as sensors on the upper leg, Case 2 vs. Case 4 (large effect). The difference here is likely because the magnitude of the normal force is greater during running than walking, which supported the classification between those activities. When compared to the accuracy results from the use of all sensors (Case 1), only using a sensor set on the lower leg with the use of a force plate (Case 5) results in a small effect size (Table 3). If the sensors are limited and the motions of standing, walking, and running need to be classified, utilizing the sensors on the lower leg with force plates may not have operationally relevant differences from the full suite of sensors. When the sensors are only placed on the lower leg and the force plate is not utilized (Case 3), the median accuracy decreases by 0.82% compared to the set with all sensors (Case 1, medium effect size, Table 3). Depending on the use case, this loss in accuracy may or may not be functionally relevant.

Using accelerometers without sEMGs reduces classification accuracy. Using one accelerometer on each segment of the leg (Case 7) decreases the median accuracy when compared to Case 1 (medium effect size). The accelerometer signals provide information on changing speeds, which can inform running and walking. It is likely that the variation in acceleration profiles that comes from using a self-paced system yielded a decrease in accuracy when compared to cases when sEMG signals were present. Case 8 (only four accelerometers) has a decrease in accuracy when compared to Case 1 (large effect size). Even removing two accelerometers when no sEMGs are present yields a large decrease in accuracy, as seen by the large effect size between Cases 7 and 8. When sEMG are not present, the relative motion between the thigh and shank is captured with the accelerometers on these segments. More distal points on the leg experience greater accelerations, as linear acceleration is a function of both the change in angular velocity and the distance from the center of rotation. It follows that the accelerometers on the lower leg capture greater relative motion. These results support that there is a higher accuracy when this segment coordination is captured.

Accelerometers alone do not achieve as high accuracies from the SVM as they do when used in tandem with other sensors. In Cases 7 and 8, when only accelerometers were used for classifying all activities, the sensitivity of running sharply decreases (Figure 7). From the lower sensitivity, it can be concluded that the use of accelerometers alone hinders the accuracy of the SVM. Similarly, the walking sensitivity sharply decreases for Case 2, which means that the use of sensors only on the upper leg makes it difficult for a SVM to positively identify walking. Perhaps the lower accuracy in walking is a result of the lower-leg muscles having a greater difference in EMG profiles during running than in walking [24]. Overall, the specificities were much higher than the sensitivities, with most of them being over 90%, which implied that the SVM method of classification does not lead to many false positives in the data, but mostly false negatives. For wearable robotic applications, high specificity might decrease potential harm to subjects. It might be more harmful to misidentify a change in action than perform no change at all, as a misidentificiation might lead to an injury if the robotic system performs in a way that the user is not expecting.

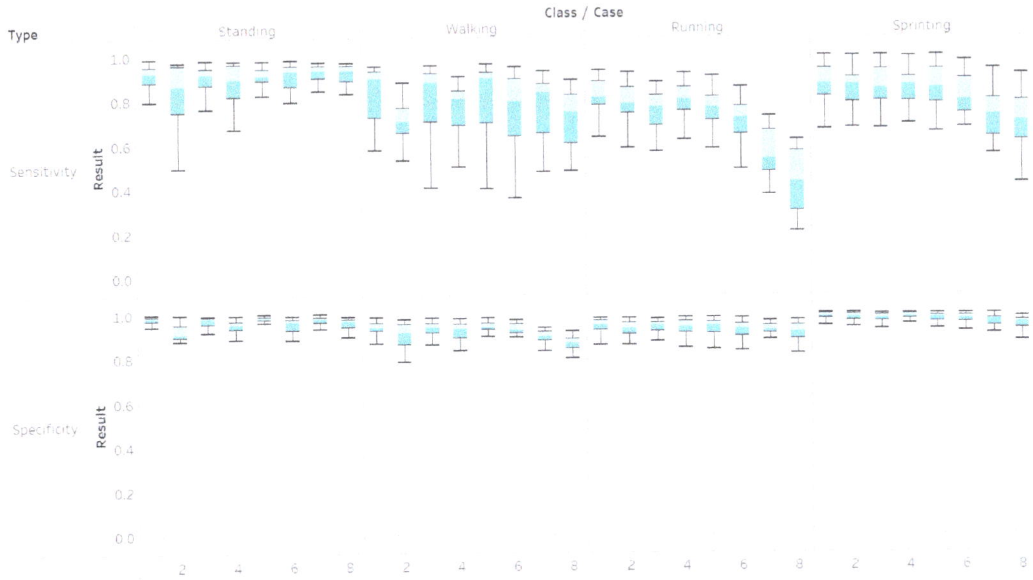

Figure 7. Sensitivities and specificities of each action per case.

This study extends the literature by examining the effect of sensor choice and placement on classifying standing, walking, and running. Through ablation, we have found that it is best to place sensors on the lower leg. Additionally, the ablation process has helped to gain an understanding regarding how different features contribute to the classification, such as the strong contributions of the lower leg when compared to the upper leg. Although other studies have used PCA for HAR, we have demonstrated that three principal components of this type of data are sufficient for a high degree of classification accuracy.

3.3. Applications of the Study

There are many applications to the results of this study. The results here can be used as a heuristic in an exoskeleton controller in order to determine when the wearer shifts from walking to running and vice-versa, supporting different controller needs. For applications, such as fitness tracking, sensors can be placed closer to the ankle than the thigh for the best results, using an EMG if possible. The ablated sensor sets can be used for optimal sensor selection in future studies. The reduced sensor suites and PCA approach can be dimensionally reduced for a faster computation time without compromising accuracy.

This paper demonstrated the successful classification of these data into standing, walking, and running, and enhanced the understanding about which sensor types and placements drive the classification. However, there are opportunities for further exploration. First, other classifiers can be explored. This paper chose to explore the effects of an SVM, which is a type of supervised learning, but there are other supervised learning techniques, such as k-nearest neighbors approximation, which can be utilized. Unsupervised techniques can also be evaluated. Additionally, the techniques that are listed here can be used to classify other types of human activities, and then ablation can be used to study which features drive those classifications.

3.4. Limitations of the Study

The small number of subjects was one major limitation of this study. The methods of this paper can be repeated with the use of a greater number of subjects, but, due to global health concerns at the time this paper was written, this was not a possibility for this particular study.

The ability of the subjects to perform the activities as another limitation of the study. Some of the subjects experienced more fatigue than others and slowed during run and sprint, leading to fewer data points for that specific activity to analyze for these cases. Consequently, we have fewer instances of sprinting and running. Because of the space that the study was completed in, we were unable to collect data for other common human activities, such as climbing stairs, but there is opportunity in the future to extend this study to additional activities.

The threshold speed values of each activity were manually set for truth labelling, so some data points might have been mislabelled near transitions. The definition of the truth states was based on the speed that the user was moving at, which was a self-selected process. Occasionally, users transitioned unintentionally. Smoothing the treadmill speeds with a Butterworth filter made the range of speeds for each task more clearly defined, but it did not entirely smooth away every unintentional transition.

The sEMG readings vary from session to session. This study's data were limited to a single session per subject and, therefore, less variability was encountered in sEMG readings than there would be multi-day study, which would be more similar to a daily use case. It is recommended to collect more data longitudinally in order to assess this variability.

During the study, the subjects were instructed to strike the left force plate with the left foot and the right force plate with the right foot. During analysis, it was observed that cross-plate strikes occurred (e.g., a left foot striking the right force plate), with a greater count towards the end of the study. These cross-strikes may occur, as subjects may have placed less attention on lateral velocity control and foot placement with study duration or fatigue onset. These cross-plate strikes were kept in the dataset. Cross-plate strikes do not affect biomechanical modeling during single-support, but they do affect the modeling of human joint torques during double-support due to the ambiguity in the length of the joints' moment arm. We do not analyze moments directly in this work, although there would be an effect on our force place sensor inputs.

4. Conclusions

This paper demonstrated which sensors were the most important for classifying standing, walking, and running through ablated data sets and an SVM that was trained on principal components. Although the ablated sets did not contain enough information to distinguish between running and sprinting, the results show that using a reduced sensor set on the lower legs will result in a classification accuracy similar to a classification accuracy when all of the sensors are used. Although including the force plates resulted in a higher classification accuracy than when the force plates were not included, the difference might not be operationally relevant. It was also shown that only using accelerometers decreases the sensitivity of the classification algorithm. Our methods involved a greater variation in speeds of activities when compared to using a fixed-speed treadmill, and yet our classification accuracy remained over 90% with all sensors included. This high accuracy indicates that the SVM used on PCA data is an effective tool for HAR and relevant sets of muscles for sensor placements have been found. Moving forward, it would be interesting to learn more regarding why dissimilarities between principal components occur.

Author Contributions: Conceptualization, S.G. and H.C.S.; Methodology, S.G., H.C.S., L.S.; Software, S.G. and H.C.S.; Formal Analysis, S.G.; Investigation, H.E. and H.C.S.; Writing—Original draft preparation, S.G.; Writing—Review and editing, H.C.S., P.S., L.S.; Supervision, H.C.S., P.S., L.S.; Funding acquisition, L.S. All authors have read and agreed to the published version of the manuscript.

Funding: Air Force Contract No. FA8702-15-D-0001.

Institutional Review Board Statement: The experimental protocol was approved by the Institutional Review Board of the Massachusetts Institute of Technology, protocol number 1703875483, approved on 9 March 2017.

Informed Consent Statement: Informed consent was obtained from all subjects involved in the study.

Data Availability Statement: The data presented in this study are available on request from the corresponding author. The data are not publicly available due to continuing study by the authors.

Acknowledgments: The authors of this paper would like to thank all the participants of this study, as well as the Air Force for providing funding.

Conflicts of Interest: The authors declare no conflict of interest.

Appendix A

The following figures are histograms of the angles of separations between all combinations of the principal components.

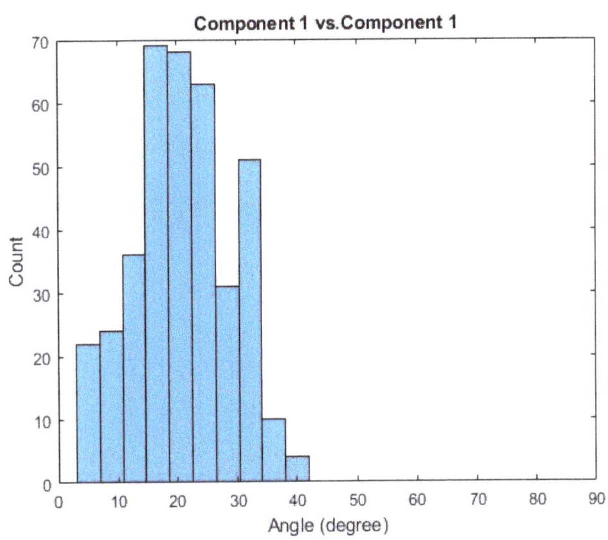

Figure A1. Histogram of the comparisons between the first principal component of some Trial A and the first principal components of some Trial B, where these comparisons are made across all trials and subjects.

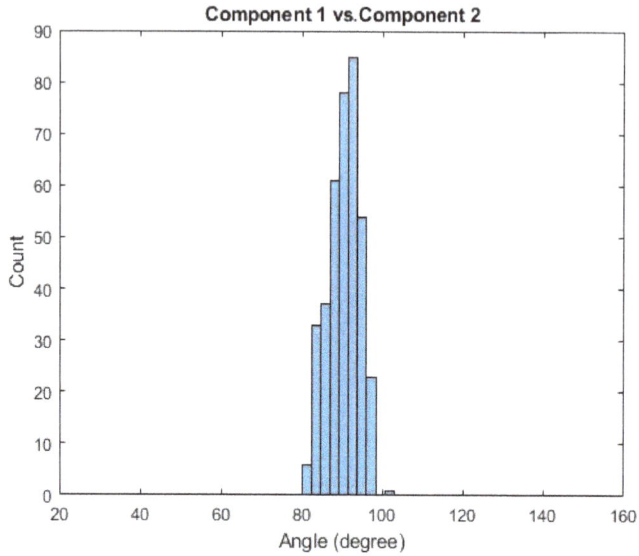

Figure A2. Histogram of the comparisons between the first principal component of some Trial A and the second principal component of some Trial B, where these comparisons are made across all trials and subjects.

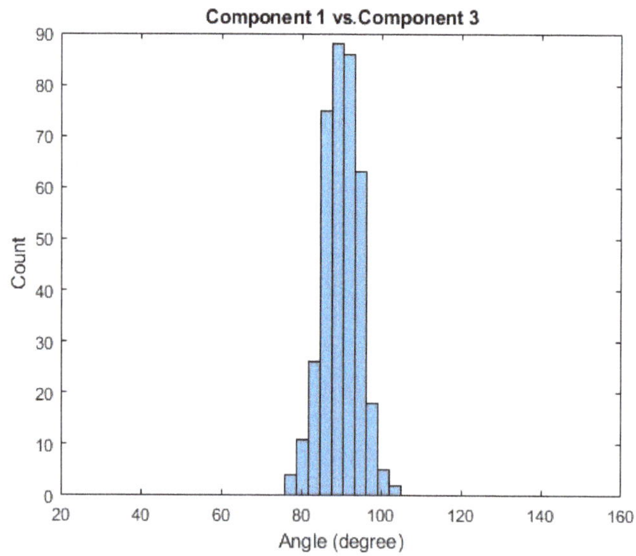

Figure A3. Histogram of the comparisons between the first principal component of some Trial A and the third principal component of some Trial B, where these comparisons are made across all trials and subjects.

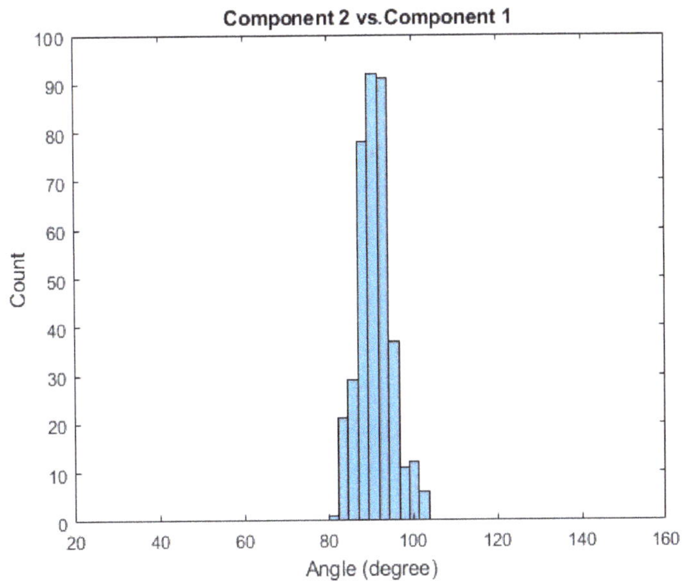

Figure A4. Histogram of the comparisons between the second principal component of some Trial A and the first principal components of some Trial B, where these comparisons are made across all trials and subjects.

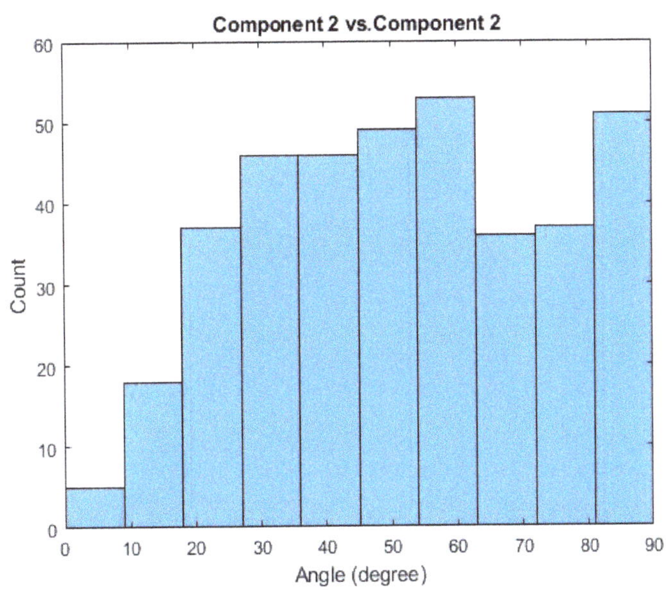

Figure A5. Histogram of the comparisons between the second principal component of some Trial A and the second principal component of some Trial B, where these comparisons are made across all trials and subjects.

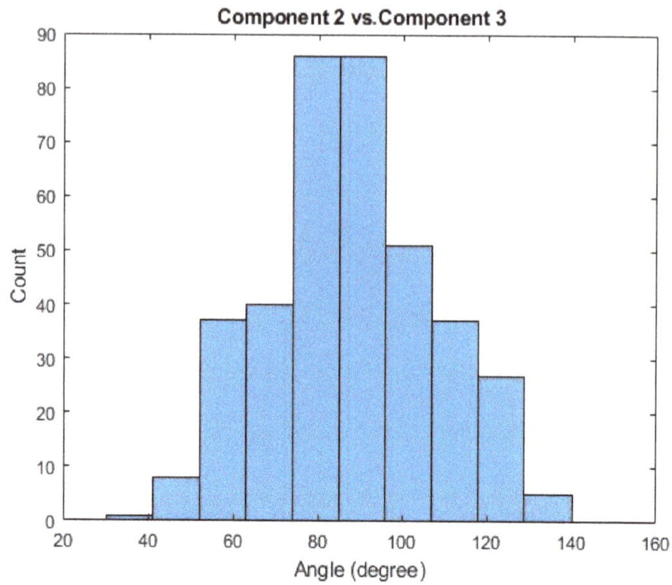

Figure A6. Histogram of the comparisons between the second principal component of some Trial A and the third principal component of some Trial B, where these comparisons are made across all trials and subjects.

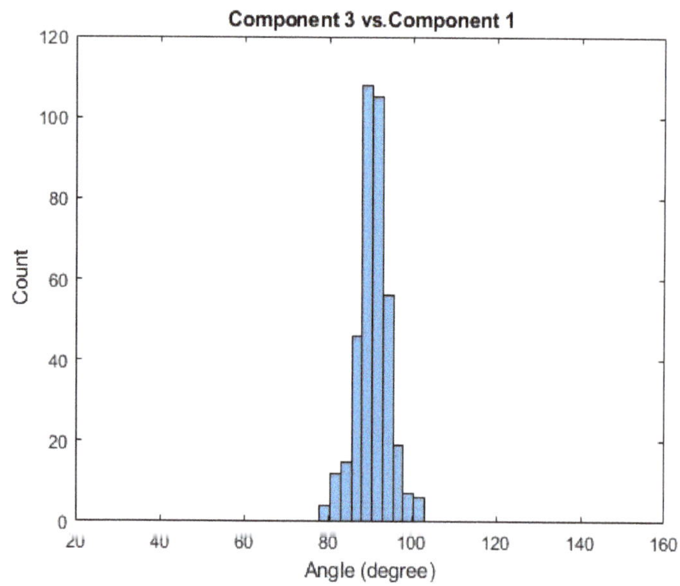

Figure A7. Histogram of the comparisons between the third principal component of some Trial A and the first principal components of some Trial B, where these comparisons are made across all trials and subjects.

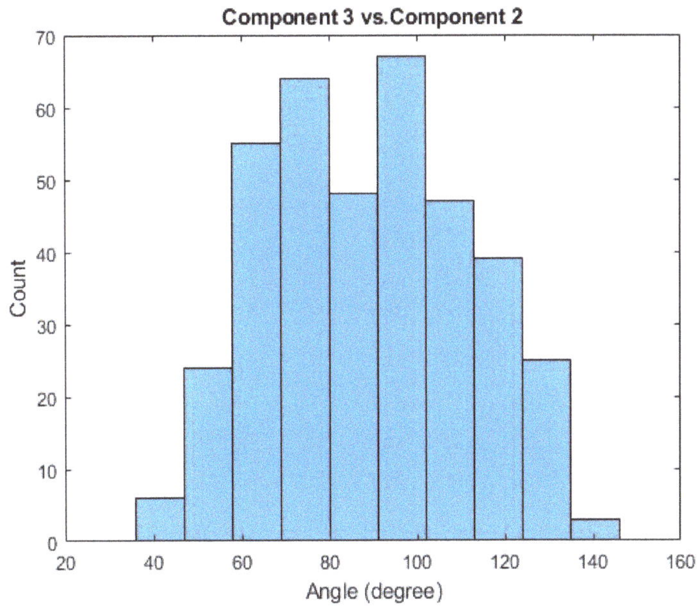

Figure A8. Histogram of the comparisons between the third principal component of some Trial A and the second principal component of some Trial B, where these comparisons are made across all trials and subjects.

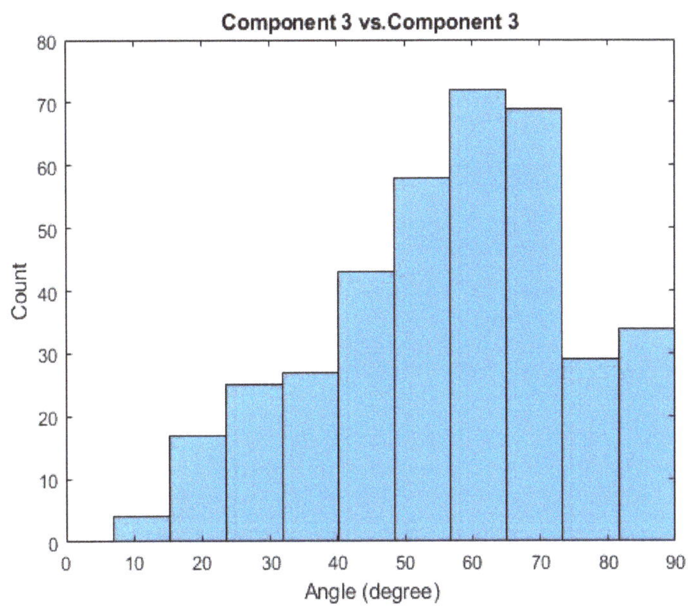

Figure A9. Histogram of the comparisons between the third principal component of some Trial A and the third principal component of some Trial B, where these comparisons are made across all trials and subjects.

References

1. Lara, O.D.; Labrador, M.A. A Survey on Human Activity Recognition Using Wearable Sensors. *IEEE Commun. Surv. Tutor.* **2013**, *15*, 1192–1209. [CrossRef]
2. Andreas, B.; Blanke, U.; Schiele, B. A Tutorial on Human Activity Recognition Using Body-Worn Inertial Sensors. *ACM Comput. Surv.* **2014**, *46*, 33:1–33:33. [CrossRef]
3. Aggarwal, J.K.; Xia, L. Human Activity Recognition from 3D Data: A Review. *Pattern Recognit. Lett. Celebr. Life Work. Maria Petrou* **2014**, *48*, 70–80. [CrossRef]
4. Munguia, T.E.; Intille, S.S.; Larson, K. Activity Recognition in the Home Using Simple and Ubiquitous Sensors. In *Pervasive Computing*; Lecture Notes in Computer Science; Ferscha, A., Mattern, F., Eds.; Springer: Berlin/Heidelberg, Germany, 2004; pp. 158–175. doi:10.1007/978-3-540-24646-6_10. [CrossRef]
5. Ferhat, A.; Mohammed, S.; Dedabrishvili, M.; Chamroukhi, F.; Oukhellou, L.; Amirat, Y. Physical Human Activity Recognition Using Wearable Sensors. *Sensors* **2015**, *15*, 31314–31338. [CrossRef]
6. Cappellini, G.; Ivanenko, Y.P.; Poppele, R.E.; Lacquaniti, F. Motor Patterns in Human Walking and Running. *J. Neurophysiol.* **2006**, *95*, 3426–3437. [CrossRef] [PubMed]
7. Wege, A.; Zimmermann, A. Electromyography Sensor Based Control for a Hand Exoskeleton. In Proceedings of the 2007 IEEE International Conference on Robotics and Biomimetics (ROBIO), Sanya, China, 15–18 December 2007; pp. 1470–1475. [CrossRef]
8. Mantyjarvi, J.; Himberg, J.; Seppanen, T. Recognizing Human Motion with Multiple Acceleration Sensors. In Proceedings of the 2001 IEEE International Conference on Systems, Man and Cybernetics. e-Systems and e-Man for Cybernetics in Cyberspace (Cat.No.01CH37236), Tucson, AZ, USA, 7–10 October 2001; Volume 2, pp. 747–752 . [CrossRef]
9. Xu, Z.; Chen, X.; Li, Y.; Lantz, V.; Wang, K.; Yang, J. A Framework for Hand Gesture Recognition Based on Accelerometer and EMG Sensors. *IEEE Trans. Syst. Man, Cybern. Part A Syst. Hum.* **2011**, *41*, 1064–1076. [CrossRef]
10. Kadrolkar, A.; Sup, F. Classification of Trunk Motion for a Backbone Exoskeleton Using Inertial Data and Surface Electromyography. In Proceedings of the 2013 IEEE International Conference on Systems, Man, and Cybernetics, Manchester, UK, 13–16 October 2013; pp. 3978–3983. [CrossRef]
11. Fabien, M.; Gonzenbach, R.R.; Arami, A.; Paraschiv-Ionescu, A.; Luft, A.R.; Aminian, K. Improving Activity Recognition Using a Wearable Barometric Pressure Sensor in Mobility-Impaired Stroke Patients. *J. Neuroeng. Rehabil.* **2015**, *12*, 72. [CrossRef]
12. Khokhar, O.; Xiao, Z.G.; Menon, C. Surface EMG Pattern Recognition for Real-Time Control of a Wrist Exoskeleton. *Biomed. Eng. Online* **2010**, *9*, 41. [CrossRef] [PubMed]
13. Andreas, B.; Roggen, D. Recognition of Visual Memory Recall Processes Using Eye Movement Analysis. In Proceedings of the 13th International Conference on Ubiquitous Computing—UbiComp'11, Beijing, China, 17–21 September 2011; ACM Press: Beijing, China. [CrossRef]
14. Naik, G.R.; Pendharkar, G.; Nguyen, H.T. Wavelet PCA for Automatic Identification of Walking with and without an Exoskeleton on a Treadmill Using Pressure and Accelerometer Sensors. In Proceedings of the 2016 38th Annual International Conference of the IEEE Engineering in Medicine and Biology Society (EMBC), Orlando, FL, USA, 16–20 August 2016; pp. 1999–2002. [CrossRef]
15. Guo, M.; Wang, Z. Segmentation and Recognition of Human Motion Sequences Using Wearable Inertial Sensors. *Multimed. Tools Appl.* **2018**, *77*, 21201–21220. [CrossRef]
16. Wolf, M.T.; Assad, C.; Stoica, A.; You, K.; Jethani, H.; Vernacchia, M.T.; Fromm, J.; Iwashita, Y. Decoding Static and Dynamic Arm and Hand Gestures from the JPL BioSleeve. In Proceedings of the 2013 IEEE Aerospace Conference, Big Sky, MT, USA, 2–9 March 2013; pp. 1–9. [CrossRef]
17. Fogelson, B.; Stirling, L.; Siu, H. Anticipation of Speed Transitions Using Electromyography. In Proceedings of the IEEE RAS/EMBS International Conference on Biomedical Robotics and Biomechatronics (BIOROB), New York, NY, USA, 29 November–2 December 2020.
18. Zardoshti-Kermani, M.; Wheeler, B.C.; Badie, K.; Hashemi, R.M. EMG feature evaluation for movement control of upper extremity prostheses. *IEEE Trans. Rehabil. Eng.* **1995**, *3*, 324–333. [CrossRef]
19. Phinyomark, A.; Thongpanja, S.; Hu, H.; Phukpattaranont, P.; Limsakul, C. The Usefulness of Mean and Median Frequencies in Electromyography Analysis. *Comput. Intell. Electromyogr. Anal. Perspect. Curr. Appl. Future Chall.* **2012**, *2012*, 195–220. doi:10.5772/50639. [CrossRef]
20. Svante, W.; Esbensen, K.; Geladi, P. Principal Component Analysis. *Chemom. Intell. Lab. Syst.* **1987**, *2*, 37–52. [CrossRef]
21. Noble, W. What is a support vector machine? *Nat. Biotechnol.* **2006**, *24*, 1565–1567. [CrossRef] [PubMed]
22. Meyes, R.; Lu, M.; de Puiseau, C.W.; Meisen, T. Ablation Studies in Artificial Neural Networks. *arXiv* **2019**, arXiv:1901.08644.
23. Kim, J.; Heimgartner, R.; Lee, G.; Karavas, N.; Perry, D.; Ryan, D.L.; Eckert-Erdheim, A.; Murphy, P.; Murphy, P.; Galiana, I.; et al. Autonomous and Portable Soft Exosuit for Hip Extension Assistance with Online Walking and Running Detection Algorithm. In Proceedings of the 2018 IEEE International Conference on Robotics and Automation (ICRA), Brisbane, Australia, 21–25 May 2018; pp. 5473–5480. [CrossRef]
24. Gazendam, M.G.J.; Hof, A.L. Averaged EMG Profiles in Jogging and Running at Different Speeds. *Gait Posture* **2007**, *25*, 604–614. [CrossRef] [PubMed]

sensors

MDPI

Communication

Arduino-Based Myoelectric Control: Towards Longitudinal Study of Prosthesis Use

Hancong Wu [1,*], Matthew Dyson [2] and Kianoush Nazarpour [1,*]

1 School of Informatics, The University of Edinburgh, Edinburgh EH8 9YL, UK
2 School of Engineering, Newcastle University, Newcastle upon Tyne NE1 7RU, UK; matthew.dyson@ncl.ac.uk
* Correspondence: hancong.wu@ed.ac.uk (H.W.); kianoush.nazarpour@ed.ac.uk (K.N.)

Abstract: Understanding how upper-limb prostheses are used in daily life helps to improve the design and robustness of prosthesis control algorithms and prosthetic components. However, only a very small fraction of published research includes prosthesis use in community settings. The cost, limited battery life, and poor generalisation may be the main reasons limiting the implementation of home-based applications. In this work, we introduce the design of a cost-effective Arduino-based myoelectric control system with wearable electromyogram (EMG) sensors. The design considerations focused on home studies, so the robustness, user-friendly control adjustments, and user supports were the main concerns. Three control algorithms, namely, direct control, abstract control, and linear discriminant analysis (LDA) classification, were implemented in the system. In this paper, we will share our design principles and report the robustness of the system in continuous operation in the laboratory. In addition, we will show a first real-time implementation of the abstract decoder for prosthesis control with an able-bodied participant.

Keywords: surface electromyogram; prosthesis control; wearable; low-cost

Citation: Wu, H.; Dyson, M.; Nazarpour, K. Arduino-Based Myoelectric Control: Towards Longitudinal Study of Prosthesis Use. *Sensors* **2021**, *21*, 763. https://doi.org/10.3390/s21030763

Academic Editor: Ernest N. Kamavuako

Received: 28 December 2020
Accepted: 20 January 2021
Published: 24 January 2021

1. Introduction

Myoelectric prostheses have the potential to restore the functionality of missing limbs. First developed in 1948 [1], myoelectric prostheses are now widely used by amputees in their daily lives. Although research and development have led to significant technological advances in the laboratory, conventional control algorithms are still the mainstream in clinical trials and commercialised applications. A literature search shows that among the 1716 articles that mention daily use of prostheses since 1990, only 69 include the original experiments outside the laboratory [2]. Limited beyond-the-laboratory testing might explain why the prosthesis rejection rate has remained as high as 50% [3,4].

One main reason for the status quo is that, in community settings, advanced prosthetic control algorithms do not perform as well as they do in the laboratory [5]. Factors such as electrode displacement [6,7], changes in limb positions [8,9], extra load on the limb [10], time between adaptation and application [11], and user learning [12,13] can degrade the performance of the prosthesis. Expansion of the training dataset helps to increase the robustness of the control, but training data in community settings are limited, and the data collected in laboratory settings can be different from user behaviour in daily life. In addition, the recalibration procedures for advanced algorithms are more complicated than the conventional control methods, and therefore, it will be challenging, if not impossible, to retune these algorithms in home trials without the support of specialists [14,15].

The limitations of portable devices are another barrier to the translational studies of advanced prosthetic technology. Currently, most of the home trials for prosthesis studies were based on commercialised prosthetic devices [16–18]. These products included self-contained control systems in their embedded microcontrollers so that they could plug and play directly in the experiments. Having the merits of convenience and robustness, these systems have limited functions and flexibility. For instance, they relied on basic prosthesis control methods, such as on–off or direct control [19], and the users could not change settings during use [20,21]. Add-on controller kits [22,23] provide platforms for the implementation of advanced control algorithms. Due to the high costs involved and the constraints on adaptable prostheses, users in developing countries have limited access to these devices.

Arduino is an open-source computing platform that can integrate data collection, signal analysis, and control on a single microcontroller. It has different series to match various peripheral devices, and the users can develop systems on built-in libraries. Preliminary studies have reported development of prosthetic control systems on Arduino boards [22,24]. With an open-source integrated development environment (IDE), developers can build the systems easily on the basis of previous designs and utilise modules developed in other projects [25–27]. Currently, the Arduino-based prosthesis research still focuses on applications with basic control functions [28,29].

In this paper, we describe the development of a cost-effective myoelectric control system for achieving multiple hand tasks with selectable control algorithms. We demonstrate its functionality and portability and show its feasibility for long-term prosthetic control and fair comparison between control algorithms. The features proposed in this work include the following: (a) a robust system structure that guarantees the accuracy of control signals during long-term operation; (b) the implementation of a clinical controller, human-learning-based controller, and pattern-recognition-based controller on a single system, which allows fair comparisons in prosthetic studies; (c) a friendly user interface that helps users to learn the system control and to adjust the control preferences in take-home studies; and (d) extendable data storage for hand-task recording with time stamps. To guarantee the generalisation, the system was developed on off-the-shelf sensors and development boards (Arduino).

2. Methods

2.1. System Features

2.1.1. System Overview

The control system consists of five major components (Figure 1). These are: (1) two sensors for surface electromyogram (EMG) recording (Gravity analog EMG sensors; OYMotion, China); (2) a development board for EMG signal processing, feature extraction, and control command generation (Arduino MKR ZERO; Arduino LLC, USA); (3) a connector module for device communication (Arduino MKR CAN shield; Arduino LLC, USA); (4) a prosthetic hand with six degrees of freedom (DoFs) (Robo-limb hand; Össur, Iceland); and (5) a data-recording unit (32-GB Kingston micro-SD card; Kingston Technology, USA). The Gravity analog EMG sensors amplify the surface EMG signals 1000 times and depress noises through differential input and an analog filter circuit. The amplified EMG signal is sampled by a 10-bit analog-to-digital converter (ADC) through the Arduino analog inputs at a 500 Hz sampling frequency. After processing the EMG signals, the corresponding motor commands are transmitted to the prosthetic hand to drive the motors through the Controller Area Network (CAN) communication.

One reason to develop the system on Arduino is because of its compatibility with a variety of analog front ends and radio frequency (RF) front ends, which is very useful in bridging the EMG sensors with prosthetic devices. It can independently power and interface with varies types of EMG sensors, including MyoWare muscle sensors (Sparkfun, USA), the Grove EMG detector (Seeed Technology Inc., China), and Gravity analog EMG sensors (OYMotion, China). Control signals can be sent to different prosthetic devices, such as the Robo-limb prosthetic hand (Össur, Iceland) and the COVVI hand (COVVI Ltd, UK) through a CAN bus or Bluetooth communication. We selected the Gravity analog EMG sensors in this study because their dry electrode design was simple to use and was more robust in long-term studies and real-world environments [30,31].

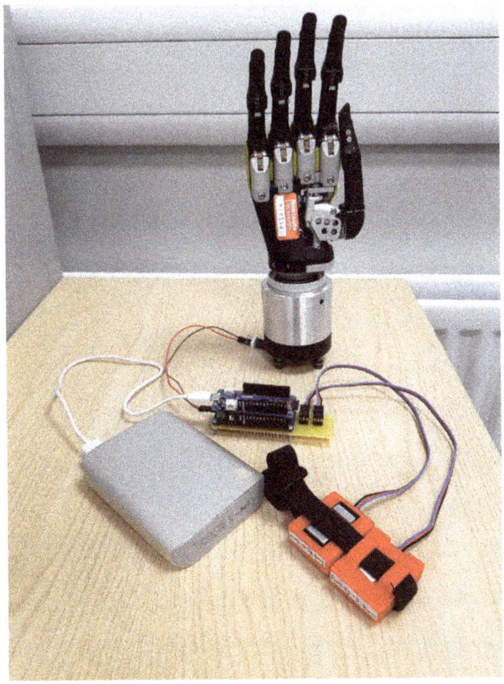

Figure 1. The prototype of the proposed myoelectric control system.

2.1.2. Signal Recording and Pre-Processing

To accurately control the prosthesis, the system should record and process the EMG signals quickly so that the delay between muscle contraction and hand movement can be minimised. Our system applied a 500 Hz sampling rate for signal recording. The sampled signals were filtered by an EMG filter module developed by OYMotion (OYMotion, Shanghai, China), which was designed to work synergistically with the analog filters on the Gravity analog EMG sensors. It was implemented by cascading an anti-hum notch filter, a second-order Butterworth lowpass filter, and a second-order Butterworth highpass filter to filter out 50 Hz power line noise, noise above 150 Hz, and noise below 20 Hz, respectively. Considering the variation in muscle strength among users, a calibration module was added to normalise the signals based on individual comfortable contraction levels [32].

2.1.3. User-Friendly Control Interface

A customised control interface was developed on a personal computer (PC) to support users in learning myoelectric control and updating control parameters. As can be seen in Figure 2, the interface is composed of a data visualisation unit and a control panel. The data visualisation unit displays the on-the-fly muscle contraction levels or selected features on the screen. This module helps users develop their control patterns for human-learning-based controllers [33], and it may also support data collection for pattern recognition [34]. The control panel allows users to select the controller based on their unique preferences and to recalibrate the system at home. Recalibration requires the users to rest their muscles and then contract them sequentially at a comfortable contraction level within ten seconds. The EMG signals in the system will be normalised by the calibration data through a normalisation equation [32]. This design aims to simplify the protocols for prosthesis adjustments so that the disruption to the user's daily life can be minimised.

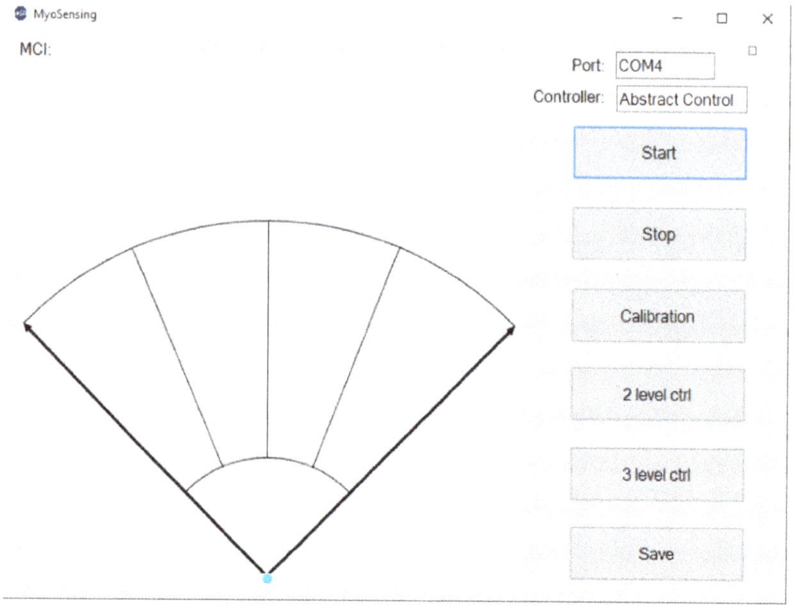

Figure 2. User interface for the proposed system.

2.2. *Controller Modules*

Three control algorithms, including the direct control, abstract control, and linear discriminant analysis (LDA) classification, were implemented in the system using two EMG channels. For each channel, the control signal was extracted by the mean absolute value (MAV) of the EMG signals over a 760 ms window with an update rate of 50 Hz. The window size was selected to smooth the control output and maintain an acceptable effector movement [33].

2.2.1. Direct Controller

Direct control is a conventional myoelectric control method that generates a single command (hand open, hand close, or grip switching) at a time [12]. It normally applies threshold-based criteria to determine the activation of control signals [35]. In our two-channel direct controller, the activation of a single control signal directed the prosthetic hand to open and close, respectively. The co-activation of both control signals allowed the user to switch the hand function between power grip, tripod grip, and lateral grip. The hand gesture persisted when neither channel was activated.

2.2.2. LDA Classifier

Classification is the most popular method in myoelectric control. It utilises the training data to develop a classifier that maps the control signals extracted from the input signals onto a discrete output variable that encodes prosthetic activities [36,37]. Because of the robustness, simplicity of implementation, and ease of training, LDA classifiers have frequently been used in previous studies [38].

In this study, the linear discriminant coefficient matrix W was trained on a PC to estimate the posterior probability of grips that the user intended to select based on two EMG channels. The trained matrix was uploaded to the system for real-time classification using $y = Wx$, where x is the $< 2 \times 1 >$ dimensional vector for the MAVs and y is the $< 5 \times 1 >$ dimensional vector encoding the posterior probability. The grip with the highest posterior probability was considered as the output of the classifier.

2.2.3. Abstract Controller

Abstract control is a relatively novel approach for myoelectric control [33]. The mechanism of abstract control is based on the fact that the motor system is able to learn multiple novel muscle synergies to achieve specific motor goals [32,39]. To facilitate the learning, a myoelectric–computer interface (MCI) is used to map the muscle activities to a non-representational multidimensional control space (Figure 2), which provides continuous visual feedback for motor system adaption [40]. The position of the cursor (blue dot in Figure 2) is controlled by the features of the EMG signals, such as the MAV. Users are asked to practise cursor control in a centre-out task to develop abstract muscle synergies [41], and each task will be mapped to a grip or a command of the prosthetic hand. More discussions of how the MCI relates to learning prosthesis control can be found in our previous studies [13,32]. In this research, the cursor position was controlled by the MAVs of two channels, and the MCI was separated into five sections, which were mapped to five different prosthetic activities. When the cursor stayed at the basket (target 0), the prosthetic hand persisted at the previous grip. Moving the cursor from the basket to target 1 to 4 (from left to right) changed the hand to a power grip, tripod grip, lateral grip, or open hand, respectively.

2.3. System Evaluation

Ethical approval was granted by the local ethics committee at Newcastle University (Ref: 17-NAZ-056). To show the feasibility of our cost-effective control system, an online analysis of the control signals and a simple pick-and-place experiment were conducted with an able-bodied participant. The participant read an information sheet and gave written consent prior to the experimental sessions. Sensors were connected to the participant's flexor carpi radialis (FCR) and extensor carpi radialis (ECR) muscles of the forearm to record the EMG activity. To connect the sensor, the skin was cleaned before the experiment. The participant sat on a chair with his elbow near his body and held his forearm at a horizontal position. We found the belly of the ECR and the FCR by asking the participant to repetitively flex and extend the wrist at a comfortable level. After fixing the sensors on the belly with an elastic band, we marked the sensor locations and orientations to ensure that they were at the same position in the multi-day experiment.

For training, the system was connected to the PC user interface shown in Figure 2 through USB, which allowed the participant to control the cursor in the MCI through muscle contraction. The participant first performed a series of contractions to learn the control of the cursor. To be specific, he was required to develop a repeatable muscle strategy that could move the cursor from the basket to the desired targets. This muscle strategy was used for the control of the direct controller and the abstract controller. To train the LDA classifier, the participant was asked to move the cursor to different targets and hold the cursor position within one target for six seconds. The MAVs during the holding period were considered as the patterns for one grip so that the dissimilarity between patterns for different grips could be guaranteed. After training the participant and the classifier, the system was connected to the prosthetic hand, and the participant started to control the hand instead of the cursor.

We first evaluated the controllability and the effector movement by analysing the real-time control signal. The participant tried to make grasps sequentially with all possible grips and to open the hand between two grasps. The MAVs, equivalent cursor positions, controller states, and the selected grips were recorded.

To evaluate the feasibility of the system in completing daily activities, the participant fitted the prosthetic hand on his right arm (Figure 3A) to carry out a pick-and-place experiment. Three objects, including a bottle, a roll tape, and a credit card simulator, were placed on a computer desk in front of the participant (Figure 3B). The participant was asked to lift the objects with a power grip, tripod grip, and lateral grip, respectively, and place them on the right-hand side of the red line by opening the hand. Videos of the experiment were recorded and are supplied as supplementary material.

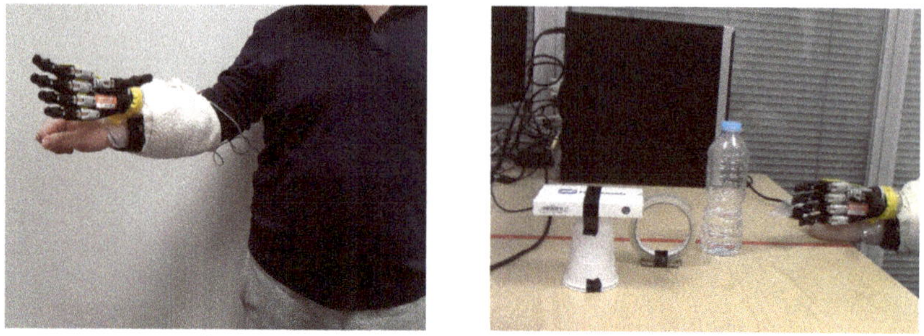

A B

Figure 3. The pick-and-place experiment setup. (**A**) The participant wore the Robo-limb prosthetic hand on his right arm with a customised socket. The prosthetic hand was powered by the battery mounted on the socket and the Arduino system was connected to the computer so that the user could switch the controller during use through the user interface. (**B**) The participant was instructed to grasp and relocate three objects using the prosthetic hand.

To verify the system performance during a long-term study, we recorded and compared the control signals before and after continuous operation. The myoelectric control system was loaded onto the Arduino board at the beginning of the experiment for a five-day test. On each day, the participant switched on and recalibrated the system based on the calibration protocol. A trial was conducted after the recalibration to ensure that the participant could easily move the cursor from the basket to each target. We kept the system working for eight hours, and the participant was required to repeat the trial at the end of the day. The MAV signals and the cursor traces were recorded and compared.

3. Results

3.1. Control Signal Analysis

Samples of the control signals and the corresponding prosthetic hand movements are presented in Figure 4. These demonstrate the relationship between the MAV and state machines within the system.

In the direct control, contracting the FCR and ECR individually moved the cursor from the basket to target 1 and target 4, respectively, which closed and opened the prosthetic hand with the selected grip. Co-contracting the muscles from resting switched the grip selection, and the hand was reset to the pre-set position. In the abstract control and the LDA classification, a grasp was completed by directly selecting the desired grip through muscle contraction/activity. Once a grip was selected, the system reset the hand to the pre-set position and then completed the grasp after a short waiting period.

Because of the continuous updating, the changes in EMG started to take effect on the MAV even though we applied a 760 ms averaging window. The delay was calculated between the time point when the MAV exceeded 5% of the maximum voluntary contraction and the time point when the cursor state changed. Based on this criterion, the average time from the participant's intention for a grasp until the cursor reached the desired target was 224 ms. The delays for the LDA classification and the abstract control were slightly longer, which were 262 ms and 268 ms, respectively. There was a 240 ms dwell time between the changes in cursor state and the controller state, which required the participant to retain the cursor within the target to make a grasp. The aim of this design was to avoid unintended hand activities due to false triggering or unconscious muscle contraction (Figure 4B).

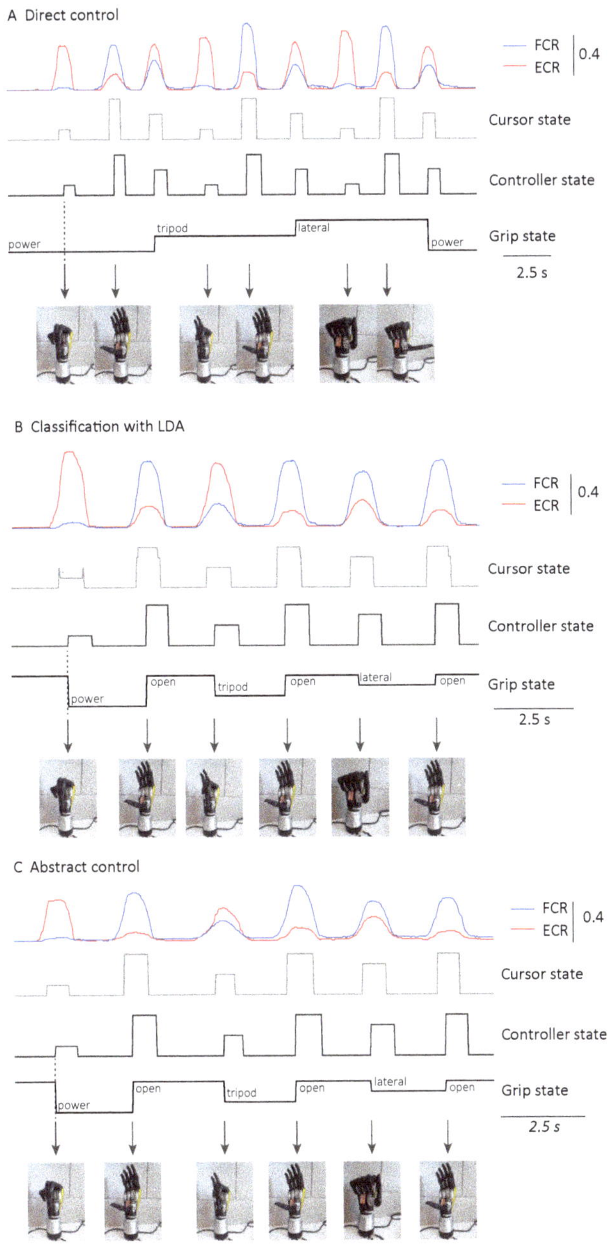

Figure 4. Control signals for the (**A**) direct controller, (**B**) linear discriminant analysis (LDA) classifier, and (**C**) abstract controller.

3.2. Pick-and-Place Experiment

The time series of the pick-and-place experiment are illustrated in Figure 5. The prosthetic hand was preset to the open-hand position at the beginning of the test. The participant sequentially moved the bottle, the roll of tape, and the credit card simulator with the corresponding grips, and opened the hand again to finish the test. The real-time performance results for the direct controller, the abstract controller, and the LDA classifier are shown in supplementary material. It took the participant 23 s to complete the test with the abstract controller and the LDA classifier, and it required 28 s with the direct controller.

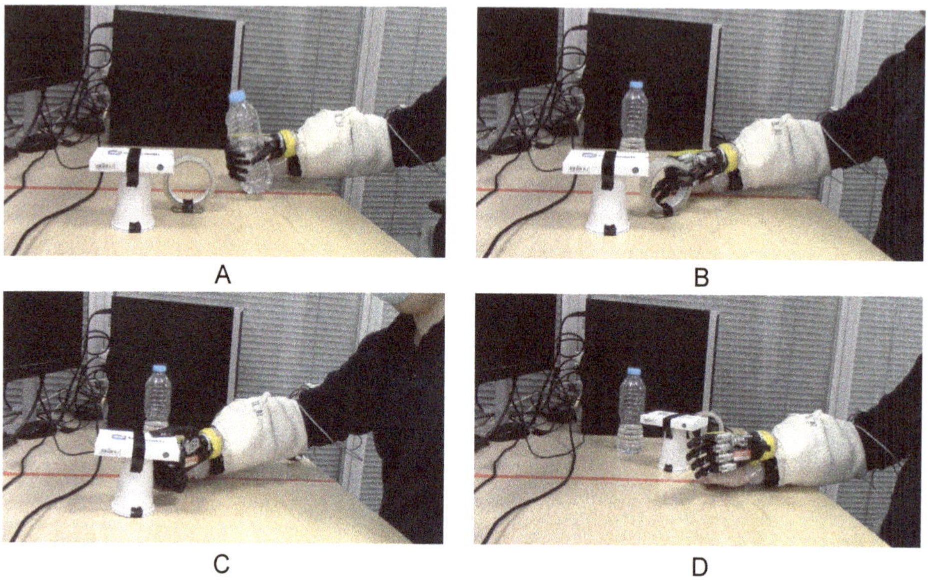

Figure 5. Real-time pick-and-place experiment with three different control algorithms. The participant (**A**) lifted the bottle with a power grip, (**B**) lifted the roll of tape with a tripod grip, (**C**) lifted the credit card simulator with a lateral grip, and (**D**) opened the prosthetic hand to relocate the objects.

3.3. System Performance in a Day-Long Study

The baselines of the control signals are presented in Figure 6A. Continuous system operation did not introduce obvious zero drift to the MAV for the filtered EMG signals. The average MAVs of the two channels at the relaxed level were increased by 0.13 and 0.12 mV, respectively, which was much smaller than the muscle contraction levels in Figure 6B. Possible reasons for the outlier in Ch1 (Test) include the sensor position and the contact between the skin and electrode.

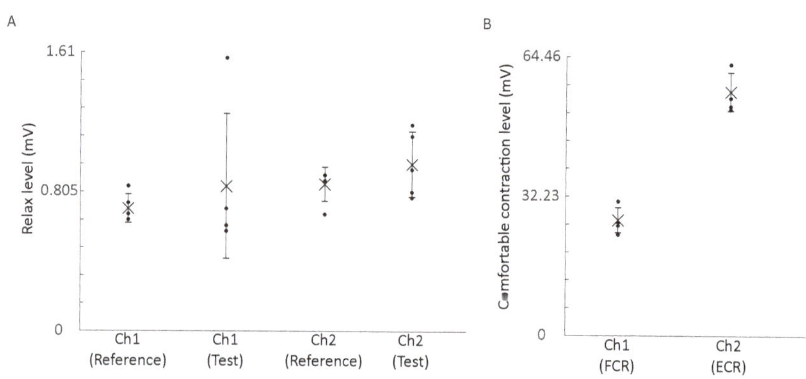

Figure 6. Mean absolute value (MAV) comparisons. (**A**) Comparison of the relaxed levels in the calibration (reference) and at the end of the day (test). (**B**) The comfortable contraction levels for two channels. Cross: mean; error bar: standard deviation; dot: individual data points (n = 5).

Figure 7 compares the cursor movement between the trial after calibration and the trial at the end of the day. No significant differences can be observed at the baseline of the cursor. The sensitivity of both sensors remained unchanged so that the participant could repeat the cursor trace without recalibration. Some user habits, such as contacting the FCR while the participant was trying to relax the ECR, could be observed in both trials.

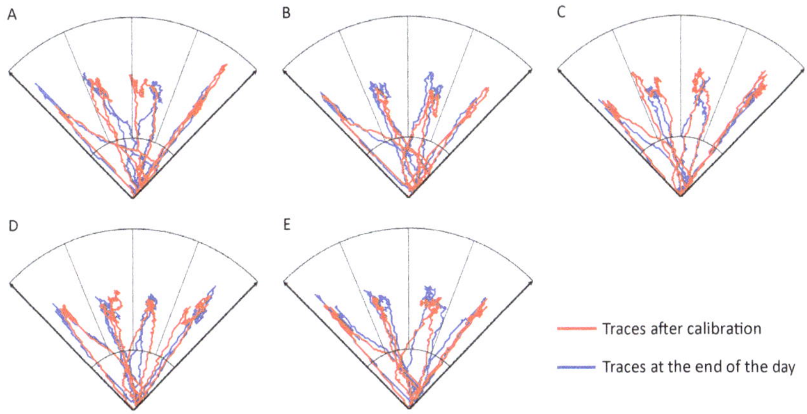

Figure 7. Representative cursor traces on (**A**) Day 1, (**B**) Day 2, (**C**) Day3, (**D**) Day 4, and (**E**) Day 5.

4. Discussion

We have shown the development of an Arduino-based myoelectric control system with three independent controllers for achieving multiple hand tasks. The system can be realised with a total cost of GBP 120, and the sensors can be replaced with any of their equivalents. The muscle contraction signals were collected by wearable EMG sensors with a 500 Hz sampling rate and could be processed within 20 ms, which allowed the system to provide movement updates in real time.

Considering the portability, we soldered the Arduino boards and connectors to a single printed circuit board (PCB) and removed unnecessary cables in the Gravity analog EMG sensors when designing the prototype (Figure 1). The total weight of the modified system was 67.8 g, which would not add significant load to an arm compared to the weight of a prosthetic limb. The sizes of the PCB and the EMG sensor are 96 × 29 × 29 mm (length × width × height) and 38 × 32 × 11 mm, respectively. We 3D-printed wearable shells for the PCB and the sensors (the red casing in Figure 1) so they could be worn on the arm with straps. When both sensors are actively measuring, the system consumes 16 mA at 3.7 V. Therefore, a rechargeable LiPo battery with 1000 mAh capacity will allow the system to work continuously for more than 48 h, in theory.

Regarding the effector movement, three controllers had similar delays. This meant that the latency was mainly introduced by filtering and the analysis window [42]. With a 50 Hz update rate, the additional latency introduced by the analysis window was below 300 ms and could not be perceivable by the user [43]. We introduced the dwell time to avoid false triggering, but the period was adjustable. It may be reduced for an experienced prosthesis user who is less likely to generate unintended commands. The controller delay of the direct controller was slightly shorter than that of the abstract controller and LDA classifier. One possible reason is that the difficulty of retaining the cursor on the MCI target is higher than for the direct control. This latency may be reduced by systematic training [13]. Meanwhile, it was noticed that the time to achieve three grips with the direct control was longer than those with the LDA classification and the abstract control in Figure 4. Extra pre-setting time for grip switching makes the direct controller effective only for tasks with a small number of grips. Otherwise, the grip switching would be time-consuming [44], which was identified as inadequate for task completion in previous surveys [45].

In the proposed system, the three controllers shared the same EMG sensors, signal processing and feature extraction modules, prosthetic hand drivers, and MCI. This approach allows a fair comparison between the basic performances of the controllers in achieving physical tasks. It also provides a platform to study the differences and similarities between the control of the cursor and the control of the prosthetic hand. For example, it can be observed from the supplementary videos that the extra time of completion with the direct controller is mainly because of the pre-setting time, which takes about one and a half seconds to switch the hand from one grip to the pre-set position of the next grip. The result matches the control signals in Figure 4. It shows the advantages of the abstract control and the LDA classification, and it may also verify that the scores of a controller in cursor control studies [46] can reflect its performances in physical task experiments.

A user interface was developed to visualise the control signals and send commands to the system through a desktop. It allows users to set up the system properly with visual feedback and to conduct recalibration in take-home studies. As learning the control on the MCI increases the performance [13,47], the user interface provides a platform for users to train themselves during long-term home studies. In the current version, the training of the LDA classifier requires the support of MATLAB. Planned future works include embedding the training algorithms into the interface.

To study long-term prosthesis use, system stability is an important aspect. We have shown that the system could provide stable control signals in a five-day test. Daily recalibration of the system only required the participant to provide his relaxed levels and comfortable contraction levels for each muscle for 10 s. Retraining of the LDA classifier was not necessary, as we introduced human learning [48] in the design of the patterns. Instead of retraining the system with new data, which might require several hours with the support of experienced professionals [49,50], the participant could actively map his muscle activities to the patterns.

In addition to the amount of time for which the device is turned on, the system allows the storage of EMG features and control signals, as well as the use of grips to describe prosthetic use. The data can be saved to a PC through the user interface and the on-board micro-SD card. Considering the limitation of storage space, the micro-SD only saves the use of grips with time stamps when new commands have been sent to the prosthetic hand. One possible way to overcome this barrier is the internet of things (IoT) [51,52], which allows the end devices (the proposed system) to upload data to a cloud server, such as the Arduino IoT Cloud. This not only reduces the risks that come with having a large amount of data stored in a single database, but also enables the comprehensive recording of prosthesis use in take-home studies [53], which will be useful in the analysis of prosthesis abandonment.

In this study, the system was only tested on a single able-bodied participant. However, since the EMG sensors are wearable and both the system and the prosthetic hand can be powered by batteries, it can be easily expanded into a wearable device for different-limbed users by mounting the components on the socket. Moreover, the update rate of the system is adjustable, so it is possible to increase the number of EMG channels by allowing a longer time for signal processing and command updates.

5. Conclusions

We introduce a system that offers the possibility of longitudinal experiments with advanced prostheses. We implemented it on a cost-effective Arduino-system platform and coupled it with wearable EMG sensors. The system allows users to adjust system settings, learn the control strategy, and record the prosthetic use outside the laboratory. This system enabled fair and balanced comparisons to be conducted between three different control algorithms. Home trials with people with limb differences will be carried out in planned future works.

Supplementary Materials: The following are available at https://www.mdpi.com/1424-8220/21/3/763/s1.

Author Contributions: H.W. produced the original draft and contributed to the system design, experiment design, data collection, and data analysis. M.D. contributed to the system and experiment design and manuscript review. K.N. conceptualised the study, designed the methodology, and contributed to the manuscript review and funding acquisition. All authors have read and agreed to the published version of the manuscript.

Funding: This work was funded by EPSRC grant EP/R004242/1.

Institutional Review Board Statement: Ethical approval was granted by the local ethics committee at Newcastle University (Ref: 17-NAZ-056).

Informed Consent Statement: The study was conducted in accordance with the Declaration of Helsinki. The subject gave his/her informed consent for inclusion before he/she participated in the study.

Data Availability Statement: The data that support the findings of this study are openly available in Edinburgh DataShare at https://doi.org/10.7488/ds/2977.

Conflicts of Interest: The authors declare no conflict of interest.

References

1. Zuo, K.J.; Olson, J.L. The evolution of functional hand replacement: From iron prostheses to hand transplantation. *Plast. Surg.* **2014**, *22*, 44–51. [CrossRef]
2. Chadwell, A.; Diment, L.; Micó-Amigo, M.; Ramírez, D.Z.M.; Dickinson, A.; Granat, M.; Kenney, L.; Kheng, S.; Sobuh, M.; Ssekitoleko, R.; et al. Technology for monitoring everyday prosthesis use: A systematic review. *J. Neuroeng. Rehabil.* **2020**, *17*, 1–26. [CrossRef] [PubMed]
3. Biddiss, E.A.; Chau, T.T. Upper limb prosthesis use and abandonment: A survey of the last 25 years. *Prosthet. Orthot. Int.* **2007**, *31*, 236–257. [CrossRef] [PubMed]
4. Espinosa, M.; Nathan-Roberts, D. Understanding prosthetic abandonment. *Proc. Hum. Factors Ergon. Soc. Annu. Meet.* **2019**, *63*, 1644–1648. [CrossRef]
5. Jiang, N.; Dosen, S.; Muller, K.R.; Farina, D. Myoelectric control of artificial limbs—Is there a need to change focus? [In the spotlight]. *IEEE Signal Process. Mag.* **2012**, *29*, 150–152. [CrossRef]
6. Young, A.J.; Hargrove, L.J.; Kuiken, T.A. The effects of electrode size and orientation on the sensitivity of myoelectric pattern recognition systems to electrode shift. *IEEE Trans. Biomed. Eng.* **2011**, *58*, 2537–2544. [CrossRef]
7. Scheme, E.; Englehart, K. Electromyogram pattern recognition for control of powered upper-limb prostheses: State of the art and challenges for clinical use. *J. Rehabil. Res. Dev.* **2011**, *48*, 643–659. [CrossRef]
8. Vujaklija, I.; Roche, A.D.; Hasenoehrl, T.; Sturma, A.; Amsuess, S.; Farina, D.; Aszmann, O.C. Translating research on myoelectric control into clinics-are the performance assessment methods adequate? *Front. Neurorobot.* **2017**, *11*, 7. [CrossRef]
9. Chadwell, A.; Kenney, L.; Thies, S.; Galpin, A.; Head, J. The reality of myoelectric prostheses: Understanding what makes these devices difficult for some users to control. *Front. Neurorobot.* **2016**, *10*, 7. [CrossRef]
10. Cipriani, C.; Sassu, R.; Controzzi, M.; Carrozza, M.C. Influence of the weight actions of the hand prosthesis on the performance of pattern recognition based myoelectric control: Preliminary study. In Proceedings of the 2011 Annual International Conference of the IEEE Engineering in Medicine and Biology Society, Boston, MA, USA, 30 August–3 September 2011; pp. 1620–1623.
11. Vidovic, M.M.C.; Hwang, H.J.; Amsüss, S.; Hahne, J.M.; Farina, D.; Müller, K.R. Improving the robustness of myoelectric pattern recognition for upper limb prostheses by covariate shift adaptation. *IEEE Trans. Neural Syst. Rehabil. Eng.* **2015**, *24*, 961–970. [CrossRef]
12. Hargrove, L.J.; Miller, L.A.; Turner, K.; Kuiken, T.A. Myoelectric pattern recognition outperforms direct control for transhumeral amputees with targeted muscle reinnervation: A randomized clinical trial. *Sci. Rep.* **2017**, *7*, 1–9. [CrossRef] [PubMed]
13. Dyson, M.; Dupan, S.; Jones, H.; Nazarpour, K. Learning, generalization, and scalability of abstract myoelectric Control. *IEEE Trans. Neural Syst. Rehabil. Eng.* **2020**, *28*, 1539–1547. [CrossRef] [PubMed]
14. Hahne, J.M.; Dähne, S.; Hwang, H.J.; Müller, K.R.; Parra, L.C. Concurrent adaptation of human and machine improves simultaneous and proportional myoelectric control. *IEEE Trans. Neural Syst. Rehabil. Eng.* **2015**, *23*, 618–627. [CrossRef] [PubMed]
15. Simon, A.M.; Lock, B.A.; Stubblefield, K.A. Patient training for functional use of pattern recognition–controlled prostheses. *J. Prosthet. Prthotics JPO* **2012**, *24*, 56. [CrossRef] [PubMed]
16. Brinton, M.; Barcikowski, E.; Davis, T.; Paskett, M.; George, J.A.; Clark, G. Portable take-home system enables proportional control and high-resolution data logging with a multi-degree-of-freedom bionic arm. *Front. Robot. AI* **2020**, *7*, 133. [CrossRef]
17. Cuberovic, I.; Gill, A.; Resnik, L.J.; Tyler, D.J.; Graczyk, E.L. Learning artificial sensation through long-term home use of a sensory-enabled prosthesis. *Front. Neurosci.* **2019**, *13*, 853. [CrossRef] [PubMed]
18. Resnik, L.; Acluche, F.; Borgia, M. The DEKA hand: A multifunction prosthetic terminal device-patterns of grip usage at home. *Prosthet. Orthot. Int.* **2018**, *42*, 446–454. [CrossRef]

19. Farina, D.; Jiang, N.; Rehbaum, H.; Holobar, A.; Graimann, B.; Dietl, H.; Aszmann, O.C. The extraction of neural information from the surface EMG for the control of upper-limb prostheses: Emerging avenues and challenges. *IEEE Trans. Neural Syst. Rehabil. Eng.* **2014**, *22*, 797–809. [CrossRef]
20. Graczyk, E.L.; Resnik, L.; Schiefer, M.A.; Schmitt, M.S.; Tyler, D.J. Home use of a neural-connected sensory prosthesis provides the functional and psychosocial experience of having a hand again. *Sci. Rep.* **2018**, *8*, 1–17. [CrossRef]
21. Hahne, J.M.; Schweisfurth, M.A.; Koppe, M.; Farina, D. Simultaneous control of multiple functions of bionic hand prostheses: Performance and robustness in end users. *Sci. Robot.* **2018**, *3*, 1–9. [CrossRef]
22. Sreenivasan, N.; Ulloa Gutierrez, D.F.; Bifulco, P.; Cesarelli, M.; Gunawardana, U.; Gargiulo, G.D. Towards ultra low-cost myoactivated prostheses. *BioMed Res. Int.* **2018**, *2018*, 9634184. [CrossRef] [PubMed]
23. George, J.A.; Davis, T.S.; Brinton, M.R.; Clark, G.A. Intuitive neuromyoelectric control of a dexterous bionic arm using a modified Kalman filter. *J. Neurosci. Methods* **2020**, *330*, 108462. [CrossRef] [PubMed]
24. Pai, U.J.; Sarath, N.; Sidharth, R.; Kumar, A.P.; Pramod, S.; Udupa, G. Design and manufacture of 3D printed myoelectric multi-fingered hand for prosthetic application. In Proceedings of the 2016 International Conference on Robotics and Automation for Humanitarian Applications (RAHA), Kollam, India, 18–20 December 2016; pp. 1–6.
25. Nguyen, N. Developing a Low-Cost Myoelectric Prosthetic Hand. Bachelor's Thesis, Metropolia University of Applied Sciences, Helsinki, Finland, May 2018.
26. Schorger, K.; Simon, J.P.; Clark, D.; Williams, A. Pneumatic Hand Prosthesis Project. Bachelor's Thesis, Cal Maritime, Vallejo, CA, USA, April 2018.
27. Curline-Wandl, S.A.; Azam Ali, M. Single channel myoelectric control of a 3D printed transradial prosthesis. *Cogent Eng.* **2016**, *3*, 1245541. [CrossRef]
28. Ariyanto, M.; Ismail, R.; Setiawan, J.D.; Yuandi, E.P. Anthropomorphic transradial myoelectric hand using tendon-spring mechanism. *Telkomnika* **2019**, *17*, 537–548. [CrossRef]
29. Canizares, A.; Pazos, J.; Benítez, D. On the use of 3D printing technology towards the development of a low-cost robotic prosthetic arm. In Proceedings of the 2017 IEEE International Autumn Meeting on Power, Electronics and Computing (ROPEC), Ixtapa, Mexico, 8–10 November 2017; pp. 1–6.
30. Gevins, A.S.; Durousseau, D.; Libove, J. Dry Electrode Brain Wave Recording System. U.S. Patent 4,967,038, 30 October 1990.
31. Forvi, E.; Bedoni, M.; Carabalona, R.; Soncini, M.; Mazzoleni, P.; Rizzo, F.; O'Mahony, C.; Morasso, C.; Cassarà, D.G.; Gramatica, F. Preliminary technological assessment of microneedles-based dry electrodes for biopotential monitoring in clinical examinations. *Sens. Actuators A Phys.* **2012**, *180*, 177–186. [CrossRef]
32. Dyson, M.; Barnes, J.; Nazarpour, K. Myoelectric control with abstract decoders. *J. Neural Eng.* **2018**, *15*, 056003. [CrossRef]
33. Pistohl, T.; Cipriani, C.; Jackson, A.; Nazarpour, K. Abstract and proportional myoelectric control for multi-fingered hand prostheses. *Ann. Biomed. Eng.* **2013**, *41*, 2687–2698. [CrossRef]
34. Fang, Y.; Zhou, D.; Li, K.; Liu, H. Interface prostheses with classifier-feedback-based user training. *IEEE Trans. Biomed. Eng.* **2016**, *64*, 2575–2583.
35. Moritz, C.T.; Perlmutter, S.I.; Fetz, E.E. Direct control of paralysed muscles by cortical neurons. *Nature* **2008**, *456*, 639–642. [CrossRef]
36. Krasoulis, A.; Vijayakumar, S.; Nazarpour, K. Multi-Grip Classification-Based Prosthesis Control with Two EMG-IMU Sensors. *IEEE Trans. Neural Syst. Rehabil. Eng.* **2020**, *28*, 508–518. [CrossRef]
37. Krasoulis, A.; Nazarpour, K. Myoelectric digit action decoding with multi-label, multi-class classification: An offline analysis. *Sci. Rep.* **2020**, *10*, 16872. [CrossRef] [PubMed]
38. Phinyomark, A.; Quaine, F.; Charbonnier, S.; Serviere, C.; Tarpin-Bernard, F.; Laurillau, Y. EMG feature evaluation for improving myoelectric pattern recognition robustness. *Expert Syst. Appl.* **2013**, *40*, 4832–4840. [CrossRef]
39. Wolpert, D.M.; Kawato, M. Multiple paired forward and inverse models for motor control. *Neural Netw.* **1998**, *11*, 1317–1329. [CrossRef]
40. Liu, X.; Mosier, K.M.; Mussa-Ivaldi, F.A.; Casadio, M.; Scheidt, R.A. Reorganization of finger coordination patterns during adaptation to rotation and scaling of a newly learned sensorimotor transformation. *J. Neurophysiol.* **2011**, *105*, 454–473. [CrossRef] [PubMed]
41. Barnes, J.; Dyson, M.; Nazarpour, K. Comparison of hand and forearm muscle pairs in controlling of a novel myoelectric interface. In Proceedings of the 2016 IEEE International Conference on Systems, Man, and Cybernetics (SMC), Budapest, Hungary, 9–12 October 2016; p. 002846.
42. Farrell, T. Analysis window induced controller delay for multifunctional prostheses. In Proceedings of the Myoelectric Controls Symp, Fredericton, NB, Canada, 13–15 August 2008; pp. 225–228.
43. Farrell, T.R.; Weir, R.F. The optimal controller delay for myoelectric prostheses. *IEEE Trans. Neural Syst. Rehabil. Eng.* **2007**, *15*, 111–118. [CrossRef]
44. Portnova-Fahreeva, A.A.; Rizzoglio, F.; Nisky, I.; Casadio, M.; Mussa-Ivaldi, F.A.; Rombokas, E. Linear and Non-linear Dimensionality-Reduction Techniques on Full Hand Kinematics. *Front. Bioeng. Biotechnol.* **2020**, *8*, 429. [CrossRef]
45. Pylatiuk, C.; Schulz, S.; Döderlein, L. Results of an Internet survey of myoelectric prosthetic hand users. *Prosthet. Orthot. Int.* **2007**, *31*, 362–370. [CrossRef]

46. Dyson, M.; Nazarpour, K. Data Driven Spatial Filtering Can Enhance Abstract Myoelectric Control in Amputees. In Proceedings of the 2018 40th Annual International Conference of the IEEE Engineering in Medicine and Biology Society (EMBC), Honolulu, HI, USA, 18–21 July 2018; pp. 3770–3773.
47. Hahne, J.M.; Wilke, M.A.; Koppe, M.; Farina, D.; Schilling, A.F. Longitudinal case study of regression-based hand prosthesis control in daily life. *Front. Neurosci.* **2020**, *14*, 600. [CrossRef]
48. Mussa-Ivaldi, F.A.; Casadio, M.; Danziger, Z.C.; Mosier, K.M.; Scheidt, R.A. Sensory motor remapping of space in human–machine interfaces. In *Progress in Brain Research*; Elsevier: Amsterdam, The Netherlands, 2011; Volume 191, pp. 45–64.
49. Powell, M.A.; Thakor, N.V. A training strategy for learning pattern recognition control for myoelectric prostheses. *J. Prosthet. Orthot. JPO* **2013**, *25*, 30. [CrossRef]
50. Resnik, L.J.; Acluche, F.; Lieberman Klinger, S. User experience of controlling the DEKA Arm with EMG pattern recognition. *PLoS ONE* **2018**, *13*, e0203987. [CrossRef]
51. Domingo, M.C. An overview of the Internet of Things for people with disabilities. *J. Netw. Comput. Appl.* **2012**, *35*, 584–596. [CrossRef]
52. Sethi, P.; Sarangi, S.R. Internet of things: Architectures, protocols, and applications. *J. Electr. Comput. Eng.* **2017**, *2017*, 9324035. [CrossRef]
53. Baker, S.B.; Xiang, W.; Atkinson, I. Internet of things for smart healthcare: Technologies, challenges, and opportunities. *IEEE Access* **2017**, *5*, 26521–26544. [CrossRef]

sensors

MDPI

Article

Flexible Recruitments of Fundamental Muscle Synergies in the Trunk and Lower Limbs for Highly Variable Movements and Postures

Hiroki Saito [1,2], Hikaru Yokoyama [1], Atsushi Sasaki [1,3], Tatsuya Kato [1,3] and Kimitaka Nakazawa [1,*]

1 Department of Life Sciences, Graduate School of Arts and Sciences, The University of Tokyo, 3-8-1 Komaba, Meguro, Tokyo 153-8902, Japan; saitohhrk@stf.teu.ac.jp (H.S.); yokoyama@idaten.c.u-tokyo.ac.jp (H.Y.); atsushi-sasaki@g.ecc.u-tokyo.ac.jp (A.S.); t-kato@g.ecc.u-tokyo.ac.jp (T.K.)
2 Department of Physical Therapy, Tokyo University of Technology, Ota, Tokyo 144-8535, Japan
3 Japan Society for the Promotion of Science, 5-3-1 Kojimachi, Chiyoda, Tokyo 102-0083, Japan
* Correspondence: nakazawa@idaten.c.u-tokyo.ac.jp; Tel.: +81-3-5454-4317

Abstract: The extent to which muscle synergies represent the neural control of human behavior remains unknown. Here, we tested whether certain sets of muscle synergies that are fundamentally necessary across behaviors exist. We measured the electromyographic activities of 26 muscles, including bilateral trunk and lower limb muscles, during 24 locomotion, dynamic and static stability tasks, and we extracted the muscle synergies using non-negative matrix factorization. Our results show that 13 muscle synergies that may have unique functional roles accounted for almost all 24 tasks by combinations of single and/or merging of synergies. Therefore, our results may support the notion of the low dimensionality in motor outputs, in which the central nervous system flexibly recruits fundamental muscle synergies to execute diverse human behaviors. Further studies are required to validate the neural representation of the fundamental components of muscle synergies.

Keywords: muscle synergies; movements; postures; the central nervous system; motor control; the neural control

Citation: Saito, H.; Yokoyama, H.; Sasaki, A.; Kato, T.; Nakazawa, K. Flexible Recruitments of Fundamental Muscle Synergies in the Trunk and Lower Limbs for Highly Variable Movements and Postures. *Sensors* **2021**, *21*, 6186. https://doi.org/10.3390/s21186186

Academic Editor: Ernest N. Kamavuako

Received: 21 July 2021
Accepted: 13 September 2021
Published: 15 September 2021

1. Introduction

To execute human movements, the central nervous system (CNS) must control many degrees of freedom from thousands of motor units within hundreds of skeletal muscles [1]. Several studies that applied factorization algorithms to complex muscle activity have identified limited sets of motor modules called muscle synergies [2]. Although the question of whether muscle synergies resulting from matrix factorization represent a neural origin or are merely a numerical artifact is still being debated [3,4], muscle synergy theory assumes that the CNS combines a few activation sets to build muscle activation commands, which simplify the control of movement [5]. Evidence of a limited set of muscle synergies has been found in various human motor behaviors, such as locomotion [6–9], reaching tasks [10,11], and sports activities [12–14].

It has been proposed that muscle synergies are shared across various motor tasks [5,10,15]. Shared synergies facilitate the robustness of the neuromuscular system, which is thought to be beneficial for stable postural control [15,16], development [17], and expert motor skills [18]. In contrast, studies have also discovered the existence of task-specific synergies to meet each biomechanical demand of the motor tasks [19,20]. An experimental study in frogs investigated muscle synergies during natural behaviors such as walking, jumping, and swimming, indicating that each motor behavior is the consequence of a combination of both synergies shared between behaviors and synergies specific to each or a few behaviors [21].

While the low dimensionality and robustness of muscle synergy models account for the advantages of redundant control of movement, the flexibility of combining muscle synergies to adapt to various mechanical demands is also considered necessary to achieve efficient movement execution [18,22,23]. A study found that during isometric force generation, six global muscle synergies extracted from pooled EMG datasets with various joint angles, directions, and force intensities were selectively recruited and changed during peak activation patterns depending on different conditions [22]. In this study, several forms of the merging of global synergies were identified, which may contribute to stabilizing force generation [22]. Furthermore, other studies found that all three muscle synergies for cycling can be well reconstructed by merging muscle synergies extracted from walking with the best reconstruction value found for higher speeds [23] and that specific merging patterns correlate with an increase in running efficiency, while other merging synergies correlate with decreased running efficiency [18]. Thus, the interpretation of the existing literature suggests that the CNS may flexibly select appropriate subsets of muscle synergies, either independently or merged, from a large set that is established to execute motor tasks [10,23,24]. Although a substantial number of in-born and learned human movements and postures share some common control variable of importance even if the tasks have some distinct biomechanical features [25], muscle synergies extracted from a diverse range of behaviors have not been comprehensively investigated.

The aim of the present study was to test the hypothesis that different subsets of muscle synergies are used in a variety of movement and postural tasks. To test this hypothesis, we first extracted muscle synergies from an EMG recording dataset made from 24 motor tasks based on the assumption that these muscle synergies represent the basic units for explaining 24 motor tasks that share common neuromechanical requirements such as supporting the body and maintaining balance statically, dynamically, and during locomotion [25,26]. We then examined whether combinations of these synergies explain each motor task.

2. Materials and Methods

2.1. Experimental Protocol

Ten healthy volunteers (aged 21–35 years, all men) participated in the study. Each participant provided written informed consent for participation in the study. The study was conducted in accordance with the Declaration of Helsinki and was approved by the local ethics committee of the University of Tokyo.

We focused on fundamental movement and postural tasks that serve as the building blocks for efficient and effective execution of a variety of daily-living activities and high-skilled performances, such as sports [25,27,28]. Specifically, we used tasks that required movements through space (locomotion) and controls against gravity (stability) in any plane [25]. Thus, all participants were asked to perform the 24 tasks described in Table 1.

Table S1 presents the details of each movement and postural task. The order of tasks was randomly assigned.

2.2. Data Collection

EMG activity was recorded from the following 26 muscles distributed across the trunk and lower limbs (13 bilateral muscles): tibialis anterior (TA), gastrocnemius medialis (MG), vastus medialis (VM), rectus femoris (RF), biceps femoris (long head, BF), gluteus maximus (GM), gluteus medius (Gmed), rectus abdominis (RA), oblique externus (OE), erector spinae at L2 (ESL2), erector spinae at Th9 (EST9), erector spinae at Th1 (EST1), and latissimus dorsi (LD). EMG activity was recorded using a wireless EMG system (Trigno Wireless System; DELSYS, Boston, MA, USA). The EMG signals were bandpass filtered (20–450 Hz), amplified (with a 300-gain preamplifier), and sampled at 1000 Hz. Three-dimensional ground reaction force (GRF) data were recorded at 1000 Hz from the force plates under each belt of the treadmill. GRF data were used to define each trial/stride data point in the walking and running tasks (i.e., the period between the initial right foot contact and the next right foot contact). We also used an electrical trigger to define each trial of the other tasks as the period between the start and end of the movement. The starting point was initiated with the verbal cue 'go' with the electrical trigger manually pressed once by the examiner. After participants completed the tasks and returned to a resting posture for about 1 s, the end of the movement was indicated by a 1-second verbal cue 'end' with the electrical trigger pressed twice.

Table 1. Movement and postural tasks.

		1	Walk (1.5 m/s)
		2	Run (2.7 m/s)
		3	Bilateral jump
Locomotion		4	Rt single leg jump
		5	Lt single leg jump
		6	Sit to stand to sit
		7	Rt single leg stance
	Static Postures	8	Lt single leg stance
		9	Deep squat
		10	Rt single leg squat
		11	Lt single leg squat
		12	Rt lunge
	Dynamic Postures	13	Lt lunge
		14	Rocking backward
		15	Rocking forward
Stability		16	Rt cross extension
		17	Lt cross extension
		18	Cat-and-dog
		19	Forward bend
		20	Rt side bend
		21	Lt side bend
	Axial	22	Backward bend
		23	Rt rotation
		24	Lt rotation

Shown are the order of 24 locomotion and stability tasks. Stability tasks are divided into three subcategories: static postures, dynamic postures and axial. Rt: right; Lt: left.

2.3. EMG Processing

The low-pass cut-off frequency influences the smoothing of EMG patterns and thus impacts the number of extracted modules [29]. To adequately compare EMG envelopes (i.e., EMG patterns with the same smoothing) of movements performed for various tasks that had different features of dynamic activities, the low-pass cut-off frequency had to be adjusted for each task. Thus, an iterative adaptive algorithm was used to extract the optimal EMG envelopes [30]. This algorithm utilized information theory to find a sample-by-sample optimal root-mean-square window for envelope estimation [30]. The algorithm allowed the filter to adequately follow fast changes in EMG activity while maintaining optimal extraction when the EMG amplitude was changing slowly [30]. A previous study used this algorithm and successfully reconstructed muscle synergies during walking in individuals with and without transfemoral amputation [31]. The smoothed EMG envelopes were time-interpolated to generate 200 time points for each trial, except for the right and left single-leg stance tasks, where the period of 15 s was time-interpolated to generate 1400 timepoints.

We created the following two types of EMG matrices for each subject to examine the repertoire of fundamental muscle synergies that could explain all datasets and how these synergies are combined in each task. Similar to previous studies [10,22,32], we pooled the EMG matrices of all 24 tasks to create an "all-task" EMG matrix for each subject (i.e., the matrix was composed of the 26 muscles × the summation of time points of the 24 single-task EMG matrices) to extract fundamental muscle synergies across all tasks. We also created a "single-task" EMG matrix composed of the 26 muscles × 1400 timepoints (seven trials × 200 timepoints for each task other than the right and left single-leg stance tasks) for each of the 24 tasks to extract muscle synergies.

2.4. Muscle Synergy Analysis

In our analysis, we first identified the muscle synergies of each task for each subject using a factorization algorithm of single-task EMG matrices, and then synergies of all tasks were extracted from all-task EMG matrices using the same algorithm. We then proceeded to characterize representative muscle synergies of individual tasks and all tasks across all participants using a hierarchical clustering algorithm. Lastly, we analyzed the similarity between synergy cluster centroids of each individual task and single or merged synergies of the all-task matrix to investigate how muscle synergies utilized by all tasks contribute to the execution of each individual movement.

To explore muscle synergies, nonnegative matrix factorization (NMF) was used for each subject from the single-task EMG matrices and the all-task EMG matrix. NMF has previously been described as a linear decomposition technique [33,34] according to Equation (1):

$$M = W{\cdot}C + e \tag{1}$$

where M ($m \times t$ matrix, where m is the number of muscles and t is the number of samples, i.e., the spatiotemporal profiles of muscle activity) is a linear combination of the following muscle weighting components: W ($m \times n$ matrix, where n is the number of muscle synergies) and C ($n \times t$ matrix, representing temporal pattern components). e is the residual error matrix. Each EMG vector in the matrix corresponding to each muscle activity was normalized to the maximum amplitude across all tasks so that all muscle scales ranged from 0 to 1. Prior to extracting muscle synergies, each muscle vector in the data matrix was standardized to have unit variance, thus ensuring that the activity in all muscles was equally weighted. However, after each synergy extraction, the unit variance scaling was removed from the data so that each muscle variable ranged from 0 to 1 for data inspection and interpretation [35]. To determine the number of muscle synergies, NMF was applied to extract each possible n from 1 to 26 from each dataset. The variance accounted for (VAF) by the reconstructed EMG (M) was calculated at each iteration to extract the optimal number of muscle synergies. VAF was defined as a 100 × square of the uncentered Pearson's correlation coefficient [36,37]. To prevent the extracted synergies from assuming a suboptimal local minimum, each synergy extraction was repeated 100 times. Thus, the iteration with the highest VAF was maintained [8]. We defined the optimal number n as the number fulfilling the following two criteria: First, n was selected as the smallest number of modules that accounted for >90% of the VAF [36]. Second, n was the smallest number to which adding another module did not increase VAF by >5% [38].

2.5. Clustering the Modules across Participants

We identified the representative synergy vectors across participants using hierarchical clustering analysis (Ward's method, Euclidian distance) of muscle synergies for each task and all tasks [8,39]. The optimal number of clusters was determined using the gap statistic [40]. Subsequently, the muscle synergies in each cluster were averaged across participants. To assess variability, we calculated the degree of similarity within each cluster to measure the consistency of each extracted cluster [10]. Here, we assessed similarity by averaging the values of all the pairwise scalar products (SPs) between the muscle synergies of each cluster.

2.6. Contributions of the Muscle Synergy of All Tasks to the Execution of Each Task

To explore whether the muscle synergy defined by the all-task matrix contributes to executing each task of movements and postures, the similarity between muscle synergies of single-task and all-task matrices was quantified by using the SP between these centroids of the synergy clusters (normalized to unit vectors). For every comparison, each of the synergy cluster centroids of all-task was matched to a synergy cluster centroid of each task by maximizing the total scalar product values. Synergy clusters that could not be matched with SP ≥ 0.75, were classified as unmatched [41].

2.7. Contributions of Merging Muscle Synergy of All Tasks towards Single-Task Execution

We also expected that all-task muscle synergies can be merged to execute each single task of movement and posture [40]. Thus, the merged synergies as a linear combination of the contributing synergies were modeled by the following formula [18,41]:

$$W_k \approx \sum_{i=0}^{N_b} D_i W_i,\ k = 1, \ldots,\ n_b \tag{2}$$

where W_k is the kth muscle synergy vector from each individual task, W_i is the ith muscle synergy vector derived from an all-task matrix, N_b is the number of synergies that contribute to the merging, and D_i is a non-negative coefficient that scales the ith synergy in the merging. D_i was obtained from a non-negative least-squares fit, implemented using MATLAB (function lsqnonneg). W_k and W_i were normalized as unit vectors. Following criteria from previous studies [18,41], the synergy merging was identified when $N_b \geq 2$, $D_i \geq 0.2$ for all i, and the SP between $\sum_{i=0}^{N_b} D_i W_i$ and W_k was $\geq$0.75.

To assess whether the synergies from each task can be explained as merging of synergies from all tasks, we first identified the synergy cluster centroids of single-task and synergies of the all-task (described above) and reconstructed each synergy cluster centroid of each individual task by merging every possible combination of the synergy cluster centroids of all tasks.

3. Results

3.1. Muscle Synergies Extracted from All-Task EMG Matrices

All of the muscle activity was accounted for by 10.6 ± 1.58 muscle synergies of all-task EMG matrices, and the mean VAF was 0.91 ± 0.001 in ten participants. Figure 1 presents 13 muscle synergies of an all-task matrix incorporating 24 trunk and lower limb movement tasks (W1 to W13), which were grouped by cluster analysis across participants, Table 2 presents the degree of similarity within each synergy cluster of all tasks, and Table 3 summarizes the characteristics of the muscle synergies.

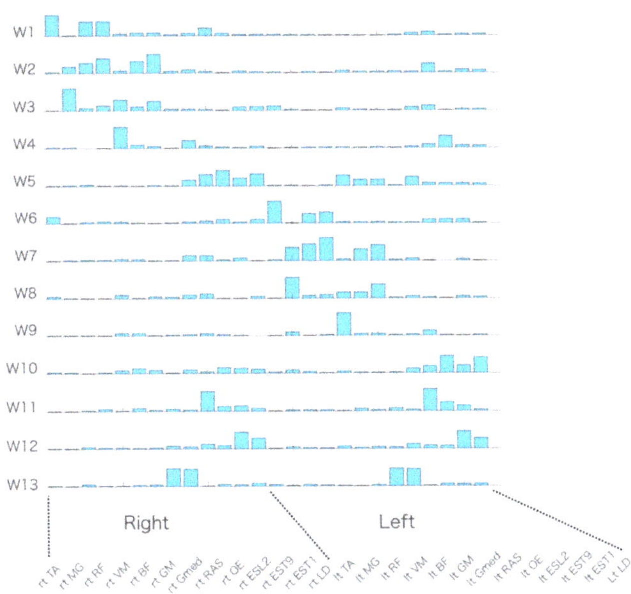

Figure 1. Muscle synergies of all tasks. Centroids of the hierarchical clustering performed on the muscle synergies of all tasks across ten participants.

Table 2. The results for the degree of similarity within each synergy cluster of all tasks.

	Similarity (Normalized)	The Number of Samples
W1	0.77 (±0.13)	8
W2	0.70 (±0.18)	8
W3	0.74 (±0.12)	8
W4	0.76 (±0.14)	7
W5	0.62 (±0.16)	13
W6	0.77 (±0.12)	10
W7	0.81 (±0.08)	6
W8	0.74 (±0.13)	7
W9	0.84 (±0.08)	4
W10	0.58 (±0.17)	8
W11	0.86 (±0.08)	5
W12	0.68 (±0.15)	11
W13	0.82 (±0.17)	11

Table 3. Characteristics of muscle synergy clusters of all tasks.

Unilateral Patterns		Major Muscles	Minor Muscles
Right Patterns	Left Patterns		
W1	W6	ispTA, ispRF, ispVM	(ispESL2, ispEST9, ispEST1, conTA, conESL2, conEST9, conLD)
W2	W7	ispVM, ispRF, ispGM, ispGmed	(ispMG, ispOE, conBF, conOE, conESL2)
W3	W8	ispMG, ispGmed	(ispRF, ispVM, ispBF, ispGM, ispEST1, ispLD, conTA, conBF, contOE, conESL2)
W4	W9	ispBF	(ispMG, ispGM, ispOE, ispESL2, conESL2, conEST9)
W5	W10	ispEST9, ispLD	(ispOE, ispESL2, ispEST1, conBF, conGM, conGmed, conOE, conESL2, conEST9, conEST1, conLD)
Bilateral patterns			
M11		bilESL2	(bilEST9, bilEST1)
M12		bilEST1	(bilLD)
M13		bilRAS, bilOE	

The following pairs showed high similarity when the muscles in W6 to W10 were reordered so that muscles on the left side of W6 to W10 corresponded to the same muscles on the right side of W1 to W5: W1 and W6 (SP = 0.93), W2 and W7 (SP = 0.97), W3 and W8 (SP = 0.97), W4 and W9 (SP = 0.85), W5, and W10 (SP = 0.93). We categorized W1 to W5 as muscle synergies with right-side dominant patterns and W6 to W10 as muscle synergies with left-side dominant patterns. W11, W12, and W13 were categorized as bilateral patterns. Muscles that account for >0.5 of activation levels are classified as major muscles and between 0.1 to 0.5 were as minor muscles. isp: ipsilateral, con: contralateral, bil: bilateral.

Visual inspection revealed that muscle synergies W1 to W5 were largely composed of the right-side muscles, while muscle synergies W6 to W10 were mainly composed of the left-side muscles. Thus, we categorized W1 to W5 as muscle synergies with right-side dominant patterns and W6 to W10 as muscle synergies with left-side dominant patterns. The following pairs showed high similarity when the muscles in W6 to W10 were reordered so that muscles on the left side of W6 to W10 corresponded to the same muscles on the right side of W1 to W5: W1 and W6 (SP = 0.93), W2 and W7 (SP = 0.97), W3 and W8 (SP = 0.97), W4 and W9 (SP = 0.85), W5, and W10 (SP = 0.93). Others, such as W11, W12, and W13, were categorized as bilateral patterns.

3.2. Relationship between Muscle Synergies Extracted from All-Task EMG Matrices and Those Extracted from Single-Task Matrices

For each task, two to four synergy clusters were identified using cluster analysis. Variability in the number of muscle synergies and VAF in subjects, and the degree of similarity within each synergy cluster of each task are presented in Table S2. Table 4 presents the number of muscle synergies in each task, which were well explained (SP > 0.75) by independent and merged muscle synergies from the all-task EMG matrices.

Of note, all synergies of each task except the one for the left single-leg stance could be explained by either single or linear combination of multiple synergies from the all-task EMG matrices (SP > 0.75). The details of the contributions of muscle synergies of all tasks to each task execution are presented in Table S3.

Figures 2 and 3 present examples of relationships between muscle synergies from the all-task EMG matrices and those from the single-task EMG matrices: locomotion tasks including walking, running, bilateral jump and sit-to-stand-to-sit (Figure 2), and stability tasks including left lunge, cat-and-dog, forward bend, and left rotation (Figure 3).

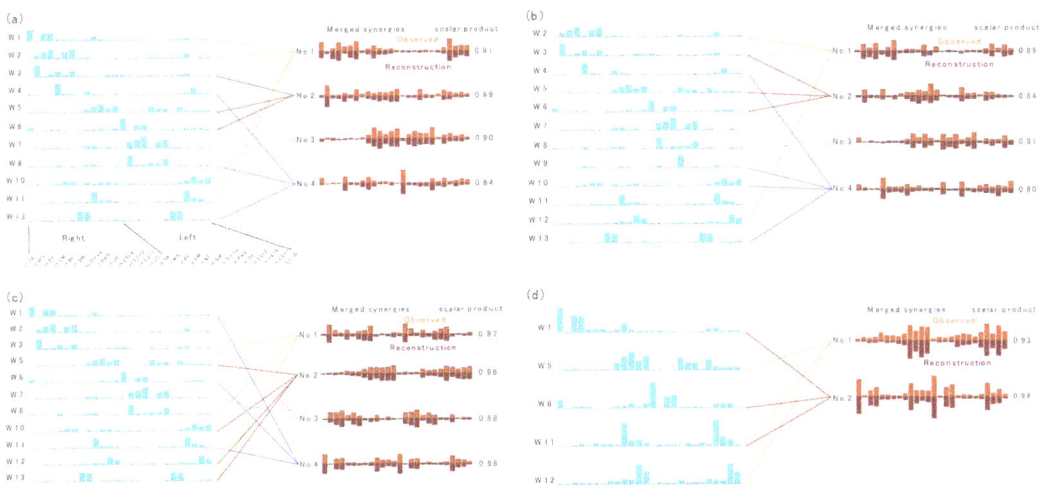

Figure 2. The relationship between muscle synergies of all tasks and muscle synergies of locomotion tasks including (**a**) walk, (**b**) run, (**c**) bilateral jump and (**d**) sit-to-stand-to-sit. The figures show the synergy cluster centroids of these tasks that could be explained by either a single or linearly combined multiple synergy cluster centroids of all tasks (synergies in blue) matched by maximizing scalar product > 0.75. Observed muscle synergies extracted from the single-task EMG (orange) and their reconstructions by merging their respective W1-combinations (dark orange) are further presented.

The relationships between muscle synergies from the all-task EMG matrices and those from the other single-task EMG matrices are shown in Figures S1 and S2.

Table 4. The relationship between synergy clusters of each task and synergy clusters of all tasks.

Movement and Postural Tasks	Walk	Run	BilJP SLS	RtSJP	Lt SJP	STS	Rt SLS	Lt SLS	DS	Rt SS	Lt SS	Rt LG	Lt LG	RB	RF	Rt CE	Lt CE	CD	FB	Rt SB	Lt SB	BB	Rt RT	Lt RT
Number of total synergy clusters	4	4	4	3	3	2	2	2	2	2	2	2	2	2	2	2	2	3	2	2	2	2	2	2
Number of synergy clusters that are well matched by a single synergy cluster of all tasks	0	0	0	0	0	0	0	0	0	0	0	0	0	1	1	1	0	1	0	0	0	1	1	1
Number of synergy clusters that are well matched by merging synergy clusters of all tasks	4	4	4	3	3	2	2	1	2	2	2	2	2	1	1	1	2	2	2	2	2	1	1	1
Number of synergy clusters that are unmatched by synergies of all tasks	0	0	0	0	0	0	0	1	0	0	0	0	0	0	0	0	0	0	0	0	0	0	0	0

The number of synergy clusters for each task and the number of a single or merged synergy cluster centroids of all tasks that were well matched (scalar product > 0.75) or unmatched to a synergy cluster centroid of each task. Bil: bilateral; Rt: right; Lt left; JP: jump; SJP: single leg jump; STS: sit-to-stand-to-sit; SLS: single leg stance; DS: deep squat; SS: single leg squat; LG: lunge; RB: rocking backward; RF: rocking forward; CE: cross extension; CD: cat-and-dog; FB; forward bend; SB: side bend; BB: backward bend; RT: rotation.

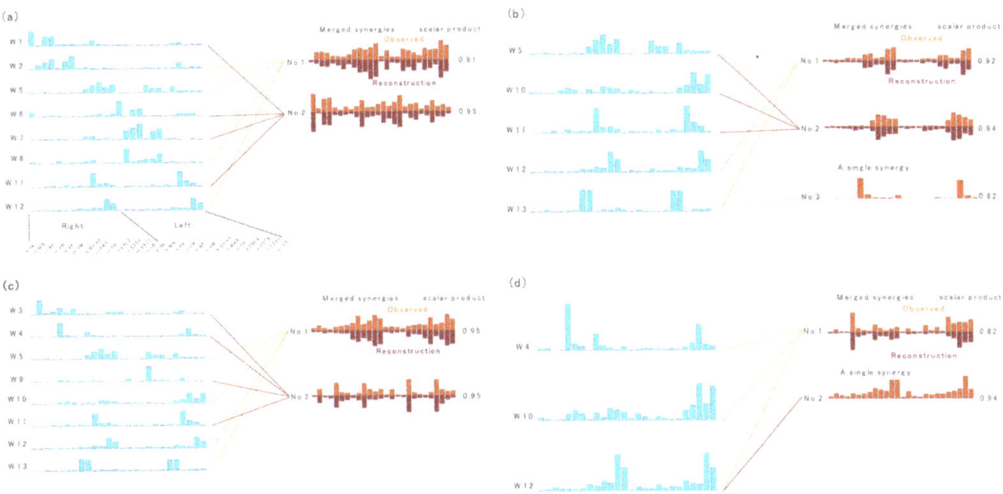

Figure 3. The relationship between muscle synergies of all tasks and muscle synergies of stability tasks including (**a**) left lunge, (**b**) cat-and-dog, (**c**) forward bend and (**d**) left rotation. The figures show the synergy cluster centroids of these task that could be explained by either a single or linearly combined multiple synergy cluster centroids of all tasks (synergies in blue) matched by maximizing scalar product > 0.75. Observed muscle synergies extracted from the single-task EMG (orange) and their reconstructions by merging their respective W1-combinations (dark orange) were further presented.

4. Discussion

Several studies have investigated shared or merged muscle synergies across different tasks, such as walking and running [8,42], walking and cycling [23], various directions of reaching [10,11] and stepping and non-stepping postural controls [35]. Their results indicated that different human behaviors use the fundamental motor modules that reflect the functional control units as a neural constraint on motor outputs. However, the extent to which muscle synergies represent the control of diverse human behaviors has not been comprehensively investigated in previous studies. In our study, we extracted muscle synergies from a large set of EMG (26 muscles) activities across bilateral locations of the trunk and lower limbs during 24 locomotion and stability tasks that were fundamental for a variety of physical activities. We found that 13 clusters of fundamental muscle synergies accounted for almost all synergy clusters of each of the 24 tasks. When we compared the synergy clusters extracted from individual tasks across participants, we found a high similarity (SP > 0.75) of a single or multiple linear combinations from the 13 fundamental muscle synergy clusters extracted from all tasks across participants. Although the question of whether muscle synergies derived from factorization algorithms are of neural origin is still controversial, some recent empirical studies using methods such as probing into neuroanatomical substrates [43] and stimulating the CNS [44] have provide direct evidence that muscle synergies observed in motor behaviors may have cortical and subcortical neural underpinnings [45]. In the following sections, we discuss the possible neural mechanism underlying a diverse set of human behaviors based on the assumptions that muscle synergies represent motor modules to coordinate patterns utilized by the CNS [45].

4.1. Characteristics of Muscle Synergies across 24 Tasks

We applied cluster analysis to the muscle synergies from the all-task EMG matrix across participants and identified 13 synergy clusters. As shown in Table 2, we broadly categorized muscle synergies into three sets based on the major contributions of the muscles (i.e., right muscle patterns, left muscle patterns, and bilateral muscle patterns). In the right and left muscle patterns, W1 and W6 were dominated by muscles around the ankle and knee joints (i.e., TA, RF, and VM). W2 and W7 were mainly composed of muscles related to the knee and hip joints (i.e., RF, VM, Gmed, and GM), and W3 and W8 employed the ankle and hip joints (i.e., MG and Gmed). Furthermore, BF mainly contributed to W4 and W9. While all four pairs were predominantly composed of extensor muscles that can move and stabilize the body during locomotion and postural tasks, they may have a distinct functional feature because the different tasks require different combinations of muscle synergies (Table S2). In contrast, the pairs of W5 and W10, W11, W12, and W13 were composed of back muscles (i.e., ES, LD) and abdominal muscles (i.e., RAS and OE) either in unilateral or bilateral patterns (Table 2). Notably, they were widely observed across 24 tasks (Table S2) and may be used for bilateral trunk movements or stabilization of the body accompanied by W1 to W10 with relatively low levels of trunk muscle activities when the lower limbs are moving [46]. Although we still do not know how muscle synergies in our study arise and whether they reflect neural structure for motor outputs, 13 muscle synergies extracted from our study may form a repertoire of whole lower limb and trunk muscle activation patterns, which can be shaped by biomechanical interactions and constrain the environment through a lifetime [18,47].

4.2. Hypothetical Neural Mechanisms Underlying Muscle-Synergy Controlling Diverse Behavior

If we assume that the muscle synergy extracted from the whole-task EMG matrices in our data may have a unique set of networks in which each synergy provides functionally necessary compositions in muscle activities, then one can expect that any combinations of these synergies may provide stable and predictable motor outputs in a diverse range of human behaviors [47]. The strength of our finding is that it indicates that there is a set of fundamental muscle synergies with different combinations of these synergies in single and/or merging states to produce 24 locomotion and stability tasks. Here, we hypothesize the existence of neural mechanisms underlying the flexible recruitment of muscle synergies in various combinations to meet the mechanical demands for each movement and posture. Figure S3 presents the functional contributions of fundamental muscle synergies for walking and running tasks revealed by merging the coefficient of synergy vectors and temporal activations of each task. For example, in walking (a), the synergy cluster centroid of No. 1 was recruited during the early stance phase of the right leg. It has been suggested that W1 with the right ankle and knee stabilizers and W2 with the right knee and hip stabilizers are merged for impact deceleration and body support [48]. Likewise, the synergy cluster centroid of No. 2 was recruited during the late stance phase of the right leg, in which W3 with the right ankle extensor and hip stabilizer largely contributed to the progression of the body [48]. Although the lack of kinematic data that define the movement phase precluded assessment of functional contributions of fundamental muscle synergies for other tasks, we show that temporal activations of fundamental muscle synergies were well reconstructed in all tasks (Table S3).

Interestingly, we found that muscle synergies in 24 locomotion and stability tasks were predominantly reconstructed by merging various combinations of fundamental muscle synergies (Table 4). A study reported that muscle synergies of cycling can result from merging synergies of walking [23]. Another recent study showed the merging of original muscle synergies during running through running training [18]. It is suggested that merged synergies were the result of the co-recruitment of multiple muscle synergies by neural networks driving the muscle synergies represented as C in Equation (1) [45,47]. Based on previous studies, we speculate that the upstream driving layer (e.g., C_{task} in Figure 4) may flexibly recruit the fundamental muscle synergies (e.g., W′ in Figure 4) located at different levels from the driving layers in the motor hierarchy to execute highly variable tasks (the schematic structure in Figure 4).

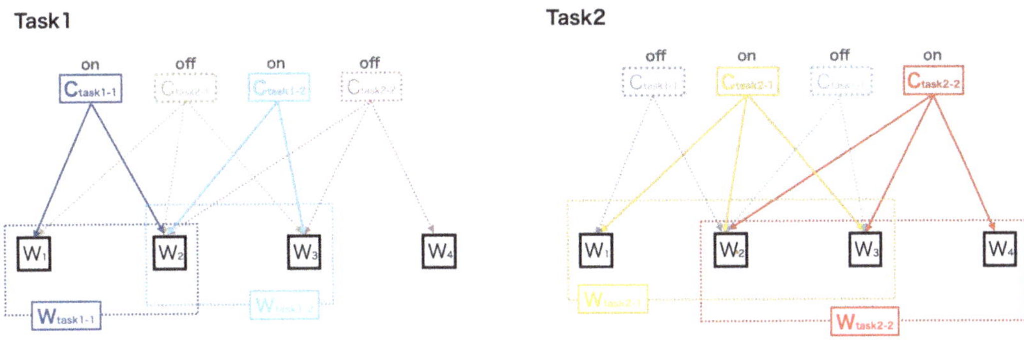

Figure 4. A hypothetical neural mechanism of merged fundamental muscle synergies with its temporal patterns in a diverse range of human behaviors. This model shows that the CNS flexibly recruits multiple synergies for different tasks. For example, $C_{task1-1}$ with W_1 and W_2 and $C_{task1-2}$ with W_2 and W_3 turn on while $C_{task2-1}$ and $C_{task2-2}$ turn off to execute task 1. Similarly, $C_{task2-1}$ with W_1, W_2, and W_3 and $C_{task2-2}$ with W_2, W_3, and W_4 turn on, ceasing to be active in $C_{task1-1}$ and $C_{task1-2}$ for task 2.

Our hypothesis is possibly equivalent to a generalized two-level CPG model for the control of locomotor muscle activity [49]. The model consists of two distinct neural network layers: (1) a pattern formation (PF) network layer that defines groups of synergistic and antagonistic motoneuron pools and (2) a rhythm generation layer that controls the activity of PF networks. However, it should be noted that the exact neural substrates encoding muscle synergies and their driving networks in humans remain largely unknown.

Since we propose that upstream driver C presents synchronous recruitments of the fundamental muscle synergies that have distinct functional roles in organizing muscle synergies for the 24 locomotion and stability tasks, it is possible that the CNS may also coordinate other simple or complex human behaviors using certain combinations of these synergies. Thus, muscle synergies during human behaviors found in previous extensive research may reflect layered structures composed of the fundamental muscle synergies extracted from our study. The advantage of these hypothetical mechanisms is that it prevents the sum of all muscle synergies from exceeding the number of relevant muscles utilized during diverse human behaviors, supporting the premise of compendium in coordinative patterns to execute several movements under different biomechanical conditions [5]. Further research is needed to investigate the muscle synergies identified by factorization algorithms coupled with CNS manipulations and/or neural recordings (e.g., CNS stimulations, spinalization, and electroencephalogram) to validate the neural representation of the fundamental muscle synergies observed in our study [45].

4.3. Clinical Implications

The results of this study may have several clinical implications. First, several studies have investigated muscle synergies in individuals with different characteristics, such as musculoskeletal and neurological disorders [50–52] as well as athletes [8,18,53]. Since we identified the fundamental muscle synergies that may underlie diverse human behaviors in healthy individuals, investigating the changes in muscle synergies such as the number of synergies as well as their compositions in a population of interest may facilitate the understanding of distinct features in motor controls that are associated with severity of symptoms [41,50] or that profile myriad skills and performance in athletes [27,28]. Second, previous studies reported that the inner structure of synergies in stroke patients was not necessarily altered, but that their recruitment patterns due to compromising descending signals were [41,54]. Since our study found that the flexible recruitments of fundamental muscle synergies play an important role in executing diverse motor tasks that related to daily activity livings in healthy individuals, it can be speculated that there is a significant relationship between altered recruitments of fundamental muscle synergies and poor motor function in stroke. Furthermore, if this is the case, it may provide a rationale for designing interventions targeting each fundamental muscle synergy with simple tasks before practicing complex tasks at the beginning of rehabilitation. Potential interventions include muscle synergy-based approaches using functional electrical stimulations to optimize the structure of fundamental synergies and temporal recruitments to improve motor performance in stroke [55–58]. Lastly, we found that different tasks with various biomechanical demands and constraints may largely share the same muscle synergies with different combinations of synergies to be merged. Thus, clinicians may choose to intensively train a particular task to transfer the effectiveness to other tasks [59], given that the transfer of motor learning effects among tasks will be high when muscle synergies involved in different motor tasks are shared [60].

4.4. Limitations

Our study had several limitations. First, it has been reported that the number of recording muscles may affect the amount and structure of muscle synergies [61]. Although EMG recordings in our study were relatively large (i.e., 26 EMG channels), we limited the recording of EMGs from only the major muscles in the trunk and lower limbs. Similarly, we were also limited to 24 fundamental tasks that involved only locomotion and postural tasks. As such, tasks that accompany coordination between the upper limbs, trunk, and lower limbs were not considered [5]. Thus, it is conceivable that some relevant muscle synergies may have been missed in our study. Second, because we used a larger set of EMG recordings and tasks, our time constraint during experiments precluded the measurement of kinematic data such as joint angles as well as velocities, and allowed for the variability of movements in each task, which may impact muscle synergy extractions. The lack of availability of kinematic data prevented the separation of the movement phase and we were thus unable to investigate the contributions of the fundamental muscle synergies for each phase of each task other than walking and running [47].Third, we extracted fundamental muscle synergies from pooled EMGs of all tasks similar to that performed in several studies [10,22,32], assuming the presence of minimum or basic units of muscle synergies that explain all the datasets [22]. It is possible that this method may overestimate data and may cause the over-fitting (i.e., 2-4 task-specific synergies are likely to be well reconstructed using any combination of 13 muscle synergies) [22,62]. Fourth, while the optimal number of healthy participants recruited in studies involving muscle synergies is unknown [63], it may be argued that ten participants in our study were not sufficient to describe data. Lastly, although we extract the fundamental muscle synergies using NMF that may present neural mechanisms for diverse human behaviors, whether the factorization-derived synergies reflect neural organization to coordinate human behaviors remains questionable [45]. This can be due to the possibility that extracted muscle synergies represent biomechanical constraints of tasks rather than neural constraints [3] and the nonlinearity in magnitude summations of the EMG or force vectors [64,65].

5. Conclusions

In this paper, we extracted a repertoire of fundamental muscle synergies from the EMGs applying a factorization algorithm during a variety of human behaviors that involve trunk and lower limb movements in healthy individuals. We found that the 13 fundamental muscle synergies in either the independent or merging state can account for almost all 24 behaviors, including locomotion and stability tasks. Our findings may support the notion that the CNS may flexibly recruit the fundamental muscle synergies to meet the various mechanical demands in a diverse range of human behaviors.

Supplementary Materials: The following are available online at https://www.mdpi.com/article/10.3390/s21186186/s1, Figure S1: Relationship between muscle synergies of all tasks and muscle synergies for other locomotion tasks; Figure S2: Relationship between muscle synergies of all tasks and muscle synergies for other stability tasks; Figure S3: Functional contributions of muscle synergies of all tasks (fundamental muscle synergies) for walking and running; Table S1: Full descriptions of movement and postural tasks; Table S2: Summary of results for the mean number of synergies, mean VAF of each task in the subjects, and the degree of similarity within each synergy cluster of each task across subjects; Table S3: Recruitment coefficients of 13 synergy clusters of all tasks for each task execution.

Author Contributions: Conceptualization and design of the work H.S., H.Y. and A.S.; data acquisition H.S., A.S. and T.K.; data analysis H.S. and H.Y.; data interpretation H.S., H.Y., A.S., T.K. and K.N.; writing—original draft preparation, H.S., H.Y., A.S., T.K. and K.N. All authors have read and agreed to the published version of the manuscript.

Funding: This project was supported by Grant-in-Aid (KAKENHI) from the Japan Society for the Promotion of Science (JSPS) for the Grand-in-Aid for Scientific Research (A) awarded to K.N. (#18H04082).

Institutional Review Board Statement: The study was conducted according to the guidelines of the Declaration of Helsinki and approved by the local ethics committee of the University of Tokyo (746).

Informed Consent Statement: Informed consent was obtained from all subjects involved in the study.

Data Availability Statement: Data will be available upon corresponding author request.

Conflicts of Interest: The authors declare no conflict of interest.

References

1. Bernstein, N.A. *The Co-Ordination and Regulation of Movements*; Pergamon Press: Oxford, UK, 1967.
2. Bizzi, E.; Cheung, V.C.K.; d'Avella, A.; Saltiel, P.; Tresch, M. Combining modules for movement. *Brain Res. Rev.* **2008**, *57*, 125–133. [CrossRef]
3. Tresch, M.C.; Jarc, A. The case for and against muscle synergies. *Curr. Opin. Neurobiol.* **2009**, *19*, 601–607. [CrossRef]
4. Kutch, J.J.; Valero-Cuevas, F.J. Challenges and New Approaches to Proving the Existence of Muscle Synergies of Neural Origin. *PLOS Comput. Biol.* **2012**, *8*, e1002434. [CrossRef]
5. Bizzi, E.; Cheung, V.C.K. The neural origin of muscle synergies. *Front. Comput. Neurosci.* **2013**, *7*, 2013. [CrossRef] [PubMed]
6. Meyer, A.J.; Eskinazi, I.; Jackson, J.N.; Rao, A.V.; Patten, C.; Fregly, B.J. Muscle Synergies Facilitate Computational Prediction of {Subject-Specific} Walking Motions. *Front. Bioeng. Biotechnol.* **2016**, *4*, 77. [CrossRef] [PubMed]
7. Cappellini, G.; Ivanenko, Y.P.; Poppele, R.E.; Lacquaniti, F. Motor Patterns in Human Walking and Running. *J. Neurophysiol.* **2006**, *95*, 3426–3437. [CrossRef]
8. Yokoyama, H.; Ogawa, T.; Kawashima, N.; Shinya, M.; Nakazawa, K. Distinct sets of locomotor modules control the speed and modes of human locomotion. *Sci. Rep.* **2016**, *6*, 36275. [CrossRef]
9. Ivanenko, Y.P.; Poppele, R.E.; Lacquaniti, F. Five basic muscle activation patterns account for muscle activity during human locomotion. *J. Physiol.* **2004**, *556*, 267–282. [CrossRef]
10. Scano, A.; Dardari, L.; Molteni, F.; Giberti, H.; Tosatti, L.M.; d'Avella, A. A Comprehensive Spatial Mapping of Muscle Synergies in Highly Variable {Upper-Limb} Movements of Healthy Subjects. *Front. Physiol.* **2019**, *10*, 1231. [CrossRef]
11. D'Avella, A.; Fernandez, L.; Portone, A.; Lacquaniti, F. Modulation of phasic and tonic muscle synergies with reaching direction and speed. *J. Neurophysiol.* **2008**, *100*, 1433–1454. [CrossRef] [PubMed]
12. Shaharudin, S.; Agrawal, S. Muscle synergies during incremental rowing {VO2max} test of collegiate rowers and untrained subjects. *J. Sports Med. Phys. Fit.* **2016**, *56*, 980–989.
13. Kristiansen, M.; Samani, A.; Madeleine, P.; Hansen, E.A. Muscle synergies during bench press are reliable across days. *J. Electromyogr. Kinesiol.* **2016**, *30*, 81–88. [CrossRef]
14. Vaz, J.R.; Olstad, B.H.; Cabri, J.; Kjendlie, P.-L.; Pezarat-Correia, P.; Hug, F. Muscle coordination during breaststroke swimming: Comparison between elite swimmers and beginners. *J. Sports Sci.* **2016**, *34*, 1941–1948. [CrossRef] [PubMed]
15. Nazifi, M.M.; Yoon, H.U.; Beschorner, K.; Hur, P. Shared and {Task-Specific} Muscle Synergies during Normal Walking and Slipping. *Front. Hum. Neurosci.* **2017**, *11*, 40. [CrossRef] [PubMed]
16. Santuz, A.; Ekizos, A.; Eckardt, N.; Kibele, A.; Arampatzis, A. Challenging human locomotion: Stability and modular organisation in unsteady conditions. *Sci. Rep.* **2018**, *8*, 2740. [CrossRef] [PubMed]
17. Yang, Q.; Logan, D.; Giszter, S.F. Motor primitives are determined in early development and are then robustly conserved into adulthood. *Proc. Natl. Acad. Sci. USA* **2019**, *116*, 12025–12034. [CrossRef]
18. Cheung, V.C.K.; Cheung, B.M.F.; Zhang, J.H.; Chan, Z.Y.S.; Ha, S.C.W.; Chen, C.-Y.; Cheung, R.T.H. Plasticity of muscle synergies through fractionation and merging during development and training of human runners. *Nat. Commun.* **2020**, *11*, 4356. [CrossRef] [PubMed]
19. Chvatal, S.A.; Ting, L.H. Common muscle synergies for balance and walking. *Front. Comput. Neurosci.* **2013**, *7*, 48. [CrossRef] [PubMed]

20. Martino, G.; Ivanenko, Y.P.; d'Avella, A.; Serrao, M.; Ranavolo, A.; Draicchio, F.; Cappellini, G.; Casali, C.; Lacquaniti, F. Neuromuscular adjustments of gait associated with unstable conditions. *J. Neurophysiol.* **2015**, *114*, 2867–2882. [CrossRef] [PubMed]
21. D'Avella, A.; Bizzi, E. Shared and specific muscle synergies in natural motor behaviors. *Proc. Natl. Acad. Sci. USA* **2005**, *102*, 3076–3081. [CrossRef]
22. Hagio, S.; Kouzaki, M. The flexible recruitment of muscle synergies depends on the required force-generating capability. *J. Neurophysiol.* **2014**, *112*, 316–327. [CrossRef]
23. Barroso, F.O.; Torricelli, D.; Moreno, J.C.; Taylor, J.; Gomez-Soriano, J.; Bravo-Esteban, E.; Piazza, S.; Santos, C.; Pons, J.L. Shared muscle synergies in human walking and cycling. *J. Neurophysiol.* **2014**, *112*, 1984–1998. [CrossRef] [PubMed]
24. Sartori, M.; Gizzi, L.; Lloyd, D.G.; Farina, D. A musculoskeletal model of human locomotion driven by a low dimensional set of impulsive excitation primitives. *Front. Comput. Neurosci.* **2013**, *7*, 79. [CrossRef]
25. David, L.; Gallahue, O.; Goodway, J.C.; Jacqueline, D. *Understanding Motor Development: Infants, Children, Adolescents, Adults*; McGraw-Hill: New York, NY, USA, 2011.
26. Huang, B.; Xiong, C.; Chen, W.; Liang, J.; Sun, B.-Y.; Gong, X. Common kinematic synergies of various human locomotor behaviours. *R. Soc. Open Sci.* **2021**, *8*, 210161. [CrossRef]
27. Cook, G.; Burton, L.; Hoogenboom, B.J.; Voight, M. Functional movement screening: The use of fundamental movements as an assessment of function—Part 2. *Int. J. Sports Phys. Ther.* **2014**, *9*, 549–563. [PubMed]
28. Cook, G.; Burton, L.; Hoogenboom, B.J.; Voight, M. Functional movement screening: The use of fundamental movements as an assessment of function—Part 1. *Int. J. Sports Phys. Ther.* **2014**, *9*, 396–409. [PubMed]
29. Hug, F. Can muscle coordination be precisely studied by surface electromyography? *J. Electromyogr. Kinesiol.* **2011**, *21*, 1–12. [CrossRef]
30. Ranaldi, S.; De Marchis, C.; Conforto, S. An automatic, adaptive, information-based algorithm for the extraction of the {sEMG} envelope. *J. Electromyogr. Kinesiol.* **2018**, *42*, 1–9. [CrossRef]
31. De Marchis, C.; Ranaldi, S.; Serrao, M.; Ranavolo, A.; Draicchio, F.; Lacquaniti, F.; Conforto, S. Modular motor control of the sound limb in gait of people with trans-femoral amputation. *J. Neuroeng. Rehabil.* **2019**, *16*, 132. [CrossRef] [PubMed]
32. Geng, Y.; Deng, H.; Samuel, O.W.; Cheung, V.; Xu, L.; Li, G. Modulation of muscle synergies for multiple forearm movements under variant force and arm position constraints. *J. Neural Eng.* **2020**, *17*, 026015. [CrossRef]
33. Tresch, M.C.; Cheung, V.C.K.; d'Avella, A. Matrix factorization algorithms for the identification of muscle synergies: Evaluation on simulated and experimental data sets. *J. Neurophysiol.* **2006**, *95*, 2199–2212. [CrossRef] [PubMed]
34. Lee, D.D.; Seung, H.S. Learning the parts of objects by non-negative matrix factorization. *Nature* **1999**, *401*, 788–791. [CrossRef] [PubMed]
35. Chvatal, S.A.; Torres-Oviedo, G.; Safavynia, S.A.; Ting, L.H. Common muscle synergies for control of center of mass and force in nonstepping and stepping postural behaviors. *J. Neurophysiol.* **2011**, *106*, 999–1015. [CrossRef] [PubMed]
36. Torres-Oviedo, G.; Macpherson, J.M.; Ting, L.H. Muscle synergy organization is robust across a variety of postural perturbations. *J. Neurophysiol.* **2006**, *96*, 1530–1546. [CrossRef] [PubMed]
37. Zar, J.H. *Biostatistical Analysis*; Pearson Education: Karnataka, India, 1999.
38. Frère, J.; Hug, F. Between-subject variability of muscle synergies during a complex motor skill. *Front. Comput. Neurosci.* **2012**, *6*, 99. [CrossRef]
39. Delis, I.; Hilt, P.M.; Pozzo, T.; Panzeri, S.; Berret, B. Deciphering the functional role of spatial and temporal muscle synergies in whole-body movements. *Sci. Rep.* **2018**, *8*, 8391. [CrossRef]
40. Tibshirani, R.; Walther, G.; Hastie, T. Estimating the number of clusters in a data set via the gap statistic. *J. R. Stat. Soc. Ser. B Stat. Methodol.* **2001**, *63*, 411–423. [CrossRef]
41. Cheung, V.C.K.; Turolla, A.; Agostini, M.; Silvoni, S.; Bennis, C.; Kasi, P.; Paganoni, S.; Bonato, P.; Bizzi, E. Muscle synergy patterns as physiological markers of motor cortical damage. *Proc. Natl. Acad. Sci. USA* **2012**, *109*, 14652–14656. [CrossRef]
42. Hagio, S.; Fukuda, M.; Kouzaki, M. Identification of muscle synergies associated with gait transition in humans. *Front. Hum. Neurosci.* **2015**, *9*, 48. [CrossRef]
43. Takei, T.; Confais, J.; Tomatsu, S.; Oya, T.; Seki, K. Neural basis for hand muscle synergies in the primate spinal cord. *Proc. Natl. Acad. Sci. USA* **2017**, *114*, 8643–8648. [CrossRef]
44. Overduin, S.A.; d'Avella, A.; Carmena, J.M.; Bizzi, E. Microstimulation Activates a Handful of Muscle Synergies. *Neuron* **2012**, *76*, 1071–1077. [CrossRef] [PubMed]
45. Cheung, V.C.K.; Seki, K. Approaches to Revealing the Neural Basis of Muscle Synergies: A Review and A Critique. *J. Neurophysiol.* **2021**, *125*, 1580–1597. [CrossRef]
46. Hodges, P.W.; Richardson, C.A. Contraction of the Abdominal Muscles Associated With Movement of the Lower Limb. *Phys Ther.* **1997**, *77*, 132–142. [CrossRef]
47. Ting, L.H.; Chiel, H.J.; Trumbower, R.D.; Allen, J.L.; McKay, J.L.; Hackney, M.E.; Kesar, T.M. Neuromechanical principles underlying movement modularity and their implications for rehabilitation. *Neuron* **2015**, *86*, 38–54. [CrossRef]
48. Jacquelin Perry, J.M.B. *Gait Analysis: Normal and Pathological Function*, 2nd ed.; SLACK Incorporated: West Deptford, NJ, USA, 2010; ISBN 9781617114304.

49. McCrea, D.A.; Rybak, I.A. Organization of mammalian locomotor rhythm and pattern generation. *Brain Res. Rev.* **2008**, *57*, 134–146. [CrossRef] [PubMed]
50. Liew, B.X.W.; Del Vecchio, A.; Falla, D. The influence of musculoskeletal pain disorders on muscle synergies—A systematic review. *PLoS ONE* **2018**, *13*, e0206885. [CrossRef]
51. Bekius, A.; Bach, M.M.; van der Krogt, M.M.; de Vries, R.; Buizer, A.I.; Dominici, N. Muscle Synergies During Walking in Children With Cerebral Palsy: A Systematic Review. *Front. Physiol.* **2020**, *11*, 632. [CrossRef] [PubMed]
52. Van Criekinge, T.; Vermeulen, J.; Wagemans, K.; Schröder, J.; Embrechts, E.; Truijen, S.; Hallemans, A.; Saeys, W. Lower limb muscle synergies during walking after stroke: A systematic review. *Disabil. Rehabil.* **2020**, *42*, 2836–2845. [CrossRef] [PubMed]
53. Taborri, J.; Agostini, V.; Artemiadis, P.K.; Ghislieri, M.; Jacobs, D.A.; Roh, J.; Rossi, S. Feasibility of muscle synergy outcomes in clinics, robotics, and sports: A systematic review. *Appl. Bionics Biomech.* **2018**, *2018*, 3934698. [CrossRef]
54. Irastorza-Landa, N.; García-Cossio, E.; Sarasola-Sanz, A.; Brötz, D.; Birbaumer, N.; Ramos-Murguialday, A. Functional synergy recruitment index as a reliable biomarker of motor function and recovery in chronic stroke patients. *J. Neural Eng.* **2021**, *18*, 046061. [CrossRef] [PubMed]
55. Bao, S.-C.; Leung, W.-C.; Cheung, V.C.K.; Zhou, P.; Tong, K.-Y. Pathway-specific modulatory effects of neuromuscular electrical stimulation during pedaling in chronic stroke survivors. *J. NeuroEng. Rehabil.* **2019**, *16*, 1–15. [CrossRef]
56. Niu, C.M.; Bao, Y.; Zhuang, C.; Li, S.; Wang, T.; Cui, L.; Xie, Q.; Lan, N. Synergy-Based FES for Post-Stroke Rehabilitation of Upper-Limb Motor Functions. *IEEE Trans. Neural Syst. Rehabil. Eng.* **2019**, *27*, 256–264. [CrossRef]
57. Ambrosini, E.; Parati, M.; Peri, E.; De Marchis, C.; Nava, C.; Pedrocchi, A.; Ferriero, G.; Ferrante, S. Changes in leg cycling muscle synergies after training augmented by functional electrical stimulation in subacute stroke survivors: A pilot study. *J. Neuroeng. Rehabil.* **2020**, *17*, 35. [CrossRef]
58. Niu, C.; Bao, Y.; Zhuang, C.; Li, S.; Wang, T.; Zhang, X.; Ma, Y.; Xuan, Z.; Gu, L.; Lan, N.; et al. Effectiveness of short-term training with a synergy-based FES paradigm on motor function recovery post-stroke. *Ann. Phys. Rehabil. Med.* **2018**, *61*, e33. [CrossRef]
59. Kantak, S.S.; Zahedi, N.; McGrath, R. Complex Skill Training Transfers to Improved Performance and Control of Simpler Tasks After Stroke. *Phys. Ther.* **2017**, *97*, 718–728. [CrossRef]
60. De Marchis, C.; Di Somma, J.; Zych, M.; Conforto, S.; Severini, G. Consistent visuomotor adaptations and generalizations can be achieved through different rotations of robust motor modules. *Sci. Rep.* **2018**, *8*, 12657. [CrossRef] [PubMed]
61. Steele, K.M.; Tresch, M.C.; Perreault, E.J. The number and choice of muscles impact the results of muscle synergy analyses. *Front. Comput. Neurosci.* **2013**, *7*, 105. [CrossRef] [PubMed]
62. Cheung, V.C.K.; d'Avella, A.; Bizzi, E. Adjustments of motor pattern for load compensation via modulated activations of muscle synergies during natural behaviors. *J. Neurophysiol.* **2009**, *101*, 1235–1257. [CrossRef] [PubMed]
63. Turpin, N.A.; Uriac, S.; Dalleau, G. How to improve the muscle synergy analysis methodology? *Eur. J. Appl. Physiol.* **2021**, *121*, 1009–1025. [CrossRef] [PubMed]
64. Lemay, M.A.; Galagan, J.E.; Hogan, N.; Bizzi, E. Modulation and vectorial summation of the spinalized frog's hindlimb end-point force produced by intraspinal electrical stimulation of the cord. *IEEE Trans. Neural Syst. Rehabil. Eng.* **2001**, *9*, 12–23. [CrossRef] [PubMed]
65. Yaron, A.; Kowalski, D.; Yaguchi, H.; Takei, T.; Seki, K. Forelimb force direction and magnitude independently controlled by spinal modules in the macaque. *Proc. Natl. Acad. Sci. USA* **2020**, *117*, 27655–27666. [CrossRef] [PubMed]

sensors

MDPI

Article

A Novel Muscle Synergy Extraction Method Used for Motor Function Evaluation of Stroke Patients: A Pilot Study

Yehao Ma [1,2], Changcheng Shi [1,2,*], Jialin Xu [1,2], Sijia Ye [1], Huilin Zhou [1,2] and Guokun Zuo [1,2]

1 Cixi Institute of Biomedical Engineering, Ningbo Institute of Materials Technology and Engineering, Chinese Academy of Sciences, Ningbo 315300, China; mayehao@nimte.ac.cn (Y.M.); xujialin@nimte.ac.cn (J.X.); yesijia258@163.com (S.Y.); zhouhuilin@nimte.ac.cn (H.Z.); moonstone@nimte.ac.cn (G.Z.)

2 Zhejiang Engineering Research Center for Biomedical Materials, Ningbo Institute of Materials Technology and Engineering, Chinese Academy of Sciences, Ningbo 315300, China

* Correspondence: changchengshi@nimte.ac.cn

Abstract: In this paper, we present a novel muscle synergy extraction method based on multivariate curve resolution–alternating least squares (MCR-ALS) to overcome the limitation of the nonnegative matrix factorization (NMF) method for extracting non-sparse muscle synergy, and we study its potential application for evaluating motor function of stroke survivors. Nonnegative matrix factorization (NMF) is the most widely used method for muscle synergy extraction. However, NMF is susceptible to components' sparseness and usually provides inferior reliability, which significantly limits the promotion of muscle synergy. In this study, MCR-ALS was employed to extract muscle synergy from electromyography (EMG) data. Its performance was compared with two other matrix factorization algorithms, NMF and self-modeling mixture analysis (SMMA). Simulated data sets were utilized to explore the influences of the sparseness and noise on the extracted synergies. As a result, the synergies estimated by MCR-ALS were the most similar to true synergies as compared with SMMA and NMF. MCR-ALS was used to analyze the muscle synergy characteristics of upper limb movements performed by healthy (n = 11) and stroke (n = 5) subjects. The repeatability and intra-subject consistency were used to evaluate the performance of MCR-ALS. As a result, MCR-ALS provided much higher repeatability and intra-subject consistency as compared with NMF, which were important for the reliability of the motor function evaluation. The stroke subjects had lower intra-subject consistency and seemingly had more synergies as compared with the healthy subjects. Thus, MCR-ALS is a promising muscle synergy analysis method for motor function evaluation of stroke patients.

Keywords: muscle synergy; MCR-ALS; sparseness; electromyography; motor function; stroke

Citation: Ma, Y.; Shi, C.; Xu, J.; Ye, S.; Zhou, H.; Zuo, G. A Novel Muscle Synergy Extraction Method Used for Motor Function Evaluation of Stroke Patients: A Pilot Study. *Sensors* **2021**, *21*, 3833. https://doi.org/10.3390/s21113833

Academic Editor: Ernest N. Kamavuako

Received: 26 April 2021
Accepted: 28 May 2021
Published: 1 June 2021

1. Introduction

How the central nervous system (CNS) controls the musculoskeletal system to solve the redundancy problem of degree of freedom (DOF) is an important research topic. One strategy widely recognized by a significant number of scholars is that the CNS accomplishes a variety of behaviors through statistical regularities involving biomechanical properties of the human body, and then synergistically applies these regularities to perform different motor tasks [1–3]. These regularities are called "muscle synergies". The limb movements are accomplished by activating these synergies coordinately [3,4].

Although the physiological origin and meaning of muscle synergies are still debated [5], it has been confirmed that motor task execution can be described by the coordination of a limited number of muscle synergies. A muscle synergy represents the relative activation strengths of a group of muscles, simplifying the control of the musculoskeletal system. Muscle synergy provides a new method for studying the motor control mechanism during a movement [6]. Ivanenko et al. found five muscle synergies accounted for muscle activity during human locomotion [7]. Scano et al. proved that a large variety of grasps can be produced by a limited subset of muscle synergies [8]. Aoi et al. developed a motor model with 69 parameters based on muscle synergy, which produced both walking and running of a human musculoskeletal model by changing only seven motor control parameters and concluded a human could change walking and running speed through seven key motor control parameters [9]. Pan et al. found that the similarity of muscle synergies of subacute stroke survivors was significantly correlated with the Brunnstrom stage [10]. Cheung et al. considered muscle synergy to be physiological markers of motor cortical damage, and the muscle synergies of stroke survivors had three distinct patterns, i.e., preservation, merging, and fractionation [11]. Thus, a growing number of studies have been focused on the synergy characteristics of subjects with nervous system diseases, such as stroke [11–13], spinal cord injury [14,15], cerebral palsy [16], and Parkinson's disease [17]. The results have shown that muscle synergy is a promising approach for motor function evaluation.

Various matrix decomposition algorithms have been applied to extract muscle synergy from recorded and processed electromyographic (EMG) signals of dynamic motor tasks. Nonnegative matrix factorization (NMF) [11–21], factor analysis (FA) [22,23], independent component analysis (ICA) [24,25], and principal component analysis (PCA) [26,27] are the four common synergy extraction algorithms. PCA produces the basis vectors (muscle synergies) with the best variance description of EMG data through singular value decomposition (SVD) [6]. Similar to PCA, FA extracts muscle synergy weights by calculating the eigenvectors of the data's covariance matrix; the eigenvectors with eigenvalues >1 are considered to be the muscle synergies. The synergies are determined by employing singular value decomposition for analyzing the data's covariance matrix. ICA makes the data statistically independent by transforming it orthogonally. ICA is employed to analyze the data with non-Gaussian variation and extract synergies that maximize the absolute value of the fourth moment of the data. In contrast to the aforementioned methods, NMF is the most widely used algorithm because of its non-negativity constraint and simple principle [6,15–21,28]. However, NMF has some limitations. For example, it is based on maximizing Gaussian likelihood, and easily falls into local optimum for identifying dependent and non-sparse components. Thus, some modified methods have been proposed. Several algorithms based on gamma and inverse Gaussian models have been used to analyze EMG data in order to deal with the influence of signal noise on muscle synergy extraction and have provided better reconstruction quality as compared with NMF [29,30]. The sparse NMF (SNMF) has been proposed to improve the performance of NMF for identifying dependent components [31]. In addition, constrained Tucker decomposition (consTD) has been employed to deal with the muscle synergy extraction problem from EMG data of various biomechanically related tasks [32]. However, there are few studies that have focused on non-sparse synergies extraction [33].

Some human muscles are biarticular and polyarticular muscles, which contribute to more biomechanical subtasks [31]. As a result, many synergies present a high degree of coactivation in certain movements with multi-joint or multi-degree of freedom. In addition, the abnormal coactivations of some muscles may increase the non-sparseness of synergies and activations for some disease patients, such as stroke [34]. It is well known that NMF presents excellent performance for extracting sparse components [35,36]. However, the performance of NMF may decrease in non-sparse synergy extraction, which significantly restricts the application of muscle synergy for evaluating motor function of stroke patients. To deal with the problem, a method associated with the characteristics of synergy mixture system itself is a feasible solution for non-sparse synergies extraction.

Self-modeling mixture analysis (SMMA) is a common signal resolution algorithm for a linear mixture system, and it can resolve component information by analyzing the statistical characteristics of various signal variables. Thus, as compared with the randomly initialized synergy matrix calculated by NMF, the synergy matrix produced by SMMA is much more consistent with the real ones. In this study, a SMMA-based synergy extraction approach called multivariate curve resolution–alternating least squares (MCR-ALS) was proposed and its performance was compared with SMMA and NMF. We evaluated the feasibility of the proposed algorithm with simulation data, for which the properties of the synergies and activations were known. The effects of the sparseness and noise intensity on the extracted synergies were investigated. In addition, the proposed method was used to analyze the muscle synergy of upper limb movements of stroke subjects; the intra-subject consistency and the number of muscle synergies between stroke and healthy subjects were compared and investigated.

2. Theory and Experiment

2.1. Muscle Synergy Pattern Model

According to the theoretical background of muscle synergy, EMG signals are considered to be the weighted summation of primitive functions reflecting the activation time information of several muscle groups, and their synergies (muscle groups) reflect the relative activation strengths of multiple muscles. Thus, muscle synergies can be extracted through unmixing EMG data as the following bilinear model [11,37]:

$$D = CS + E \tag{1}$$

where D is the $m \times n$ matrix which consists of a set of m preprocessed EMG signals and n data points for each EMG signal, correspondingly; C is the $m \times r$ matrix of weight coefficient describing the enrollment of muscle groups (muscle synergy); r represents the number of muscle synergies; S is a $r \times n$ matrix of activations constituting the basic components or primitive functions; and E is residual matrix with size of $m \times n$. Every muscle synergy is a time-invariant module reflecting relative activation strengths of multiple muscles, which is activated by time-varying commands (activation) descending from CNS [11].

2.2. Simulated Data

A simulation study is a useful approach for synergy extraction algorithm assessment as the properties of the synergies and activations are known. In this study, various simulated data sets with different noise intensities were constructed to evaluate the accuracy and robustness of the proposed synergy extraction algorithm. According to the muscle synergy pattern model (Equation (1)), simulated data, D_{sim}, was generated according to following equation:

$$D_{sim} = C_{sim}S_{sim} + E_{sim} \tag{2}$$

where D_{sim} is a 10×1000 matrix which consists of a set of 10 simulated EMG signals (muscles or channels) and 1000 data points for each EMG signal, correspondingly; C_{sim} is a 10×4 weight matrix containing 4 synergies; S_{sim} is a 4×1000 matrix of activation; and E_{sim} is the signal-dependent noise matrix with a size of 10×1000.

In past studies, researchers have proven that the noise in neural control signal increases with an increase in the magnitude of signal [38]. Thus, in this study, the noise, whose standard deviation (SD) is positive proportional to noiseless signal value, was added to the EMG signal [39]. The scale factor a was used to change the noise intensities of EMG signals; its value varied from 0.05 to 0.15. The signal-dependent noises with a = 0.05, 0.10, 0.15 (SNR = 26, 20, 16 dB) were added to noiseless signal to generate different simulated data sets, according to Equation (2).

In this study, to evaluate the effect of the sparseness on the proposed synergy extraction method, the synergies and activation profiles with six degrees of sparseness (0.1, 0.2, 0.3, 0.4, 0.5, 0.6) were used to create the simulated data. The synergies with various degrees of sparseness were created randomly and their values were between [0, 1]. The activation profiles with various degrees of sparseness were generated by linear combination of Gaussian and Lorentz functions by control of the full width at half maximum. The degree of sparseness of vector x (synergy or activation profile) was evaluated by Equation (3) [40], which is:

$$\text{sparseness}(x) = \frac{\sqrt{n} - \left(\sum |x_i|\right) / \sqrt{\sum x_i^2}}{\sqrt{n} - 1} \tag{3}$$

where n is the dimension of x, and x_i is the ith variable of x. In the following paragraphs, the sparseness of synergies and activations represents the average sparseness of all synergies and all activations in muscle synergy pattern model, respectively.

2.3. Experimental Data

Eleven healthy subjects (age range 25–37 years, mean age 26.7 years) and five stroke subjects (Brunnstrom 3–5, upper limb Fugl-Meyer score of 22–60, arm Fugl-Meyer score of 16–36, age range 52–79 years, mean age 67.8 years) volunteered to participate in this study. The healthy subjects had no neurological disorders and limb surgeries in their clinical histories. The stroke subjects had no limb surgeries in their clinical histories. The subjects were asked not to drink alcohol and caffeine 24 h before the experiment. The experiment was performed based on the Declaration of Helsinki and confirmed by the Ethics Committee of Cixi Institute of Biomedical Engineering.

The EMG data used in this study were collected from upper limb movements based on an upper limb rehabilitation robot. Subjects sat on a chair with their back straight and perpendicular to the ground. The robot resistance was adjusted to a uniform and comfortable level for each subject. Each subject performed upper limb movements according to the trajectory (semicircle) on the screen of the rehabilitation robot, shown in Figure 1. In the experiment, subjects sat on the side of the upper limb rehabilitation robot and operated a shot stick using their evaluated upper limb (dominant hand for healthy subject and affected hand for patients) to accomplish the task repeatedly (six trials for each subject). To improve the repeatability of the experiment, the motion range of the shot stick was limited by a computer program that ensured it moved along a semicircle trajectory. In any trial, each subject put his hand on the robot's tray and grasped the shot stick with five fingers on the starting position. The experimenter started the data acquisition and give the "go" signal. The subjects did the movement with the shot stick after the "go" signal and stopped at the ending position shown on the screen. The consumed time of the movement was about 2 s. Data collection stopped automatically after 3 s.

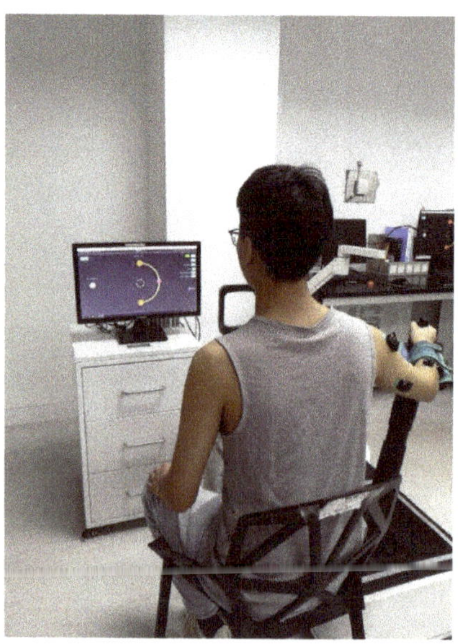

Figure 1. The upper limb movement experiment based on the rehabilitation robot.

In this study, the activities of seven muscles, including the anterior deltoid (DA), posterior deltoid (DP), triceps brachii (TI), biceps brachii (BI), extensor carpi radialis (ECR), flexor carpi radialis (FCR), and brachioradialis (BIO), were recorded through a 16-channel

EMG system (Delsys Inc., Boston, MA, USA). Before placing the EMG sensors, excessive hair was shaved from the skin and alcohol was applied to wipe the skin to remove oils and surface residues. Each EMG sensor was placed on the muscle belly under the guidance of a therapist. The raw EMG signals were recorded with a sampling frequency of 1926 Hz. Figure 2 presents the raw EMG signals of seven muscles recorded during a trial of one healthy subject. To attenuate DC offset and high-frequency noise, the raw EMG signals were band-pass filtered between 40 and 400 Hz (3th order zero-lag Butterworth) [41]. The filtered EMG signals were full-wave rectified and low-pass filtered at 5 Hz (3th order zero-lag Butterworth) to calculate the envelopes of EMG signals [41,42]. The envelopes normalized by the maximum of each envelope itself were used to extract muscle synergy [12].

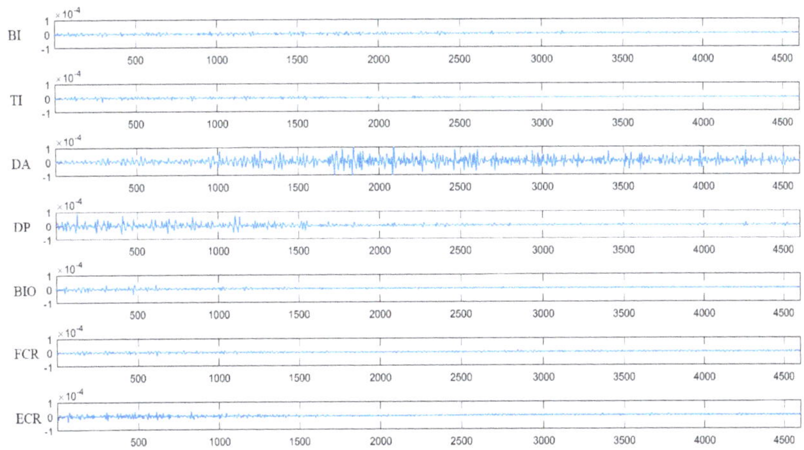

Figure 2. Raw EMG data recorded during the trial of one healthy subject. The unit of the Y coordinate is V.

3. Methodology

3.1. NMF

NMF, which was first proposed by Lee and Seung [35,36], is usually used to reduce data dimension. Because of its nonnegative constraint, NMF is also applied to analyze medical [43] and space object [44] data.

NMF is the most widely used muscle synergy extraction method [15–21], which is usually based on the multiplicative update rules [36]. After creation of the random initial matrices (*C* and *S*), the iteration is to minimize the Frobenius norm of the residual matrix (preprocessed EMG matrix *D* minus multiplication of the matrix *S* and *C*) illustrated by Equation (4). The stop criterion is set based on the parameter Q, the percentage of change in the lack of fit between two iterations ($f_l(S,C)$ and $f_{l+1}(S,C)$), which is obtained through Equation (5). The stop criterion is: (1) Q equal to 0.01% and (2) max number of iterations equal to 1000. Equations (4) and (5) are as follows:

$$f(S,C) = \frac{1}{2}\|D - CS\|_F^2 \quad (4)$$

$$Q = 100 \cdot \left(\frac{f_{l+1}(S,C) - f_l(S,C)}{f_l(S,C)} \right) \quad (5)$$

3.2. SMMA

SMMA, which is also called simple-to-use interactive self-modeling mixture analysis (SIMPLISMA), is a matrix decomposition method for a linear mixture system. SMMA is proposed based on pure variables, which contribute from one component [45]. Thus, SMMA resolves the signal of mixtures by analyzing the characteristics of various variables of signal, and then extracts the values of the variables containing quantitative information to resolve the linear mixture system. SMMA is usually used in a signal analysis of a chemical mixture system [46]. According to Equation (1), muscle synergy is also a linear mixture system and the EMG signals are considered to be weighted summation of time-variant activation. The nonstationarity of the activation profile demonstrates that it is feasible to find pure variables in EMG data. The pure variables, which contain the relative content information of components (synergies), are used to resolve the components (activations).

3.3. MCR-ALS

MCR-ALS, which evolved from SMMA, is a popular matrix factorization algorithm [47,48]. For MCR-ALS, the initial synergy and activation matrices, containing relative content and component information, respectively, are obtained from SMMA. In addition, alternating least squares (ALS) is used to optimize the resolution according to Equations (6) and (7) [49], and nonnegative constraint is imposed to ensure that the decomposition results have physical significance. In the iterative process, if the elements of S and C are negative, it is set to zero forcibly. The stop criterion of MCR-ALS is the same as NMF (Q equal to 0.01%, max number of iteration equal to 1000). Equations (6) and (7) are as follows:

$$C = (DS^T)(SS^T)^{-1} \tag{6}$$

$$S = (C^T C)^{-1} C^T D \tag{7}$$

3.4. Algorithm Evaluation

To compare the performances of the three algorithms, they were applied to analyze the simulated data created by synergies and activations with different sparseness and noise levels. The Pearson's correlation coefficient between true and extracted synergies was applied to assess the three muscle synergy extraction algorithms. The match between extracted and true synergies was performed based on their similarity evaluated by the Pearson's correlation coefficient. We paired the true and extracted synergies with highest similarity. If more than one extracted synergy had the greatest similarity to the same true synergies, simultaneously, we achieved full math by studying the total correlation of all synergy pairs. The match pattern with the max sum of correlation coefficient was used as the final match result. The statistical results (performances on 25 simulated data containing randomly created synergies and activations for each sparseness combination and noise level) of the correlation coefficients of full matched synergies were used to assess the accuracy and robustness of different synergy extraction methods.

In the experimental study, the repeatability of different algorithms was evaluated by analyzing the variation of repeated estimated synergies (25 times) utilizing Pearson's correlation coefficient. The evaluation was based on the Pearson's correlation coefficient between matched synergies extracted from any two performances. Similar to the simulation study, the first step was to match the synergies extracted from two calculations for the same data set (trial). The pair of synergies with the highest correlation were matched together. If two or more synergies had the maximum correlation coefficient with the same synergy, a full match was realized by analyzing the overall correlation of all synergy pairs. The match pattern with the max sum of correlation coefficient was considered to be the final match result. The correlation coefficients of all possible combinations of full matched synergies extracted from 25 repeated performances were used to assess the repeatability of the three synergy extraction algorithms.

In this study, the intra-subject consistency of synergies across multiple trials of each subject was applied to assess the robustness of the proposed synergy extraction algorithm. The intra-subject consistency was calculated by analyzing Pearson's correlation coefficient between each pair of trials for each subject. For each pair of trials, the correlation coefficients of all possible combinations of full matched synergies from the two trial data sets, respectively, were calculated. For each subject, the average of the correlation coefficients of all trial combinations was used to represent intra-subject consistency. A paired t-test was used to analyze the difference in intra-subject consistency for various algorithms. One-way ANOVA was performed to detect the difference in intra-subject consistencies between healthy and stroke subjects.

3.5. Choose the Number of Synergies

In the analysis described above, we assumed that the correct number of synergies was known before decomposition. However, in many situations, the number of synergies, r, is unknown. In this study, we applied variance account for (VAF) to determine the number of muscle synergies r [50–52]. VAF was calculated according to the following equation:

$$VAF = 1 - (\|D - M\|^2 / \|D - mean(D)\|^2) \tag{8}$$

where $M = CS$, which represents the reconstruction matrix of synergy extraction algorithm; the operator "mean" is to produce a matrix with the same size as D, whose columns are made up of the mean of corresponding column in D. Here, r is defined as the minimum when VAF exceeded 80% [11,13].

4. Results

4.1. Evaluation with Simulated Data

In order to evaluate the feasibilities of the three muscle synergy extraction algorithms, we employed various simulated EMG datasets generated from components (synergies and activation profiles) with different sparseness combinations. The similarity between estimated and true components was used to assess various synergy extraction methods. As a result, the performance of NMF was affected by both the sparseness of synergies and the sparseness of activations. NMF became less accurate as the degree of sparseness decreased. In contrast, SMMA and MCR-ALS were mainly affected by activation sparseness, because the purities of pure variables were easily influenced by the non-sparseness of activations. However, the influence of the non-sparseness of synergies on the two methods was relatively small. Figure 3 presents the performances of NMF, SMMA, and MCR-ALS versus synergy sparseness and activation sparseness under the circumstance of noise level 0.05. Obviously, the performance of NMF was different from SMMA and MCR-ALS.

To validate the robustness of each synergy extraction method, we analyzed their performances under circumstances of different noise intensities. Figure 4 shows the average correlation coefficients for the fully identified synergies compared across two settings (sparseness, noise) for the three methods. From the figure, it can be seen that the noise was an important influence factor for synergy extraction. The estimated synergies became less accurate as the noise level increased. In the three methods, MCR-ALS provided the best performance, especially when sparseness of synergy and activation was low. For example, the average correlation coefficient of synergy estimated by MCR-ALS was 0.96 when the synergy sparseness and noise level were 0.1 and 0.05, while the average correlation coefficients of synergies estimated by SMMA and NMF were 0.92 and 0.83, respectively. As the synergy sparseness increased, the performances of NMF and MCR-ALS became close. Lacking iterative optimization restricted the performance of SMMA. In addition, the non-negativity of its decomposing results was hard to be ensured. For the data containing non-sparse components (synergy or activation), NMF easily fell into local optimum and its result was non-unique, which lead to its poor performance. However, MCR-ALS could supply accurate and unique decomposing results.

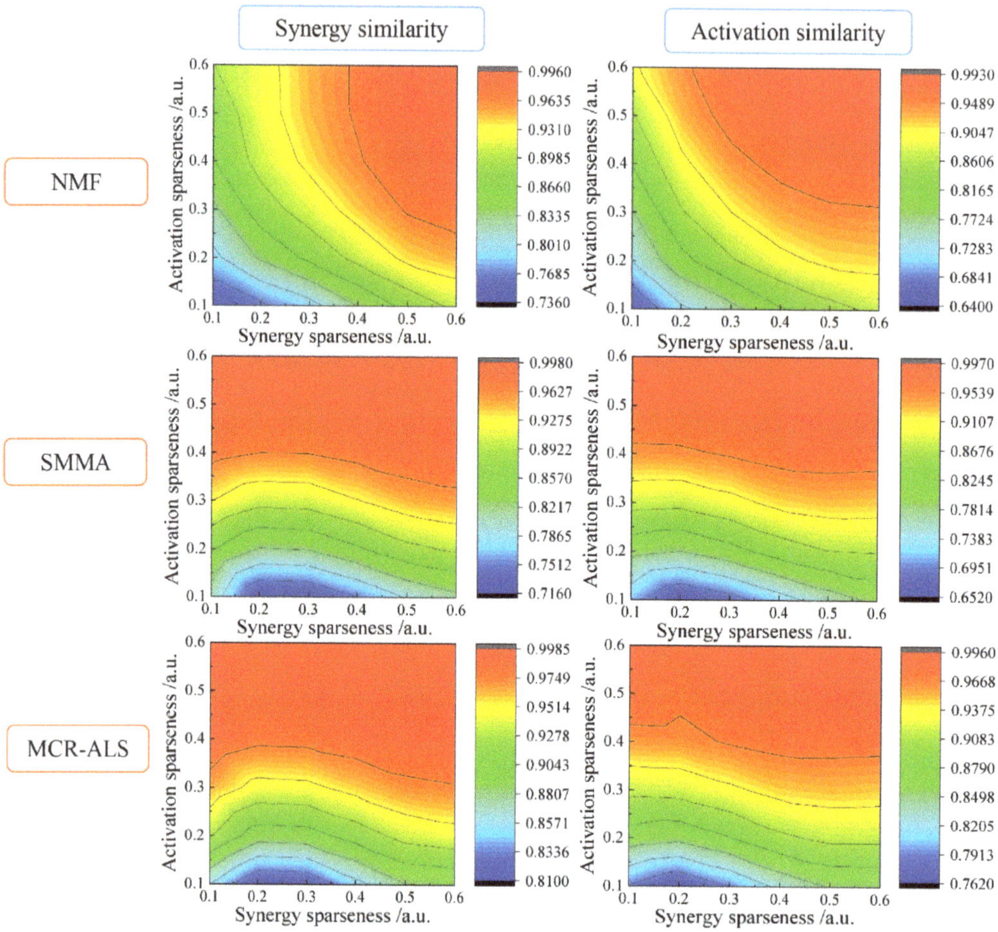

Figure 3. The performances of NMF, SMMA, and MCR-ALS versus synergy sparseness and activation sparseness under the circumstance of noise level 0.05. (The color represents the degree of similarity evaluated by the Pearson's correlation coefficient between the synergies and activations extracted by the matrix decomposition algorithms and the true synergies and activations; the isolines represent the same value of average correlation coefficients; a.u. represents arbitrary unit).

4.2. Results of Motor Function Evaluation by Muscle Synergy

In the simulation study, SMMA provided unique decomposition as compared with NMF, but it still could not present satisfactory performance. MCR-ALS, a developed algorithm which evolved from SMMA, optimized the resolution through ALS iteration. Thus, MCR-ALS was the most reliable and accurate synergy extraction. In this study, MCR-ALS was also applied to analyze the muscle synergies of upper limb movements of stroke subjects, to improve the performance of muscle synergy on motor function evaluation.

For each healthy subject, three muscle synergies were needed to ensure the reconstruction accuracy for the EMG data. For the conventional method (NMF), muscle synergies were extracted with an average VAF value of 85.67 ± 5.64%. However, for the proposed method (MCR-ALS), the muscle synergies were identified with an average VAF value of 86.32 ± 4.97%. The number of synergies of stroke subjects was different from healthy subjects. Three muscle synergies were enough to reconstruct the EMG data accurately for three stroke subjects, while for the other two stroke subjects, four muscle synergies were needed to realize accurate data reconstruction. Muscle synergies were extracted by NMF with an average VAF value of 86.35 ± 5.44%. For MCR-ALS, the muscle synergies were identified with an average VAF value of 86.47 ± 5.39%. Obviously, the two methods have close reconstruction ability.

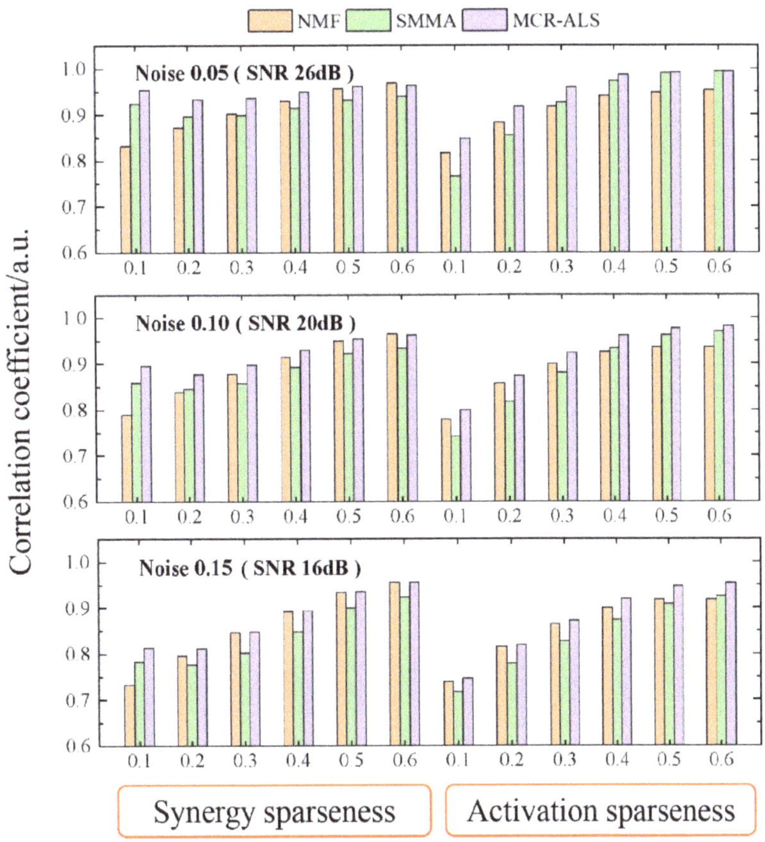

Figure 4. Average correlation coefficients for the fully identified synergies compared across two settings (sparseness, noise) for the three methods.

To compare the repeatability of MCR-ALS and NMF, we analyzed 25 repeated decomposition results for each trial dataset. As a result, the average repeatability of NMF was 0.86 ± 0.11 and 0.74 ± 0.13 for all healthy and stroke subjects, respectively. Figure 5 is the boxplot of the correlation coefficients of muscle synergies extracted from a typical trial EMG data of healthy subject. As seen in Figure 5, the average correlation coefficients of three synergies were 0.92, 0.80, and 0.65, respectively, with box lengths of 0.08, 0.25, and 0.52. Obviously, NMF presented inferior repeatability for the upper limb movement data, and it usually fell into local optimum. However, MCR-ALS could extract relative concentration information through pure variables in the initialization phase instead of random initialization, thus, its repeatability was one. In other words, the initialized synergy and activation matrix is unique.

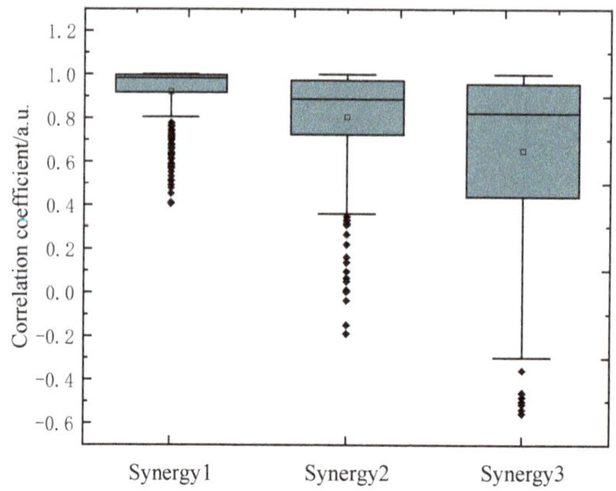

Figure 5. The boxplots of the correlation coefficients of muscle synergies extracted from a typical trial data of healthy subject using NMF.

Figure 6 shows the muscle synergies extracted by NMF and MCR-ALS from one healthy subject. As seen in the figure, the synergies produced by NMF had much larger standard deviations as compared with the ones identified by MCR-ALS, especially for Synergy 2 and Synergy 3. As shown in Figure 6, the synergies extracted by NMF and MCR-ALS all had low sparseness, especially Synergy 2 and Synergy 3. Thus, the performance of NMF might be affected by the non-sparseness of synergy. A similar conclusion can be obtained through the synergy analysis of stroke subjects. As Figure 7 shows, the standard deviations of the synergies extracted by MCR-ALS were much smaller than the ones identified by NMF.

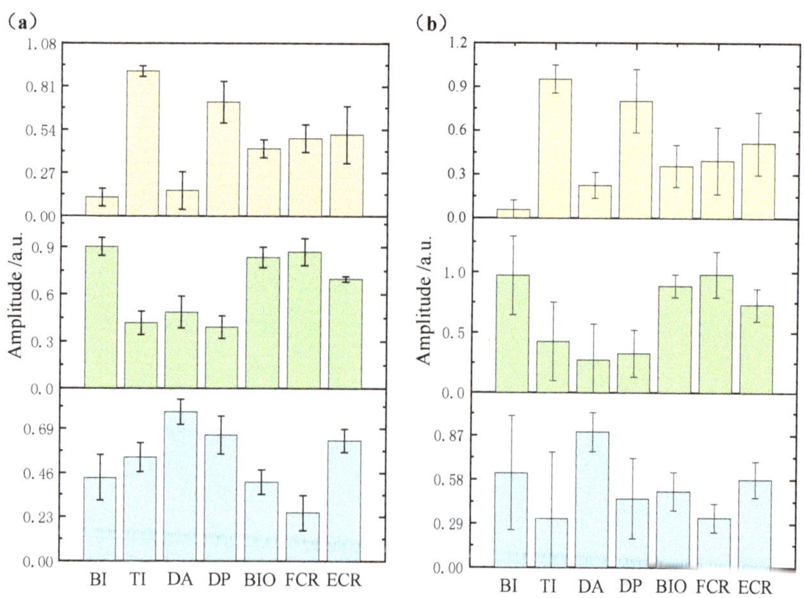

Figure 6. The statistical chart of synergies extracted by MCR-ALS (**a**) and NMF (**b**) from a typical healthy subject.

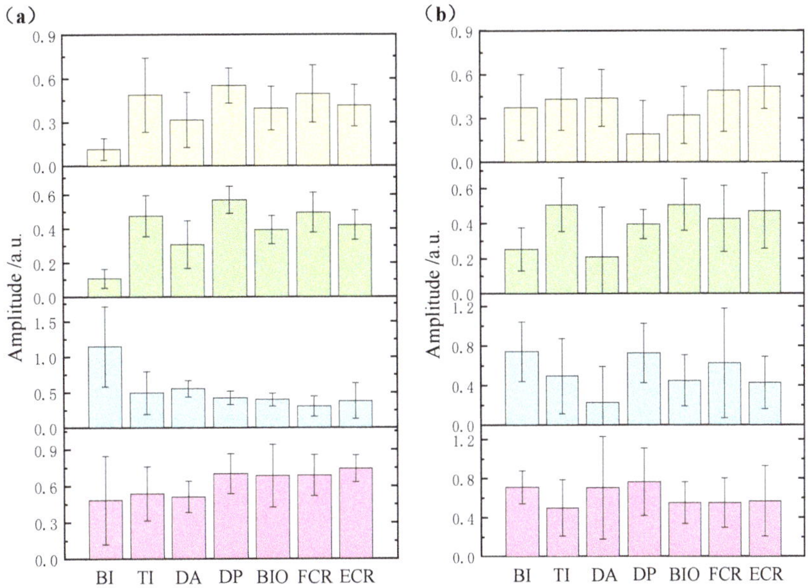

Figure 7. The statistical chart of synergies extracted by MCR-ALS (**a**) and NMF (**b**) from a typical stroke subject.

To further evaluate the performance of the two methods, intra-subject consistency was calculated. Figure 8 represents the statistical results of intra-subject consistency computed to compare the standard approach (NMF) and the novel approach (MCR-ALS) for healthy and stroke subjects. For the healthy subjects, there was a significant difference in intra-subject consistency between the two methods ($p < 0.001$). The consistencies of MCR-ALS and NMF were equal to 0.86 ± 0.04 and 0.69 ± 0.09, respectively. For the stroke subjects, there was also a significant difference in intra-subject consistency between the two methods ($p = 0.01$). The consistencies of MCR-ALS and NMF were equal to 0.56 ± 0.11 and 0.38 ± 0.15, respectively. Obviously, NMF had inferior consistency as compared with MCR-ALS. The main reason may be that NMF could not obtain optimum in non-sparseness component extraction. From the experimental study, MCR-ALS could provide more reliable synergy identification as compared with NMF.

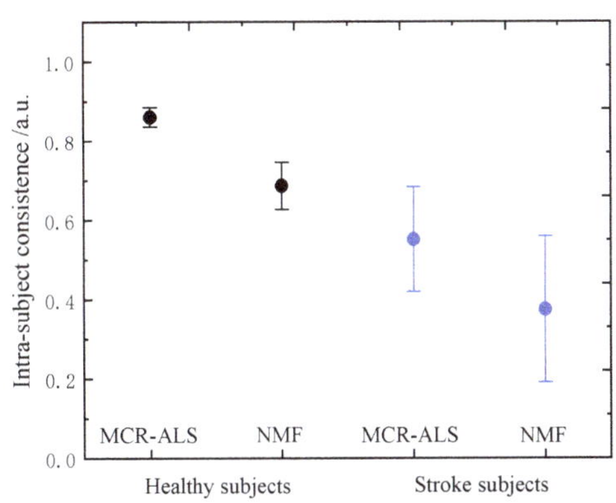

Figure 8. The statistical results of intra-subject consistency computed to compare the standard approach (NMF) and the novel approach (MCR-ALS) for all healthy and stroke subjects.

Compared with healthy subjects, some chronic stroke subjects had more synergies, which was consistent with those reported in the literature [11]. The two groups of subjects had a significant difference in intra-subject consistency (F = 78.275, $p < 0.001$ for MCR-ALS and F = 27.835, $p < 0.001$ for NMF). Stroke subjects had inferior synergy consistencies as compared with healthy subjects (0.38 ± 0.15 vs. 0.69 ± 0.09 for NMF and 0.56 ± 0.11 vs. 0.86 ± 0.04 for MCR-ALS), which demonstrated that stroke patients had poorer motor control ability as compared with healthy subject. Thus, the number of synergies and intra-subject consistency are promising indexes used for motor function evaluation for stroke patients.

5. Discussion

In the simulation study, NMF showed good performances for identifying sparse synergies. However, for non-sparse synergy extraction, NMF could not provide satisfactory performance. The estimated synergies had a large error as compared with true synergies. Many human muscles are biarticular and polyarticular, which induces the phenomenon of coactivation and non-sparse synergies in many human movements with multiple degrees of freedom. The experiment of this study is a great example, the synergies of upper limb movements involving digital, wrist, elbow, and shoulder joints are non-sparse. In addition, the abnormal coactivations of stroke patients' muscles also affects the sparseness of synergies. The analysis results of the experimental data showed that NMF easily fell into local optimum in extracting non-sparse components.

From the results of the simulation study, SMMA was a feasible method for identifying muscle synergy. In addition, its decomposing result was unique. However, the non-sparse activations affected the purities of pure variables. In other words, it was hard to find several data points completely contributed from one synergy. Thus, SMMA could not provide satisfying resolution. However, MCR-ALS, as a developed algorithm, optimized the purities of pure variables through ALS. As a result, MCR-ALS was the most reliable synergy identification method, even though the EMG data contained non-sparse synergies and non-sparse activations simultaneously.

Predicting the number of synergies is an important task for muscle synergy analysis. We applied VAF to predict the number of synergy vectors ensuring enough reconstruction accuracy of the matrix decomposition algorithm for EMG data. In the analysis of the EMG study (healthy and stroke subjects), the VAF values of NMF and MCR-ALS were very close. Thus, MCR-ALS also had enough learning ability for EMG data, ensuring a robust prediction of the number of synergies.

The non-negativity is also a key index for the matrix decomposition algorithm. The SMMA usually produces negative values resulting from the amplitude differences between the pure variables and weight of synergies. Thus, non-negative constraint is necessary for SMMA to ensure the physical significance of decomposition. However, the relative relationship between pure variables and weight of synergies would remain if the non-sparseness of activation profiles were slight. Thus, the synergies were still similar to real ones. MCR-ALS evolved from SMMA, and its non-negative constraint ensures the positive decomposing results. As compared with the pure mathematical method (NMF), MCR-ALS and SMMA are strongly associated with system theory and the structure of the data itself, thus, they have greater robustness.

For stroke patients, their synergies have a low degree of sparseness because of the abnormal muscle activations and movements with multiple degrees of freedom. The poor reliability of NMF severely restricts the application of muscle synergy to motor function evaluation. MCR-ALS supplied more reliable muscle synergy information as compared with NMF, contributing to the robust motor function evaluation. In addition, chronic stoke subjects might have more synergies and inferior intra-subject consistency as compared with healthy subjects. Therefore, the two parameters are two promising indexes used for motor function evaluation of stroke survivors. In future study, we plan to analyze the muscle synergies of more stroke subjects with the novel EMG analysis method.

6. Conclusions

In this study, a novel muscle synergy extraction method called MCR-ALS was proposed. Its performance was compared with two other matrix decomposition algorithms (NMF and SMMA). The results showed the following: (1) The problem of non-unique decomposition of NMF was resolved through pure variable extraction (SMMA and MCR-ALS); (2) As a developed algorithm evolved from SMMA, MCR-ALS presented the greatest reliability in synergy identification as compared with NMF and SMMA, especially for the data containing non-sparse components. In addition, MCR-ALS was used to analyze the muscle synergy characteristics of stroke subjects. Through comparative study, stoke subjects have more synergies and inferior intra-subject consistency as compared with healthy subjects. Therefore, MCR-ALS is a promising muscle synergy extraction method. The results of this study are of great significance for promoting the application of muscle synergy for evaluating motor function of stroke patients.

Author Contributions: Conceptualization, Y.M. and C.S.; methodology, Y.M., J.X. and C.S.; resource, Y.M. and C.S.; software, Y.M. and S.Y.; validation, J.X.; formal analysis, Y.M. and H.Z.; investigation, Y.M. and S.Y.; supervision, C.S.; data curation, S.Y.; writing—original draft preparation, Y.M. and S.Y.; writing—review and editing, C.S.; visualization, Y.M., H.Z. and J.X.; funding acquisition, G.Z., C.S., Y.M. and J.X. All authors have read and agreed to the published version of the manuscript.

Funding: This work was supported by the China Postdoctoral Science Foundation Grant (2019M662128), the Major Scientific and Technological Projects in Ningbo City (2018B10073 and 2019B10034), the Key Research and Development Program of Zhejiang Province (2019C03090), the Zhejiang Provincial Natural Science Foundation of China (LQ20F030003 and LQ19F030002), the Ningbo Natural Science Foundation (2019A610089).

Institutional Review Board Statement: The study was conducted according to the guidelines of the Declaration of Helsinki, and approved by the Research Ethics Board of Cixi Institute of Biomedical Engineering, Ningbo Institute of Materials Technology and Engineering, Chinese Academy of Sciences.

Informed Consent Statement: Informed consent was obtained from all subjects involved in the study.

Data Availability Statement: The data presented in this study are available on request from the corresponding author.

Acknowledgments: The authors would like to thank Lin Cao, Haiyan Gu, and Xiaoguang Liu at Ningbo Rehabilitation Hospital for their clinical guidance and support.

Conflicts of Interest: The authors declare no conflict of interest.

References

1. Safavynia, S.A.; Torres-Oviedo, G.; Ting, L.H. Muscle synergies: Implications for clinical evaluation and rehabilitation of movement. *Top. Spinal Cord Inj. Rehabil.* **2011**, *17*, 16–24. [CrossRef]
2. Ting, L.H.; Chiel, H.J.; Trumbower, R.D.; Allen, J.L.; McKay, J.L.; Hackney, M.E.; Kesar, T.M. Neuromechanical principles underlying movement modularity and their implications for rehabilitation. *Neuron* **2015**, *86*, 38–54. [CrossRef] [PubMed]
3. Tresch, M.C.; Saltiel, P.; d'Avella, A.; Bizzi, E. Coordination and localization in spinal motor systems. *Brain Res. Rev.* **2002**, *39*, 66–79. [CrossRef]
4. Li, Z.; Liu, H.; Yin, Z.; Chen, K. Muscle synergy alteration of human during walking with lower limb exoskeleton. *Front. Neurosci.* **2019**, *12*, 1050. [CrossRef]
5. Tresch, M.C.; Jarc, A. The case for and against muscle synergies. *Curr. Opin. Neurobiol.* **2009**, *19*, 601–607. [CrossRef] [PubMed]
6. Lambert-Shirzad, N.; Van der Loos, H.F.M. On identifying kinematic and muscle synergies: A comparison of matrix factorization methods using experimental data from the healthy population. *J. Neurophysiol.* **2017**, *117*, 290–302. [CrossRef]
7. Ivanenko, Y.P.; Poppele, R.E.; Lacquaniti, F. Five basic muscle activation patterns account for muscle activity during human locomotion. *J. Physiol.* **2004**, *556*, 267–282. [CrossRef]
8. Scano, A.; Chiavenna, A.; Tosatti, L.M.; Müller, H.; Atzori, M. Muscle synergy analysis of a hand-grasp dataset: A limited subset of motor modules may underlie a large variety of grasps. *Front. Neurorobot.* **2018**, *12*, 57. [CrossRef] [PubMed]
9. Aoi, S.; Ohashi, T.; Bamba, R.; Fujiki, S.; Tamura, D.; Funato, T.; Senda, K.; Ivanenko, Y.; Tsuchiya, K. Neuromusculoskeletal model that walks and runs across a speed range with a few motor control parameter changes based on the muscle synergy hypothesis. *Sci. Rep.* **2019**, *9*, 1–13.

10. Pan, B.; Sun, Y.; Xie, B.; Huang, Z.; Wu, J.; Hou, J.; Liu, Y.; Huang, Z.; Zhang, Z. Alterations of muscle synergies during voluntary arm reaching movement in subacute stroke survivors at different levels of impairment. *Front. Comput. Neurosci.* **2018**, *12*, 69. [CrossRef]
11. Cheung, V.C.K.; Turolla, A.; Agostini, M.; Silvoni, S.; Bennis, C.; Kasi, P.; Paganoni, S.; Bonato, P.; Bizzi, E. Muscle synergy patterns as physiological markers of motor cortical damage. *Proc. Acad. Natl. Sci. USA* **2012**, *109*, 14652–14656. [CrossRef]
12. Clark, D.J.; Ting, L.H.; Zajac, F.E.; Neptune, R.R.; Kautz, S.A. Merging of healthy motor modules predicts reduced locomotor performance and muscle coordination complexity poststroke. *J. Neurophysiol.* **2010**, *103*, 844–857. [CrossRef]
13. Li, S.; Zhuang, C.; Niu, C.M.; Bao, Y.; Xie, Q.; Lan, N. Evaluation of functional correlation of task-specific muscle synergies with motor performance in patients postsroke. *Front. Neurol.* **2017**, *8*, 337. [CrossRef]
14. Chvatal, S.A.; Macpherson, J.M.; Torres-Oviedo, G.; Ting, L.H. Absence of postural muscle synergies for balance after spinal cord transection. *J. Neurophysiol.* **2013**, *110*, 1301–1310. [CrossRef]
15. Hayes, H.B.; Chvatal, S.A.; French, M.A.; Ting, L.H.; Trumbower, R.D. Neuromuscular constraints on muscle coordination during overground walking in persons with chronic incomplete spinal cord injury. *Clin. Neurophysiol.* **2014**, *125*, 2024–2035. [CrossRef]
16. Tang, L.; Chen, X.; Cao, S.; Wu, D.; Zhao, G.; Zhang, X. Assessment of upper limb motor dysfunction for children with cerebral palsy based on muscle synergy analysis. *Front. Hum. Neurosci.* **2017**, *11*, 130. [CrossRef]
17. Mileti, I.; Zampogna, A.; Santuz, A.; Asci, F.; Prete, Z.D.; Arampatzis, A.; Palermo, E.; Suppa, A. Muscle synergies in Parkinson's disease. *Sensors* **2020**, *20*, 3209. [CrossRef] [PubMed]
18. Cheung, V.C.K.; Cheung, B.M.F.; Zhang, J.H.; Chan, Z.Y.S.; Ha, S.C.W.; Chen, C.Y.; Cheung, R.T.H. Plasticity of muscle synergies through fractionation and merging during development and training of human runners. *Nat. Commun.* **2020**, *11*, 4356. [CrossRef]
19. Sabzevari, V.R.; Jafari, A.H.; Boostani, R. Muscle synergy extraction during arm reaching movements at different speeds. *Technol. Health Care* **2017**, *25*, 123–136. [CrossRef] [PubMed]
20. Ghislieri, M.; Agostini, V.; Knaflitz, M. How to improve robustness in muscle synergy extraction. In Proceedings of the 41st Annual International Conference of the IEEE Engineering in Medicine and Biology Society, Berlin, Germany, 23–27 July 2019; pp. 1525–1528.
21. Barradas, V.R.; Kutch, J.J.; Kawase, T.; Koike, Y.; Schweighofer, N. When 90% of the variance is not enough: Residual EMG from muscle synergy extraction influences task performance. *J. Neurophysiol.* **2020**, *93*, 2180–2190.
22. Kieliba, P.; Tropea, P.; Pirondini, E.; Coscia, M.; Micera, S.; Artoni, F. How are muscle synergies affected by electromyography pre-processing? *IEEE Trans. Neural Syst. Rehabilit. Eng.* **2018**, *26*, 882–893. [CrossRef] [PubMed]
23. Saito, A.; Watanabe, K.; Akima, H. Coordination among thigh muscles including the vastus intermedius and adductor magnus at different cycling intensities. *Hum. Mov. Sci.* **2015**, *40*, 14–23. [CrossRef]
24. Kargo, W.J.; Nitz, D.A. Early skill learning is expressed through selection and tuning of cortically represented muscle synergies. *J. Neurosci.* **2003**, *23*, 11255–11269. [CrossRef] [PubMed]
25. Hart, C.B.; Giszter, S.F. Distinguishing synchronous and time-varying synergies using point process interval statistics: Motor primitives in frog and rat. *Front. Comput. Neurosci.* **2013**, *7*, 52. [CrossRef]
26. Ting, L.H.; Macpherson, J.M. A limited set of muscle synergies for force control during a postural task. *J. Neurophysiol.* **2005**, *93*, 609–613. [CrossRef]
27. Falaki, A.; Huang, X.; Lewis, M.M.; Latash, M.L. Motor equivalence and structure of variance: Multi-muscle postural synergies in Parkinson's disease. *Exp. Brain Res.* **2017**, *235*, 2243–2258. [CrossRef]
28. Rabbi, M.F.; Pizzolato, C.; Lloyd, D.G.; Carty, C.P.; Devaprakash, D.; Diamond, L.E. Non-negative matrix factorisation is the most appropriate method for extraction of muscle synergies in walking and running. *Sci. Rep.* **2020**, *10*, 8266. [CrossRef]
29. Devarajan, K.; Cheung, V.C.K. On nonnegative matrix factorization algorithms for signal-dependent noise with application to electromyography data. *Neural Comput.* **2014**, *26*, 1128–1168. [CrossRef]
30. Santuz, A.; Ekizos, A.; Janshen, L.; Baltzopoulos, V.; Arampatzis, A. On the methodological implications of extracting muscle synergies from human locomotion. *Int. J. Neural Syst.* **2017**, *27*, 1750007. [CrossRef] [PubMed]
31. Soomro, M.H.; Conforto, S.; Giunta, G.; Ranaldi, S.; De Marchis, C. Comparison of initialization techniques for the accurate extraction of muscle synergies from myoelectric signals via nonnegative matrix factorization. *Appl. Bionics Biomech.* **2018**, *2018*, 3629347. [CrossRef]
32. Ebied, A.; Kinney-Lang, E.; Spyrou, L.; Escudero, J. Muscle activity analysis using higher-order tensor decomposition: Application to muscle synergy extraction. *IEEE Access* **2019**, *7*, 27257–27271. [CrossRef]
33. Ebied, A.; Kinney-Lang, E.; Spyrou, L.; Escudero, J. Evaluation of matrix factorisation approaches for muscle synergy extraction. *Med. Eng. Phys.* **2018**, *57*, 51–60. [CrossRef]
34. Wright, Z.A.; Rymer, W.Z.; Slutzky, M.W. Reducing abnormal muscle coactivation after stroke using a myoelectric-computer interface: A pilot study. *Neurorehabilit. Neural Repair* **2014**, *28*, 443–451. [CrossRef]
35. Lee, D.D.; Seung, H.S. Learning the parts of objects by non-negative matrix factorization. *Nature* **1999**, *401*, 788–791. [CrossRef]
36. Lee, D.D.; Seung, H.S. Algorithms for non-negative matrix factorization. In Proceedings of the 14th Annual Neural Information Processing Systems Conference, Denver, CO, USA, 27 November–2 December 2000; pp. 556–562.
37. Hagio, S.; Fukuda, M.; Kouzaki, M. Identification of muscle synergies associated with gait transition in humans. *Front. Hum. Neurosci.* **2015**, *9*, 48. [CrossRef]
38. Harris, C.M.; Wolpert, D.M. Signal-dependent noise determines motor planning. *Nature* **1998**, *394*, 780–784. [CrossRef]

39. Tresch, M.C.; Cheung, V.C.K.; d'Avella, A. Matrix factorization algorithms for the identification of muscle synergies: Evaluation on simulated and experimental data sets. *J. Neurophysiol.* **2006**, *95*, 2199–2212. [CrossRef] [PubMed]
40. Hoyer, P.O. Non-negative matrix factorization with sparseness constraints. *J. Mach. Learn. Res.* **2004**, *5*, 1457–1469.
41. Yang, N.; An, Q.; Kogami, H.; Yamakawa, H.; Tamura, Y.; Takahashi, K.; Kinomoto, M.; Yamasaki, H.; Itkonen, M.; Shibata-Alnajjar, F.; et al. Temporal features of muscle synergies in sit-to-stand motion reflect the motor impairment of post-stroke patients. *IEEE Trans. Neural Syst. Rehabilit. Eng.* **2019**, *27*, 2118–2127. [CrossRef] [PubMed]
42. Barroso, F.O.; Torricelli, D.; Moreno, J.C.; Taylor, J.; Gomez-Soriano, J.; Bravo-Esteban, E.; Piazza, S.; Santos, C.; Pons, J.L. Shared muscle synergies in human walking and cycling. *J. Neurophysiol.* **2014**, *112*, 1984–1998. [CrossRef]
43. Diehn, S.; Zimmermann, B.; Tafintseva, V.; Bagcioglu, M.; Kohler, A.; Ohlson, M.; Fjellheim, S.; Kneipp, J. Discrimination of grass pollen of different species by FTIR spectroscopy of individual pollen grains. *Anal. Bioanal. Chem.* **2020**, *412*, 6459–6474. [CrossRef]
44. Geng, X.R.; Ji, L.Y.; Sun, K. Non-negativematrix factorization based unmixing for principal component transformed hyperspectral data. *Front. Inform. Technol. Electron. Eng.* **2016**, *17*, 403–412. [CrossRef]
45. Windig, W.; Guilment, J. Interactive self-modeling mixture analysis. *Anal. Chem.* **1991**, *63*, 1425–1432. [CrossRef]
46. Vrielynck, L.; Dupuy, N.; Coustillier, G.; Merlin, J.C. Self-modelling analysis applied to nanosecond transient absorption spectroscopy of flavone: An aid to elucidate and characterise reaction intermediates. *Spectrochim. Acta A* **2002**, *58*, 2633–2645. [CrossRef]
47. Liu, Z.; Huang, X.; Jiang, Z.; Tuo, X. Investigation of the binding properties between levamlodipine and HSA based on MCR-ALS and computer modeling. *Spectrochim. Acta A* **2021**, *245*, 118929. [CrossRef] [PubMed]
48. Horii, S.; Ando, M.; Samuel, A.Z.; Take, A.; Nakashima, T.; Matsumoto, A.; Takahashi, Y.; Takeyama, H. Detection of penicillin G produced by penicillium chrysogenum with raman microspectroscopy and multivariate curve resolution-alternating least-squares methods. *J. Nat. Prod.* **2020**, *83*, 3223–3229. [CrossRef] [PubMed]
49. Paatero, P.; Tapper, U. Positive matrix factorization: A non-negative factor model with optimal utilization of error estimates of data values. *Environmetrics* **1994**, *5*, 111–126. [CrossRef]
50. Cheung, V.C.K.; Piron, L.; Agostini, M.; Silvoni, S.; Turolla, A.; Bizzi, E. Stability of muscle synergies for voluntary actions after cortical stroke in humans. *Proc. Acad. Natl. Sci. USA* **2009**, *106*, 19563–19568. [CrossRef]
51. Chiovetto, E.; Berret, B.; Delis, I.; Panzeri, S.; Pozzo, T. Investigating reduction of dimensionality during single-joint elbow movements: A case study on muscle synergies. *Front. Comput. Neurosci.* **2013**, *7*, 11. [CrossRef] [PubMed]
52. Torres-Oviedo, G.; Macpherson, J.M.; Ting, L.H. Muscle synergy organization is robust across a variety of postural perturbations. *J. Neurophysiol.* **2006**, *96*, 1530–1546. [CrossRef]

sensors

MDPI

Article

Electromyographic Assessment of the Efficacy of Deep Dry Needling versus the Ischemic Compression Technique in Gastrocnemius of Medium-Distance Triathletes

María Benito-de-Pedro [1], César Calvo-Lobo [1], Daniel López-López [2], Ana Isabel Benito-de-Pedro [1], Carlos Romero-Morales [3], Marta San-Antolín [3,*], Davinia Vicente-Campos [4] and David Rodríguez-Sanz [1]

1 Faculty of Nursing, Physiotherapy and Podiatry, Universidad Complutense de Madrid, 28040 Madrid, Spain; mariabenito1309@gmail.com (M.B.-d.-P.); cescalvo@ucm.es (C.C.-L.); anabenito1912@gmail.com (A.I.B.-d.-P.); davidrodriguezsanz@ucm.es (D.R.-S.)

2 Research, Health and Podiatry Group, Department of Health Sciences, Faculty of Nursing and Podiatry, Universidade da Coruña, 15403 Ferrol, Spain; daniellopez@udc.es

3 Villaviciosa de Odón Campus, Universidad Europea de Madrid, 28670 Madrid, Spain; carlos.romero@universidadeuropea.es

4 Facultad de Ciencias de la Salud, Universidad Francisco de Vitoria, Pozuelo de Alarcon, 28223 Madrid, Spain; Davinia.vicente@ufv.es

* Correspondence: marta.sanantolin@gmail.com or marta.sanantolin@universidadeuropea.es

Citation: Benito-de-Pedro, M.; Calvo-Lobo, C.; López-López, D.; Benito-de-Pedro, A.I.; Romero-Morales, C.; San-Antolín, M.; Vicente-Campos, D.; Rodríguez-Sanz, D. Electromyographic Assessment of the Efficacy of Deep Dry Needling versus the Ischemic Compression Technique in Gastrocnemius of Medium-Distance Triathletes. *Sensors* **2021**, *21*, 2906. https://doi.org/10.3390/s21092906

Academic Editor: Ernest N. Kamavuako

Received: 17 March 2021
Accepted: 15 April 2021
Published: 21 April 2021

Abstract: Several studies have shown that gastrocnemius is frequently injured in triathletes. The causes of these injuries are similar to those that cause the appearance of the myofascial pain syndrome (MPS). The ischemic compression technique (ICT) and deep dry needling (DDN) are considered two of the main MPS treatment methods in latent myofascial trigger points (MTrPs). In this study superficial electromyographic (EMG) activity in lateral and medial gastrocnemius of triathletes with latent MTrPs was measured before and immediately after either DDN or ICT treatment. Taking into account superficial EMG activity of lateral and medial gastrocnemius, the immediate effectiveness in latent MTrPs of both DDN and ICT was compared. A total of 34 triathletes was randomly divided in two groups. The first and second groups (n = 17 in each group) underwent only one session of DDN and ICT, respectively. EMG measurement of gastrocnemius was assessed before and immediately after treatment. Statistically significant differences (p = 0.037) were shown for a reduction of superficial EMG measurements differences (%) of the experimental group (DDN) with respect to the intervention group (ICT) at a speed of 1 m/s immediately after both interventions, although not at speeds of 1.5 m/s or 2.5 m/s. A statistically significant linear regression prediction model was shown for EMG outcome measurement differences at V1 (speed of 1 m/s) which was only predicted for the treatment group (R^2 = 0.129; β = 8.054; F = 4.734; p = 0.037) showing a reduction of this difference under DDN treatment. DDN administration requires experience and excellent anatomical knowledge. According to our findings immediately after treatment of latent MTrPs, DDN could be advisable for triathletes who train at a speed lower than 1 m/s, while ICT could be a more advisable technique than DDN for training or competitions at speeds greater than 1.5 m/s.

Keywords: myofascial pain syndrome; trigger points; electromyography; deep dry needling; ischemic pressure technique

1. Introduction

The triceps surae complex includes both gastrocnemius and soleus muscles with attachments proximally at the posterior part of the knee and distally at the posterior part of the ankle [1], respectively. The triceps surae assists in control of the knee and ankle and thus, plays a role in all lower extremity activities including triathlon [2,3]. Currently, a triathlon (swimming, cycling, running) is a growing sport activity [4], with an exponential increase in their annual licenses [5]. Several studies have shown that the area that suffers the greatest number of injuries in this sport is the triceps surae [6,7].

Myofascial pain syndrome (MPS) is a regional pain condition associated with the presence of myofascial trigger points (MTrPs) [8]. The presence of MTrPs is considered to be the first sign of muscle overload [9]. This disorder can affect any skeletal muscle in the body and it usually accounts for 21% of orthopaedic clinic visits, 30% of general medicine visits, and around 85%–93% of patients who are referred to pain management clinics [10]. MPS usually occurs in unconditioned muscles, under high tension conditions or after direct local trauma or strains [11]. The presence of MTrPs is linked to overuse and inappropriate muscle use [8].

MTrPs which may be considered as the primary source of pain in MPS were defined as hyperirritable spots within a taut band of skeletal muscle [12]. Two different kinds of MTrPs were divided according to their classification by clinical activity: (1) latent and (2) active MTrPs [9,13]. Latent MTrPs can develop after maintaining muscle activation for long periods of time, exaggerated muscle contractions, or repeated physical activity. Latent MTrPs pain can be triggered by digital compression, stretching, and/or overload [10].

A very high percentage of injuries in triathlon usually occur mainly for two reasons: (1) micro-trauma of repetition or traumatisms and (2) muscular overuse [14], both of which are the main reason for the appearance of MPS [15].

Treatment of this musculoskeletal condition includes deactivation of MTrPs by procedures such as deep dry needling (DDN) or ischemic compression technique (ICT), in order to evaluate their effects on pain and functionality. DDN is considered as a safe and effective method for decreasing pain and improving function by eliciting a local twitch response (LTR) in the MTrPs [16]. Active MTrPs are associated with higher motor end plate noise than latent MTrPs, which means that latent MTrPs can display less irritability than active MTrPs and local twitch responses could be elicited with more difficulty [17]. Eliciting LTR during treatment with DDN would modulate a motor-neuron activity and disrupt the abnormal motor end plante activity [18]. The evidence suggests that this method of treatment performed by physical therapists was more effective than no treatment, sham dry needling and/or other types of treatment [19].

Another efficient technique for treating MTrPs is manual local ICT [20], whereby a pressure for 90 s using the thumb is applied to the MTrP [21]. ICT is considered the most common, non-invasive therapy currently used for the treatment of MTrPs [22].

Diverse studies report alterations in muscle function such as, electromyographic activity increased in the antagonist and synergists muscles of subjects with latent MTrPs [23,24], muscle fatigue increased and overload of motor units close to the latent MTrPs [25]. Moreover, latent MTrPs contribute to accelerated fatigability [26].

When performing a surface electromyography (EMG) of the muscles with MTrPs, resting activity is rarely recorded in them, but they usually show increased motor activity during contraction [10]; however, when the test is performed by means of an intramuscular needle EMG; of an unstimulated MTrP, greater spontaneous activity is observed in MTrP locations than in sites located outside of these MTrPs [27,28]. Spontaneous electrical activity (SEA) is an indication of spontaneous release of acetylcholine (ACh) at the neuromuscular junction (NMJ) [29]. This activity is characteristic of both active and latent MTrPs, which is recorded with an EMG intramuscular needle, when the muscle is at rest. The potential of the terminal motor plate in dysfunction of the extrafusal fibre of the muscle, may present two characteristics under the presence of an MTrP: (1) On one hand, a low degree of continuous electrical activity is present and represents an abnormal activity of end plate or end plate noise and (2) on the other hand, higher amplitude peaks or end plate peaks are accompanied by end plate noise of latent MTrPs [28].

Stimulation of a latent MTrPs (either by ICT or DDN) induces local pain and/or referred pain and presents a lower degree of pain in latent MTrPs than in active ones, without finding any type of end plate noise in areas of non-MTrPs [27].

The mode of action of the DDN is associated withs both mechanical and neurophysiological mechanisms, but still the specific mechanism of action of the DDN is not yet known exactly [30,31]. It is intended that the mechanical effects of the needle interrupt the integrity of the terminal motor plate in dysfunction [32] and alterations in the length and tension of muscle fibres [33] increase blood flow in the muscle and therefore oxygenation there in [33,34]. Kubo et al. found that DDN caused an increase in blood flow at the needling site, for 30 min starting at the moment of needling onwards [33]. Sato et al. described that the release of vasoactive substance, such as calcitonin gene-related peptide (CGRP) and substance *p* (SP) leads to vasodilatation in small vessels and increased blood flow [35].

Until now we have not found any article that measures the muscle activity in the gastrocnemius at an electromyographic level immediately after the latent PGM puncture is carried out, for this reason we have considered it relevant to carry out this investigation, in order to demonstrate whether the proven acceleration of muscle fatigue typical of latent PGM is reduced immediately after treatment with PSP and/or ICT, as measured by surface electromyography.

The aim of the study was to compare the immediate effectiveness in latent medial and lateral gastrocnemius MTrPs of both DDN and ICT assessed by superficial EMG. We hypothesized a greater improvement in EMG activity in triathletes who received DDN than in those receiving ICT.

2. Methods

2.1. Design

The current study was a secondary analysis of a single blinded clinical trial [36] that used a randomized between-group design to investigate the immediate superficial electromiographic effects of DDN treatment with a single session of this technique versus a single treatment session of ICT. Both treatments were applied to latent medial and lateral gastrocnemius MTrPs of medium-distance triathletes. The principal outcome of this clincal trial was identification of EMG activity over the latent treated MTrPs. This superficial EMG measurement was carried out for both groups of triathletes according to our previously published randomized clinical trial protocol [36]. Measurements were made in all participants by a blinded evaluator before and 5, 10, 15, 20 and 25 min after treatment session and those measurements were made 24 h after the last sport session. The study was approved by the human research committee of the Hospital Clinico San Carlos, Madrid-Spain (CEIC Hospital Clínico San Carlos 02/17), and before participation in the study all subjects signed an informed consent form. This clinical trial was prospectively registered at Clinicaltrials.gov with the number NCT03273985.

2.2. Participants

Participants in this study were medium distance triathletes ((1.9 km of swimming, 90 km of cycling, and 21 km of running) [4] who were recruited from the podiatric and physiotherapy clinic, Fisiofuenla s.l.p, from September to December 2017, according to a randomized sampling method. All of them had $\geq$3 years of experience participating in triathlons and they trained around 15 to 20 h weekly. They were selected after identifying them via a clinical exploration conducted by the principal investigator in which, the presence in gastrocnemius of latent MTrPs in the places determined by Travell and Simons [37] as the most common and symptomatic, and identification based on selection criteria were used to select the participants. All participants were evaluated by the same principal investigator who has extensive experience, at least three years, in MTrPs diagnosis and its treatment, which increased confidence in the identification of MTrPs [38]. Inclusion criteria necessary for the participation of triathletes in the study consisted of two parameters: (1) presence of a knot or hypersensitive point in the taut band of skeletal muscle and (2) presence of local or referred pain after mechanical stimulation in the superficial area of the latent MTrP. Several criteria were used to exclude the triathlete from the study: (1) Age > 75 and <18 years; (2) score $\geq$ 4 on the Douler Neuropathique 4 (DN4) questionnaire, which denotes lower limb neurological disorders [39]; (3) cognitive alterations according to positive results on the Pfeiffer questionnaire [40]; (4) taking anticoagulant or antiaggregant medication; (5) existence of prosthesis in the lower limb; (6) presence of systemic or local infection in the lower limb; (7) fibromyalgia, autoimmune disease, iron deficiency, or hypothyroidism; and (8) fear of needles.

2.3. Simple Size Calculation

A sample size calculation using the G*Power 3.1.9.2 software (G*Power©; University of Dusseldorf, Dusseldorf, Germany) was obtained according to the difference between two independent groups. This calculation was based on the EMG percentage (%) at a medium speed of 1.5 m/s (V2) for voluntary isometric maximum force contraction (MVC) of the latent MTrP of medial and lateral gastrocnemius muscle measured immediately after interventions. A pilot study (n = 12) with two groups consisting of six triathletes in the experimental group who received DDN (expressed as mean ± standard deviation [SD] of 22.96% ± 8.38%) and six triathletes in the intervention group who received ICT (35.41% ± 16.24%) [41] was initially conducted. In addition, a one-tailed hypothesis, effect size of 0.96, α error probability of 0.05, power (1-β error probability) of 0.80, and allocation ratio $N_2/N_1 = 1$ were used in order to determine the suitable sample size calculation. Therefore, a sample size of 30 triathletes was calculated with an actual power of 0.82. Finally, regarding possible 10% loss during follow-up, a total sample size of 34 triathletes with ± 17 triathletes in each of the experimental groups (DDN and ICT, respectively), was used.

2.4. Principal Outcome: Electromyographic Measurement

A double channel surface bipolar EMG was used to measure muscular activity in the gastrocnemius, (Verity Medical Ltd., Hans, UK), EMG ranged from 0.2 to 2000 µV root mean square (RMS prolonged) with a sensitivity of 0.1 µV RMS, accuracy of 4% of indications µV 0.3 µV at 200 Hz, and selective band filter of bandwidth 3 dB differential instrumentation amplifier.

Electrode preparation and placement of was carried out following the Surface EMG Recommendations for Non-invasive Assessment of Muscles (SENIAM) recommendations [42]. The two electrodes were placed longitudinally after careful preparation of the skin in which each muscle was cleaned with 70% alcohol, marked with and indelible marker and then shaved. The skin was shaved and scratched with fine sandpaper and then cleaned with ethyl alcohol [43]. The electrodes were secured with hypoallergenic adhesive plaster [42] and an elastic bandage was placed on them [44] in order to prevent their movement.

The location of surface electrodes was established by SENIAM protocol. According to the protocol, the first electrode should be located in external gastrocnemius near the junction line between the fibula head and heel. We placed the electrode at 1/3 of the fibula head following the line of junction [45]. The second one was located in internal gastrocnemius the electrodes were placed on the most promising area of the muscle [45].

Silver or silver/chloride electrodes with rectangular shape and a measure of 4 × 5 mm embedded in a round gel pad of 10 mm diameter (MedicotestGmbH, Andernach, Germany) were used and the reference ground electrode was placed in a free area of the muscle over the femur [42].

This study was divided into two parts. First, the patient was asked for a voluntary isometric maximum force contraction (MVC) [46] after receiving precise instructions on how to perform the MVC of the muscle in question. Three maximal attempts of 6 s each were performed, separated by 3 min to recover from medial and lateral gastrocnemius fatigue, and only the best performance was chosen for the statistical analysis [47]. The participants were instructed to exert maximum effort against resistance. Patients were placed in prone position, and a plantar flexion movement of the tibio-fibular talar joint, with the knee extended (a specialized therapist exerted maximum resistance on the sole of the foot) [48] was required. Patients were verbally encouraged while performing the evaluation. The purpose of this test is to allow the researcher to compare the maximum amplitudes with submaximal tasks, such as walking [44].

Certain studies with intramuscular EMG indicate that muscle with latent MTrP fatigues faster than normal muscles [26]. In 2011 Hong and Arendt Nielsen showed that reduction in mean power spectral frequency of intramuscular EMG activity occurred earlier in latent MTrPs than in normal muscle fibers [26].

Therefore, according to the SENIAM protocol, a comparison was carried out with EMG measurement on the same reference points in the affected leg [42], before and after treatment. The measurements were taken in order to evaluate muscle activity in the gastrocnemius with the superficial EMG selected for this study showing high intra and inter-examiner reliabilities (intraclass correlation coefficient from [ICC] ranging from 0.94 to 0.98). This measurement was performed in the second part of the test protocol at three different speeds on a treadmill before and after treatment (V1: 1 m/sg, V2: 1.5 m/sg, and V3: 2.5 m/sg). To determine the fastest speed that the subject could successfully complete, they were urged to begin the test on the treadmill for the lowest speed before and after treatment. If the patient could walk comfortably at this speed during the assigned 2 min period, the speed of the treadmill was increased to the next level. The study continued until maximum speed data were collected on the treadmill or until the subject could not maintain the established speed [48].

To carry out the data analysis (Software Watscope, Digital North Inc. Waterloo, ON, Canada), the first and last 5 s of the EMG study signals were selected, and the mean power frequency (FPM) was calculated in each segment [49].

2.5. Treatment Allocation

First, an external researcher collected the necessary subject data who participated in this study. Afterwards, the distribution of the same was carried out in both groups, DDN or ICT. This distribution was determined with the statistical and epidemiological analysis system called Epidat 4.2 (Consejería de Sanidad, Xunta de Galicia (España); Organización Panamericana de Salud (OPS-OMS); Universidad CES Colombia). Individual and numbered sheets were prepared sequentially with randomized assignment and placed in sealed opaque envelopes. A second investigator was in charge of opening the envelope. Each variable was measured five times before and after the intervention by an independent investigator.

2.6. Experimental Group: DDN

The experimental group of triathletes of this study received only one session of DDN for which disposable stainless-steel needles (0.3 × 50 mm, Agupunt, Madrid, Spain) were used. These needles were introduced into the MTrP after its location within the taut band [50]. The "fast in, fast out" technique was selected to carry out the DDN treatment [16]. First, the area was cleaned with alcohol, and the therapist wore sterile gloves. The needle was then held firmly between the thumb and the index finger and was directed with deepening penetration through the MTrP [51]. The needle moved up and down without exiting the skin. Finally, the DDN technique was applied for a maximum number of 8–10 insertions or up to the limit of patient tolerance [52].

2.7. Intervention Group: ICT

Ischemic compression is a habitually manual pressure technique. In the present study this technique was applied to the intervention group of triathletes. In order to carry out this treatment, pressure was applied with the thumb in the latent MTrP until the pain reached its maximal tolerable level. This pressure was sustained for 1 min [53].

2.8. Statistical Analysis

Statistical analyses were performed by the software IBM SPSS (version 19.0, IBM Corp., Armonk, NY, USA). The Shapiro-Wilk test was performed to assess normality. Group differences were examined using, the Student's t-test for independent variables for the variables that were adjusted to normal ($p > 0.05$). For the variables that were not adjusted to normal, the non-parametric Mann–Whitney U test ($p < 0.05$) was used. These statistical analyses were carried out according to the provided sample size calculation and following prior published secondary analyses of this randomized clinical trial [36,54]. The mean ± SD, range (minimum – maximum), mean differences and the 95% confidence intervals (CI) for each outcome measurement were calculated. Furthermore, bar graphs including 95% CI error bars were added in order to illustrate statistically significant differences for parametric data. All analyses were considered statistically significant at $p < 0.05$, with a CI of 95%.

In addition, for outcome measurements, reliability analyses were performed. Calculation of ICCs, minimum detectable changes (MDCs), standard errors of measurement (SEM) and lower and upper limits of the 95% CI were obtained. The formula SEM = SD × $\sqrt{}$(1-ICC) was used to calculate SEM values and thus measure the error range of the parameter. The MDC values were calculated to determine the change in magnitude necessary so that without being influenced by random variations or measurement errors, confidence changes are provided. SEM and MDC were analysed according to Bland and Altman. The classification of ICC values was as follows: the value was considered poor when ICC < 0.40, ICC was fair with an ICC value of 0.40–0.59. ICC was considered good with value range from 0.60 to 0.74 and ICC value ranging from 0.75–1 was categorized as excellent [55].

Finally, multivariate regression analyses were carried out in order to predict the EMG differences after treatments. Linear regression models were carried out by the stepwise selection method to predict the outcome measurements differences for EMG values at V1 (speed of 1 m/s), V2 (speed of 1.5 m/s) and V3 (speed of 2.5 m/sg) as dependent variables. Descriptive data, such as sex, height, weight, BMI and foot length, as well as treatment group (DDN = 0; ICT = 1) were included as independent variables. R^2 coefficients were used to determine the quality of adjustment. Pre-stablished P_{in} and P_{out} F-probability values were set at 0.05 and 0.10, respectively.

3. Results

A total of 46 triathletes were recruited initially to participate in this study. Of all participants, 12 participants were excluded from the study. Ten were excluded because they did not present latent MTrPs at the time of the evaluation and two because they took the medication at the time of study completion. Finally, 34 subjects (20 males and 14 females) agreed to participate in the study. None of the subjects presented any adverse effects (Figure 1).

3.1. Sociodemographic Characteristic by Treatment Groups.

The demographic data of the sample studied were divided by type of treatment. The experimental group was treated with DDN, and the intervention group was treated with ICT. Ages, both in the experimental and intervention groups (35.29 ± 5.39 and 33.76 ± 76 years, respectively) did not present statistically significant differences between them ($p = 0.215$). Weight did not present presented statistically significant differences between experimental and intervention groups 65.17 ± 10.71 and 69.17 ± 10.66 kg, respectively; $p = 0.141$. Height did not present statistically significant differences between both experimental and intervention groups 170.35 ± 12.94 and 174.94 ± 6.96 cm, respectively; $p = 0.103$. Body mass index (BMI) did not present statistically significant differences in the experimental and intervention groups 22.37 ± 1.92 and 22.48 ± 2.35 kg/cm^2, respectively; $p = 0.443$. Foot length did not show any statistically significant differences between both groups ($p = 0.421$) with values in the experimental and intervention groups of 41.55 ± 3.26 and 41.35 ± 2.73 cm, respectively; These results matched the randomization of the sample when selecting for one treatment or another.

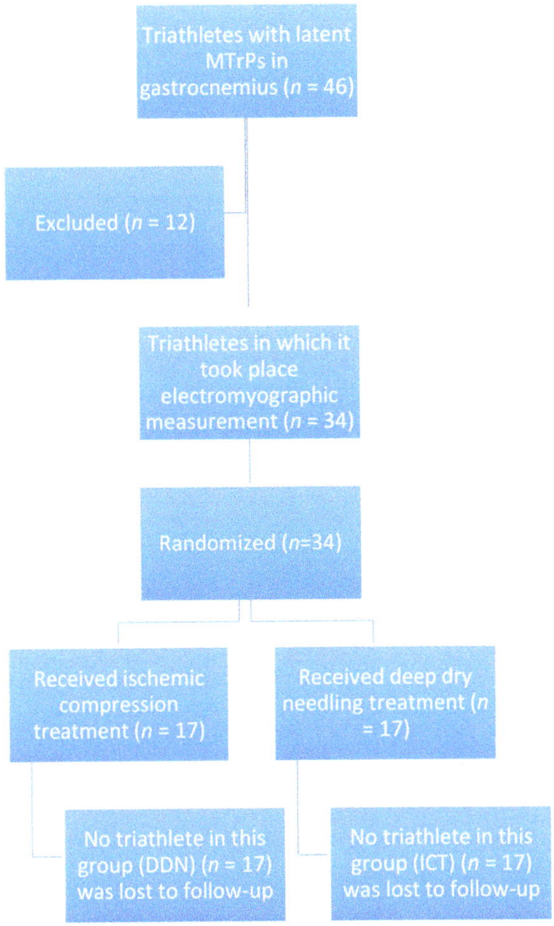

Figure 1. Flow diagram of patients throughout the course of the study.

3.2. *Electromyographic Measurement*

According to Tables 1 and 2 and Figure 2, statistically significant differences (p = 0.037) were shown for a reduction of superficial EMG measurements differences (%) of the experimental group (DDN) with respect to intervention group (ICT) at a speed of 1 m/s (V1) immediately after both interventions, although not at speeds of 1.5 m/s (V2) or 2.5 m/s (V3).

Table 1. Sociodemographic characteristics according to the division by treatment groups.

	Experimental Group (DDN) (n = 17)	**Control Group (ICT) (n = 17)**	***p* Value**
Age (years)	35.29 ± 5.39 (32.73–37.85)	33.76 ± 5.76 (31.02–36.50)	0.215
Weight (kg)	65.17 ± 10.71(60.08–70.27)	69.17 ± 10.66 (64.10–74.24)	0.141
Height (cm)	170.35 ± 12.94 (164.19–176.50)	174.94 ± 6.96 (171.62–178.25)	0.103
BMI (kg/m^2)	22.37 ± 1.92 (21.46–23.29)	22.48 ± 2.35 (21.36–23.6)	0.443
Foot length (cm)	41.55 ± 3.26 (40.00–43.11)	41.35 ± 2.73 (40.05–42.65)	0.421

Abbreviations; DDN, deep dry needling; IC, ischemic compression; m^2, meter^2; cm, centimetres; kg, kilograms; BMI, body mass index; 95% CI, confidence interval at 95%. Statistical significance for a $p < 0.05$ value, with a confidence interval of 95%.

Table 2. Electromyographic measurement of muscle activity in medial or lateral gastrocnemius.

Variable	Before Treatment: Experimental Group (DDN)	Before Treatment: Intervention Group (ICT)	*p* Value	After Treatment: Experimental Group (DDN)	After Treatment: Intervention Group (ICT)	Mean Difference (95% CI)	*p* Value
EMG FPM (μV). V1 (%)	33.75 ± 14.84 (26.69–40.81)	26.79 ± 9.40 (22.32–31.26)	0.056 **	−4.46 ± 11.53 (−30.74–11.08)	3.59 ± 10.00 (−9.29–30.33)	−8.05 (−15.60; −0.51)	**0.037** * ($t = -2.176$)
EMG FPM (μV). V2 (%)	34.84 ± 14.43 (27.98–41.70)	30.36 ± 11.34 (24.97–35.76)	0.160 **	−1.00 ± 9.63 (−16.56–19.67)	4.45 ± 10.62 (−13.86–35.56)	−5.45 (−12.53; 1.63)	0.127 * ($t = -1.167$)
EMG FPM (μV). V3 (%)	38.69 ± 15.19 (31.46–45.91)	34.68 ± 11.69 (29.12–40.24)	0.197 **	2.13 ± 7.42 (−8.07–24.15)	3.23 ± 11.46 (−22.83–31.29)	−1.10 (−7.85; 5.84)	0.734 † ($U = 155.00$)

Abbreviations: DDN, deep dry needling; ICT, ischemic compression technique; FPM, mean power frequency (μV), % percentage with respect to the MVC; EMG, electromyography; V1, speed of 1 m/sg; V2, speed of 1.5 m/sg; V3, speed 2.5 m/sg; * Student's parametric t test for independent samples. † Non-parametric Mann–Whitney U test for independent samples. Statistical significance for a *p*-value < 0.05 (bold).

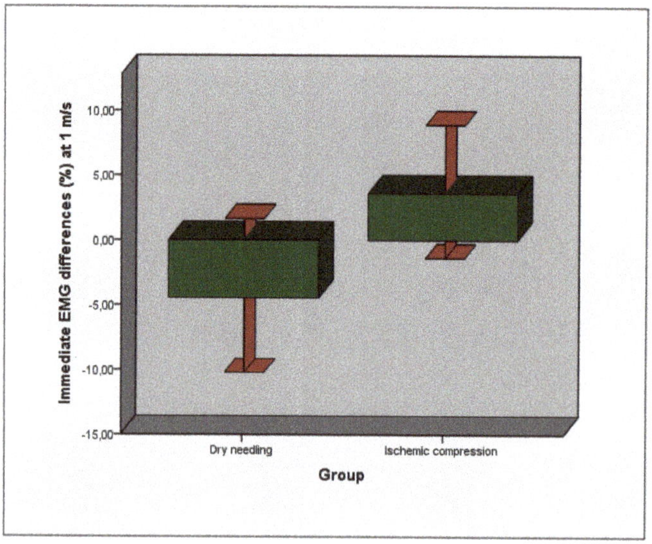

Figure 2. Bars graph with 95% confidence intervals (CI) error bars of immediate electromyograp- ic measurement differences after experimental (DDN) and control (ICT) interventions of muscle activity in medial or lateral gastrocnemius at a speed of 1 m/s (V1).

3.3. *Reliability Analysis*

V1 EMG measurements showed excellent reliability with an ICC value of 0.969 (lower and upper limits of the 95% CI ranging from 0.946 to 0.984), SEM of 2.61% and MDC of 3.19%, both expressed in percentage with respect to the MVC. V2 EMG measurements also showed excellent reliability with an ICC value of 0.957 (lower and upper limits of the 95% CI ranging from 0.924 to 0.977), SEM of 3.07% and MDC of 3.46%, both expressed in percentage with respect to the MVC. V3 EMG measurements showed excellent reliability with an ICC value of 0.952 (lower and upper limits of the 95% CI from 0.914 to 0.975), SEM of 3.32% and MDC of 3.60%, both expressed in percentage with respect to the MVC.

3.4. *Multivariate Prediction Analysis*

A statistically significant linear regression prediction model was shown for EMG outcome measurement differences at V1 (speed of 1 m/s) which was only predicted for the treatment group ($R^2 = 0.129$; β = 8.054; F [1,32] = 4.734; $p = 0.037$) showing a reduction of this difference under DDN treatment. The rest of descriptive data, such as sex, height, weight, BMI and foot length, were excluded from this linear regression model according to the pre-stablished $P_{in} = 0.05$ and $P_{out} = 0.10$ values. Thus, descriptive data did not influence nor predict the EMG outcome measurement differences at V1 (speed of 1 m/s). In addition, linear regression models for EMG outcomes measurements differences at V2 (speed of 1.5 m/s) and V3 (speed of 2.5 m/sg) did not show any valid prediction model.

4. Discussion

DDN is an invasive treatment procedure that requires a thorough understanding of human anatomy in order to be properly performed [56]. It contains a risk of skin infection as a continuity solution occurs with this technique. Several cases of possible infection are described [57] and even Lee et al. described the development of an acute cervical epidural hematoma as a result of needling therapy in the area [43]. Therefore, a thorough knowledge of the anatomy is suggested to try to avoid these complications and to be thorough with hygiene measures. Treatment of latent MTrPs with DDN in gastrocnemius is related to intramuscular oedema, produced by the needling [50], procedure was associated with acute pain postneedling [50].

DDN of latent MTrPs leads to a temporary increase in muscle tone in the needling area, possibly due to intramuscular oedema in this area [58]. Regarding adverse effects of DDN, in 2014, Cummings et al. reported [59] a case of pneumothorax complications after DDN in the iliocostalis muscle. A deep spine infection [60] and infected of hip prothesis were also described too [57]. In addition, a cervical epidural hematoma [61] has been reported after DDN treatment.

Common adverse effects included bruising, bleeding and pain during and after treatment. Correct technique, proper hygiene and anatomical knowledge are stipulated as preventive measures to avoid risks [62].

In case of obtaining similar results, how it happens in our study in the post treatment at 1.5 and 2.5 m/s speeds, with both techniques and taking into account the possible adverse effects resulting from DDN treatment [57,58,61], we would choose the ICT technique as the better treatment option in patients with latent MTrPs in gastrocnemius considering their EMG activity.

In the triceps surae the existence of latent MTrPs is associated with future muscular dysfunction [63].

A 2013 study [64], found a high percentage of latent MTrPs in the asymptomatic population and located a high prevalence of this point in the gastrocnemius muscle [64]

In keeping with the integrated hypothesis of a trigger point proposed by Simons, the zone around an MTrP is in an ischemic state with a shortage of glucose and oxygen [65]. Therefore, treatment with compression of MTrPs helps to improve sarcomere contractions [66].

Takamoto et al. demonstrated that compression applied to the MTrPs would affect to autonomic nervous activity [67]. They published a study in 2009 [67] in which MTrP compression elevated the activity of the parasympathetic nervous system and showed that the physiological mechanisms of pain relief could be induced by the pressure application over MTrPs. That increase entailed an increase in peripheral blood [68], one of the more important factors involved in relieving muscle fatigue [67]. The increased blood flow, provides a concurrent increased in the glucose availability to the muscle [69].

The insertion of a needle at the endplate region reduced the quantity of acetylcholine (ACh) by increasing discharges, leading to a lesser SEA. Hsieh et al. they found an increase in a number of hypoxic-responsive proteins after DDN stimulation that can promote angiogenesis, vasodilation, and altered glucose metabolism in MTrP location [69].

In this study, the effects of DDN at the precise site of latent MTrPs, as compared with those of ICT in the same location shows measure differences in the superficial EMG results, immediately after treatment, at 1 m/s speed.

At a speed of 1 m/s, triathletes with latent MTPs, treated with DDN, immediately improve their muscular activity compared to those treated with ICT; in such a way that those triathletes who begin their recovery at low training speeds after injuries, they should receive as treatment, DDN.

Based on our results, from 1.5 m/s of training speed, the activity data of muscle in triceps surae of triathletes with latent MTPs are similar, so treatment with ICT is recommended, when seeking to demonstrate less adverse effects.

A study on intramuscular EMG in latent MTrPs, in this case the trapezius, associated their existence with accelerated muscle fatigue, although patients did not have painful symptoms [65]. These results did not match what we found in our study, in which it was shown that despite deactivating the latent MTrP from both treatment techniques the results of EMG measurements were very similar to those before the treatment, perhaps conditioned by the immediately measurement.

With intramuscular EMG it is evident that a reduction in the fatigue progression measurement appears earlier in latent MTrPs muscle fibres than in muscle fibres without latent MTrP during a sustained isometric contraction [26]. These results suggest that latent MTrPs are associated with accelerated muscle fatigue.

The muscles fibres of latent MTrPs showed increased motor activity during contraction [10] when the measured was made with surface EMG. Measured with intramuscular EMG the activity in latent MTrPs was much higher than in areas where there is not latent MTrPs both at rest and in isometric contraction [24]. On the other hand when the measurement is made with surface EMG, there are no notable differences between the results obtained at rest and during the contraction [24].

Hong et al. conduced a study in 2014 in which an increase in surface EMG activity in latent MTrPs in the upper trapezius muscle when the muscle was submitted to low loads (25% of MVC) and short times (less than 10 s) of isometric contraction was not observed. On the other hand when the muscle was subject to low loads (25% of MVC) and somewhat longer times (7 min) of isometric and fatigue contractions an increase in surface EMG activity in latent MTrPs was induced [24]. These results may not contradict with those in our study because the performed contraction by triathletes was concentric in the Hong study. In 2019, Baraja-Vegas et al. published a study in which they used DN over latent MTrPs of medial gastrocnemius muscle and observed that the RMS peak amplitude of each subsequent LTR decreased when compared with the initial RMS peak amplitude of previous LTRs [70].

Another study in computer workers related to upper trapezius muscle demonstrated a small decrease in superficial EMG activity after the application of ICT [52] or DN [71].

In previous studies by our research group, we observed that both interventions DDN and ICT showed similarly efficacy in treatment of latent MTrPs in the gastrocnemius of triathletes in terms of dorsiflexion of tibiofibular-talar joint in addiction to changes in static and dynamic plantar pressures [36]. In addition, when our research group measured the pressure pain threshold (PPT) and thermography in latent MTrPs of triathletes treated with both techniques, we observed that local mechanosensitivity had immediately increased after treatment with DDN while this increase did not occur when triathletes were treated with ICT [72].

ICT has demonstrated moderately strong evidence for immediate pain relief in MTrPs, but this evidence is limited in terms of long-term pain relief. A recent review of the literature demonstrated moderately strong evidence supporting the use of ischemic pressure for immediate pain relief at the TrPs, but only limited evidence for long-term pain relief [73].

The great pressure stimulus caused by the needle on the MTrP sends strong neural impulses to the posterior horn, breaking the pain-spasm-pain circle of the MTrP as described by the gate control theory [74].

Hsieh et al. [75] demonstrated that when LTRs were elicited by DDN to a MTrP region suppression of spontaneous electrical activity (SEA) occurred.

According to the various theories, deactivation of trigger points may be attributed to mechanical [32], and biochemical [76] changes around needle insertion. One study demonstrated a decrease in motor end-plate hyperactivity in MTrP, in patients treated with DDN [77].

Sympathetic system regulation can be affected by the effects of needle insertion, and DDN might cause reduction of sympathetic response after treatment with this method [61].

Several authors have theorized about the possibility that rapid movement of the needle into a MTrP might stimulate afferent fibres and could block the pain information generated in the MTrP´s nociceptors though a "gate control" mechanism [78].

In 2019, Barajas et al, conducted a study in which only superficial EMG changes, after the completion of DDN in latent MTrP were described. The decrease in local twitch response amplitudes (brief and sudden contractions of the MTrP taut band) peak after DDN with respect to before DDN treatment in the latent MTrPs [70].

In addition, treatment of latent MTrPs with DDN is related to intramuscular oedema, an improvement of muscle contraction reaction, and increase in muscle stiffness [57].

5. Limitations

As a main limitation, this study reflected only the EMG measurement results immediately after an intervention, and an intervention group with placebo treatment was not used.

Another limitation was pain measurements of were not obtained in this study because the aim in this study was measure EMG activity in latent MTrPs.

The fact that a single treatment session was carried out, and in the case of ICT, of only 60 s., could determine that the treatment was scarce in order to see results.

6. Conclusions

DDN administration requires experience and excellent anatomical knowledge. According to our findings immediately after treatment of latent MTrPs when the muscle was subdued due to a concentric contraction, DDN could be advisable for triathletes who train at a speed lower than 1 m/s, those who begin their recovery at low training speeds after injuries, they should receive as treatment, DDN, while ICT could be a more advisable technique than DDN for training or competition at speeds greater than 1.5 m/s. Further studies with longer follow-up periods and placebo interventions are suggested.

Author Contributions: Conceptualization: M.B.-d.-P., C.C.-L., D.L.-L., A.I.B.-d.-P., C.R.-M., M.S.-A., D.V.-C., and D.R.-S.; data curation: M.B.-d.-P., C.C.-L., D.L.-L., A.I.B.-d.-P., C.R.-M., and D.R.-S.; investigation: M.B.-d.-P., C.C.-L., A.I.B.-d.-P., C.R.-M., M.S.-A., D.V.-C., and D.R.-S methodology: M.B.-d.-P., C.C.-L., D.L.-L., A.I.B.-d.-P., C.R.-M., M.S.-A., D.V.-C., and D.R.-S resources: M.B.-d.-P., C.C.-L., D.L.-L., M.S.-A., D.V.-C., and D.R.-S.; software: M.B.-d.-P., C.C.-L., D.L.-L., A.I.B.-d.-P., C.R.-M., M.S.-A., D.V.-C., and D.R.-S.; supervision: M.B.-d.-P., C.C.-L., D.L.-L., A.I.B.-d.-P., C.R.-M., M.S.-A., and D.R.-S. validation: M.B.-d.-P., C.C.-L., D.L.-L., A.I.B.-d.-P., C.R.-M., M.S.-A., D.V.-C., and D.R.-S. visualization: M.B.-d.-P., C.C.-L., D.L.-L., A.I.B-d.-P., C.R.-M., M.S.-A., D.V.-C., and D.R.-S.; writing—original draft: M.B.-d.-P., C.C.-L., D.L.-L., A.I.B.-d.-P., M.S.-A., D.V.-C., and D.R.-S. writing—review and editing: M.B.-d.-P., C.C.-L., D.L.-L., A.I.B.-d.-P., C.R.-M., M.S.-A., D.V.-C., and D.R.-S. All authors have read and agreed to the published version of the manuscript.

Funding: This research received no external funding.

Institutional Review Board Statement: The study was conducted according to the guidelines of the Declaration of Helsinki, and approved by the human research committee of Hospital Clínico San Carlos, Madrid-Spain (CEIC Hospital Clínico San Carlos 02/17).

Informed Consent Statement: Informed consent was obtained from all subjects involved in the study.

Data Availability Statement: Raw data will be available upon corresponding author request.

Acknowledgments: There is no acknowledgement. This research received no specific grant from any funding agency in the public, commercial or not-for-profit sectors.

Conflicts of Interest: The authors declare no conflict of interest.

References

1. Abdulmassih, S.; Phisitkul, P.; Femino, J.E.; Amendola, A. Triceps surae contracture: Implications for foot and ankle surgery. *J. Am. Acad. Orthop. Surg.* **2013**, *21*, 398–407. Available online: http://www.jaaos.org/cgi/doi/10.5435/JAAOS-21-07-398 (accessed on 16 April 2021). [CrossRef] [PubMed]
2. Mokhtarzadeh, H.; Yeow, C.H.; Hong Goh, J.C.; Oetomo, D.; Malekipour, F.; Lee, P.V.-S. Contributions of the soleus and gastrocnemius muscles to the anterior cruciate ligament loading during single-leg landing. *J. Biomech.* **2013**, *46*, 1913–1920. [CrossRef] [PubMed]
3. Kuitunen, S.; Kyröläinen, H.; Avela, J.; Komi, P.V. Leg stiffness modulation during exhaustive stretch-shortening cycle exercise. *Scand. J. Med. Sci. Sports* **2007**, *17*, 67–75. [CrossRef] [PubMed]
4. Getzin, A.R.; Milner, C.; Harkins, M. Fueling the Triathlete: Evidence-Based Practical Advice for Athletes of All Levels. *Curr. Sports. Med. Rep.* **2017**, *16*, 240–246. [CrossRef]
5. Consejo Superior de Deportes. Licencias—Portal del Consejo Superior de Deportes. 2016. Available online: http://www.csd.gob.es/csd/asociaciones/1fedagclub/03Lic/ (accessed on 16 April 2021).
6. Andersen, C.A.; Clarsen, B.; Johansen, T.V.; Engebretsen, L. High prevalence of overuse injury among iron-distance triathletes. *Br. J. Sports Med.* **2013**, *47*, 857–861. [CrossRef]
7. Bertola, I.P.; Sartori, R.P.; Corrêa, D.G.; Zotz, T.G.G.; Gomes, A.R.S. Profile of injures prevalence in athletes who participated in SESC Triathlon Caiobá-2011. *Acta Ortop. Bras.* **2014**, *22*, 191–196. [CrossRef]
8. Bron, C.; Dommerholt, J.D. Etiology of myofascial trigger points. *Curr. Pain Headache Rep.* **2012**, *16*, 439–444. Available online: http://www.ncbi.nlm.nih.gov/pubmed/22836591 (accessed on 16 April 2021). [CrossRef]
9. Ge, H.-Y.; Fernández-de-Las-Peñas, C.; Yue, S.-W. Myofascial trigger points: Spontaneous electrical activity and its consequences for pain induction and propagation. *Chin. Med.* **2011**, *6*, 13. Available online: http://cmjournal.biomedcentral.com/articles/10.1186/1749-8546-6-13 (accessed on 16 April 2021). [CrossRef]
10. Borg-Stein, J.; Simons, D.G. Focused review: Myofascial pain. *Arch. Phys. Med. Rehabil.* **2002**, *83*, S40–S47. Available online: http://www.ncbi.nlm.nih.gov/pubmed/11973695 (accessed on 16 April 2021). [CrossRef]
11. Shah, J.P.; Thaker, N.; Heimur, J.; Aredo, J.V.; Sikdar, S.; Gerber, L. Myofascial Trigger Points Then and Now: A Historical and Scientific Perspective. *PM&R* **2015**, *7*, 746–761.
12. Simons, D.G. Understanding effective treatments of myofascial trigger points. *J. Bodyw. Mov. Ther.* **2002**, *6*, 81–88. [CrossRef]
13. Travell, J.G.; Simons, D.G. *Dolor y Disfunción Miofascial. el Manual de Los Puntos Gatillo*; Mitad Inferior del Cuerpo: Baltimore, MD, USA, 2004.
14. Zwingenberger, S.; Valladares, R.D.; Walther, A.; Beck, H.; Stiehler, M.; Kirschner, S.; Engelhardt, M.; Kasten, P. An epidemiological investigation of training and injury patterns in triathletes. *J. Sports Sci.* **2014**, *32*, 583–590. [CrossRef]
15. Dommerholt, J. Dry needling—Peripheral and central considerations. *J. Man. Manip. Ther.* **2011**, *19*, 223–227. [CrossRef]
16. Hong, C.Z. Lidocaine injection versus dry needling to myofascial trigger point. The importance of the local twitch response. *Am. J. Phys. Med. Rehabil.* **1994**, *73*, 256–263. [CrossRef]
17. Kuan, T.-S.; Hsieh, Y.-L.; Chen, S.-M.; Chen, J.-T.; Yen, W.-C.; Hong, C.-Z. The myofascial trigger point region: Correlation between the degree of irritability and the prevalence of endplate noise. *Am. J. Phys. Med. Rehabil.* **2007**, *86*, 183–189. [CrossRef]
18. Chou, L.-W.; Kao, M.-J.; Lin, J.-G. Probable mechanisms of needling therapies for myofascial pain control. *Evid. Based Complement. Altern. Med.* **2012**, *2012*, 705327. [CrossRef]
19. Gattie, E.; Cleland, J.A.; Snodgrass, S. The effectiveness of trigger point dry needling for musculoskeletal conditions by physical therapists: A systematic review and meta-analysis. *J. Orthop. Sports Phys. Ther. Mov. Sci. Media* **2017**, *47*, 133–149. [CrossRef]
20. Hains, G.; Descarreaux, M.; Hains, F. Chronic shoulder pain of myofascial origin: A randomized clinical trial using ischemic compression therapy. *J. Manip. Physiol. Ther.* **2010**, *33*, 362–369. [CrossRef]
21. Apropos of a muscles. In *Myofascial Pain and Dysfunction*; Travell, J.G.; Simons, D.G. (Eds.) Williams & Wilkins: Baltimore, MD, USA, 2002.
22. Cagnie, B.; Castelein, B.; Pollie, F.; Steelant, L.; Verhoeyen, H.; Cools, A. Evidence for the Use of Ischemic Compression and Dry Needling in the Management of Trigger Points of the Upper Trapezius in Patients with Neck Pain: A Systematic Review. *Am. J. Phys. Med. Rehabil.* **2015**, *94*, 573–583. [CrossRef]
23. Miota Ibarra, J.; You Ge, H.; Wang, W.; Martínez Vizcaíno, V.; Graven.Nielsen, T.; Arendt-Nielsen, L. Latent myofascial trigger points are associated with an increased antagonistic muscle activity during agonist muscle contraction. *J. Pain* **2011**, *12*, 1282–1288. [CrossRef]
24. Ge, H.-Y.; Monterde, S.; Graven-Nielsen, T.; Arendt-Nielsen, L. Latent Myofascial Trigger Points Are Associated with an Increased Intramuscular Electromyographic Activity During Synergistic Muscle Activation. *J. Pain* **2014**, *15*, 181–187. [CrossRef]
25. Ge, H.-Y.; Arendt-Nielsen, L.; Madeleine, P. Accelerated muscle fatigability of latent myofascial trigger points in humans. *Pain Med.* **2012**, *13*, 957–964. [CrossRef]
26. Ge, H.-Y.; Arendt-Nielsen, L. Latent Myofascial Trigger Points. *Curr. Pain Headache Rep.* **2011**, *15*, 386–392. [CrossRef]
27. Gerwin, R.D.; Dommerholt, J.; Shah, J.P. An expansion of Simons' integrated hypothesis of trigger point formation. *Curr. Pain Headache Rep.* **2004**, *8*, 468–475. [CrossRef]
28. Simons, D.G.; Hong, C.-Z.; Simons, L.S. Endplate potentials are common to midfiber myofacial trigger points. *Am. J. Phys. Med. Rehabil.* **2002**, *81*, 212–222. [CrossRef]

29. Abbaszadeh-Amirdehi, M.; Ansari, N.N.; Naghdi, S.; Olyaei, G.; Nourbakhsh, M.R. Therapeutic effects of dry needling in patients with upper trapezius myofascial trigger points. *Acupunct. Med.* **2017**, *35*, 85–92. [CrossRef]
30. Cagnie, B.; Dewitte, V.; Barbe, T.; Timmermans, F.; Delrue, N.; Meeus, M. Physiologic effects of dry needling. *Curr. Pain Headache Rep.* **2013**, *17*, 348. [CrossRef]
31. Kalichman, L.; Vulfsons, S. Dry Needling in the Management of Musculoskeletal Pain. *J. Am. Board Fam. Med.* **2010**, *23*, 640–646. [CrossRef]
32. Simons, D.G.; Travell, J.G.S.L. *Travell and Simons' Myofascial Pain and Dysfunction: The Trigger Point Manual Upper Half of Body*, 2nd ed.; Lippincott Williams & Wilkins: Baltimore, MD, USA, 1999.
33. Kubo, K.; Yajima, H.; Takayama, M.; Ikebukuro, T.; Mizoguchi, H.; Takakura, N. Effects of acupuncture and heating on blood volume and oxygen saturation of human Achilles tendon in vivo. *Eur. J. Appl. Physiol.* **2010**, *109*, 545–550. [CrossRef]
34. Ohkubo, M.; Hamaoka, T.; Niwayama, M.; Murase, N.; Osada, T.; Kime, R.; Kurosawa, Y.; Sakamoto, A.; Katsumura, T. Local increase in trapezius muscle oxygenation during and after acupuncture. *Dyn. Med.* **2009**, *8*, 2. [CrossRef]
35. Sato, A.; Sato, Y.; Shimura, M.; Uchida, S. Calcitonin gene-related peptide produces skeletal muscle vasodilation following antidromic stimulation of unmyelinated afferents in the dorsal root in rats. *Neurosci. Lett.* **2000**, *283*, 137–140. [CrossRef]
36. Benito-de-Pedro, M.; Becerro-de-Bengoa-Vallejo, R.; Elena Losa-Iglesias, M.; Rodríguez-Sanz, D.; López-López, D.; Palomo-López, P.; Mazoteras-Pardo, V.; Calvo-Lobo, C. Effectiveness of Deep Dry Needling vs Ischemic Compression in the Latent Myofascial Trigger Points of the Shortened Triceps Surae from Triathletes on Ankle Dorsiflexion, Dynamic, and Static Plantar Pressure Distribution: A Clinical Trial. *Pain Med.* **2020**, *21*, e172–e181. [CrossRef] [PubMed]
37. Travell, J.G.; Simons, D.G. *Myofascial Pain and Dysfunction: The Trigger Point Manual*; Williams & Wilkins: Baltimore, MD, USA, 1992.
38. Myburgh, C.; Larsen, A.H.; Hartvigsen, J. A Systematic, Critical Review of Manual Palpation for Identifying Myofascial Trigger Points: Evidence and Clinical Significance. *Arch. Phys. Med. Rehabil.* **2008**, *89*, 1169–1176. [CrossRef] [PubMed]
39. Llamas-Ramos, R.; Pecos-Martín, D.; Gallego-Izquierdo, T.; Llamas-Ramos, I.; Plaza-Manzano, G.; Ortega-Santiago, R.; Joshua Cleland, J.; Fernández-de-las-Peñas, C. Comparison of the Short-Term Outcomes between Trigger Point Dry Needling and Trigger Point Manual Therapy for the Management of Chronic Mechanical Neck Pain: A Randomized Clinical Trial. *J. Orthop. Sports Phys. Ther.* **2014**, *44*, 852–861. [CrossRef] [PubMed]
40. Martínez de la Iglesia, J.; Dueñas Herrero, R.; Onís Vilches, M.C.; Aguado Taberné, C.; Albert Colomer, C.; Luque Luque, R. Spanish language adaptation and validation of the Pfeiffer's questionnaire (SPMSQ) to detect cognitive deterioration in people over 65 years of age. *Med. Clin.* **2001**, *117*, 129–134.
41. Faul, F.; Erdfelder, E.; Lang, A.-G.; Buchner, A. G*Power 3: A flexible statistical power analysis program for the social, behavioral, and biomedical sciences. *Behav. Res. Methods* **2007**, *39*, 175–191. [CrossRef]
42. Hermens, H.J.; Freriks, B.; Disselhorst-Klug, C.; Rau, G. Development of recommendations for SEMG sensors and sensor placement procedures. *J. Electromyogr. Kinesiol.* **2000**, *10*, 361–374. [CrossRef]
43. De Vito, G.; McHugh, D.; Macaluso, A.; Riches, P.E. Is the coactivation of biceps femoris during isometric knee extension affected by adiposity in healthy young humans? *J. Electromyogr. Kinesiol.* **2003**, *13*, 425–431. [CrossRef]
44. Murley, G.S.; Bird, A.R. The effect of three levels of foot orthotic wedging on the surface electromyographic activity of selected lower limb muscles during gait. *Clin. Biomech.* **2006**, *21*, 1074–1080. [CrossRef]
45. Cram, J.R.; Kasman, E.S.; Holtz, J. *Introduction to Surface Electromyography*; Aspen Publishers: New York, NY, USA, 1998.
46. Pfeifer, K.; Banzer, W. Motor performance in different dynamic tests in knee rehabilitation. *Scand. J. Med. Sci. Sports.* **1999**, *9*, 19–27. [CrossRef]
47. Fernández-Carner, J.; Ge, H.-Y.; Kimura, Y.; Fernández-de-Las-Peñas, C.; Arendt-Nielsen, L. Increased spontaneous electrical activity at a latent myofascial trigger point after nociceptive stimulation of another latent trigger point. *Clin. J. Pain* **2010**, *26*, 138–143. [CrossRef]
48. Huang, S.; Ferris, D.P. Muscle activation patterns during walking from transtibial amputees recorded within the residual limb-prosthetic interface. *J. Neuroeng. Rehabil.* **2012**, *9*, 55. [CrossRef]
49. Hermens, H.J.; Merletti, R.; Rix, H.F.B. *The State of the Art on Signal Processing Methods for Surface Elektromyographie*; SENIAM-deliverable 7; Roessingh Research and Development b.v.: Enschede, The Netherlands, 1999.
50. Mayoral, O. Dry needling treatments for myofascial trigger points. *J. Musculoskelet. Pain* **2010**, *18*, 411–416. [CrossRef]
51. Baldry, P. Superficial versus deep dry needling. *Acupunct. Med.* **2002**, *20*, 78–81. [CrossRef]
52. Aguilera, F.J.M.; Martín, D.P.; Masanet, R.A.; Botella, A.C.; Soler, L.B.; Morell, F.B. Immediate effect of ultrasound and ischemic compression techniques for the treatment of trapezius latent myofascial trigger points in healthy subjects: A randomized controlled study. *J. Manip. Physiol. Ther.* **2009**, *32*, 515–520. [CrossRef] [PubMed]
53. Cagnie, B.; Dewitte, V.; Coppieters, I.; Van Oosterwijck, J.; Cools, A.; Danneels, L. Effect of Ischemic Compression on Trigger Points in the Neck and Shoulder Muscles in Office Workers: A Cohort Study. *J. Manip. Physiol. Ther.* **2013**, *36*, 482–489. [CrossRef]
54. Benito-de-Pedro, M.; Becerro-de-Bengoa-Vallejo, R.; Losa-Iglesias, M.E.; Rodríguez-Sanz, D.; López-López, D.; Cosín-Matamoros, J.; Martínez-Jiménez, E.M.; Calvo-Lobo, C. Effectiveness between Dry Needling and Ischemic Compression in the Triceps Surae Latent Myofascial Trigger Points of Triathletes on Pressure Pain Threshold and Thermography: A Single Blinded Randomized Clinical Trial. *J. Clin. Med.* **2019**, *8*, 1632. [CrossRef]

55. Lin, Y.-C.; Lai, C.-H.; Chang, W.-H.; Tu, L.-W.; Lin, J.-C.; Chou, S.-W. Immediate effects of ischemic compression on neck function in patients with cervicogenic cephalic syndrome. *J. Manip. Physiol. Ther.* **2012**, *35*, 301–307. [CrossRef]
56. Halle, J.S.; Halle, R.J. Pertinent dry needling considerations for minimizing adverse effects—Part two. *Int. J. Sports Phys. Ther.* **2016**, *11*, 810.
57. Steentjes, K.; de Vries, L.M.A.; Ridwan, B.U.; Wijgman, A.J.J. Infection of a hip prosthesis after dry needling. *Ned. Tijdschr. Geneeskd.* **2016**, *160*, A9364.
58. Baraja-Vegas, L.; Martín-Rodríguez, S.; Piqueras-Sanchiz, F.; Faundez-Aguilera, J.; Bautista, I.J.; Barrios, C.; Garcia-Escudero, M.; Fernández-de-las-Peñas, C. Localization of Muscle Edema and Changes on Muscle Contractility After Dry Needling of Latent Trigger Points in the Gastrocnemius Muscle. *Pain Med.* **2019**, *20*, 1387–1394. [CrossRef] [PubMed]
59. Cummings, M.; Ross-Marrs, R.; Gerwin, R. Pneumothorax complication of deep dry needling demonstration. *Acupunct. Med.* **2014**, *32*, 517–519. [CrossRef] [PubMed]
60. Callan, A.K.; Bauer, J.M.; Martus, J.E. Deep Spine Infection after Acupuncture in the Setting of Spinal Instrumentation. *Spine Deform.* **2016**, *4*, 156–161. [CrossRef] [PubMed]
61. Lee, J.-H.; Lee, H.; Jo, D.-J. An Acute Cervical Epidural Hematoma as a Complication of Dry Needling. *Spine* **2011**, *36*, E891–E893. [CrossRef]
62. *Trigger Point Dry Needling: An Evidence and Clinical-Based Approach*, 2nd ed.; Fernández-de-las-Peñas, C. (Ed.) Churchill Livingstone,: London, UK, 2019.
63. Grieve, R.; Barnett, S.; Coghill, N.; Cramp, F. Myofascial trigger point therapy for triceps surae dysfunction: A case series. *Man Ther.* **2013**, *18*, 519–525. [CrossRef]
64. Grieve, R.; Barnett, S.; Coghill, N.; Cramp, F. The prevalence of latent myofascial trigger points and diagnostic criteria of the triceps surae and upper trapezius: A cross sectional study. *Physiotherapy* **2013**, *99*, 278–284. [CrossRef]
65. Simons, D.G. Review of enigmatic MTrPs as a common cause of enigmatic musculoskeletal pain and dysfunction. *J. Electromyogr. Kinesiol.* **2004**, *14*, 95–107. [CrossRef]
66. Jafari, M.; Bahrpeyma, F.; Togha, M. Effect of ischemic compression for cervicogenic headache and elastic behavior of active trigger point in the sternocleidomastoid muscle using ultrasound imaging. *J. Bodyw. Mov. Ther.* **2017**, *21*, 933–939. [CrossRef]
67. Takamoto, K.; Sakai, S.; Hori, E.; Urakawa, S.; Umeno, K.; Ono, T.; Nishijo, H. Compression on trigger points in the leg muscle increases parasympathetic nervous activity based on heart rate variability. *J. Physiol. Sci.* **2009**, *59*, 191–197. [CrossRef]
68. Ishii, H.; Niioka, T.; Watanabe, H.; Izumi, H. Inhibitory effects of excess sympathetic activity on parasympathetic vasodilation in the rat masseter muscle. *Am. J. Physiol. Regul. Integr. Comp. Physiol.* **2007**, *293*, R729–R736. [CrossRef]
69. Hsieh, Y.-L.; Yang, S.-A.; Yang, C.-C.; Chou, L.-W. Dry needling at myofascial trigger spots of rabbit skeletal muscles modulates the biochemicals associated with pain, inflammation, and hypoxia. *Evid. Based Complement. Altern. Med.* **2012**, *2012*, 342165. [CrossRef]
70. Baraja-Vegas, L.; Martín-Rodríguez, S.; Piqueras-Sanchiz, F.; Martín-Ruiz, J.; Yeste Fabregat, M.; Florencio, L.L.; Fernández-de-las-Peñas, C. Electromyographic Activity Evolution of Local Twitch Responses During Dry Needling of Latent Trigger Points in the Gastrocnemius Muscle: A Cross-Sectional Study. *Pain Med.* **2019**, *21*, 1224–1229. [CrossRef]
71. De Meulemeester, K.E.; Castelein, B.; Coppieters, I.; Barbe, T.; Cools, A.; Cagnie, B. Comparing Trigger Point Dry Needling and Manual Pressure Technique for the Management of Myofascial Neck/Shoulder Pain: A Randomized Clinical Trial. *J. Manip. Physiol. Ther.* **2017**, *40*, 11–20. [CrossRef]
72. Behrangrad, S.; Abbaszadeh-Amirdehi, M.; Kordi Yoosefinejad, A.; Mokhtar Esmaeilnejadganji, S. Comparison of dry needling and ischaemic compression techniques on pain and function in patients with patellofemoral pain syndrome: A randomised clinical trial. *Acupunct Med.* **2020**, *38*, 371–379. [CrossRef]
73. Vernon, H.; Schneider, M. Chiropractic management of myofascial trigger points and myofascial pain syndrome: A systematic review of the literature. *J. Manip. Physiol. Ther.* **2009**, *32*, 14–24. [CrossRef]
74. Hong, C.Z. Myofascial pain therapy. *J. Musculoskelet. Pain* **2004**, *12*, 37–43. [CrossRef]
75. Hsieh, Y.-L.; Chou, L.-W.; Joe, Y.-S.; Hong, C.-Z. Spinal cord mechanism involving the remote effects of dry needling on the irritability of myofascial trigger spots in rabbit skeletal muscle. *Arch. Phys. Med. Rehabil.* **2011**, *92*, 1098–1105. [CrossRef]
76. Shah, J.P.; Gilliams, E.A. Uncovering the biochemical milieu of myofascial trigger points using in vivo microdialysis: An application of muscle pain concepts to myofascial pain syndrome. *J. Bodyw. Mov. Ther.* **2008**, *12*, 371–384. [CrossRef]
77. Chen, J.T.; Chung, K.C.; Hou, C.R.; Kuan, T.S.; Chen, S.M.; Hong, C.Z. Inhibitory effect of dry needling on the spontaneous electrical activity recorded from myofascial trigger spots of rabbit skeletal muscle. *Am. J. Phys. Med. Rehabil.* **2001**, *80*, 729–735. [CrossRef]
78. Abbaszadeh-Amirdehi, M.; Ansari, N.N.; Naghdi, S.; Olyaei, G.; Nourbakhsh, M.R. The neurophysiological effects of dry needling in patients with upper trapezius myofascial trigger points: Study protocol of a controlled clinical trial. *BMJ Open.* **2013**, *3*, e002825. [CrossRef]

sensors

MDPI

Article

Noninvasive Assessment of Neuromechanical Coupling and Mechanical Efficiency of Parasternal Intercostal Muscle during Inspiratory Threshold Loading

Manuel Lozano-García [1,2,3,*], Luis Estrada-Petrocelli [1,2,4], Abel Torres [1,2,3], Gerrard F. Rafferty [5], John Moxham [5], Caroline J. Jolley [5,6,†] and Raimon Jané [1,2,3,†]

1 Institute for Bioengineering of Catalonia (IBEC), The Barcelona Institute of Science and Technology (BIST), UPC Campus Diagonal-Besòs, Av. d'Eduard Maristany 10–14, 08019 Barcelona, Spain; lestrada@ibecbarcelona.eu (L.E.-P.); atorres@ibecbarcelona.eu (A.T.); rjane@ibecbarcelona.eu (R.J.)
2 Biomedical Research Networking Centre in Bioengineering, Biomaterials and Nanomedicine (CIBER-BBN), 08028 Barcelona, Spain
3 Department of Automatic Control (ESAII), Universitat Politècnica de Catalunya (UPC)-Barcelona Tech, 08028 Barcelona, Spain
4 Facultad de Ingeniería, Universidad Latina de Panamá, Panama City 0823-00923, Panama
5 Centre for Human & Applied Physiological Sciences, School of Basic & Medical Biosciences, Faculty of Life Sciences & Medicine, King's College London, King's Health Partners, London SE1 9RT, UK; gerrard.rafferty@kcl.ac.uk (G.F.R.); john.moxham@kcl.ac.uk (J.M.); caroline.jolley@kcl.ac.uk (C.J.J.)
6 King's College Hospital NHS Foundation Trust, King's Health Partners, London SE5 9RS, UK
* Correspondence: mlozano@ibecbarcelona.eu; Tel.: +34-695-119-250
† C.J.J. and R.J. contributed equally to this work.

Citation: Lozano-García, M.; Estrada-Petrocelli, L.; Torres, A.; Rafferty, G.F.; Moxham, J.; Jolley, C.J.; Jané, R. Noninvasive Assessment of Neuromechanical Coupling and Mechanical Efficiency of Parasternal Intercostal Muscle during Inspiratory Threshold Loading. *Sensors* **2021**, *21*, 1781. https://doi.org/10.3390/s21051781

Academic Editor: Marco Di Rienzo

Received: 11 January 2021
Accepted: 28 February 2021
Published: 4 March 2021

Publisher's Note: MDPI stays neutral with regard to jurisdictional claims in published maps and institutional affiliations.

Abstract: This study aims to investigate noninvasive indices of neuromechanical coupling (NMC) and mechanical efficiency (MEff) of parasternal intercostal muscles. Gold standard assessment of diaphragm NMC requires using invasive techniques, limiting the utility of this procedure. Noninvasive NMC indices of parasternal intercostal muscles can be calculated using surface mechanomyography ($sMMG_{para}$) and electromyography ($sEMG_{para}$). However, the use of $sMMG_{para}$ as an inspiratory muscle mechanical output measure, and the relationships between $sMMG_{para}$, $sEMG_{para}$, and simultaneous invasive and noninvasive pressure measurements have not previously been evaluated. $sEMG_{para}$, $sMMG_{para}$, and both invasive and noninvasive measurements of pressures were recorded in twelve healthy subjects during an inspiratory loading protocol. The ratios of $sMMG_{para}$ to $sEMG_{para}$, which provided muscle-specific noninvasive NMC indices of parasternal intercostal muscles, showed nonsignificant changes with increasing load, since the relationships between $sMMG_{para}$ and $sEMG_{para}$ were linear (R^2 = 0.85 (0.75–0.9)). The ratios of mouth pressure (P_{mo}) to $sEMG_{para}$ and $sMMG_{para}$ were also proposed as noninvasive indices of parasternal intercostal muscle NMC and MEff, respectively. These indices, similar to the analogous indices calculated using invasive transdiaphragmatic and esophageal pressures, showed nonsignificant changes during threshold loading, since the relationships between P_{mo} and both $sEMG_{para}$ (R^2 = 0.84 (0.77–0.93)) and $sMMG_{para}$ (R^2 = 0.89 (0.85–0.91)) were linear. The proposed noninvasive NMC and MEff indices of parasternal intercostal muscles may be of potential clinical value, particularly for the regular assessment of patients with disordered respiratory mechanics using noninvasive wearable and wireless devices.

Keywords: inspiratory threshold loading; neuromechanical coupling; parasternal intercostal muscles; respiratory pressure; surface electromyography; surface mechanomyography

1. Introduction

Evaluating respiratory muscle function contributes to enhancing the diagnosis, phenotyping, and monitoring of patients with respiratory symptoms and neuromuscular diseases [1]. Inspiratory pressure generation is dependent on the level of muscle electrical activation and the transduction of this electrical activation into pressure, termed neuromechanical coupling (NMC). Poor transduction of inspiratory muscle electrical activation into pressure, neuromechanical uncoupling, is common in obstructive lung disease [2–4], and has been related to the perception of dyspnea in both obstructive lung disease [5] and in neuromuscular disease [6]. Measuring inspiratory muscle NMC is therefore clinically important for the assessment of patients with disordered respiratory mechanics.

Inspiratory muscle NMC estimation requires simultaneous measurement of muscle electrical activation and mechanical output. To date, most studies reporting inspiratory muscle NMC have focused on the diaphragm, the principal inspiratory muscle [7]. The gold standard measurement of diaphragm electrical activation and mechanical output involves the use of invasive techniques, including crural diaphragm electromyography using a multipair esophageal electrode catheter ($oesEMG_{di}$) to assess neural respiratory drive [8], and the balloon-catheter technique to measure transdiaphragmatic pressure (P_{di}) [1]. However, the utility of these techniques is limited by the invasive nature of the measurements, subject tolerance, and requirement for specialist equipment and training. In this regard, noninvasive measurement of inspiratory muscle NMC would facilitate the assessment of respiratory muscle function.

Parasternal intercostal muscles are obligatory muscles of inspiration and their activation is coupled to that of the diaphragm [9]. Measurements of surface electromyography of the parasternal intercostal muscles ($sEMG_{para}$) have been shown to provide a robust measure of load on the respiratory muscle pump and to strongly correlate with measurements of $oesEMG_{di}$ [10–12]. Measurements of $sEMG_{para}$ have also been related to breathlessness in both healthy subjects [3,12] and patients with respiratory disease [3,12–15]. Therefore, and given the advantage of being noninvasive, $sEMG_{para}$ has been proposed as an alternative measure for the assessment of neural respiratory drive [10,11].

Surface mechanomyography (sMMG) measures muscle surface vibrations due to motor unit mechanical activity [16], and represents the mechanical counterpart of motor unit electrical activity, as measured by surface electromyography (sEMG). In non-respiratory skeletal muscles, the ratio of sMMG to sEMG has been proposed for the assessment of muscle electromechanical efficiency [17,18]. In the respiratory system, sMMG and pressure represent different mechanical states of inspiratory muscles, the latter being a more global measure of respiratory muscle mechanical output. We have recently used sMMG and sEMG measurements from lower chest wall intercostal spaces ($sMMG_{lic}$ and $sEMG_{lic}$, respectively) to provide noninvasive indices of lower chest wall inspiratory muscle NMC [19], which did not change significantly with increasing inspiratory load due to the linear relationship between $sMMG_{lic}$ and $sEMG_{lic}$ measurements. Measurements of $sMMG_{lic}$ have also been used, in combination with measurements of mouth pressure (P_{mo}), to calculate the mechanical efficiency (MEff) of lower chest wall inspiratory muscles, that is the transformation of muscle mechanical activation into pressure generation, in healthy subjects and patients with chronic obstructive pulmonary disease (COPD) [20]. Lower chest wall recordings are, however, less easily accessible and more contaminated by non-respiratory chest wall and abdominal muscle activation than parasternal intercostal recordings [21,22]. Nevertheless, the use of parasternal intercostal muscle sMMG ($sMMG_{para}$) as a noninvasive measure of inspiratory muscle mechanical output has not previously been investigated.

Inspiratory muscle mechanical output is usually estimated as pressure. Gold standard P_{di} measurements are derived from simultaneous measurements of gastric (P_{gas}) and esophageal (P_{oes}) pressures ($P_{di} = P_{gas} - P_{oes}$). Since P_{di} is specific to the diaphragm, P_{di} has been widely used, together with $oesEMG_{di}$, for the assessment of diaphragm NMC [4,12,19,23–26]. Measurements of P_{oes} alone are, however, not specific to the diaphragm and are influenced by the activation of all inspiratory muscles [1]. Compared to P_{di}, P_{oes} is therefore expected to better reflect the mechanical output of extra-diaphragmatic muscles, including the parasternal intercostal muscles. Moreover, P_{mo}, which provides a noninvasive approximation of P_{oes} [1], could be used to obtain noninvasive indices of NMC and MEff of parasternal intercostal muscles. However, to our knowledge, no study has evaluated the relationships among simultaneous invasive and noninvasive measurements of respiratory pressures and both $sEMG_{para}$ and $sMMG_{para}$ measurements for the assessment of NMC and MEff of parasternal intercostal muscles.

The principal aim of this study was therefore to investigate the use of $sMMG_{para}$ as a novel noninvasive index of parasternal intercostal muscle mechanical output. In addition, we wished to examine the relationship between $sMMG_{para}$ and $sEMG_{para}$ to provide muscle-specific noninvasive indices of NMC of parasternal intercostal muscles, in healthy subjects during an incremental inspiratory threshold loading protocol. We hypothesized that, as in lower chest wall inspiratory muscles, there would be a linear relationship between measurements of $sMMG_{para}$ and $sEMG_{para}$ in healthy subjects, thus resulting in nonsignificant changes in NMC of parasternal intercostal muscles (the ratio of $sMMG_{para}$ to $sEMG_{para}$ measurements) during threshold loading. We also aimed to explore the relationship between measurements of $sEMG_{para}$ and $sMMG_{para}$ and both invasive and noninvasive measurements of respiratory pressures, to obtain indices of NMC and MEff of parasternal intercostal muscles.

2. Materials and Methods

2.1. Data Acquisition and Preprocessing

2.1.1. Measurements

This study was carried out in the Respiratory Physiology Laboratory at the King's College London, King's College Hospital, London, United Kingdom. Ethics approval was obtained from the National Health Service Health Research Authority (National Research Ethics Service Committee London—Dulwich 05/Q0703) and the study was conducted in accordance with the Declaration of Helsinki.

All recordings were obtained from 12 healthy subjects, including 6 male and 6 female, with no medical history of neuromuscular or cardiorespiratory disease and with the following characteristics (median (interquartile range)): age 33 (30–39) years, body mass index (BMI) 22.2 (20.6–24.2) kg/m^2, forced expiratory volume in 1 second 98% (95–106%) % of predicted, and forced expiratory volume in 1 second/forced vital capacity 82% (74–84%) % of predicted [19]. Written consent was provided by all subjects prior to study participation.

$sEMG_{para}$ was acquired using a pair of surface electrodes in bipolar configuration located bilaterally in the second intercostal space [10,27]. The $sEMG_{para}$ recordings were amplified, high-pass filtered with a cut-off frequency of 10 Hz, and alternating current coupled. $sMMG_{para}$ was acquired using a triaxial accelerometer (TSD109C2; BIOPAC Systems Inc, Goleta, CA, USA) attached to the skin in the second intercostal space, between the $sEMG_{para}$ electrodes and close to the right one. P_{gas} and P_{oes} were acquired using a catheter tip pressure transducer (CTO-2; Gaeltec Devices Ltd., Dunvegan, UK), as previously described [12]. Airflow was acquired using a pneumotachometer (4830; Hans Rudolph Inc, Shawnee, KS, USA). P_{mo} was acquired using a differential pressure transducer (MP45; Validyne Engineering, Northridge, CA, USA) connected to a side port of the pneumotachometer.

All respiratory signals were acquired using a 16-bit analog-to-digital converter (PowerLab 16/35; ADInstruments Ltd, Oxford, UK) at a sampling frequency of 4000 Hz ($sEMG_{para}$), 2000 Hz ($sMMG_{para}$), and 100 Hz (pressures and airflow). LabChart software (Version 7.2, ADInstruments Pty, Colorado Springs, CO, USA) was used to manage the signal acquisition.

2.1.2. Acquisition Protocol

Initially, each participant carried out two types of maximal volitional respiratory maneuvers: maximal static inspiratory pressure (PImax) [1] and maximal inspirations to total lung capacity [10,28]. These maneuvers were performed several times to ensure each subject's maximal volitional effort. All signals were recorded during two minutes of resting breathing (L0). After that, each participant carried out an inspiratory threshold loading protocol at inspiratory threshold loads ranging from 12% (L1) to 60% (L5) of PImax, in increments of 12%, as we previously described [19]. Inspiratory loads were generated using an electronic inspiratory muscle trainer (POWERbreathe K5; POWERbreathe International Ltd, Southam, UK) attached to the distal end of the pneumotachometer. Subjects breathed through the pneumotachometer and performed 30 breaths against each inspiratory load, with a resting period in between to allow all measurements to return to baseline.

2.1.3. Data Preprocessing

All data were processed and analyzed in MATLAB (The MathWorks, Inc., vR2020a, Natick, MA, USA). $sEMG_{para}$ signals were down-sampled at 2000 Hz and filtered between 10 and 600 Hz. Removal of 50 Hz interference was performed by means of a comb filter. $sMMG_{para}$ signals were down-sampled at 500 Hz and filtered between 5 and 40 Hz. The three acceleration components recorded by the accelerometer were root sum squared to calculate the magnitude of the $sMMG_{para}$ signal ($|sMMG_{para}|$).

A zero-crossing threshold-based detector was applied to P_{mo} in order to detect inspiratory and expiratory segments during resting breathing and each threshold load for subsequent inspiratory muscle activity estimation. For each respiratory cycle, the following parameters were calculated: inspiratory time, inspiratory time/total respiratory cycle time, inspiratory volume (i.e., the area under the curve of the inspiratory flow), and breathing frequency. The median values of all respiratory cycles were then calculated separately for L0–L5, and 10 respiratory cycles that contained the four parameters nearest to the median values were automatically selected, resulting in 60 respiratory cycles per subject.

The moving minimum of the P_{di} signal and the moving maximum of the P_{oes} signal were calculated using a moving window of 1.5 times the maximum inspiratory time of each load, and subtracted from the P_{di} and P_{oes} signals, respectively.

2.2. Data Processing

2.2.1. Inspiratory Muscle Activity Estimation

Inspiratory muscle activity estimation has usually been based on conventional amplitude estimators, such as the average rectified value (ARV) or the root mean square (RMS) [28,29]. However, when applied to myographic respiratory signals, these parameters are greatly affected by cardiac activity, and therefore prior rejection of signal segments that contain cardiac noise is required. Fixed sample entropy (fSampEn) has been demonstrated to reduce cardiac activity in myographic respiratory signals [30,31], allowing inspiratory muscle activity to be estimated without the need for prior rejection of cardiac artefacts. fSampEn consists in calculating sample entropy within a moving window over a signal, using a fixed tolerance value for all windows [30]. In this way, fSampEn is sensitive to variations in both signal complexity and signal amplitude. Therefore, like ARV and RMS, fSampEn can track amplitude variations in myographic respiratory signals. Furthermore, since cardiac artifacts exhibit a more deterministic pattern compared to respiratory EMG and MMG signals, fSampEn also inherently reduces cardiac activity, and therefore it is less influenced by cardiac artefacts than ARV or RMS.

In this study, fSampEn was calculated for the $sEMG_{para}$ ($fSEsEMG_{para}$) and $|sMMG_{para}|$ ($fSE|sMMG_{para}|$) signals, using the optimal fSampEn parameters that we previously described for respiratory muscle activity estimation (i.e., a moving window of 500 ms with a 50 ms step, m equal to 2, and r equal to 0.3 ($sEMG_{para}$) and 0.5 ($|sMMG_{para}|$) times the global standard deviation of each signal) [32]. Inspiratory muscle activity was then estimated for each respiratory cycle as the mean inspiratory P_{di}, P_{oes}, P_{mo}, $fSE|sMMG_{para}|$, and $fSEsEMG_{para}$. The mean values of $fSE|sMMG_{para}|$ and $fSEsEMG_{para}$ were expressed as percentages of their respective largest mean values obtained during the inspiratory threshold loading protocol and the two maximal volitional maneuvers ($fSE|sMMG_{para}|_{\%max}$ and $fSEsEMG_{para\%max}$).

2.2.2. Neuromechanical Coupling and Mechanical Efficiency

NMCs of parasternal intercostal muscles were calculated as the ratios of fSE|sMMG$_{para}$|$_{\%max}$, mean P_{di}, mean P_{oes}, and mean P_{mo} to fSEsEMG$_{para\%max}$ (NMC$_{MMG\text{-}EMGpara}$, NMC$_{Pdi\text{-}EMGpara}$, NMC$_{Poes\text{-}EMGpara}$, and NMC$_{Pmo\text{-}EMGpara}$, respectively). MEffs of parasternal intercostal muscles were calculated as the ratios of mean P_{di}, mean P_{oes}, and mean P_{mo} to fSE|sMMG$_{para}$|$_{\%max}$ (MEff$_{Pdi\text{-}MMGpara}$, MEff$_{Poes\text{-}MMGpara}$, and MEff$_{Pmo\text{-}MMGpara}$, respectively). The average value of the ten respiratory cycles selected for resting breathing and each load was calculated for all NMCs and MEffs.

2.3. *Statistical Analysis*

All data correspond to median and interquartile range. Changes in respiratory pressures, fSE|sMMG$_{para}$|$_{\%max}$, and fSEsEMG$_{para\%max}$ with increasing threshold load were assessed using Friedman tests followed by multiple Wilcoxon signed-rank tests with Bonferroni adjusted *p*-values.

An increasing NMC with increasing load can be well explained by an exponential relationship between the measurements involved [19]. By contrast, a NMC that remains almost constant or increases slightly with increasing load indicates that the relationship between the measurements involved can be well explained by a linear model. Therefore, in this study, the relationships between measurements of respiratory pressures, fSE|sMMG$_{para}$|$_{\%max}$, and fSEsEMG$_{para\%max}$ were assessed individually using both linear and exponential regression models, together with Spearman's rank correlation coefficients (ρ). Moreover, changes in NMCs and MEffs during threshold loads L1–L5 were assessed using Friedman tests, followed by multiple Wilcoxon signed-rank tests with Bonferroni adjustment for multiple comparisons, with the same method that we previously described [19]. The significance level for all statistical tests was set at 0.05.

3. Results

Figure 1 shows an example of respiratory signals recorded in a healthy subject during resting breathing and the inspiratory threshold loading protocol.

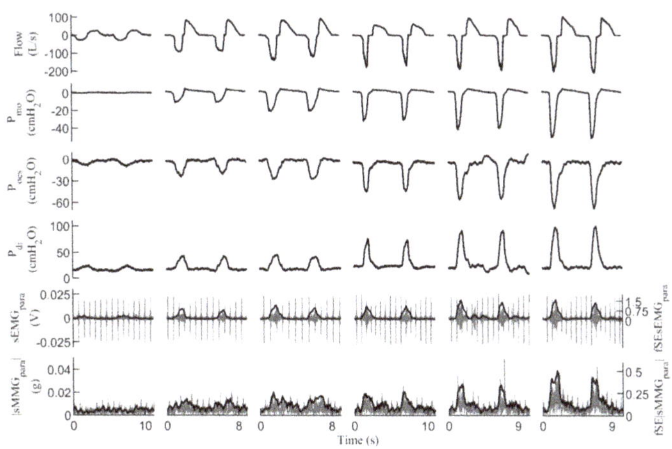

Figure 1. Signals recorded during the inspiratory threshold loading protocol in a healthy subject. Two respiratory cycles are shown for quiet resting breathing and inspiratory threshold loads at 12%, 24%, 36%, 48%, and 60% of maximal static inspiratory pressure. For the sEMG$_{para}$ and |sMMG$_{para}$| signals, the corresponding fixed sample entropy time-series (fSEsEMG$_{para}$ and fSE|sMMG$_{para}$| respectively) are also shown. P_{mo} = mouth pressure, P_{oes} = esophageal pressure, P_{di} = transdiaphragmatic pressure, sEMG$_{para}$ = surface electromyography of the parasternal intercostal muscles, |sMMG$_{para}$| = surface mechanomyography of the parasternal intercostal muscles.

3.1. Respiratory Pressures

Median (interquartile range (IQR)) PImax for the group was 87.0 (78.0–116.5) cmH_2O. The inspiratory threshold loads increased from 10.5 (9.5–14.0) cmH_2O during load L1 to 52.0 (47.0–70.0) cmH_2O during load L5. Mean P_{di}, mean P_{oes}, and mean P_{mo} increased significantly between successive loads during the inspiratory threshold loading protocol (Figure 2). Although all pressures increased in parallel during threshold loading, mean P_{mo} was consistently lower than mean P_{oes} and mean P_{di}.

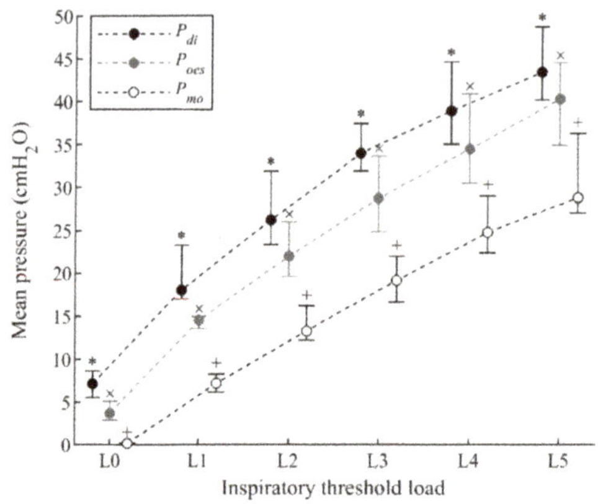

Figure 2. Evolution of respiratory pressures during progressive inspiratory threshold loading. Data points represent median and interquartile (IQR) range of the 12 subjects for each load. All data points with the same symbol (*, ×, or +) were significantly different to each other.

3.2. sMMG and sEMG of Parasternal Intercostal Muscles during Threshold Loading

Measurements of fSE | $sMMG_{para}$ | $_{\%max}$ increased progressively from 10.9% (9.4–14.8%) during resting breathing to 64.7% (54.2–75.9%) during load L5 (Figure 3). Similarly, measurements of $fSEsEMG_{para\%max}$ increased from 13.5% (10.4–17.6%) during resting breathing to 72.2% (67.2–80.9%) during load L5. Increases in fSE | $sMMG_{para}$ | $_{\%max}$ and $fSEsEMG_{para\%max}$ were statistically significant between successive loads, except between loads L3 and L4 in $fSEsEMG_{para\%max}$, and between loads L4 and L5 in both fSE | $sMMG_{para}$ | $_{\%max}$ and $fSEsEMG_{para\%max}$ measurements. Moreover, although fSE | $sMMG_{para}$ | $_{\%max}$ and $fSEsEMG_{para\%max}$ increased in parallel during threshold loading, fSE | $sMMG_{para}$ | $_{\%max}$ was consistently lower than $fSEsEMG_{para\%max}$.

3.3. Noninvasive Measurements of Neuromechanical Coupling and Mechanical Efficiency of Parasternal Intercostal Muscles

Having an increasing or almost constant pattern of NMC and MEff indices with increasing load depends on whether the relationship between the measurements involved is exponential or linear respectively, as we previously described [19]. Accordingly, the relationships between measurements of respiratory pressures, fSE | $sMMG_{para}$ | $_{\%max}$, and $fSEsEMG_{para\%max}$ were firstly assessed individually using both linear and exponential regression models. Secondly, the evolution of the group NMC and MEff indices during threshold loads L1–L5 was assessed using Friedman tests, followed by multiple Wilcoxon signed rank tests with Bonferroni adjustment for multiple comparisons.

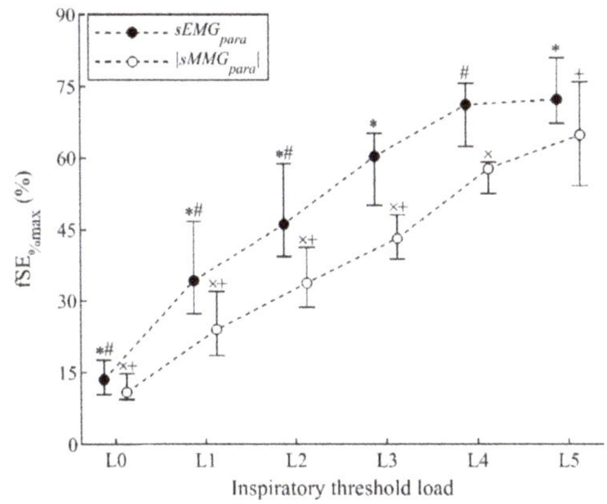

Figure 3. Fixed sample entropy measurements of surface mechanomyography (fSE | sMMG$_{para}$ | $_{\%max}$) and surface electromyography (fSEsEMG$_{para\%max}$) of the parasternal intercostal muscles during inspiratory threshold loading. Data points represent median and interquartile range of the 12 subjects for each load. All data points with the same symbol (*, #, ×, or +) were significantly different to each other.

The relationship between fSE | sMMG$_{para}$ | $_{\%max}$ and fSEsEMG$_{para\%max}$ measurements shown in Figure 4a was firstly assessed. Very strong positive correlations between fSE | sMMG$_{para}$ | $_{\%max}$ and fSEsEMG$_{para\%max}$ were found individually (Table 1). Individual linear and exponential regression results shown in Table 2 indicate that the linear model, besides being simpler, performed slightly better than the exponential model to describe the relationships between fSE | sMMG$_{para}$ | $_{\%max}$ and fSEsEMG$_{para\%max}$. The evolution of the group NMC$_{MMG\text{-}EMGpara}$ during progressive inspiratory threshold loading is shown in Figure 4b. Slight and nonsignificant increases were found in NMC$_{MMG\text{-}EMGpara}$ between loads L1 and L5, confirming that the relationship between fSE | sMMG$_{para}$ | $_{\%max}$ and fSEsEMG$_{para\%max}$ can be well explained by a linear model.

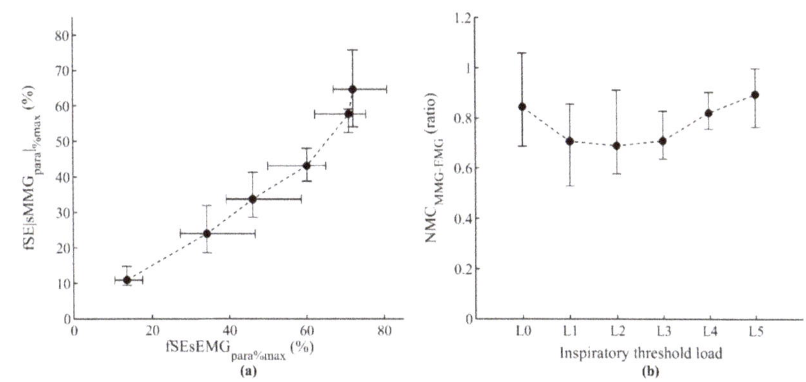

Figure 4. Relationship between fixed sample entropy measurements of surface mechanomyography (fSE | sMMG$_{para}$ | $_{\%max}$) and surface electromyography (fSEsEMG$_{para\%max}$) of the parasternal intercostal muscles (**a**) and the corresponding neuromechanical coupling ratio (NMC$_{MMG\text{-}EMGpara}$) (**b**), during the incremental inspiratory threshold loading protocol. Data points represent median and interquartile range of the 12 subjects for each load.

Table 1. Spearman's ρ between measurements of respiratory pressures, fSE | $sMMG_{para\,|\,\%max}$, and $fSEsEMG_{para\%max}$.

ID	$fSEsEMG_{para\%max}$ fSE | $sMMG_{para\,\|\,\%max}$	$fSEsEMG_{para\%max}$ Mean P_{di}	$fSEsEMG_{para\%max}$ Mean P_{oes}	$fSEsEMG_{para\%max}$ Mean P_{mo}	fSE | $sMMG_{para\,\|\,\%max}$ Mean P_{di}	fSE | $sMMG_{para\,\|\,\%max}$ Mean P_{oes}	fSE | $sMMG_{para\,\|\,\%max}$ Mean P_{mo}
1	0.88	0.90	0.89	0.89	0.89	0.90	0.92
2	0.93	0.94	0.97	0.95	0.89	0.93	0.93
3	0.95	0.87	0.95	0.97	0.82	0.92	0.95
4	0.95	0.92	0.95	0.92	0.94	0.96	0.95
5	0.96	0.96	0.94	0.94	0.96	0.96	0.95
6	0.93	0.88	0.89	0.87	0.92	0.93	0.92
7	0.98	0.98	0.96	0.96	0.97	0.97	0.97
8	0.97	0.98	0.98	0.98	0.96	0.96	0.97
9	0.88	0.91	0.91	0.92	0.92	0.91	0.91
10	0.91	0.82	0.85	0.84	0.91	0.94	0.94
11	0.80	0.76	0.77	0.74	0.96	0.97	0.96
12	0.87	0.91	0.90	0.88	0.91	0.92	0.92
Median (IQR)	0.93 (0.88–0.96)	0.91 (0.88–0.95)	0.92 (0.89–0.96)	0.92 (0.87–0.96)	0.92 (0.90–0.96)	0.94 (0.92–0.96)	0.94 (0.92–0.95)

IQR = interquartile range, P_{di} = transdiaphragmatic pressure, P_{oes} = esophageal pressure, P_{mo} = mouth pressure, $fSEsEMG_{para\%max}$ = fixed sample entropy measurements of surface electromyography of the parasternal intercostal muscles, fSE | $sMMG_{para\,|\,\%max}$ = fixed sample entropy measurements of surface mechanomyography of the parasternal intercostal muscles. All correlations were statistically significant.

Table 2. Adjusted R^2 of the linear and exponential regression models to describe the relationships between measurements of respiratory pressures, fSE | $sMMG_{para\,|\,\%max}$, and $fSEsEMG_{para\%max}$.

ID	$fSEsEMG_{para\%max}$ fSE | $sMMG_{para\,\|\,\%max}$		$fSEsEMG_{para\%max}$ Mean P_{di}		$fSEsEMG_{para\%max}$ Mean P_{oes}		$fSEsEMG_{para\%max}$ Mean P_{mo}		fSE | $sMMG_{para\,\|\,\%max}$ Mean P_{di}		fSE | $sMMG_{para\,\|\,\%max}$ Mean P_{oes}		fSE | $sMMG_{para\,\|\,\%max}$ Mean P_{mo}	
	Lin.	Exp.	Lin.	Exp.	Lin.	Exp.	Lin.	Exp.	Lin.	Exp.	Lin.	Exp.	Lin.	Exp.
1	0.74	0.69	0.81	0.78	0.77	0.72	0.83	0.73	0.80	0.69	0.85	0.75	0.88	0.73
2	0.81	0.80	0.91	0.84	0.96	0.87	0.92	0.82	0.70	0.60	0.76	0.64	0.80	0.65
3	0.86	0.91	0.74	0.81	0.90	0.83	0.94	0.82	0.74	0.76	0.75	0.67	0.79	0.65
4	0.76	0.86	0.91	0.90	0.93	0.89	0.79	0.84	0.81	0.70	0.83	0.67	0.89	0.76
5	0.88	0.91	0.93	0.90	0.91	0.88	0.93	0.86	0.92	0.83	0.90	0.81	0.91	0.79
6	0.91	0.87	0.78	0.74	0.87	0.79	0.75	0.67	0.83	0.78	0.91	0.82	0.83	0.73
7	0.96	0.93	0.95	0.89	0.95	0.88	0.94	0.87	0.95	0.88	0.95	0.87	0.96	0.88
8	0.95	0.86	0.98	0.92	0.98	0.91	0.97	0.90	0.93	0.91	0.94	0.92	0.94	0.92
9	0.84	0.78	0.82	0.71	0.81	0.70	0.86	0.73	0.87	0.79	0.84	0.76	0.89	0.79
10	0.86	0.80	0.74	0.66	0.80	0.74	0.76	0.68	0.84	0.77	0.88	0.85	0.91	0.86

Table 2. *Cont.*

ID	fSEsEMG$_{para\%max}$ fSE∣sMMG$_{para}$∣$_{\%max}$		fSEsEMG$_{para\%max}$ Mean P_{di}		fSEsEMG$_{para\%max}$ Mean P_{oes}		fSEsEMG$_{para\%max}$ Mean P_{mo}		fSE∣sMMG$_{para}$∣$_{\%max}$ Mean P_{di}		fSE∣sMMG$_{para}$∣$_{\%max}$ Mean P_{oes}		fSE∣sMMG$_{para}$∣$_{\%max}$ Mean P_{mo}	
	Lin.	Exp.	Lin.	Exp.	Lin.	Exp.	Lin.	Exp.	Lin.	Exp.	Lin.	Exp.	Lin.	Exp.
11	0.70	0.74	0.83	0.83	0.73	0.73	0.56	0.57	0.91	0.81	0.95	0.87	0.91	0.87
12	0.73	0.76	0.76	0.81	0.83	0.84	0.82	0.81	0.81	0.71	0.87	0.75	0.88	0.74
Median (IQR)	0.85 (0.75–0.90)	0.83 (0.77–0.89)	0.83 (0.77–0.92)	0.82 (0.76–0.90)	0.88 (0.80–0.94)	0.84 (0.74–0.88)	0.84 (0.77–0.93)	0.81 (0.71–0.85)	0.83 (0.80–0.91)	0.77 (0.71–0.82)	0.88 (0.84–0.93)	0.78 (0.71–0.86)	0.89 (0.85–0.91)	0.77 (0.73–0.87)

IQR = interquartile range, Lin. = linear regression model, Exp. = exponential regression model, P_{di} = transdiaphragmatic pressure, P_{oes} = esophageal pressure, P_{mo} = mouth pressure, fSEsEMG$_{para\%max}$ = fixed sample entropy measurements of surface electromyography of the parasternal intercostal muscles, fSE∣sMMG$_{para}$∣$_{\%max}$ = fixed sample entropy measurements of surface mechanomyography of the parasternal intercostal muscles.

Secondly, the relationships between mean P_{di}, mean P_{oes}, and mean P_{mo} and fSEsEMG $_{para\%max}$ measurements shown in Figure 5a were analyzed. Individual analyses showed very strong positive correlations between mean P_{di}, mean P_{oes}, and mean P_{mo} and $fSEsEMG_{para\%max}$ (Table 1). Moreover, the linear and exponential regression results shown in Table 2 indicated that these relationships are better described by linear models than by exponential models. The group $NMC_{Pdi\text{-}EMGpara}$ and $NMC_{Poes\text{-}EMGpara}$ increased very little, not significantly, from load L1 to load L5 (Figure 5b). The group $NMC_{Pmo\text{-}EMGpara}$ increased significantly from load L1 to load L2, but not significantly from load L3 to load L5, thus tending to stabilize around a constant value as the load increases. These results are consistent with the fact that the relationships between mean P_{di}, mean P_{oes}, and mean P_{mo} and $fSEsEMG_{para\%max}$ are better explained by linear than by exponential models.

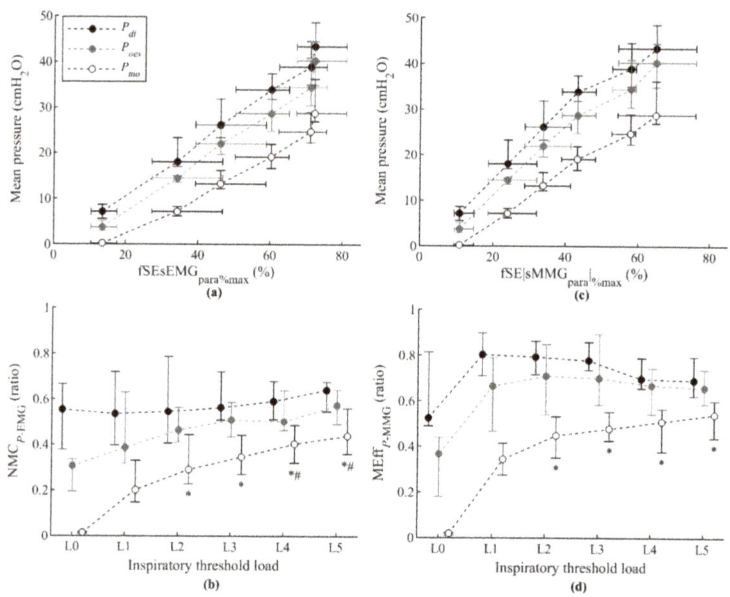

Figure 5. Relationship between respiratory pressures and fixed sample entropy measurements of both surface electromyography ($fSEsEMG_{para\%max}$) (**a**) and surface mechanomyography ($fSE\,|\,sMMG_{para\,|\,\%max}$) (**c**) of the parasternal intercostal muscles, and the corresponding neuromechanical coupling ($NMC_{P\text{-}EMGpara}$) (**b**) and mechanical efficiency ($MEff_{P\text{-}MMGpara}$) (**d**) ratios, during the incremental inspiratory threshold loading protocol. Data points represent median and interquartile range of the 12 subjects for each load. Symbols ∗ and # indicate statistically significant differences with respect to inspiratory threshold loads L1 and L2, respectively.

Finally, the relationships between mean P_{di}, mean P_{oes}, and mean P_{mo} and $fSE\,|\,sMMG_{para\,|\,\%max}$ measurements shown in Figure 5c were analyzed. Individual analyses showed very strong positive correlations between mean P_{di}, mean P_{oes}, and mean P_{mo} and $fSE\,|\,sMMG_{para\,|\,\%max}$ (Table 1). The linear and exponential regression results indicated that these relationships are better described by linear models than by exponential models (Table 2). Accordingly, the group $MEff_{Pdi\text{-}MMGpara}$ and $MEff_{Poes\text{-}MMGpara}$ changed very little, not significantly, from load L1 to load L5 (Figure 5d), and the group $MEff_{Pmo\text{-}MMGpara}$ increased significantly from load L1 to load L2, but not significantly from load L2 to load L5, thus tending to change little as the load increases.

4. Discussion

This study describes for the first time the use of $sMMG_{para}$ as a measure of the mechanical activity of parasternal intercostal muscles, allowing calculation of noninvasive indices of NMC of parasternal intercostal muscles ($NMC_{MMG\text{-}EMGpara}$, i.e., the ratio of fSE | $sMMG_{para}$ | $_{\%max}$ to $fSEsEMG_{para\%max}$). $NMC_{MMG\text{-}EMGpara}$ showed little and not significant changes with progressive increases in inspiratory load between 12% and 60% of PImax. This is due to the mostly linear increase in fSE | $sMMG_{para}$ | $_{\%max}$ relative to $fSEsEMG_{para\%max}$ during threshold loading.

The aforementioned results are equivalent to those found in our previous study using $sMMG_{lic}$ and $sEMG_{lic}$ recordings from the lower chest wall inspiratory muscles [19]. In that study, we showed that both $sEMG_{lic}$ and $sMMG_{lic}$ measurements reflect, in part, the activation of the diaphragm, but also extra-diaphragmatic lower chest wall and abdominal muscle activation [22,33,34], which progressively increases with increasing threshold load to optimize the functioning of the diaphragm [35]. The contribution of the activation of extra-diaphragmatic muscles to $sMMG_{lic}$ and $sEMG_{lic}$ in a similar way was the reason for the parallel increase of fSE | $sMMG_{lic}$ | $_{\%max}$ and $fSEsEMG_{lic\%max}$ during incremental threshold loading, and therefore for the nonsignificant changes in lower chest wall inspiratory muscle NMC from load L1 to load L5.

In the present study, both fSE | $sMMG_{para}$ | $_{\%max}$ and $fSEsEMG_{para\%max}$ measurements have also been found to increase progressively and mostly in parallel with increasing threshold load. The increasing pattern of parasternal intercostal muscle activation during inspiratory threshold loading was previously reported by Reilly et al. [10] using $sEMG_{para}$ measurements in healthy subjects. They reported that, although root-mean-square-based measurements of $oesEMG_{di\%max}$ were consistently greater than those of $sEMG_{para\%max}$, there was a strong relationship between them and they increased mostly linearly during threshold loading. Such coupling between parasternal intercostal muscles and the diaphragm had previously been suggested by De Troyer and Sampson [9], who indicated that parasternal intercostals are involuntarily activated, together with the diaphragm, during inspiratory breathing efforts. Accordingly, $sEMG_{para}$ has been proposed as an alternative noninvasive measure of neural respiratory drive [10,12,14,36].

In non-respiratory skeletal muscles, the relationship between measurements of sMMG and sEMG has been used to characterize some neuromuscular diseases. Orizio et al. [37] analyzed the ratio of sMMG to sEMG, which they called electromechanical coupling efficiency, of finger flexors in patients with myotonic dystrophy, who presented lower values as compared to those of control subjects. Barry et al. [38] used the ratio of sMMG to sEMG, recorded from the biceps brachii, to study several pediatric muscle diseases. They found a significant reduction in the ratio in affected subjects. The same ratio was used in patients with spastic cerebral palsy by Akataki et al. [39], who found significantly lower ratios of sMMG to sEMG in the patients than in the normal group.

The sMMG signal provides information about muscle contractile properties, reflects the mechanical properties of motor unit activity, and serves as an estimate of muscle force generation [16]. Therefore, relationships between sMMG and sEMG measurements, as expressed by $NMC_{MMG\text{-}EMGpara}$ in this study, provide muscle-specific noninvasive indices of NMC, which depend mainly on muscle mechanics. In the respiratory system, however, such indices reflect only the first step in the transformation of neural respiratory drive into ventilation. Next steps include the translation of respiratory muscle force into pressure, and the translation of pressure into ventilation. These steps depend on several aspects, such as chest wall geometry, airways resistance, or lung compliance, which can be altered in patients with disordered ventilatory mechanics, such as in COPD or in restrictive lung disease, thus causing neuromechanical and neuroventilatory uncoupling [40]. Therefore, it is desirable to have noninvasive indices of NMC, other than $NMC_{MMG\text{-}EMGpara}$, capable of reflecting the uncoupling that may occur in the different steps of the transformation of neural respiratory drive into ventilation. Different combinations of measurements of the electrical activation of respiratory muscles, respiratory pressures, and lung volumes have been proposed in previous studies as indices of NMC and neuroventilatory coupling [2,6,19,41–43]. However, the indices proposed in those previous studies involved at least one invasive measurement (i.e., P_{oes}, P_{di}, or $oesEMG_{di}$).

This is the first study to explore the relationships between measurements of $sEMG_{para}$ and $sMMG_{para}$, and measurements of P_{di}, P_{oes}, and P_{mo} to propose noninvasive indices of NMC and MEff of parasternal intercostal muscles. We found progressive and mostly linear increases in $fSEsEMG_{para\%max}$ relative to mean P_{di}, mean P_{oes}, and mean P_{mo} during threshold loading. Analogous results were found between $fSEsEMG_{lic\%max}$ and P_{di}, expressed as a percentage of maximum, in our previous study [19]. Respiratory pressures are considered measurements of global respiratory mechanical output [1], thus reflecting the action of several inspiratory muscles. The linear relationships found between $fSEsEMG_{para\%max}$ and the three different mean pressures (mean P_{di}, mean P_{oes}, and mean P_{mo}) may result therefore from the contribution of progressive activation of parasternal intercostal muscles to $sEMG_{para}$, and of parasternal and other extra-diaphragmatic inspiratory muscles to respiratory pressure measurements. Accordingly, $NMC_{Pdi\text{-}EMGpara}$, $NMC_{Poes\text{-}EMGpara}$, and $NMC_{Pmo\text{-}EMGpara}$ remained almost constant or increased slightly during threshold loading. It is noteworthy, however, that $NMC_{Pmo\text{-}EMGpara}$ increased significantly at the onset of inspiratory loading and from load L1 to load L2, which was due to the low values of P_{mo} during quiet resting breathing. Nevertheless, $NMC_{Pmo\text{-}EMGpara}$ behaved similarly to $NMC_{Poes\text{-}EMGpara}$ and $NMC_{Pdi\text{-}EMGpara}$ as load increased. Progressive and mostly linear increases were also found in $fSE|sMMG_{para}|_{\%max}$ relative to mean P_{di}, mean P_{oes}, and mean P_{mo} during threshold loading. These results are in accordance with those previously found between mean $fSE|sMMG_{lic}|$ and mean P_{di} in our previous study [33]. Accordingly, $MEff_{Pdi\text{-}MMGpara}$ and $MEff_{Poes\text{-}MMGpara}$ showed nonsignificant changes during threshold loading. $MEff_{Pmo\text{-}MMGpara}$, as $NMC_{Pmo\text{-}EMGpara}$, increased significantly at the onset of inspiratory loading and from load L1 to load L2, but behaved similarly to $MEff_{Poes\text{-}MMGpara}$ and $MEff_{Pdi\text{-}MMGpara}$ at higher loads. The $NMC_{Pmo\text{-}EMGpara}$ and $MEff_{Pmo\text{-}MMGpara}$ indices proposed in this study therefore provide noninvasive measurements of the contribution of parasternal intercostal muscle activation to the generation of respiratory pressures.

Parasternal intercostal recordings have the advantage, over lower chest wall inspiratory muscle recordings, of being less affected by the limitations generally associated with surface recordings, such as the difficulty in finding the optimal sensor position or the strong influence of chest wall thickness and subcutaneous fat [44,45]. Moreover, parasternal intercostal recordings are less susceptible to crosstalk from postural chest wall and abdominal muscle activity [21,22]. The noninvasive indices of $NMC_{MMG\text{-}EMGpara}$, $NMC_{Pmo\text{-}EMGpara}$, and $MEff_{Pmo\text{-}MMGpara}$ proposed in this study would therefore make the evaluation of respiratory muscle function easier and faster to perform, and thus more acceptable in patients with altered respiratory mechanics, such as in obstructive lung disease and neuromuscular disease. Neuroventilatory uncoupling resulting from respiratory muscle weakness and an increased elastic load of the lungs has been related to the degree of dyspnea in patients with neuromuscular disease [6]. Also, in chronic pulmonary diseases, neuromechanical uncoupling has been associated with the perception of breathlessness and limited exercise tolerance [3,5]. The proposed indices would therefore be of potential value to the clinical assessment of these patients.

This study may provide a basis for future research. The clinical utility of the proposed noninvasive NMC and MEff indices requires testing in disease states, since our study dataset was recorded from twelve healthy subjects. The size of the dataset reflects the difficulty in recruiting subjects for studies using invasive measures of diaphragmatic function. The study participants had body mass index values within the normal range. However, it is well known that sEMG and sMMG are affected by the thickness of subcutaneous fat [46,47]. The effect of body mass index on $sMMG_{para}$ and $sEMG_{para}$ measurements should therefore be a focus of future investigation.

5. Conclusions

We have proposed the combined use of P_{mo} and parasternal intercostal sEMG and sMMG recordings to obtain noninvasive indices of NMC and MEff of parasternal intercostal muscles, in healthy adults during an incremental inspiratory threshold loading protocol. The combination of $sMMG_{para}$ and $sEMG_{para}$ measurements ($NMC_{MMG\text{-}EMGpara}$) provides a muscle-specific noninvasive index of NMC of parasternal intercostal muscles, whose pattern during threshold loading is similar to that previously found in lower chest wall inspiratory muscle NMC. Global noninvasive indices of NMC and MEff of parasternal intercostal muscles have also been proposed by combining P_{mo} measurements and both $sEMG_{para}$ and $sMMG_{para}$ measurements ($NMC_{Pmo\text{-}EMGpara}$ and $MEff_{Pmo\text{-}MMGpara}$, respectively), reflecting the contribution of parasternal intercostal muscles to global respiratory mechanical output. Similar patterns have been found in $NMC_{Pmo\text{-}EMGpara}$ and $MEff_{Pmo\text{-}MMGpara}$, and their analogous invasive indices, calculated using P_{oes} and P_{di}.

The proposed noninvasive indices derived from P_{mo}, $sMMG_{para}$, and $sEMG_{para}$ may prove to be useful indices of NMC and MEff of parasternal intercostal muscles, particularly for the assessment of respiratory muscle function using wearable devices. Advances in sensor technologies have led to an increasing trend and interest in the use of wearable and wireless physiological monitoring devices in medicine [48–50]. These devices may contribute to improving the assessment of patients with chronic respiratory diseases by allowing home monitoring of respiratory muscle function in a wireless and noninvasive manner. In this regard, the proposed noninvasive indices of NMC could be easily implemented in a portable device capable of acquiring sEMG and sMMG signals, allowing regular monitoring of patients with impaired respiratory mechanics.

Author Contributions: Conceptualization, M.L.-G., J.M., C.J.J. and R.J.; methodology, M.L.-G., L.E.-P., A.T., G.F.R., C.J.J. and R.J.; software, M.L.-G.; validation, M.L.-G., L.E.-P., A.T., G.F.R., C.J.J. and R.J.; formal analysis, M.L.-G.; investigation, M.L.-G., L.E.-P., A.T., G.F.R., C.J.J. and R.J.; writing—original draft preparation, M.L.-G.; writing—review and editing, M.L.-G., L.E.-P., A.T., G.F.R., C.J.J. and R.J.; visualization, M.L.-G.; supervision, J.M., C.J.J. and R.J. All authors have read and agreed to the published version of the manuscript.

Funding: This work was supported in part by the Generalitat de Catalunya (CERCA Programme and GRC 2017 SGR 01770), the Gobierno de España (RTI2018-098472-B-I00 MCIU/AEI/FEDER, UE), and the Biomedical Research Networking Centre in Bioengineering, Biomaterials and Nanomedicine (CIBER-BBN, Instituto de Salud Carlos III/FEDER). Manuel Lozano-García and Luis Estrada-Petrocelli were the recipients of two European Respiratory Society Fellowships (ERS LTRF 2015-5185 and ERS LTRF 2017 01-00086, respectively).

Institutional Review Board Statement: The study was conducted according to the guidelines of the Declaration of Helsinki, and approved by the National Health Service Health Research Authority (National Research Ethics Service Committee London—Dulwich 05/Q0703).

Informed Consent Statement: Written informed consent was obtained from all subjects involved in the study.

Data Availability Statement: The data presented in this study are available from the corresponding author on reasonable request.

Conflicts of Interest: The authors declare no conflict of interest. The funders had no role in the design of the study; in the collection, analyses, or interpretation of data; in the writing of the manuscript, or in the decision to publish the results.

References

1. Laveneziana, P.; Albuquerque, A.; Aliverti, A.; Babb, T.; Barreiro, E.; Dres, M.; Dubé, B.P.; Fauroux, B.; Gea, J.; Guenette, J.A.; et al. ERS statement on respiratory muscle testing at rest and during exercise. *Eur. Respir. J.* **2019**, *53*, 1801214. [CrossRef] [PubMed]
2. Gorini, M.; Spinelli, A.; Ginanni, R.; Duranti, R.; Gigliotti, F.; Scano, G. Neural Respiratory Drive and Neuromuscular Coupling in Patients with Chronic Obstructive Pulmonary Disease (COPD). *Chest* **1990**, *98*, 1179–1186. [CrossRef] [PubMed]
3. Duiverman, M.L.; de Boer, E.W.J.; van Eykern, L.A.; de Greef, M.H.G.; Jansen, D.F.; Wempe, J.B.; Kerstjens, H.A.M.; Wijkstra, P.J. Respiratory muscle activity and dyspnea during exercise in chronic obstructive pulmonary disease. *Respir. Physiol. Neurobiol.* **2009**, *167*, 195–200. [CrossRef] [PubMed]
4. Druz, W.S.; Sharp, J.T. Electrical and mechanical activity of the diaphragm accompanying body position in severe chronic obstructive pulmonary disease. *Am. Rev. Respir. Dis.* **1982**, *125*, 275–280. [PubMed]

5. O'Donnell, D.E.; Ora, J.; Webb, K.A.; Laveneziana, P.; Jensen, D. Mechanisms of activity-related dyspnea in pulmonary diseases. *Respir. Physiol. Neurobiol.* **2009**, *167*, 116–132.
6. Lanini, B.; Misuri, G.; Gigliotti, F.; Iandelli, I.; Pizzi, A.; Romagnoli, I.; Scano, G. Perception of dyspnea in patients with neuromuscular disease. *Chest* **2001**, *120*, 402–408. [CrossRef]
7. Mead, J.; Loring, S.H. Analysis of volume displacement and length changes of the diaphragm during breathing. *J. Appl. Physiol.* **1982**, *53*, 750–755. [CrossRef] [PubMed]
8. Luo, Y.M.; Moxham, J.; Polkey, M.I. Diaphragm electromyography using an oesophageal catheter: Current concepts. *Clin. Sci.* **2008**, *115*, 233–244. [CrossRef] [PubMed]
9. De Troyer, A.; Sampson, M.G. Activation of the parasternal intercostals during breathing efforts in human subjects. *J. Appl. Physiol. Respir. Environ. Exerc. Physiol.* **1982**, *52*, 524–529. [CrossRef]
10. Reilly, C.C.; Jolley, C.J.; Ward, K.; MacBean, V.; Moxham, J.; Rafferty, G.F. Neural respiratory drive measured during inspiratory threshold loading and acute hypercapnia in healthy individuals. *Exp. Physiol.* **2013**, *98*, 1190–1198. [CrossRef]
11. Lin, L.; Guan, L.; Wu, W.; Chen, R. Correlation of surface respiratory electromyography with esophageal diaphragm electromyography. *Respir. Physiol. Neurobiol.* **2019**, *259*, 45–52. [CrossRef]
12. Reilly, C.C.; Ward, K.; Jolley, C.J.; Lunt, A.C.; Steier, J.; Elston, C.; Polkey, M.I.; Rafferty, G.F.; Moxham, J. Neural respiratory drive, pulmonary mechanics and breathlessness in patients with cystic fibrosis. *Thorax* **2011**, *66*, 240–246. [CrossRef] [PubMed]
13. Schmidt, M.; Kindler, F.; Gottfried, S.B.; Raux, M.; Hug, F.; Similowski, T.; Demoule, A. Dyspnea and surface inspiratory electromyograms in mechanically ventilated patients. *Intensive Care Med.* **2013**, *39*, 1368–1376. [CrossRef] [PubMed]
14. Murphy, P.B.; Kumar, A.; Reilly, C.; Jolley, C.; Walterspacher, S.; Fedele, F.; Hopkinson, N.S.; Man, W.D.-C.; Polkey, M.I.; Moxham, J.; et al. Neural respiratory drive as a physiological biomarker to monitor change during acute exacerbations of COPD. *Thorax* **2011**, *66*, 602–608. [CrossRef]
15. Suh, E.-S.; Mandal, S.; Harding, R.; Ramsay, M.; Kamalanathan, M.; Henderson, K.; O'Kane, K.; Douiri, A.; Hopkinson, N.S.; Polkey, M.I.; et al. Neural respiratory drive predicts clinical deterioration and safe discharge in exacerbations of COPD. *Thorax* **2015**, *70*, 1123–1130. [CrossRef] [PubMed]
16. Orizio, C.; Gobbo, M. Mechanomyography. In *Wiley Encyclopedia of Biomedical Engineering*; John Wiley & Sons, Inc.: Hoboken, NJ, USA, 2006.
17. Madeleine, P.; Bajaj, P.; Søgaard, K.; Arendt-Nielsen, L. Mechanomyography and electromyography force relationships during concentric, isometric and eccentric contractions. *J. Electromyogr. Kinesiol.* **2001**, *11*, 113–121. [CrossRef]
18. Ebersole, K.T.; Malek, D.M. Fatigue and the electromechanical efficiency of the vastus medialis and vastus lateralis muscles. *J. Athl. Train.* **2008**, *43*, 152–156. [CrossRef] [PubMed]
19. Lozano-Garcia, M.; Estrada-Petrocelli, L.; Moxham, J.; Rafferty, G.F.; Torres, A.; Jolley, C.J.; Jane, R. Noninvasive Assessment of Inspiratory Muscle Neuromechanical Coupling During Inspiratory Threshold Loading. *IEEE Access* **2019**, *7*, 183634–183646. [CrossRef]
20. Sarlabous, L.; Torres, A.; Fiz, J.A.; Gea, J.; Martínez-Llorens, J.M.; Jané, R. Efficiency of mechanical activation of inspiratory muscles in COPD using sample entropy. *Eur. Respir. J.* **2015**, *46*, 1808–1811. [CrossRef]
21. De Troyer, A.; Estenne, M.; Ninane, V.; Van Gansbeke, D.; Gorini, M. Transversus abdominis muscle function in humans. *J. Appl. Physiol.* **1990**, *68*, 1010–1016. [CrossRef]
22. Sinderby, C.; Friberg, S.; Comtois, N.; Grassino, A. Chest wall muscle cross talk in canine costal diaphragm electromyogram. *J. Appl. Physiol.* **1996**, *81*, 2312–2327. [CrossRef] [PubMed]
23. Finucane, K.E.; Panizza, J.A.; Singh, B. Efficiency of the normal human diaphragm with hyperinflation. *J. Appl. Physiol.* **2005**, *99*, 1402–1411. [CrossRef] [PubMed]
24. Finucane, K.E.; Singh, B. Human diaphragm efficiency estimated as power output relative to activation increases with hypercapnic hyperpnea. *J. Appl. Physiol.* **2009**, *107*, 1397–1405. [CrossRef] [PubMed]
25. Finucane, K.E.; Singh, B. Diaphragm efficiency estimated as power output relative to activation in chronic obstructive pulmonary disease. *J. Appl. Physiol.* **2012**, *113*, 1567–1575. [CrossRef] [PubMed]
26. Laghi, F.; Shaikh, H.S.; Morales, D.; Sinderby, C.; Jubran, A.; Tobin, M.J. Diaphragmatic neuromechanical coupling and mechanisms of hypercapnia during inspiratory loading. *Respir. Physiol. Neurobiol.* **2014**, *198*, 32–41. [CrossRef] [PubMed]
27. MacBean, V.; Hughes, C.; Nicol, G.; Reilly, C.C.; Rafferty, G.F. Measurement of neural respiratory drive via parasternal intercostal electromyography in healthy adult subjects. *Physiol. Meas.* **2016**, *37*, 2050–2063. [CrossRef] [PubMed]
28. Jolley, C.J.; Luo, Y.M.; Steier, J.; Reilly, C.; Seymour, J.; Lunt, A.; Ward, K.; Rafferty, G.F.; Polkey, M.I.; Moxham, J. Neural respiratory drive in healthy subjects and in COPD. *Eur. Respir. J.* **2009**, *33*, 289–297. [CrossRef] [PubMed]
29. Ibitoye, M.O.; Hamzaid, N.A.; Zuniga, J.M.; Hasnan, N.; Wahab, A.K.A. Mechanomyographic parameter extraction methods: An appraisal for clinical applications. *Sensors* **2014**, *14*, 22940–22970. [CrossRef] [PubMed]
30. Sarlabous, L.; Torres, A.; Fiz, J.A.; Jané, R. Evidence towards improved estimation of respiratory muscle effort from diaphragm mechanomyographic signals with cardiac vibration interference using sample entropy with fixed tolerance values. *PLoS ONE* **2014**, *9*, e88902. [CrossRef]
31. Estrada, L.; Torres, A.; Sarlabous, L.; Jané, R. Improvement in neural respiratory drive estimation from diaphragm electromyographic signals using fixed sample entropy. *IEEE J. Biomed. Heal. Informatics* **2016**, *20*, 476–485. [CrossRef]

32. Lozano-García, M.; Estrada, L.; Jané, R. Performance evaluation of fixed sample entropy in myographic signals for inspiratory muscle activity estimation. *Entropy* **2019**, *21*, 183.
33. Lozano-García, M.; Sarlabous, L.; Moxham, J.; Rafferty, G.F.; Torres, A.; Jané, R.; Jolley, C.J. Surface mechanomyography and electromyography provide non-invasive indices of inspiratory muscle force and activation in healthy subjects. *Sci. Rep.* **2018**, *8*, 16921.
34. Demoule, A.; Verin, E.; Locher, C.; Derenne, J.P.; Similowski, T. Validation of surface recordings of the diaphragm response to transcranial magnetic stimulation in humans. *J. Appl. Physiol.* **2003**, *94*, 453–461. [CrossRef] [PubMed]
35. Goldman, M.D.; Grassino, A.; Mead, J.; Sears, T.A. Mechanics of the human diaphragm during voluntary contraction: Dynamics. *J. Appl. Physiol.* **1978**, *44*, 840–848. [CrossRef] [PubMed]
36. Reilly, C.C.; Jolley, C.J.; Elston, C.; Moxham, J.; Rafferty, G.F. Measurement of parasternal intercostal electromyogram during an infective exacerbation in patients with cystic fibrosis. *Eur. Respir. J.* **2012**, *40*, 977–981. [CrossRef]
37. Orizio, C.; Esposito, F.; Sansone, V.; Parrinello, G.; Meola, G.; Veicsteinas, A. Muscle surface mechanical and electrical activities in myotonic dystrophy. *Electromyogr. Clin. Neurophysiol.* **1997**, *37*, 231–239. [PubMed]
38. Barry, D.T.; Gordon, K.E.; Hinton, G.G. Acoustic and surface EMG diagnosis of pediatric muscle disease. *Muscle Nerve* **1990**, *13*, 286–290. [CrossRef] [PubMed]
39. Akataki, K.; Mita, K.; Itoh, K.; Suzuki, N.; Watakabe, M. Acoustic and electrical activities during voluntary isometric contraction of biceps brachii muscles in patients with spastic cerebral palsy. *Muscle Nerve* **1996**, *19*, 1252–1257. [CrossRef]
40. Jensen, D.; Pattinson, K.; Jolley, C.J. Mechanisms of Breathlessness. In *European Respiratory Society Monograph*; Bausewein, C., Currow, D., Johnson, M., Eds.; European Respiratory Society Journals: Lausanne, Switzerland, 2016; pp. 111–113.
41. Laghi, F.; Jubran, A.; Topeli, A.; Fahey, P.J.; Garrity, E.R.; Arcidi, J.M.; de Pinto, D.J.; Edwards, L.C.; Tobin, M.J. Effect of Lung Volume Reduction Surgery on Neuromechanical Coupling of the Diaphragm. *Am. J. Respir. Crit. Care Med.* **1998**, *157*, 475–483. [CrossRef]
42. O'Donnell, D.E.; Bertley, J.C.; Chau, L.K.; Webb, K.A. Qualitative aspects of exertional breathlessness in chronic airflow limitation: Pathophysiologic mechanisms. *Am. J. Respir. Crit. Care Med.* **1997**, *155*, 109–115.
43. Jolley, C.J.; Luo, Y.M.; Steier, J.; Rafferty, G.F.; Polkey, M.I.; Moxham, J. Neural respiratory drive and breathlessness in COPD. *Eur. Respir. J.* **2015**, *45*, 355–364. [CrossRef] [PubMed]
44. Glerant, J.C.; Mustfa, N.; Man, W.D.; Luo, Y.M.; Rafferty, G.; Polkey, M.I.; Moxham, J. Diaphragm electromyograms recorded from multiple surface electrodes following magnetic stimulation. *Eur. Respir. J.* **2006**, *27*, 334–342. [CrossRef]
45. Dos Reis, I.M.M.; Ohara, D.G.; Januário, L.B.; Basso-Vanelli, R.P.; Oliveira, A.B.; Jamami, M. Surface electromyography in inspiratory muscles in adults and elderly individuals: A systematic review. *J. Electromyogr. Kinesiol.* **2019**, *44*, 139–155. [CrossRef] [PubMed]
46. Scheeren, E.M.; Mineiro, L.; Neves, E.B.; Krueger, E.; Nogueira Neto, G.N.; Nohama, P.; Scheeren, E.M.; Mineiro, L.; Neves, E.B.; Krueger, E.; et al. Influence of subcutaneous fat on mechanomyographic signals at three levels of voluntary effort. *Res. Biomed. Eng.* **2017**, *32*, 307–317. [CrossRef]
47. Nordander, C.; Willner, J.; Hansson, G.-A.; Larsson, B.; Unge, J.; Granquist, L.; Skerfving, S. Influence of the subcutaneous fat layer, as measured by ultrasound, skinfold calipers and BMI, on the EMG amplitude. *Eur. J. Appl. Physiol.* **2003**, *89*, 514–519. [CrossRef] [PubMed]
48. Ràfols-De-Urquía, M.; Estrada, L.; Estévez-Piorno, J.; Sarlabous, L.; Jané, R.; Torres, A. Evaluation of a Wearable Device to Determine Cardiorespiratory Parameters from Surface Diaphragm Electromyography. *IEEE J. Biomed. Heal. Informatics* **2019**, *23*, 1964–1971.
49. Xu, J.; Konijnenburg, M.; Ha, H.; Van Wegberg, R.; Song, S.; Blanco-Almazán, D.; Van Hoof, C.; Van Helleputte, N. A 36 μW 1.1 mm2 Reconfigurable Analog Front-End for Cardiovascular and Respiratory Signals Recording. *IEEE Trans. Biomed. Circuits Syst.* **2018**, *12*, 774–783. [CrossRef]
50. Song, S.; Konijnenburg, M.; Van Wegberg, R.; Xu, J.; Ha, H.; Sijbers, W.; Stanzione, S.; Biswas, D.; Breeschoten, A.; Vis, P.; et al. A 769 μW Battery-Powered Single-Chip SoC with BLE for Multi-modal Vital Sign Monitoring Health Patches. *IEEE Trans. Biomed. Circuits Syst.* **2019**, *13*, 1506–1517. [CrossRef]

sensors

MDPI

Article

Using a System-Based Monitoring Paradigm to Assess Fatigue during Submaximal Static Exercise of the Elbow Extensor Muscles

Kaci E. Madden, Dragan Djurdjanovic and Ashish D. Deshpande *

Department of Mechanical Engineering, The University of Texas at Austin, Austin, TX 78712, USA; kaci.madden@utexas.edu (K.E.M.); dragand@me.utexas.edu (D.D.)
* Correspondence: ashish@austin.utexas.edu; Tel.: +1-512-475-7773

Abstract: Current methods for evaluating fatigue separately assess intramuscular changes in individual muscles from corresponding alterations in movement output. The purpose of this study is to investigate if a system-based monitoring paradigm, which quantifies how the dynamic relationship between the activity from multiple muscles and force changes over time, produces a viable metric for assessing fatigue. Improvements made to the paradigm to facilitate online fatigue assessment are also discussed. Eight participants performed a static elbow extension task until exhaustion, while surface electromyography (sEMG) and force data were recorded. A dynamic time-series model mapped instantaneous features extracted from sEMG signals of multiple synergistic muscles to extension force. A metric, called the Freshness Similarity Index (FSI), was calculated using statistical analysis of modeling errors to reveal time-dependent changes in the dynamic model indicative of performance degradation. The FSI revealed strong, significant within-individual associations with two well-accepted measures of fatigue, maximum voluntary contraction (MVC) force ($r_{rm} = -0.86$) and ratings of perceived exertion (RPE) ($r_{rm} = 0.87$), substantiating the viability of a system-based monitoring paradigm for assessing fatigue. These findings provide the first direct and quantitative link between a system-based performance degradation metric and traditional measures of fatigue.

Keywords: human fatigue monitoring; neuromuscular fatigue; surface electromyography time-frequency signal analysis; time-series modeling; autoregressive moving average model with exogenous inputs; isometric contraction; elbow extension

Citation: Madden, K.E.; Djurdjanovic, D.; Deshpande, A.D. Using a System-Based Monitoring Paradigm to Assess Fatigue during Submaximal Static Exercise of the Elbow Extensor Muscles. *Sensors* **2021**, *21*, 1024. https://doi.org/10.3390/s21041024

Academic Editor: Ernest N. Kamavuako

Received: 31 December 2020
Accepted: 29 January 2021
Published: 3 February 2021

1. Introduction

1.1. Background

Fatigue, commonly defined as "any exercise-induced reduction in the ability of a muscle to generate force or power" [1], is a complex accumulation of psychological and physiological processes that impair muscle function and diminish the capacity of the central nervous system to activate muscles [1–3]. Neuromuscular fatigue presents a major obstacle for achieving desired performance in a variety of circumstances. For healthy individuals in physically demanding professions (e.g., astronauts, soldiers, athletes, etc.), prolonged periods of training and operations are known to adversely affect task efficiency [4], movement accuracy [5], and performance [4], while also increasing susceptibility to overuse injuries [4]. For patients with neurological or cerebrovascular diseases, such as stroke, multiple sclerosis, and Parkinson's disease, fatigue is also a typical and potentially debilitating symptom [6,7]. Thus, assessing fatigue has important implications for preventing neuromuscular injury [8], optimizing training loads [9], and guiding effective, individualized treatment strategies for rehabilitation [7].

In a clinical setting, standard methods for assessing fatigue rely upon self-reported questionnaires or rating scales [1,10] that capture how an individual experiences fatigue. Mental fatigue can be experienced as an increase in the perceived effort to complete a task [11] or a reduction in motivation and concentration [1]. Ratings of perceived exertion (RPE) [12–14] are used to study mental fatigue in healthy and affected populations. A higher perception of effort is known to limit exercise tolerance [13] and adversely affect physical performance during endurance tasks [11,14]. Although subjective rating scales contain valuable information, they are indirect measures of fatigue that provide qualitative information with low-resolution [10]. Moreover, self-perceived fatigue is not always accompanied by a loss of force-producing capacity [6,11,14] or changes in physiological variables [13,14], especially during endurance tasks.

A decline in maximum voluntary contraction (MVC) force has become a "gold standard" [15] indicator for confirming the occurrence of fatigue in the physiological sciences [1,14,15] because it can directly quantify a loss in force-generating capacity. Despite their value as objective assessment tools, MVC force measures are often taken immediately before and after a bout of exercise and only capture the overall mechanical manifestation of fatigue. Consequently, they lack valuable insight regarding the progression of fatigue during the task itself, including the underlying physiological processes that contribute to the degraded performance of the neuromuscular system. Neuromuscular fatigue can be identified by measuring the evoked force from twitch responses after electrically stimulating muscles during maximal or submaximal voluntary contractions [16]. However, this technique is also applied before and after a fatiguing exercise.

Surface electromyography (sEMG) has been widely used to address this issue by enabling the continuous measurement of muscle activity during exercise. Since fatigue begins to accumulate at the start of a muscle contraction and continuously progresses throughout the exercise [17], changes in the sEMG signal can reveal indications of localized muscle fatigue long before a decline in force or power output occurs [1,2,18]. For instance, during sustained contractions at submaximal force levels, a progressive increase in sEMG amplitude and compression of the sEMG signal spectrum can be detected [2,17,18]. Fourier-based spectral features extracted from the sEMG signal, such as the mean or median frequency, are the most widely used indices of localized muscle fatigue and have been employed in numerous applications [2,18,19].

Extensive work has been devoted to developing more advanced spectral estimation and signal processing techniques that can accommodate the non-stationary behavior in sEMG signals [20,21]. The majority of these efforts, which are thoroughly discussed elsewhere [21–24], were devoted to developing fatigue assessment metrics that reflect the localized manifestations of fatigue within a muscle. Thus, these metrics are often univariate, monitored independently for each muscle, and analyzed separately from associated changes in joint movement. Less attention has been paid to developing multivariate metrics that utilize more information from the sEMG signal, aggregate activity from all contributing muscles, and establish a relationship with kinematic or kinetic movement variables. Such metrics would be beneficial for assessing how the neuromuscular system fatigues as a whole during exercise.

1.2. Related Literature

Model-based methods that relate sEMG parameters to movement variables have shown success in producing a single, unified metric for monitoring fatigue, overcoming some of the aforementioned issues. Previous studies have applied linear regressions [23], artificial neural networks [23,25,26], linear projection methods [27], and correlations [28] to map net changes in sEMG parameters to overall reductions in power [23] or force [28]. Although promising, these approaches do not continuously monitor changes in the dynamic relationship between sEMG and movement output over time—a relationship that is significantly altered in the presence of fatigue [29]. They also require (i) *a priori* assumptions about the linearity of fatigue progression [23,25,26], (ii) extensive data sets containing the entire time-course of fatigue to train models [23,25–28], and (iii) reference contractions to probe for fatigue-induced changes in parameters at the beginning and end of an endurance task [28].

Recent studies have approached human performance monitoring using a system-based monitoring paradigm, which is relatively well-known in the machine monitoring community [30]. The system-based approach monitors how the performance of the human neuromusculoskeletal (NMS) degrades during prolonged exercise by continuously tracking changes in the dynamic relationship between sEMG and movement output over time. Musselman et al. were the first to pursue this direction [31]. The dynamic relationship was described using vectorial autoregressive models with exogenous inputs (vARX), which took instantaneous intensity and frequency features from upper-arm sEMG signals as inputs and related them to joint angular velocities as model outputs. The methodology was tested on data from participants performing a repetitive sawing movement until voluntary exhaustion. Xie and Djurdjanovic [32], Madden et al. [33], and Yang et al. [34] modified this work by instead using autoregressive moving average models with exogenous inputs (ARMAX) with second-order muscle dynamics to describe the NMS system during both constant force and repetitive movement tasks. Two additional sEMG features, namely instantaneous variance and entropy, were incorporated as model inputs with either force [32,33], joint velocity [32], or limb displacement [33] serving as outputs, depending on the task.

The models in all four studies [31–34] were trained with data from the initial portion of the task before fatigue onset to capture the system dynamics during a normal, unfatigued state. Progressive changes in system behavior were evaluated by tracking the divergence of model prediction error distributions between the unfatigued state and subsequent periods of time. Statistically significant trends in a divergence measure, referred to as either the freshness similarity index (FSI) [32,33], fatigue index [34], or global freshness index [31], provided evidence that performance degradation occurred during the exercises.

The system-based monitoring paradigm overcomes the limitations imposed by alternative, model-based approaches to monitoring fatigue [23,25,26,28]. However, the previous system-based monitoring efforts [31–34] have not established a formal association between performance degradation and fatigue. Although the studies verified their findings using trends in sEMG features to reveal indications of localized muscle fatigue in individual muscles, the indices in [31–34] are constructed as global measures of how the performance of the entire NMS system changes over time. Thus, to claim that a system-based monitoring approach is a viable method for monitoring fatigue, further research is needed to formally associate the performance degradation index with well-established fatigue measures that quantify a global reduction in force-producing capacity [1,15] and heightened perception of exertion [12], rather than changes intramuscular mechanisms. Furthermore, modifications can be made to the system-based paradigm used in these previous works [31–34] to produce sEMG features that are more representative of neural activation signals to the NMS system, provide a complete representation of the NMS system by incorporating all contributing muscles, and facilitate online performance assessment.

To this end, the primary aim of this work is to firmly establish the viability of the system-based monitoring paradigm for assessing fatigue by relating the performance degradation index to well-accepted measures of fatigue that capture changes in force-generating capacity (MVC force) and self-perceived fatigue (RPE). We present a methodology, modified from previous works, to generate a sensitive and concise index of performance degradation (FSI) occurring across multiple muscles and sensor sources during a submaximal static exercise. We then substantiate its viability for assessing fatigue by evaluating within-individual associations between the FSI and measures of MVC force and RPE. We discuss the improvements made to the paradigm to facilitate its use as an online assessment tool and more accurately represent changes occurring in the NMS. The results of this work have promising implications for informing new methods of monitoring fatigue. Tracking fatigue-related changes in performance may lead to more personalized training regimens and therapeutic modalities for rehabilitation. Interventions involving robotic exoskeletons present an especially promising application of the system-based monitoring paradigm because these devices possess high-resolution sensors that can collect physiological, dynamic, and kinematic measures in real-time.

2. Materials and Methods

2.1. Participants

Eight healthy, right-handed men (26.6 ± 6.1 years, 76.2 ± 12.4 kg, 178.9 ± 6.6 cm) with no known neurological disorders were recruited from the university population to participate in the study. All participants were fully informed of any risks associated with the experiments before giving their informed written consent to participate in the investigation. The study was conducted in accordance with the Declaration of Helsinki [35], and the experimental procedure was approved by the Internal Review Board organized by the Office of Research Support at The University of Texas at Austin under the protocol number 2013-05-0126.

2.2. Experimental Setup

Participants were seated in a high-back chair with a five-point harness that restrained their waist and shoulders (Figure 1). A single-degree-of-freedom exoskeleton testbed was grounded to the base of the chair and used for testing. The device consists of an upper arm linkage, capstan drive elbow joint, and lower arm linkage with a wrist cuff. The chair and link lengths were adjusted to accommodate each participant. The participants' upper arm was positioned at 90° of flexion and 45° of horizontal abduction, and their elbow placed in 90° of flexion. The medial epicondyle of the participants' humerus was aligned with the exoskeleton elbow joint axis, and the forearm was placed in a neutral position. The wrist cuff was positioned below the participants' ulnar styloid process and securely attached their forearm to the exoskeleton. The lower arm linkage of the exoskeleton was grounded to the chair using a mechanical structure to prevent the elbow joint from rotating during the isometric contractions described in Section 2.3. As a result, the robot actuator remained unpowered during experimentation. A multi-axis force/torque sensor mounted to a linear sliding joint was housed between the wrist cuff and exoskeleton linkage and used to measure the participants' elbow extension force. The linear slider allowed passive travel in the direction parallel to the ulna bone to minimize off-axis forces due to robot-human misalignment [36].

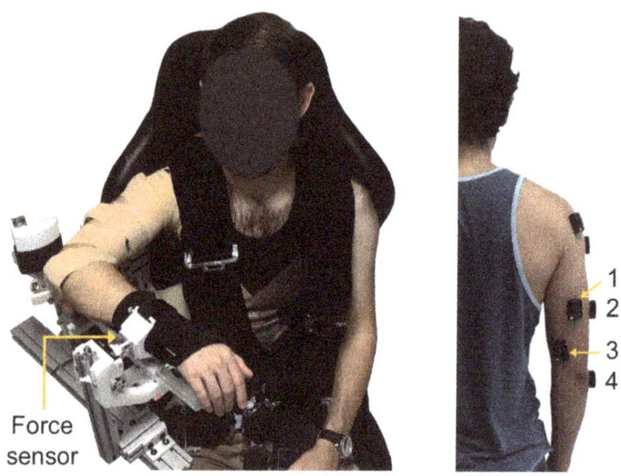

Figure 1. Experimental setup. (**Left**) Exoskeleton testbed. (**Right**) sEMG sensor placement: (1) long, (2) lateral, and (3) medial heads of the triceps brachii, and (4) anconeus muscles.

2.3. Experimental Protocol

Experiments were carried out in the ReNeu Robotics Laboratory at the University of Texas at Austin. All participants performed the same experiment on two days separated by 72 h of rest [37,38] in a temperature-controlled room set to 70°. Both sessions were performed at the same time of day and followed the same general protocol, which consisted of three elbow extension tasks: (1) baseline maximum voluntary contractions (MVCs) (2) a constant-force endurance task sustained at 30% MVC until exhaustion, and (3) a follow-up MVC. Only results from the first session are reported in this paper. Participants were instructed to refrain from consuming caffeine on the day of testing [39] and exercising 24 h before the experiment.

Before testing, the participants performed isometric elbow extension, elbow flexion, shoulder flexion, shoulder abduction, and shoulder extension contractions for which they were asked to maximally and submaximally exert force. During testing, participants were provided with real-time visual feedback of their elbow extension force, in the form of a gauge display, on a computer monitor placed at eye-level. For the MVCs, participants were instructed to gradually increase extension force output from zero to maximum over a 3 s period and maintain their maximal force for an additional 2–3 s. The participants were given strong verbal encouragement to provide maximal effort during each contraction. At baseline, a minimum of three MVCs separated by one minute of rest were performed. If peak forces from two of the three MVCs were not within 5%, additional trials were performed until this criterion was met. The trial consisting of the highest value was retained and considered the MVC force. Participants then rested for at least eight minutes to minimize residual fatigue from the MVC tasks.

Before the endurance task, each participant was familiarized with their MVC levels by performing brief elbow extension contractions at various force levels (i.e., 30% and 60% MVC). For the endurance task, participants performed a sustained, isometric contraction at 30% MVC until their force fell below 10–15% of the target value [39,40]. In related works examining fatigue, the MVC thresholds of the isometric contractions vary between 25–35% [28,39–41]. The contraction level for evaluation was set to 30% MVC for this study, as it is the average between these ranges. The target force (30% MVC) and the participant's actual extension force were displayed on the computer monitor. Participants matched and tracked the target line for as long as possible and were verbally encouraged to maintain a steady force output. Every 30 s, participants reported a rating of perceived exertion (RPE) using the Borg CR-10 scale [12]. These ratings ranged from 0 ("no exertion at all") to 10 ("maximal exertion"). Immediately after termination of the endurance task, participants reported a final RPE and performed a follow-up MVC to determine the amount of fatigue induced.

2.4. Data Acquisition

A Delsys Trigno Wireless EMG system (Delsys Inc., Boston, MA, USA) was used to collect sEMG activity from the triceps brachii (long, lateral, and medial heads), anconeus, biceps brachii, brachioradialis, and deltoid (anterior, middle, and posterior) muscles. The scope of this paper requires analysis of only the muscles that extend the elbow, that is, the triceps brachii and anconeus (Figure 1). Participants' body hair was shaved, and skin lightly abraded with a pumice stone then cleansed with isopropyl alcohol to ensure good skin-to-electrode contact before sEMG sensor placement. Electrodes were positioned over each muscle according to European recommendations for Surface Electromyography for Non-Invasive Assessment of Muscles (SENIAM) [42]. Elbow extension forces were measured with a multi-axis force/torque sensor (ATI, Nano25). An xPC Target (Mathworks, MATLAB module) running Simulink Real-Time and hosting NI data acquisition (NI DAQ) boards (National Instruments, Inc., Austin, TX, USA) synchronously recorded all data at 1 kHz.

2.5. Data Processing

Raw sEMG signals were bandpass filtered from 10 to 400 Hz [43,44] using a 4th order Butterworth filter (zero-lag, non-causal) [45], then demeaned [46] to remove the DC offset. Data from the force/torque sensor was low-pass filtered using a 4th order Butterworth filter (zero-lag, non-causal) with a 6 Hz cutoff frequency. The processed sEMG and force measures are used in Sections 2.6.1 and 2.6.2.

2.6. System-Based Monitoring

2.6.1. sEMG Feature Extraction

The first step in the system-based monitoring workflow (Figure 2) involves extracting features from the filtered sEMG signals [47] that capture how the signal energy changes in both the time and frequency domains. Cohen's class of time-frequency distributions (TFD) was used to obtain a two-dimensional probability density function, $C(t,\omega)$, describing the joint distribution of energy of the sEMG signal, $s(t)$, over time, t, and frequency, ω, where

$$C(t,\omega) = \frac{1}{4\pi^2}\cdot$$

$$\iiint_{-\infty}^{+\infty} s^*(u-\frac{1}{2}\tau)s(u+\frac{1}{2}\tau)\phi(\theta,\tau)e^{-j(\theta(t-u)+\tau\omega)}d\tau du d\theta, \quad (1)$$

with $s^*(t)$ signifying the complex conjugate of $s(t)$ and $\phi(\theta,\tau)$ denoting the so-called TFD kernel. The binomial kernel, a signal independent member of the reduced interference distribution family of kernels, was used for this analysis due to its desirable mathematical properties [31].

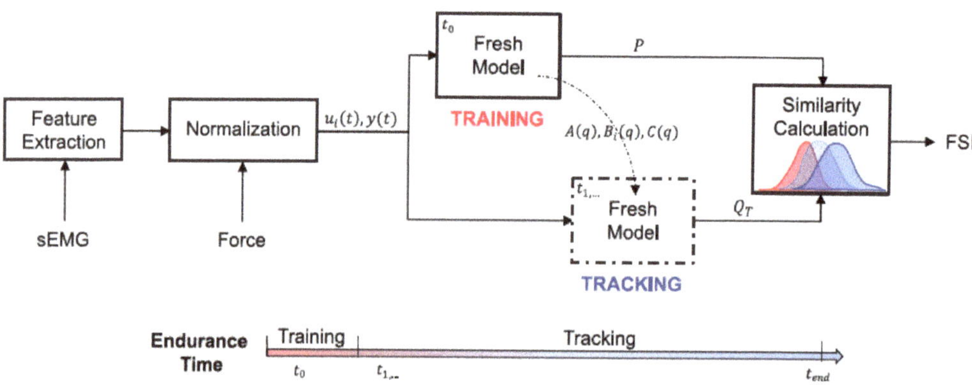

Figure 2. System-based monitoring workflow. Features are extracted from the surface electromyography (sEMG) signals of each muscle. The sEMG features and elbow extension force are then normalized and used as the inputs ($u_i(t)$) and output ($y(t)$) to a dynamic time-series model. Training data from the start of the endurance task (t_0) is used to identify the polynomial coefficients ($A(q), B_i(q), C(q)$) of the "Fresh Model" and calculate a reference distribution (P) of one-step ahead prediction errors. The remaining endurance task data ($t_{1,\ldots,end}$) is incrementally introduced to the tuned "Fresh Model" for which updated prediction error distributions (Q_T) are calculated at each time step, T. The overlap between P and Q_T is evaluated to obtain a time-series of freshness similarity index (FSI) values that quantify performance degradation.

Calculation of the zero- and first-order moments (i.e., $< f^0|t >$ and $< f^1|t >$) of $C(t,\omega)$ provide the instantaneous energy and instantaneous mean frequency of the sEMG signal, respectively, with

$$< f^0|t > = \int_{-\infty}^{+\infty} C(t,\omega)d\omega = |a_i(t)|^2 \tag{2}$$

$$< f^1|t > = \int_{-\infty}^{+\infty} \frac{C(t,\omega)}{< f^0|t >}\omega d\omega = f_{im}(t), \tag{3}$$

where $a_i(t)$ is the instantaneous amplitude—a parameter that is approximately equal to the RMS amplitude of the sEMG signal [48,49]. The instantaneous mean frequency, labeled as $f_{im}(t)$, and instantaneous amplitude, $a_i(t)$, are widely used as myoelectric indicators of fatigue. As a result, significant decreasing trends in $f_{im}(t)$ and increasing trends in $a_i(t)$ during the constant-force endurance task would substantiate the presence of localized muscle fatigue [2,18,19].

Previous system-based monitoring studies [31–33] used the instantaneous energy ($< f^0|t >$), rather than $a_i(t)$, as an input to the dynamic model described in Section 2.6.3. However, we adopted $a_i(t)$ because it is analogous to the RMS amplitude of the sEMG signal that reflects changes in "neural drive" due to fatigue [1]. Moreover, the square root calculation in (2) attenuates the high magnitude spikes produced when computing the zero-order moment, which is apparent in [31]. Previous works also extracted two additional sEMG features, representing the second-order moment and entropy of the signal, to be used as model inputs [32–34]. When including these features in our dynamic model, the performance degradation metric described in Section 2.6.4 did not significantly change. Therefore, we reduced the complexity of our model by restricting the number of model inputs to include only $a_i(t)$ and $f_{im}(t)$ for each muscle.

2.6.2. Normalization

Data from the MVC and endurance tasks were smoothed using 10 ms and 1.5 ms sliding windows, respectively. Maximal values obtained over a 1.5 s period around the peak MVC reference force were determined for each muscle and used to normalize the corresponding $a_i(t)$ signals from the endurance task. Force and $f_{im}(t)$ signals from the endurance task were normalized to their average values during the initial 10 s of the endurance task. All signals were then downsampled to 100 Hz. This procedure prepared the data to be used in the model described in Section 2.6.3 and shown in Figure 2. Figure 3 depicts the force and sEMG features after normalization for one representative participant.

The normalization strategy presented in this work was another improvement made to previous system-based monitoring attempts, which used data from the entire endurance task to normalize the signals [32,33]. By scaling $a_i(t)$ to MVC values and $f_{im}(t)$ to initial values, our normalization approach produced signals that are more representative of neural activation signals and the frequency-based sEMG indices found in the literature for assessing localized muscle fatigue. Moreover, our approach could be employed for online performance assessment because the only data needed for normalization was collected at the beginning of the experiment (i.e., baseline MVC contractions performed before testing and the initial few seconds of the endurance task).

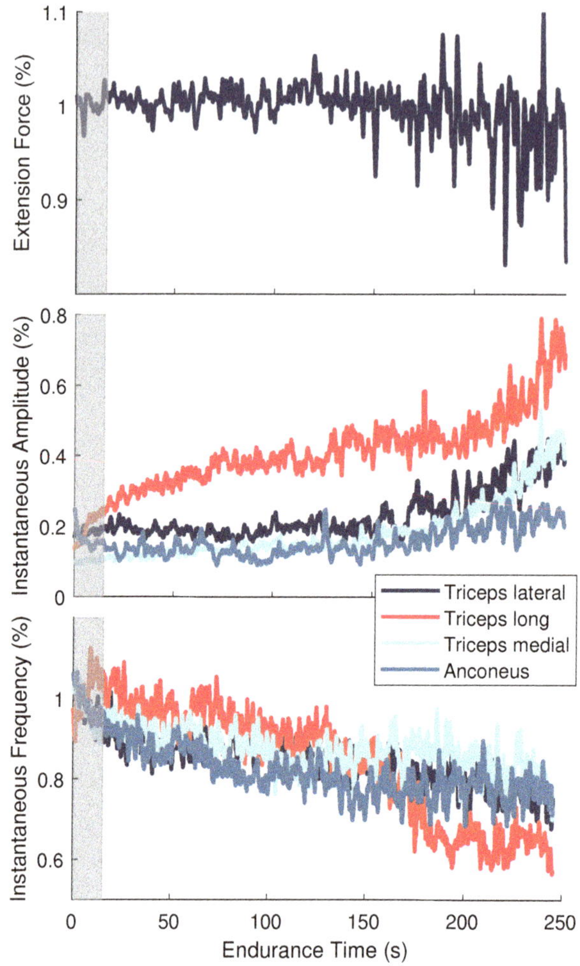

Figure 3. Normalized signals for a single representative participant during the endurance task. (**Top**) Elbow extension force. (**Middle**) Instantaneous amplitude ($a_i(t)$) and (**Bottom**) instantaneous frequency ($f_{im}(t)$) features for the elbow extensor muscles. Gray shaded area signifies the training data set.

2.6.3. Modeling

Human skeletal muscle can be considered a viscoelastic system whose physiological input is a neural signal and output response is a generated force [50]. Thus, the normalized sEMG features extracted from the triceps brachii (long, lateral, and medial heads) and anconeus muscles were used as neural inputs to a dynamic model whose output is elbow extension force. The dynamics were represented using an autoregressive moving average model with exogenous inputs (ARMAX). This form of parametric system identification approximates force as a linear transformation of sEMG features and noise terms and can be expressed as

$$A(q)y(t) = \sum_{i=1}^{n_u} B_i(q)u_i(t) + C(q)e(t), \quad (4)$$

where the system output, $y(t)$, is the elbow extension force, the system input, $u_i(t)$, is an $n_u \times 1$ vector of the normalized sEMG features, and $e(k)$ is the model disturbance considered to be zero mean Gaussian process noise. Since two sEMG features ($a_i(k)$ and $f_{im}(k)$) were extracted from each muscle, $n_u = 8$. The polynomials A, B_i, and C are expressed in terms of the time-shift operator, q^{-1}, and can be written as

$$\begin{aligned} A(q) &= 1 + a_1 q^{-1} + \ldots + a_{n_a} q^{-n_a} \\ B_i(q) &= b_1 + b_2 q^{-1} + \ldots + b_{n_b} q^{-n_b+1} \\ C(q) &= 1 + c_1 q^{-1} + \ldots + c_{n_c} q^{-n_c}, \end{aligned} \tag{5}$$

where n_a, n_b, and n_c are their respective orders. The model was structured such that each muscle is considered a second-order dynamic system [32]. This approach is in line with Gottlieb and Agarwal [50] and Thelen et al. [51] who found that a second-order system can adequately describe the functional relationship between sEMG and force [50] or joint torque [51]. Thus, the orders of the polynomials were selected to be 8 for $A(q)$ and $B_i(q)$ and 7 for $C(q)$. Separate models were trained for each user with data selected from the initial 15 s of the endurance task (Figures 2 and 3). This training data set captures the state of the users before significant fatigue could develop. Thus, the trained model, referred to as the "fresh model" (Figure 3), captures the system dynamics corresponding to the user's least degraded, or least fatigued, state.

2.6.4. Performance Tracking

Using the training data set, a reference distribution, P, of 1-step ahead prediction errors was generated by the "fresh model" (Figure 2). The remaining data from the endurance task was segmented into T epochs that were 4 s in length. The endurance time for each participant determined the total number of epochs. These data segments were sequentially presented to the "fresh model" to calculate the latest 1-step ahead prediction error distributions, Q_T. The Fidelity similarity metric [52,53] was then calculated to evaluate the amount of overlap between the reference and updated distributions over time. The metric, which is referred to as the Freshness Similarity Index (FSI), is defined as

$$FSI = 1 - \sum_{i=1}^{N} \sqrt{P(i) Q_T(i)} \tag{6}$$

and ranges from 0 to 1, where values near 0 indicate a high degree of similarity and those close to 1 suggest little similarity. For context, if the dynamic system remains unaltered with time, the updated distributions will be comparable to the fresh distribution. However, if the system dynamics change due to fatigue or injury, for example, the updated distribution will shift or change shape, reducing the amount of overlap with the fresh distribution. Thus, the FSI is a metric that reflects how the ARMAX approximation of the system dynamics degrades over time with respect to a normal, unfatigued state.

Previous system-based monitoring studies used similarity/divergence measures, including Matusita's overlap coefficient measure [31–33] and the Kullback-Leibler divergence measure [33]. However, the Fidelity similarity metric was used in this work due to its superior sensitivity to changes in modeling errors for the data in this study. All data processing and modeling was conducted using MATLAB software (R2017b) [54].

2.7. Statistical Analysis

A paired samples t-test was used to test for differences between baseline (pre-endurance task) and follow-up (post-endurance task) MVC forces, and Cohen's d was used to calculate the effect size between time points. A one-factor repeated measures analysis of variance (RM-ANOVA) was used to evaluate mean differences in RPE scores collected after the first, middle, and last 30 s of the endurance task. For each sEMG feature, a two-factor RM-ANOVA was used to test for differences across time and within muscles using average values over the first, middle, and last 30 s of the endurance task. FSI was quantified in two ways. For statistical analysis, averages over the first, middle, and last 30 s of the endurance task were used in a one-factor RM-ANOVA to evaluate mean differences over time. For graphical representation, average FSI values over each 1% of the endurance time were presented. A Greenhouse-Geisser correction was applied to correct for violations of sphericity when Mauchly's test was significant. Significant main effects were further examined using estimated marginal means with a Tukey-Kramer adjustment for multiple comparisons.

Within-subject correlations [55] were performed using repeated-measures correlation (*rmcorr*) [56] analysis to evaluate the associations between FSI and measures of force-generating capacity (MVC force) and self-perceived fatigue (RPE scores). Although associations between parameters may typically be analyzed using simple correlations that quantify between-subject associations, within-subject associations are more important to this study because FSI is an individual-specific metric. *Rmcorr* analysis also provides benefits over simple correlation techniques when considering the change in variables over time. Multiple data points per participant can be used in a *rmcorr*, whereas simple correlations require time-series data to be aggregated so that all observations are independent of each other. As a result, *rmcorr* can yield much greater power than simple correlation methods and detect relationships between variables that might otherwise be masked when using aggregated data. Two *rmcorr* analyses were used to estimate linear models with subject-specific intercepts relating FSI to MVC force and FSI to RPE scores. Paired data from the start (i.e., pre-endurance task/first 30 s) and end (i.e., post-endurance task/last 30 s) of the endurance task was used for the *rmcorr* between MVC force and FSI. Paired data from the first, middle, and last 30 s of the task was used for the *rmcorr* between RPE and FSI. The resulting *rmcorr* coefficient (r_{rm}) quantified the common within-individual association between variables.

Although the results from the *rmcorr* analyses were used to evaluate the FSI metric, between-subject associations were also reported based on simple correlations. Pearson's product-moment correlation coefficient (r) was used to assess the association between FSI and MVC force. The Spearman rank correlation coefficient (r_S) was used to evaluate the relationship between FSI and RPE because the RPE scores were treated as ordinal data. To minimize biases introduced by the time-dependency among data points, the paired data was aggregated into difference scores representing the overall change in measures from the start (i.e., pre-endurance task/first 30 s) to the end (i.e., post-endurance task/last 30 s) of the endurance task. Shapiro-Wilk tests verified that all difference scores were normally distributed. We hypothesized that FSI would be negatively correlated with MVC force and positively correlated with RPE.

Using the guidelines presented in [57], correlation coefficients were interpreted as very strong ($r \geq 0.9$), strong ($0.7 \leq r < 0.9$), moderate ($0.5 \leq r < 0.7$), weak ($0.3 \leq r < 0.5$), and negligible ($r < 0.3$). All statistical analyses were conducted using R software (3.6.1) [58]. RM-ANOVAs and follow-up tests were analyzed using the *afex* and *emmeans* packages. Within-subject correlations were determined using the *rmcorr* package [56]. Statistical significance was set at $p < 0.05$ for all testing. Data are reported as mean ± standard error of the mean (SE) unless stated otherwise.

3. Results

3.1. Confirmation of Fatigue

The average endurance time across participants was 287.4 ± 28.0 s. The average MVC force at baseline was 139.8 ± 10.1 N and significantly declined by 49.5 ± 8.8 N, or 35.6 ± 6.1%, ($t(7) = -5.63, p < 0.001, d = -1.99$; Figure 4a) at follow-up. This substantial decline in MVC force from baseline to follow-up verifies that the experimental protocol successfully induced fatigue across participants.

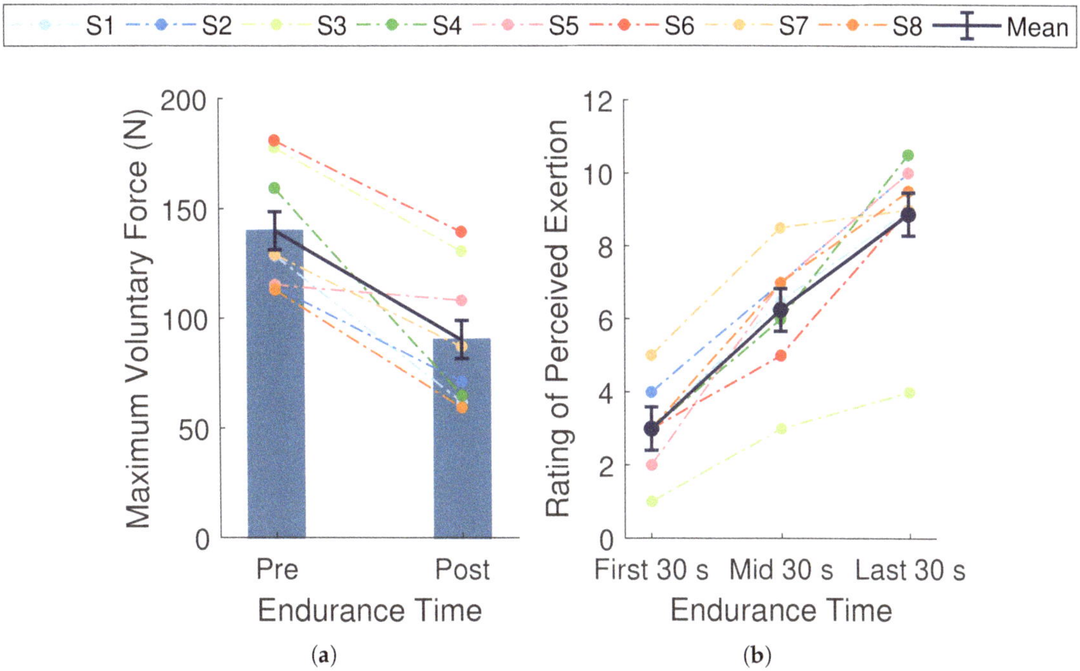

Figure 4. Measures of fatigue. (**a**) Maximal voluntary contraction (MVC) forces taken at baseline (pre-endurance task) and follow-up (post-endurance task). (**b**) Ratings of perceived exertion (RPE) during the first, middle, and last 30 s of the endurance task. Dark blue bars and data points connected by solid lines are means ± SE. Dotted lines represent data from a single participant ($n = 8$) whose assigned color is consistent across figures. MVC force significantly declined ($p < 0.001$, $d = -1.99$) and RPE significantly increased over time ($p < 0.001$, $\eta_p^2 = 0.91$).

A significant change in mean RPE scores occurred during the endurance task ($F(2, 14) = 74.15$, $p < 0.001$, $\eta_p^2 = 0.91$; Figure 4b). Post hoc pairwise comparisons revealed significant differences between all measured time points (all p-values < 0.001). There was an overall mean increase of 5.9 ± 0.5 across participants, with slightly higher changes in scores during the first half (3.2 ± 0.5) compared to the last half (2.6 ± 0.5) of the task. The overall rise in RPE scores indicates the endurance task became increasingly more difficult for the participants as time progressed, providing evidence of self-perceived fatigue.

3.2. Evidence of Localized Muscle Fatigue

A significant main effect of time was found for the instantaneous amplitude ($a_i(t)$) during the endurance task ($F(1.38, 9.68) = 116.65, p < 0.001, \eta_p^2 = 0.83$; Figure 5). No significant differences were present across muscles ($F(2.04, 14.27) = 3.48, p = 0.058$, $\eta_p^2 = 0.33$), nor was there a muscle by time interaction ($F(1.94, 13.58) = 3.26, p = 0.071$, $\eta_p^2 = 0.32$). The mean $a_i(t)$ across all muscles at the beginning, midpoint, and end of the task was 0.17 ± 0.02, 0.2 ± 0.02, and 0.34 ± 0.02, respectively. There was an average increase of $16 \pm 1\%$ ($p < 0.001$) over the course of the task, with a greater increase in $a_i(t)$ during the second half of the task ($13 \pm 1\%$, $p < 0.05$) compared to the first half ($3 \pm 1\%$, $p < 0.001$).

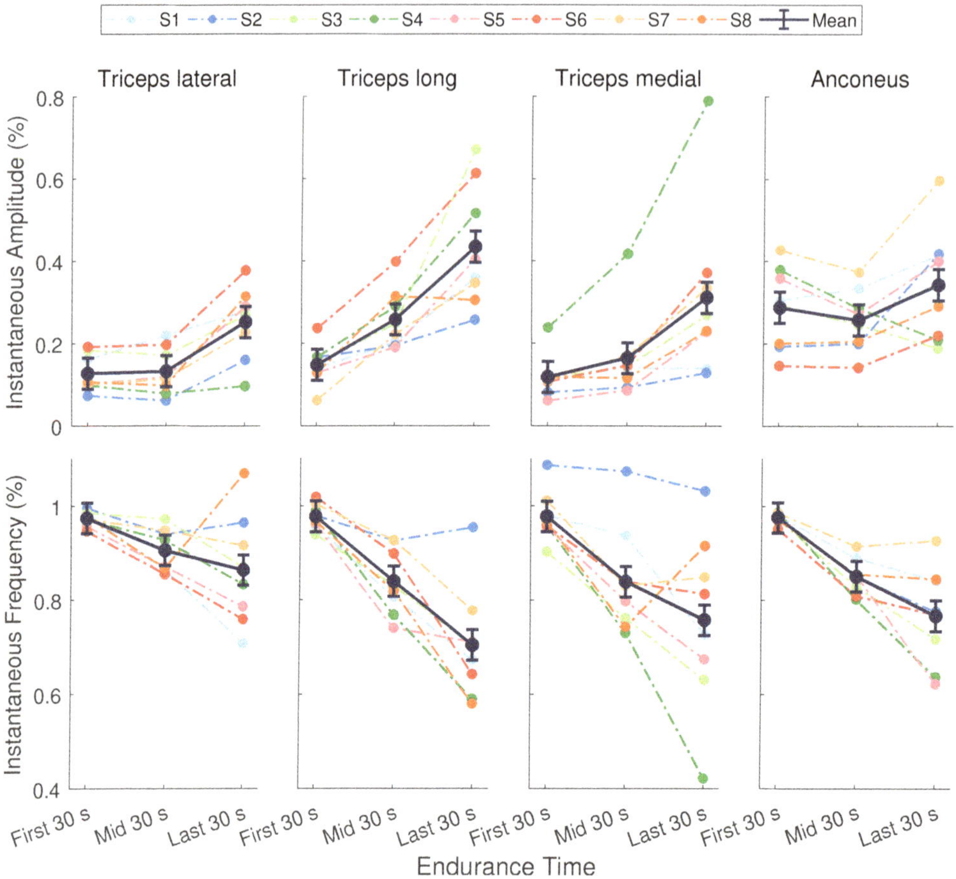

Figure 5. Normalized sEMG features from the elbow extensor muscles. (**Top**) Instantaneous amplitude ($a_i(t)$) and (**Bottom**) instantaneous mean frequency ($f_{im}(t)$). Dark points separated by solid lines are means ± SE for the first, middle, and last 30 s of the task. Dotted lines represent data from a single participant ($n = 8$) whose assigned color is consistent across figures. There was a significant main effect of time for $f_{im}(t)$ ($p < 0.001$, $\eta_p^2 = 0.83$) and $a_i(t)$ ($p < 0.001$, $\eta_p^2 = 0.83$).

There was a significant main effect of time for the instantaneous mean frequency ($f_{im}(t)$) during the endurance task ($F(1.19, 8.34) = 33.97, p < 0.001, \eta_p^2 = 0.83$; Figure 5). There were no significant differences across muscles ($F(1.88, 13.14) = 2.82, p = 0.098, \eta_p^2 = 0.29$), nor was there a muscle by time interaction ($F(2.80, 19.57) = 2.55, p = 0.089, \eta_p^2 = 0.27$). The mean $f_{im}(t)$ across all muscles during the first, middle, and last 30 s was 0.98 ± 0.02, 0.86 ± 0.02, and 0.77 ± 0.02, respectively. On average, the decrease in $f_{im}(t)$ during the first half of the task ($12 \pm 2\%$, $p < 0.001$) was slightly greater than the decrease during the second half of the task ($9 \pm 2\%$, $p < 0.05$), resulting in an overall decline from start to end of $20 \pm 2\%$ ($p < 0.001$).

The average increase in $a_i(t)$ coupled with a decrease in $f_{im}(t)$ across muscles indicates that significant localized fatigue developed in the elbow extensor muscles during the endurance task. These trends in sEMG features can be attributed to central and peripheral nervous system mechanisms and intramuscular adaptations [2,17,18]. Our results are consistent with other studies that evaluated the elbow extensor muscles in male participants during sustained isometric contractions [39,40]. For an isometric endurance task held at 25% MVC, Krogh-Lund and Jorgensen [40] found that the median frequency decreased almost linearly in the medial head of the triceps brachii. The RMS amplitude also increased in this muscle, showing greater changes in the last half of the contraction compared to the first. These results parallel the average trends across individuals in our study for $f_{im}(t)$ and $a_i(t)$, respectively, of the triceps medial head (Figure 5, third column). Davidson and Rice [39] observed significant increases in the RMS amplitude of all three triceps heads (medial, lateral, and long) during an isometric endurance task at 20% MVC. The amplitude of the anconeus muscle, however, revealed smaller increases from the start to the end of the task. Moreover, the long head of the triceps displayed the greatest increase in amplitude across participants at the end of the contraction compared to the other muscles when the participants' shoulder was in 90° of flexion [39]. The average trends in $a_i(t)$ in our study are in agreement with these findings (Figure 5, top row).

The anconeus and long, lateral, and medial heads of the triceps brachii are considered a synergistic muscle group because they all act to extend the elbow [59]. Evidence suggests that these muscles follow a general hierarchic recruitment pattern to preserve energy [60], where the order of activation depends upon the muscle's size [60], joint articulation [60,61], fiber composition [59,62,63], and level of effort required by the task [60,64]. Following these principles, the anconeus muscle will activate first at low levels of force, followed by the medial head of the triceps brachii. When effort reaches a moderate-to-high level, the lateral head will be recruited next, followed by the long head [60]. When averaged across individuals, the results from our study closely mirror this recruitment strategy (Figure 5, top row). The anconeus displayed the greatest average $a_i(t)$ of all the synergists at the start of the task. During the first half of the task, sEMG of the medial head showed a moderate increase in $a_i(t)$ and the largest decrease in $f_{im}(t)$. The $a_i(t)$ of the lateral head remained nearly unchanged, while the $f_{im}(t)$ showed a modest decrease during this period, indicating it may not have been fully recruited yet. During the second half of the endurance task, all muscles showed a steady increase in $a_i(t)$ and decrease in $f_{im}(t)$, with the long and lateral heads of the triceps brachii showing the greatest mean changes. These results show that the endurance task, whose target force was only 30% MVC, started as a low effort task but progressed to a moderate-to-high effort task that required increased recruitment of all muscles. The average rise RPE confirmed that subjects felt the level of effort required to maintain force increased during the task.

Although a hierarchic recruitment pattern [60] is evident when averaged across participants, considerable inter-individual variation in this strategy was present in our study. For example, some participants (S6) showed the largest changes in sEMG activity for the long head of the triceps, whereas others (S4) revealed more dynamic trends in the medial head (Figure 5). Moreover, trends in the sEMG amplitude of the anconeus muscle varied widely across individuals. Inter-muscular variability was also evident in our study. The fatigue response within a muscle is known to be variable over time [28,65] and often exhibits curvilinear behavior depending on the intensity of the muscle contraction [66] and activation of other synergist muscles. This type of behavior is most notable in the non-linear trends in the instantaneous amplitude of the anconeus muscle and the reversed trends in the triceps brachii heads over the last half of the endurance task for participant S8 (Figure 5).

3.3. *Trends in Performance Degradation*

There was a significant change in average FSI over the course of the endurance task ($F(2, 14) = 34.17$, $p < 0.001$, $\eta_p^2 = 0.83$; Figure 6a). Post hoc pairwise comparisons showed significant differences between all time points (all p-values < 0.001). From the first 30 s to the last 30 s of the task, FSI increased by an average of 0.45 ± 0.05. These results demonstrate that the FSI metric was sensitive to fatigue-induced changes in performance over time. The significant increase observed in the FSI metric (Figure 6) indicates that a progressive temporal change occurred in the dynamic relationship between muscle activity and force output during the endurance task. This general trend coincides with changes in force-generating capacity (MVC force), self-perceived exertion (RPE), and localized muscle fatigue ($f_{im}(t)$ and $a_i(t)$), suggesting that the phenomenon captured by the FSI metric reflects a degradation in performance over time.

The full time-series of FSI values for each participant are shown in Figure 6b. Although the average trend in FSI is close to linear when averaged across individuals, most participants displayed a non-linear degradation in performance. Moreover, inter-individual differences in the non-linear trends were also apparent. Performance degraded quickly for some participants during the first half of the experiment (S7, S8), whereas others (S2, S5, S6) showed higher rates of change during the latter half.

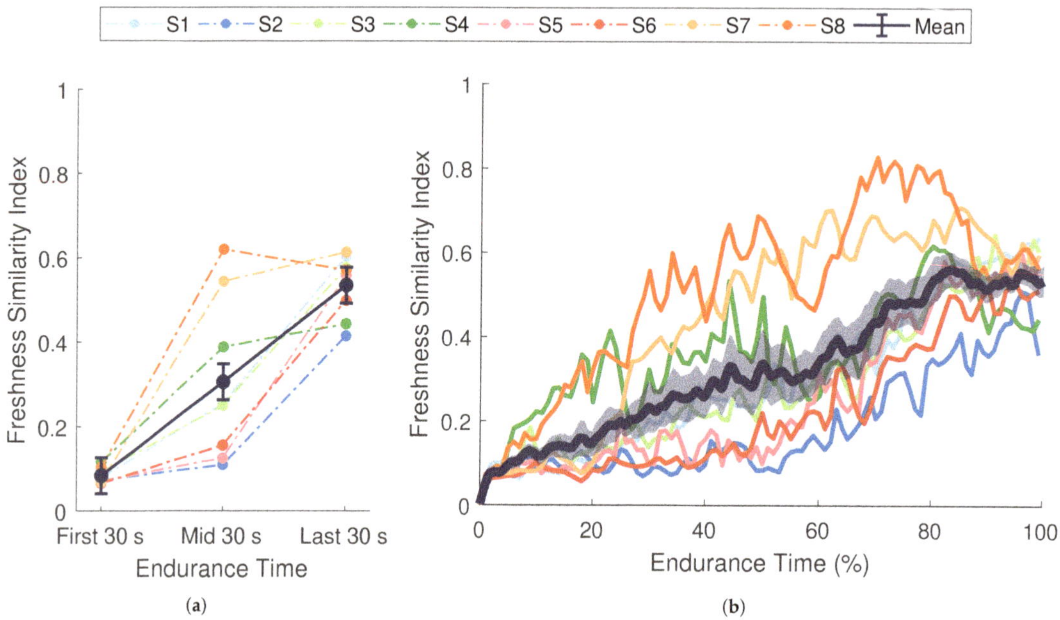

Figure 6. Freshness similarity index (FSI). (**a**) Dark blue data points separated by solid lines are means ± SE for the first, middle, and last 30 s of the task. (**b**) The dark blue line with shaded envelope represents the mean ± SE over each 1% of endurance time. Additional colored lines (dotted in (**a**), solid in (**b**)) represent data from a single participant ($n = 8$) whose assigned color is consistent across figures. FSI increased significantly over time ($p < 0.001$, $\eta_p^2 = 0.83$).

3.4. Relationship Between Measures of Performance Degradation and Fatigue

The *rmcorr* analyses revealed a strong, negative association between FSI and MVC force ($r_{rm}(7) = -0.86, 95\%$ CI $[-0.98, -0.32], p < 0.01$; Figure 7a), and a strong, positive association between FSI and RPE ($r_{rm}(15) = 0.87, 95\%$ CI $[0.64, 0.96], p < 0.001$; Figure 7b). These analyses were used to evaluate whether changes in performance degradation were paralleled by changes in mechanical and self-perceived fatigue within the individual. In other words, for a given individual, was an increase in FSI associated with a decrease in MVC force and an increase in RPE. The results indicate that participants who displayed significant performance degradation also experienced a considerable reduction in force-generating capacity and a rise in perceived effort. These strong within-subject relationships between FSI and both well-established measures of fatigue suggest that the degradation in performance captured by the FSI metric is representative of fatigue, thereby substantiating the use of an ARMAX-based monitoring paradigm for assessing fatigue

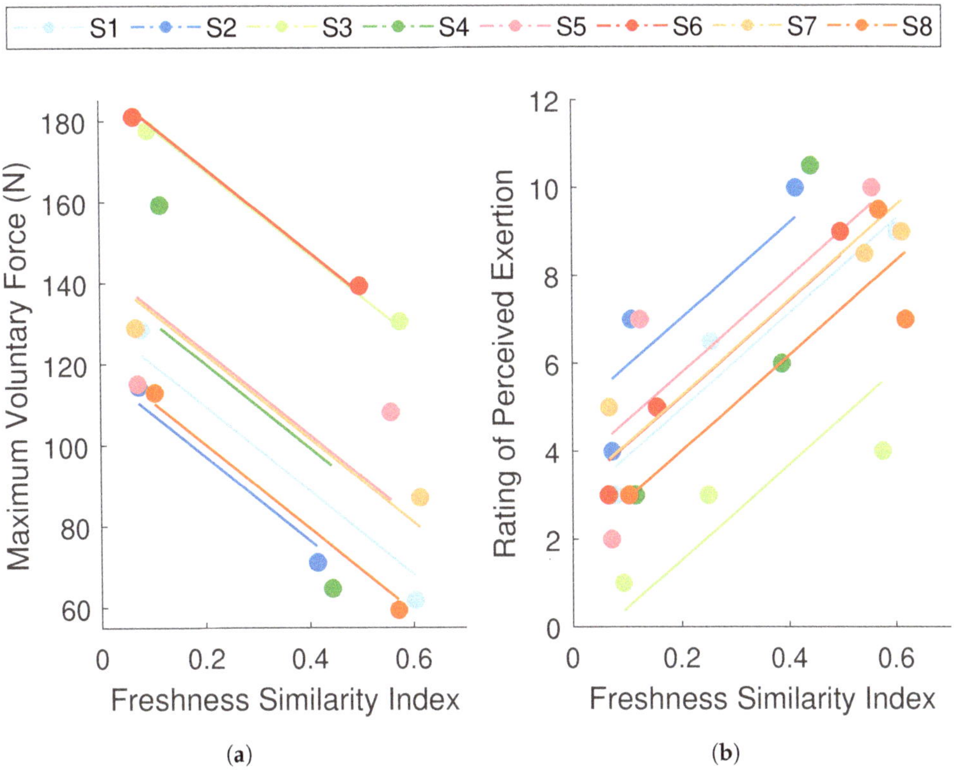

Figure 7. Repeated measures correlations between the freshness similarity index (FSI) and (**a**) maximum voluntary contraction (MVC) force and (**b**) ratings of perceived exertion (RPE). Data points are grouped by participant ($n = 8$), where each color summarizes all observations from one participant and corresponding lines represent the *rmcorr* fit for that participant. Participant color assignments are consistent with those in other figures. FSI revealed a strong, negative relationship with MVC force ($r_{rm} = -0.86, p < 0.01$) and a strong, positive relationship with RPE ($r_{rm} = 0.87, p < 0.001$).

Simple correlations between overall changes in FSI and MVC force ($r(6) = 0.41$, $p = 0.846$) and overall changes in FSI and RPE across participants ($r_s = -0.34$, $p = 0.796$) were not significant. However, we did not expect to observe between-subject associations. Between-subject associations would suggest that participants with high values of FSI also tend to have high values of RPE and low values of MVC force. However, since the FSI is an individual-specific metric, its absolute value may not be comparable across participants.

4. Discussion

4.1. Viability of a System-Based Monitoring Approach for Assessing Fatigue

The primary purpose of this study was to substantiate the viability of the system-based monitoring paradigm for assessing fatigue by relating the FSI metric to well-accepted measures of fatigue that capture a net reduction in force-generating capacity (MVC force) and self-perceived fatigue (RPE). The strong within-individual associations between FSI and these traditional measures indicate that the system-based monitoring approach captured fatigue-induced changes in performance, substantiating its use for assessing fatigue. These findings provide the first direct, quantitative link between a system-based approach to monitoring performance degradation and well-accepted measures of fatigue.

To that end, we verified that participants developed fatigue during the endurance task by observing significant reductions in MVC force and increases in RPE. Previous studies that implemented a system-based monitoring paradigm [31–33] verified their findings by identifying fatigue in individual muscles using trends in sEMG features. However, trends in the relevant sEMG features reflect localized intramuscular adaptations rather than a global reduction in force-generating capacity [15] or heightened perception of exertion [11,12], whereas the FSI metric is a global representation of system-based performance degradation. Furthermore, in these works, the sEMG features were used as inputs to the vARX and ARMAX models, so comparisons of the sEMG features to the results of the FSI metric might be biased. For these reasons, the present study sought to confirm fatigue using well-accepted global measures of fatigue that are external to the modeling paradigm (i.e., MVC force and RPE) in addition to trends in localized muscle signals. Significant changes in MVC force, RPE, and the sEMG features ($f_{im}(t)$ and $a_i(t)$) indicate that the participants fatigued during the endurance task.

4.2. Improvements to the System-Based Monitoring Paradigm

Additional novelty to the research presented in this paper is in the improvements made to the system-based monitoring paradigm presented in previous works. The modifications, which were specified throughout Section 2.6 and are discussed in more detail below, serve to more accurately represent changes occurring in the NMS and facilitate the use of the system-based monitoring paradigm as an online assessment tool.

We selected the sEMG instantaneous amplitude ($a_i(t)$) as an input to the ARMAX model to minimize the influence of high magnitude transients associated with the instantaneous energy feature used in other studies [31–34] and provide a comparable sEMG feature to the commonly used RMS amplitude. As such, $a_i(t)$ served to attenuate signal artifacts and better reflected the neural activation of the muscle [1]. To simplify our model structure, we excluded two additional sEMG features from the ARMAX formulation that were used as model inputs in [32–34]. These extra features, which capture the variance and entropy of the sEMG signal, provided redundant information and added complexity to our model without improving the sensitivity of the FSI metric to fatigue-related changes in the dynamic relationship between the sEMG features and force.

We normalized the model inputs and outputs in a way that is both consistent with how sEMG signals are processed in the literature [18,67,68] and more suitable for online fatigue assessment compared to previous works [31–33]. As a result, the magnitude of the sEMG features fell within predictable bounds, and data from only the baseline MVC contractions and the initial few seconds of the endurance task were needed for scaling. Our strategy would allow for an ARMAX model to be trained using data from short contractions performed before the endurance task, then employed for online monitoring during the endurance task itself. This offers an improvement to previous works whose normalization methods produced model input values that far exceeded the predictable bounds of 0 to 1 [31] or required data from the entire endurance task to obtain the scaling factors [32,33], which would restrict the use of the methodology to post hoc analysis.

Lastly, sEMG features from all elbow extensor muscles were incorporated as inputs to the dynamic model, providing a complete representation of the neuromuscular system responsible for elbow extension. This comprehensive approach extends the capability of previous works, which used a single synergistic calf [32] or forearm [34] muscle to represent the neuromuscular system responsible for isometric plantar flexion and hand grasping, respectively. Although evidence suggests that elbow extensor muscles follow a general hierarchic recruitment pattern, these patterns can vary considerably between individuals and muscles [60], and did vary in our study. Despite these differences, some researchers choose to monitor only one head of the triceps brachii by assuming the sEMG activity from one muscle is representative of the entire synergistic group (i.e., the "equivalent muscle" concept [59]). Although this may be true for brief static contractions [59], the concept does not apply during submaximal contractions held until failure [39]. As a result, assessment approaches that only monitor how one muscle from a synergist group fatigues could underestimate the fatiguing process as a whole. The inclusion of all contributing muscles in our model accommodates the inter-individual differences in muscle recruitment strategies without loss of information by excluding any one particular muscle. Moreover, our approach eliminates the need for *a priori* information regarding muscle fatigability. This is important because the factors contributing to the inter-individual variation (i.e., differences in muscle composition, anatomy, and fitness level) are difficult to measure, making it infeasible to know which muscles will be most fatiguable for a given participant before an experiment is performed.

4.3. Performance of the FSI Metric

The FSI metric showed sensitivity to the performance degradation occurring across multiple muscles and sensor sources during an isometric endurance task. The significant increase in FSI demonstrates that the metric was sensitive to changes in the dynamic relationship between sEMG features from the elbow extensor muscles and force that occurred over time. Alterations in this relationship between sEMG amplitude and force are known to occur in the presence of fatigue during isometric tasks [29]. Moreover, by utilizing both amplitude and frequency based sEMG features from each muscle [5], our multivariate ARMAX model effectively detected fatigue-induced changes in the muscle signals [41] and accounted for changes in muscle behavior due to fatigue and those due to altered force production [5].

As a single metric, the FSI also proved to be a concise representation of performance degradation occurring across multiple muscles and sensor sources. Typically, researchers will evaluate fatigue by using sEMG to separately assess intramuscular changes in individual muscles from corresponding alterations in force or movement output. Instead, our system-based methodology uses an ARMAX formulation to represent the neuromusculoskeletal system as an input-output dynamic model and monitors the model's residuals error over time via the FSI metric. This approach reduces the number of potential monitoring parameters from nine (eight sEMG features and one force signal) to one (FSI), thereby providing a concise representation of fatigue-related degradation in performance.

Most importantly, monitoring the FSI metric also allows for the continuous assessment of fatigue during a task. This can elucidate non-linear performance changes or adaptations that arise over time due to fatigue, as evidenced by the curvilinear evolution of the FSI metric for the majority of individuals in our study. As a result, the system-based monitoring paradigm has clear benefits over MVC-based approaches that must be performed before and after bouts of exercise.

4.4. Advantages of a System-Based Monitoring Approach over Alternative Model-Based Techniques for Fatigue Monitoring

The system-based modeling paradigm presented in this paper offers decided advantages over existing model-based fatigue monitoring strategies. First, the methodology does not restrict how performance degradation can evolve over time, thereby allowing for a non-linear progression of FSI. Compared to other model-based fatigue assessment approaches, which utilize *a priori* assumptions that fatigue will progress linearly over time [23,25,26], the methodology is less restrictive and can allow for a more accurate evolution of fatigue-induced changes in performance. Secondly, the ARMAX model used in this study need only be trained on a small data set from the initial portion of the task before fatigue onset. Alternative fatigue modeling attempts require extensive data sets containing the entire time-course of fatigue to train the models [23,25–28]. This constraint limits the practicality of these approaches due to time-consuming data collection and computationally expensive procedures. The system-based methodology also allows changes in performance to be continually tracked during the endurance task itself, in contrast with other models that use reference contractions to probe for fatigue-induced changes in parameters at discrete time points (e.g., the beginning and end of a task) [28]. Furthermore, our paradigm produces a single overall measure of fatigue, providing an advantage over a model-based technique that used multiple model kernels to evaluate fatigue in each muscle individually [69]. Lastly, our black-box modeling approach requires very few biomechanical assumptions and is capable of performing in a real-time capacity. This offers decided advantages over musculoskeletal modeling approaches that demand knowledge of anatomical parameters and involve time-consuming optimization procedures [70].

4.5. Limitations of the Study

Since the system-based modeling paradigm is in a nascent state, the meaning of the absolute value of the FSI is not yet well understood. This is a common issue shared among fatigue metrics [25,27,28,71], however, because the relative change in the parameter over time is generally of more interest than the absolute value of the parameter. The lack of between-subject associations between FSI and other measures of fatigue found in our study verified that the relative change in FSI is not reflecting the differences within individuals. However, with further investigation and participant-specific considerations, FSI values may become more interpretable.

The sample size may be a limitation of the simple Pearson and Spearman correlations used in this work. With a larger group of participants, it may be possible to observe significant between-subject associations between the FSI and both MVC force and RPE. In fact, a multimuscle fatigue score (MMFS) developed in [28] showed weak ($r = 0.31$) and moderate ($r = -0.56$) relationships with ratings of perceived fatigue (RPF) and changes in MVC force, respectively, using Pearson product-moment correlations on data from 20 participants. In our study, the sample size was sufficient to evaluate the sensitivity of the FSI to fatigue-related changes in performance using RM-ANOVAs and demonstrate the within-subject associations between FSI and both MVC force and RPE using *rmcorr* analyses. The *rmcorr* analysis can accommodate smaller sample sizes because it uses multiple data points per participant and accounts for non-independence of error between observations using analysis of covariance to statistically adjust for the inter-individual variability [56]. As a result, the degrees of freedom and power will generally be higher using *rmcorr* compared to simple correlations, which use aggregated measures to meet the assumption that data is Independent and Identically Distributed (IID) [56].

This study tested only male participants. However, it is not uncommon for fatigue studies to include only one gender in the participant group [25,28,39,40,65,72]. A related study that evaluated elbow extensor fatigability during a sustained isometric task at 15% MVC until failure reported no differences in endurance time or sEMG amplitude across men and women [73], contrary to observations from other muscle groups that exhibit sex differences [73,74]. Thus, despite the single-gender participant pool used in our study, the findings in [73] provide evidence that our system-based paradigm could account for gender in this muscle group. However, further investigation is necessary to confirm the accuracy of the proposed system-based monitoring paradigm for gender and other factors, such as age.

The ARMAX models were trained on data that was individual- and task-specific, meaning the model parameters, which were estimated for each participant individually during a specific submaximal isometric task, may not be generalizable to other participants or exercises. However, this warrants further investigation. Although model specificity is a shared limitation among other model-based fatigue assessment strategies [23,25,28], personalized models are still essential for making patient-specific clinical decisions [75] or when accurate fatigue monitoring is required, that is, during recovery after musculoskeletal injuries or rehabilitation for patients with neuromuscular disorders [7].

Lastly, insight concerning the specific muscles experiencing fatigue is not reflected in the FSI, as was the case in the model-based approach by [28]. However, the purpose of the system-based monitoring paradigm is to provide a concise measure of fatigue-related changes in performance across multiple muscles and sensor sources. Thus, condensing the number of monitoring parameters down to a single metric allows for a uniform approach to assessing how the entire NMS system responsible for the fatiguing task behaves across individuals. Although only four muscles were considered in the NMS system responsible for elbow extension in this study, the system-based monitoring paradigm is flexible to accommodate any number of inputs.

4.6. Applications of the Study

There are many practical applications of this research. The ability to characterize and track fatigue-related changes in neuromuscular system performance during exercise has the potential to inform therapeutic modalities for rehabilitation. It also can become useful when personalizing exercise regimens to target strength or endurance deficits, or by indicating when to stop exercising before significant fatigue leads to the onset of injury. More specifically, this work has the potential to improve fatigue monitoring techniques during robot-aided movement training, which typically apply traditional signal processing methods to analyze localized fatigue of individual muscles using sEMG [10]. Robotic exoskeletons are equipped with high-resolution sensors, such as force sensors and encoders, that can capture kinematic and kinetic measurements reflecting the quality of a user's movement [76]. In combination with physiological measures, such as sEMG, a system-based monitoring paradigm could fuse the data from these sensor sources and produce a single metric to assess fatigue, such as the FSI. This metric could then be used as an input to an exoskeleton controller that alters the level of robot-applied assistance or resistance to accommodate a patient's capability and needs [77].

4.7. Future Work

Several aspects of the presented methodology are ripe for further exploration to enhance its utility as a diagnostic and monitoring tool. In this work, we chose to use an isometric task to validate that the FSI captures fatigue because it is a simple contraction that does not require the muscle to change length, thereby minimizing the non-stationary behavior of the sEMG signals. Further validation using concentric and eccentric exercises will open the possibility of fatigue monitoring during dynamic movements, which are integral to various therapeutic modalities. Additionally, a formal exploration of how the FSI metric behaves across multiple days of testing and in response to periods of rest and recovery would help prove its effectiveness as a clinical tool. Further advancements to the dynamic model might also lead to improved modeling accuracy and fatigue tracking, especially when expanding the application of this work to more dynamic movements involving multiple joints. In this work, we assumed a linear dynamic relationship between muscle activity and movement output for analytical tractability. Future work could examine the appropriateness of the linear assumption by comparing its accuracy to non-linear dynamic models [78]. In the long run, the approach presented in this paper could be adapted to monitor fatigue in real-time and used to update control laws of robots, e.g., exoskeletons, for optimal human-robot performance.

5. Conclusions

This paper presented and validated a framework for continuously assessing fatigue using a system-based monitoring paradigm. The paradigm modeled the dynamic relationship between sEMG features extracted from multiple synergistic muscles to force output, then employed statistical analysis of modeling errors to reveal how performance degraded in each participant over time. The index of performance degradation (FSI) revealed strong, within-individual associations with two well-established fatigue measures, substantiating its applicability as a fatigue monitoring tool. The FSI provided a sensitive and concise representation of the temporal changes in the dynamic relationship between limb force and sEMG parameters during submaximal static exercise. Improvements were made to the system-based monitoring paradigm to facilitate online fatigue assessment and more accurately represent changes occurring in the NMS. This work presents the first step toward evaluating the clinical viability of a system-based monitoring strategy for assessing fatigue by comparing its performance with traditional fatigue measures. Ultimately, the ability to monitor and assess fatigue has important implications for preventing neuromuscular injury, optimizing training loads, and guiding effective, individualized treatment strategies for rehabilitation.

Author Contributions: Conceptualization, K.E.M., D.D., A.D.D.; methodology, K.E.M. and D.D.; software, K.E.M. and D.D.; validation, K.E.M., D.D., A.D.D.; formal analysis, K.E.M., D.D.; investigation, K.E.M.; resources, K.E.M., A.D.D.; data curation, K.E.M.; writing—original draft preparation, K.E.M.; writing—review and editing, K.E.M., D.D., A.D.D.; visualization, K.E.M.; supervision, K.E.M., D.D., A.D.D.; project administration, K.E.M., A.D.D.; funding acquisition, K.E.M. and A.D.D. All authors have read and agreed to the published version of the manuscript.

Funding: This work was supported in part by a NASA Space Technology Research Fellowship under grant NNX15AQ32H.

Institutional Review Board Statement: The study was conducted according to the guidelines of the Declaration of Helsinki. The experimental procedure was approved by the Internal Review Board organized by the Office of Research Support at The University of Texas at Austin under the protocol number 2013-05-0126 approved on 18 July 2019.

Informed Consent Statement: Informed consent was obtained from all participants involved in the study.

Data Availability Statement: The data presented in this study are available on request from the corresponding author. The data are not publicly available due to continuing study by the authors.

Acknowledgments: The authors would like to thank Job Ramirez for assistance in building the exoskeleton test-bed.

Conflicts of Interest: The authors declare no conflict of interest.

Abbreviations

The following abbreviations are used in this manuscript:

sEMG	Surface Electromyography
MVC	Maximum Voluntary Contraction
RPE	Rating of Perceived Exertion
FSI	Freshness Similarity Index
RMS	Root Mean Square
TFD	Time Frequency Distribution
ARMAX	Autoregressive Moving Average Model with Exogenous Inputs
RM-ANOVA	Repeated Measures Analysis of Variance
rmcorr	Repeated Measures Correlation
MU	Motor Unit
NMS	Neuromusculoskeletal System

References

1. Gandevia, S.C. Spinal and supraspinal factors in human muscle fatigue. *Physiol. Rev.* **2001**, *81*, 1725–1789. [CrossRef]
2. De Luca, C.J. Myoelectrical manifestations of localized muscular fatigue in humans. *Crit. Rev. Biomed. Eng.* **1984**, *11*, 251–279. [PubMed]
3. Carroll, T.J.; Taylor, J.L.; Gandevia, S.C. Recovery of central and peripheral neuromuscular fatigue after exercise. *J. Appl. Physiol.* **2017**, *122*, 1068–1076. [CrossRef] [PubMed]
4. Knapik, J.J.; Reynolds, K.L.; Harman, E. Soldier load carriage: Historical, physiological, biomechanical, and medical aspects. *Mil. Med.* **2004**, *169*, 45–56. [CrossRef] [PubMed]
5. Luttmann, A.; Jäger, M.; Laurig, W. Electromyographical indication of muscular fatigue in occupational field studies. *Int. J. Ind. Ergon.* **2000**, *25*, 645–660. [CrossRef]
6. Zwarts, M.J.; Bleijenberg, G.; Van Engelen, B.G.M. Clinical neurophysiology of fatigue. *Clin. Neurophysiol.* **2008**, *119*, 2–10. [CrossRef]
7. Lou, J.S.; Weiss, M.D.; Carter, G.T. Assessment and management of fatigue in neuromuscular disease. *Am. J. Hosp. Palliat. Med.* **2010**, *27*, 145–157. [CrossRef]
8. Gorelick, M.; Brown, J.; Groeller, H. Short-duration fatigue alters neuromuscular coordination of trunk musculature: Implications for injury. *Appl. Ergon.* **2003**, *34*, 317–325. [CrossRef]
9. Ament, W.; Verkerke, G.J. Exercise and fatigue. *Sport. Med.* **2009**, *39*, 389–422. [CrossRef]

10. Mugnosso, M.; Marini, F.; Holmes, M.; Morasso, P.; Zenzeri, J. Muscle fatigue assessment during robot-mediated movements. *J. Neuroeng. Rehabil.* **2018**, *15*, 119. [CrossRef]
11. Schiphof-Godart, L.; Roelands, B.; Hettinga, F.J. Drive in sports: How mental fatigue affects endurance performance. *Front. Psychol.* **2018**, *9*, 1383. [CrossRef] [PubMed]
12. Borg, G. Psychophysical scaling with applications in physical work and the perception of exertion. *Scand. J. Work. Environ. Health* **1990**, *16*, 55–58. [CrossRef] [PubMed]
13. Marcora, S.M.; Staiano, W.; Manning, V. Mental fatigue impairs physical performance in humans. *J. Appl. Physiol.* **2009**, *106*, 857–864. [CrossRef] [PubMed]
14. Van Cutsem, J.; Marcora, S.; De Pauw, K.; Bailey, S.; Meeusen, R.; Roelands, B. The effects of mental fatigue on physical performance: A systematic review. *Sport. Med.* **2017**, *47*, 1569–1588. [CrossRef]
15. Vøllestad, N.K. Measurement of human muscle fatigue. *J. Neurosci. Methods* **1997**, *74*, 219–227. [CrossRef]
16. Norberto, M.S.; De Arruda, T.B.; Papoti, M. A new approach to evaluate neuromuscular fatigue of extensor elbow muscles. *Front. Physiol.* **2020**, *11*, 553296. [CrossRef]
17. Bigland-Ritchie, B.; Woods, J.J. Changes in muscle contractile properties and neural control during human muscular fatigue. *Muscle Nerve* **1984**, *7*, 691–699. [CrossRef]
18. Merletti, R.; Knaflitz, M.; De Luca, C.J. Myoelectric manifestations of fatigue in voluntary and electrically elicited contractions. *J. Appl. Physiol.* **1990**, *69*, 1810–1820. [CrossRef]
19. Basmajian, J.V. Muscles alive. Their functions revealed by electromyography. *Acad. Med.* **1962**, *37*, 802.
20. Bonato, P.; Roy, S.H.; Knaflitz, M.; De Luca, C.J. Time-frequency parameters of the surface myoelectric signal for assessing muscle fatigue during cyclic dynamic contractions. *IEEE Trans. Biomed. Eng.* **2001**, *48*, 745–753. [CrossRef]
21. Farina, D. Interpretation of the surface electromyogram in dynamic contractions. *Exerc. Sport Sci. Rev.* **2006**, *34*, 121–127. [CrossRef] [PubMed]
22. Cifrek, M.; Medved, V.; Tonković, S.; Ostojić, S. Surface EMG based muscle fatigue evaluation in biomechanics. *Clin. Biomech.* **2009**, *24*, 327–340. [CrossRef] [PubMed]
23. González-Izal, M.; Malanda, A.; Gorostiaga, E.; Izquierdo, M. Electromyographic models to assess muscle fatigue. *J. Electromyogr. Kinesiol.* **2012**, *22*, 501–512. [CrossRef] [PubMed]
24. Rampichini, S.; Vieira, T.M.; Castiglioni, P.; Merati, G. Complexity Analysis of Surface Electromyography for Assessing the Myoelectric Manifestation of Muscle Fatigue: A Review. *Entropy* **2020**, *22*, 529. [CrossRef]
25. MacIsaac, D.T.; Parker, P.A.; Englehart, K.B.; Rogers, D.R. Fatigue estimation with a multivariable myoelectric mapping function. *IEEE Trans. Biomed. Eng.* **2006**, *53*, 694–700. [CrossRef]
26. Rogers, D.R.; MacIsaac, D.T. Training a multivariable myoelectric mapping function to estimate fatigue. *J. Electromyogr. Kinesiol.* **2010**, *20*, 953–960. [CrossRef]
27. Rogers, D.R.; MacIsaac, D.T. EMG-based muscle fatigue assessment during dynamic contractions using principal component analysis. *J. Electromyogr. Kinesiol.* **2011**, *21*, 811–818. [CrossRef]
28. McDonald, A.C.; Mulla, D.M.; Keir, P.J. Using EMG amplitude and frequency to calculate a multimuscle fatigue score and evaluate global shoulder fatigue. *Hum. Factors* **2019**, *61*, 526–536. [CrossRef]
29. Dideriksen, J.L.; Farina, D.; Enoka, R.M. Influence of fatigue on the simulated relation between the amplitude of the surface electromyogram and muscle force. *Philos. Trans. R. Soc. A Math. Phys. Eng. Sci.* **2010**, *368*, 2765–2781. [CrossRef]
30. Isermann, R. *Fault-Diagnosis Applications: Model-Based Condition Monitoring: Actuators, Drives, Machinery, Plants, Sensors, and Fault-Tolerant Systems*; Springer Science & Business Media: Berlin/Heidelberg, Germany, 2011.
31. Mussleman, M.; Gates, D.; Djurdjanovic, D. A System-Based Approach to Monitoring the Performance of a Human Neuromusculoskeletal System. *Int. J. Progn. Health Manag.* **2016**, *7*, 14.
32. Xie, Y.Y.; Djurdjanovic, D. Monitoring of human neuromusculoskeletal system performance through model-based fusion of electromyogram signals and kinematic/dynamic variables. *Struct. Health Monit.* **2019**, 1475921719848006. [CrossRef]
33. Madden, K.E.; Djurdjanovic, D.; Deshpande, A.D. Monitoring human neuromusculoskeletal system performance during spacesuit glove use: A pilot study. In Proceedings of the IEEE Aerospace Conference, Big Sky, MT, USA, 3–10 March 2018; pp. 1–10.
34. Yang, K.; Nicolini, L.; Kuang, I.; Lu, N.; Djurdjanovic, D. Long-term modeling and monitoring of neuromusculoskeletal system performance using tattoo-like EMG sensors. *Int. J. Progn. Health Manag.* **2019**, *10*, 004.
35. Hellmann, F.; Verdi, M.; Schlemper Junior, B.R.; Caponi, S. 50th anniversary of the Declaration of Helsinki: The double standard was introduced. *Arch. Med Res.* **2014**, *45*, 600–601. [CrossRef] [PubMed]
36. Schiele, A.; van der Helm, F.C.T. Kinematic design to improve ergonomics in human machine interaction. *IEEE Trans. Neural Syst. Rehabil. Eng.* **2006**, *14*, 456–469. [CrossRef] [PubMed]
37. Taylor, A.; Bronks, R. Reproducibility and validity of the quadriceps muscle integrated electromyogram threshold during incremental cycle ergometry. *Eur. J. Appl. Physiol. Occup. Physiol.* **1995**, *70*, 252–257. [CrossRef] [PubMed]
38. Ng, J.K.; Richardson, C.A. Reliability of electromyographic power spectral analysis of back muscle endurance in healthy subjects. *Arch. Phys. Med. Rehabil.* **1996**, *77*, 259–264. [CrossRef]
39. Davidson, A.W.; Rice, C.L. Effect of shoulder angle on the activation pattern of the elbow extensors during a submaximal isometric fatiguing contraction. *Muscle Nerve* **2010**, *42*, 514–521. [CrossRef]

40. Krogh-Lund, C.; Jørgensen, K. Changes in conduction velocity, median frequency, and root mean square-amplitude of the electromyogram during 25% maximal voluntary contraction of the triceps brachii muscle, to limit of endurance. *Eur. J. Appl. Physiol. Occup. Physiol.* **1991**, *63*, 60–69. [CrossRef] [PubMed]
41. Rogers, D.R.; MacIsaac, D.T. A comparison of EMG-based muscle fatigue assessments during dynamic contractions. *J. Electromyogr. Kinesiol.* **2013**, *23*, 1004–1011. [CrossRef]
42. Hermens, H.J.; Freriks, B.; Merletti, R.; Stegeman, D.; Blok, J.; Rau, G.; Disselhorst-Klug, C.; Hägg, G. European recommendations for surface electromyography. *Roessingh Res. Dev.* **1999**, *8*, 13–54.
43. Merletti, R. Standards for reporting EMG data. *J. Electromyogr. Kinesiol.* **1999**, *9*, 3–4.
44. Potvin, J.; Brown, S. Less is more: High pass filtering, to remove up to 99% of the surface EMG signal power, improves EMG-based biceps brachii muscle force estimates. *J. Electromyogr. Kinesiol.* **2004**, *14*, 389–399. [CrossRef] [PubMed]
45. Semmlow, J.L.; Griffel, B. *Biosignal and Medical Image Processing*; CRC Press: Boca Raton, FL, USA, 2014.
46. Dankaerts, W.; O'Sullivan, P.B.; Burnett, A.F.; Straker, L.M.; Danneels, L.A. Reliability of EMG measurements for trunk muscles during maximal and sub-maximal voluntary isometric contractions in healthy controls and CLBP patients. *J. Electromyogr. Kinesiol.* **2004**, *14*, 333–342. [CrossRef] [PubMed]
47. Bonato, P.; Gagliati, G.; Knaflitz, M. Analysis of myoelectric signals recorded during dynamic contractions. *IEEE Eng. Med. Biol. Mag.* **1996**, *15*, 102–111. [CrossRef]
48. von Tscharner, V. Intensity analysis in time-frequency space of surface myoelectric signals by wavelets of specified resolution. *J. Electromyogr. Kinesiol.* **2000**, *10*, 433–445. [CrossRef]
49. Boashash, B. *Time-Frequency Signal Analysis and Processing: A Comprehensive Reference*, 2nd ed.; Academic Press: Oxford, UK, 2016; Chapter 16.
50. Gottlieb, G.L.; Agarwal, G.C. Dynamic relationship between isometric muscle tension and the electromyogram in man. *J. Appl. Physiol.* **1971**, *30*, 345–351. [CrossRef] [PubMed]
51. Thelen, D.G.; Schultz, A.B.; Fassois, S.D.; Ashton-Miller, J.A. Identification of dynamic myoelectric signal-to-force models during isometric lumbar muscle contractions. *J. Biomech.* **1994**, *27*, 907–919. [CrossRef]
52. Cha, S.H. Comprehensive survey on distance/similarity measures between probability density functions. *Int. J. Math. Model. Methods Appl. Sci.* **2007**, *1*, 1.
53. Hernández-Rivera, E.; Coleman, S.P.; Tschopp, M.A. Using similarity metrics to quantify differences in high-throughput data sets: Application to X-ray diffraction patterns. *ACS Comb. Sci.* **2017**, *19*, 25–36. [CrossRef]
54. MATLAB. *Version 9.3.0 (R2017b)*; The MathWorks Inc.: Natick, MA, USA, 2017.
55. Bland, J.M.; Altman, D.G. Statistics notes: Calculating correlation coefficients with repeated observations: Part 1—Correlation within subjects. *BMJ* **1995**, *310*, 446. [CrossRef]
56. Bakdash, J.Z.; Marusich, L.R. Repeated measures correlation. *Front. Psychol.* **2017**, *8*, 456. [CrossRef] [PubMed]
57. Mukaka, M.M. A guide to appropriate use of correlation coefficient in medical research. *Malawi Med. J.* **2012**, *24*, 69–71. [PubMed]
58. R Core Team. *R: A Language and Environment for Statistical Computing*; R Foundation for Statistical Computing: Vienna, Austria, 2019.
59. Le Bozec, S.; Maton, B.; Cnockaert, J.C. The synergy of elbow extensor muscles during static work in man. *Eur. J. Appl. Physiol. Occup. Physiol.* **1980**, *43*, 57–68. [CrossRef] [PubMed]
60. Neumann, D.A. *Kinesiology of the Musculoskeletal System; Foundation for Rehabilitation*, 2nd ed.; Mosby/Elsevier: St. Louis, MO, USA, 2010; Chapter 6.
61. Zhang, L.Q.; Nuber, G.W. Moment distribution among human elbow extensor muscles during isometric and submaximal extension. *J. Biomech.* **2000**, *33*, 145–154. [CrossRef]
62. Elder, G.C.; Bradbury, K.; Roberts, R. Variability of fiber type distributions within human muscles. *J. Appl. Physiol.* **1982**, *53*, 1473–1480. [CrossRef]
63. Le Bozec, S.; Maton, B. Differences between motor unit firing rate, twitch characteristics and fibre type composition in an agonistic muscle group in man. *Eur. J. Appl. Physiol. Occup. Physiol.* **1987**, *56*, 350–355. [CrossRef]
64. Kuo, K.H.M.; Clamann, H.P. Coactivation of synergistic muscles of different fiber types in fast and slow contractions. *Am. J. Phys. Med. Rehabil.* **1981**, *60*, 219–238.
65. Gonzalez-Izal, M.; Falla, D.; Izquierdo, M.; Farina, D. Predicting force loss during dynamic fatiguing exercises from non-linear mapping of features of the surface electromyogram. *J. Neurosci. Methods* **2010**, *190*, 271–278. [CrossRef]
66. Merletti, R.; Conte, L.R.; Orizio, C. Indices of muscle fatigue. *J. Electromyogr. Kinesiol.* **1991**, *1*, 20–33. [CrossRef]
67. Hostens, I.; Seghers, J.; Spaepen, A.; Ramon, H. Validation of the wavelet spectral estimation technique in biceps brachii and brachioradialis fatigue assessment during prolonged low-level static and dynamic contractions. *J. Electromyogr. Kinesiol.* **2004**, *14*, 205–215. [CrossRef]
68. Woods, J.J.; Bigland-Ritchie, B. Linear and non-linear surface EMG/force relationships in human muscles. An anatomical/functional argument for the existence of both. *Am. J. Phys. Med.* **1983**, *62*, 287–299. [PubMed]
69. Asefi, M.; Moghimi, S.; Kalani, H.; Moghimi, A. Dynamic modeling of SEMG–force relation in the presence of muscle fatigue during isometric contractions. *Biomed. Signal Process. Control* **2016**, *28*, 41–49. [CrossRef]
70. Rao, G.; Berton, E.; Amarantini, D.; Vigouroux, L.; Buchanan, T.S. An EMG-driven biomechanical model that accounts for the decrease in moment generation capacity during a dynamic fatigued condition. *J. Biomech. Eng.* **2010**, *132*, 071003. [CrossRef]

71. Dimitrov, G.V.; Arabadzhiev, T.I.; Mileva, K.N.; Bowtell, J.L.; Crichton, N.; Dimitrova, N.A. Muscle fatigue during dynamic contractions assessed by new spectral indices. *Med. Sci. Sport. Exerc.* **2006**, *38*, 1971–1979. [CrossRef]
72. González-Izal, M.; Malanda, A.; Navarro-Amézqueta, I.; Gorostiaga, E.M.; Mallor, F.; Ibañez, J.; Izquierdo, M. EMG spectral indices and muscle power fatigue during dynamic contractions. *J. Electromyogr. Kinesiol.* **2010**, *20*, 233–240. [CrossRef] [PubMed]
73. Dearth, D.J.; Umbel, J.; Hoffman, R.L.; Russ, D.W.; Wilson, T.E.; Clark, B.C. Men and women exhibit a similar time to task failure for a sustained, submaximal elbow extensor contraction. *Eur. J. Appl. Physiol.* **2010**, *108*, 1089–1098. [CrossRef] [PubMed]
74. Hunter, S.K. The relevance of sex differences in performance fatigability. *Med. Sci. Sport. Exerc.* **2016**, *48*, 2247. [CrossRef] [PubMed]
75. Forkan, A.R.M.; Khalil, I. A clinical decision-making mechanism for context-aware and patient-specific remote monitoring systems using the correlations of multiple vital signs. *Comput. Methods Programs Biomed.* **2017**, *139*, 1–16. [CrossRef]
76. Nordin, N.; Xie, S.Q.; Wünsche, B. Assessment of movement quality in robot-assisted upper limb rehabilitation after stroke: A review. *J. Neuroeng. Rehabil.* **2014**, *11*, 137. [CrossRef]
77. Marchal-Crespo, L.; Reinkensmeyer, D.J. Review of control strategies for robotic movement training after neurologic injury. *J. Neuroeng. Rehabil.* **2009**, *6*, 1–15. [CrossRef]
78. Musselman, M.; Gates, D.; Djurdjanovic, D. System based monitoring of a neuromusculoskeletal system using divide and conquer type models. In Proceedings of the IEEE Aerospace Conference, Big Sky, MT, USA, 4–11 March 2017; pp. 1–12.

sensors

MDPI

Article

Wheelchair-Mounted Upper Limb Robotic Exoskeleton with Adaptive Controller for Activities of Daily Living

Bridget Schabron [1], Jaydip Desai [1,*] and Yimesker Yihun [2]

[1] Neuro-Robotics Lab, Biomedical Engineering, Wichita State University, Wichita, KS 67260, USA; bcschabron@shockers.wichita.edu
[2] Robotics and Control Lab, Mechanical Engineering, Wichita State University, Wichita, KS 67260, USA; yimesker.yihun@wichita.edu
* Correspondence: jaydip.desai@wichita.edu; Tel.: +1-316-978-7608

Abstract: Neuro-muscular disorders and diseases such as cerebral palsy and Duchenne Muscular Dystrophy can severely limit a person's ability to perform activities of daily living (ADL). Exoskeletons can provide an active or passive support solution to assist these groups of people to perform ADL. This study presents an artificial neural network-trained adaptive controller mechanism that uses surface electromyography (sEMG) signals from the human forearm to detect hand gestures and navigate an in-house-built wheelchair-mounted upper limb robotic exoskeleton based on the user's intent while ensuring safety. To achieve the desired position of the exoskeleton based on human intent, 10 hand gestures were recorded from 8 participants without upper limb movement disabilities. Participants were tasked to perform water bottle pick and place activities while using the exoskeleton, and sEMG signals were collected from the forearm and processed through root mean square, median filter, and mean feature extractors prior to training a scaled conjugate gradient backpropagation artificial neural network. The trained network achieved an average of more than 93% accuracy, while all 8 participants who did not have any prior experience of using an exoskeleton were successfully able to perform the task in less than 20 s using the proposed artificial neural network-trained adaptive controller mechanism. These results are significant and promising thus could be tested on people with muscular dystrophy and neuro-degenerative diseases.

Keywords: electromyography; artificial neural network; exoskeleton; assistive technology; robotics; hand gestures

Citation: Schabron, B.; Desai, J.; Yihun, Y. Wheelchair-Mounted Upper Limb Robotic Exoskeleton with Adaptive Controller for Activities of Daily Living. *Sensors* **2021**, *21*, 5738. https://doi.org/10.3390/s21175738

Academic Editor: Ernest N. Kamavuako

Received: 31 July 2021
Accepted: 23 August 2021
Published: 26 August 2021

Publisher's Note: MDPI stays neutral with regard to jurisdictional claims in published maps and institutional affiliations.

1. Introduction

Actively-actuated exoskeletons with joints with several degrees-of-freedom (DOF) are complex and powerful due to the use of actuators for the movement of the joint compared with both passively and non-actuated exoskeletons. The use of active actuators enhances the ability to move multiple joints at the same time, creating a greater work space to perform ADL [1] and also offering greater precision compared with the gravity-offsetting systems of springs, elastic bands, and counter-weights used in passive and non-actuated exoskeletons. Currently, there are several actively-actuated exoskeletons on the market, including the Orthojacket [2], Multimodal Neuroprosthesis for Daily Upper Limb Support (MUNDUS) [3], the robotic upper-extremity repetitive trainer (RUPERT) [4], the biomimetic orthosis for the neurorehabilitation of the elbow and shoulder (BONES) [5], and the motorized upper-limb orthotic system (MULOS) [6]. Most of these actively controlled exoskeletons require a precise control and user interface system that infers a user's input to control the actuators in real-time. There are a wide variety of options for intuiting user input; the simplest and most frequently used option is a switch, key, button, or joystick [7,8]. In fact, a survey of upper limb hybrid exoskeletons in 2017 noted that these were utilized in 50% of hybrid exoskeletons (exoskeletons with functional electrical stimulation) [8]. The signal from these kinds of sensors is easier to receive and most directly interpret. However, this also limits user movement, taking away the user's ability to perform activities with both arms while also controlling the exoskeleton [8]. Another method is the use of voice commands, which free up the user's hands but are difficult to utilize in crowded or noisy environments. There is also a visual system that monitors eye movement to infer user intent, which is limiting due to the placement of the eye-tracking unit being restricted to a table [8].

Biological signals are also used as an input to drive the exoskeletons, as well as for more general feedback to gather information on fatigue or the muscle groups utilized during use of the exoskeleton. Electroencephalography (EEG) and surface electromyography (sEMG) are the most complex inputs used for deriving user intent to control exoskeletons [8]. EEG, or electrical signals of the brain, are occasionally used for brain–machine type interfaces for exoskeleton control [8]. However, when utilized to control exoskeletons, the EEG signals are not intuitive and are often used with predefined trajectories, such as for the BCI-Controlled Wearable Robot in 2014 and the BCI-Controlled Exoskeleton in 2015, or predefined trajectories in addition to a joystick for the degree of movement for the THINK2GRASP exoskeleton in 2013 [8]. The MUNDUS is another example that uses a brain–computer interface (BCI), but it combines EMG with a USB button and eye-tracking user input options as well [3].

The use of sEMG, or surface-level electrical muscle signals, is a less complicated input than the brain but nonetheless can be quite complex. Some exoskeleton controllers have used simple thresholding on EMG signals [9]. Others take more complex approaches, offering more possibilities for using the signals, such as a neurofuzzy modifier to estimate joint torques from EMG signals [10,11]. While this concept of joint torque estimation is promising, it also requires a higher number of sensors, with 16 wired sEMG sensors along deltoid, pectoral, biceps, triceps, flexor, and extensor muscles of the arm in one study [10] and 14 wired sEMG signals covering many of the same muscles in another study [11]. Although these studies have shown the promise of such a system, commercial applicability would be difficult to attain given the number and wide-ranging placements of sensors; imagine a patient trying to put even 14 sensors on alone, or having to them put on his/her arms and shoulders, each and every day. Of course, there is the possibility of the development of a sleeve and electrodes that do not require re-placement every day, but the complexity from the sheer number of signals still remains.

Another study estimated the force of antagonist and agonist muscles of the arm (biceps—short and long head and flexor and extensor carpi radialis) using the Hill model and sEMG signal inputs from those muscles and the use of different forces as an input for a proportional–integral (PI) controller. The same study also looked at a different control method that used the same sEMG signals but instead utilized machine learning, specifically linear discriminant analysis, to classify movements [12]. A study using the ETS-MARSE related sEMG signals to muscle forces and torques using a proportional constant for each muscle [13]. These various studies show the interest and promise of EMG, as well as the limitations of the current use of EMG input. This research article proposes a compact, additive-manufactured, and actively-actuated 3 DOF wheelchair-mounted upper limb robotic exoskeleton that offers shoulder horizontal abduction and adduction, shoulder flexion and extension, and elbow flexion and extension movements based on artificial neural network-trained hand gesture recognition and an adaptive controller mechanism to ensure the user's safety while maneuvering the assistive exoskeleton for daily activities. Eight able-bodied participants without upper limb movement disabilities were recruited in this feasibility study to evaluate the response of the proposed system during individual joint control movements followed by a water bottle pick and place task using hand gestures. Quantitative data sets and subjective feedback were acquired to evaluate the proposed technology, which has the potential to be utilized in the near future on people with muscular dystrophy and neuro-degenerative diseases that limit a person's ability to move their arms in three-dimensional space.

2. Methods

The proposed artificial neural network-trained hand gesture recognition system based on sEMG and an adaptive controller mechanism was implemented on an in-house-built wheelchair-mounted upper limb robotic exoskeleton. The exoskeleton was intended to be used for both assistive and rehabilitative tasks for people with upper-limb impairment. The mechanism was designed based on human anatomical data and the desired workspace to accommodate functional ranges of motions to perform some ADLs. The system is equipped with safety features of mechanical joint limits and electronic sensors to communicate with the user through physiological data, such as EMG and force myography signals (FMG).

2.1. Wheelchair-Mounted Upper Limb Robotic Exoskeleton Design

Human anatomical data from studies on human anatomy were collected to create the computer-aided design (CAD) model to locate the joints and allow flexion/extension and abduction/adduction at the shoulder joint and flexion/extension for the elbow. To accommodate ranges of sizes of subjects while minimizing stresses at the joints, the design incorporated adjustable linkages. The computer-aided models of the wheelchair-mounted robotic exoskeleton were created using SolidWorks ®. The entire prototype was divided into three assemblies, named (1) the adjustable elbow forearm assembly, (2) the shoulder flexion/extension assembly, and (3) the shoulder horizontal abduction/adduction assembly.

The adjustable elbow forearm assembly of the prototype consists of a forearm exoshell part and an adjustable elbow link. The adjustable link allows the total length of the lower arm assembly to vary from 21.4 to 26.4 cm, depending on the user's arm length. The forearm exoshell is an open L-shaped piece and is rounded at the corner edges in order to better fit the user, with the strap channels following the rounded contours. At the interface, filleted holes are added for ventilation to prevent the arm from overheating. The assembly of the forearm shell (purple) and the adjustable link (cyan) is shown in Figure 1a.

An upper arm exoshell attaches to an upper arm link with a motor case to create the shoulder flexion/extension assembly, as shown in Figure 1b. Similar to the forearm exoshell, the upper arm exoshell has holes for ventilation and rectangular channels for straps. One link on the side of the upper arm exoshell helps to keep alignment with elbow joint. The last main assembly is the shoulder horizontal abduction/adduction assembly, as shown in Figure 1c.

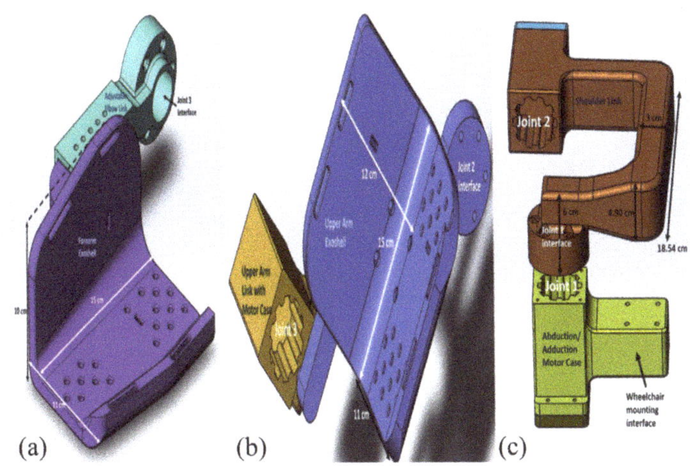

Figure 1. (**a**) Adjustable elbow forearm assembly, (**b**) shoulder flexion/extension assembly, and (**c**) shoulder horizontal abduction/adduction assembly of the exoskeleton.

The individual parts were assembled as shown in Figure 2a and mounted on a wheelchair, as shown in Figure 2b. The exoskeleton was manufactured with fusion deposition modeling using the RAISE3D Pro2 Plus printer using NylonX, a 20% carbon-fiber filled nylon material and PLA. The overall weight of the printed wheelchair-mounted upper limb robotic exoskeleton with hardware is less than 6.61 lb, and it was easily mounted on the back of the powered wheelchair using two screws and did not restrict the movement of the powered wheelchair through regular-sized doors.The printed robotic exoskeleton with motors and sensors mounted to a powered wheelchair is shown in Results Section 3.2.

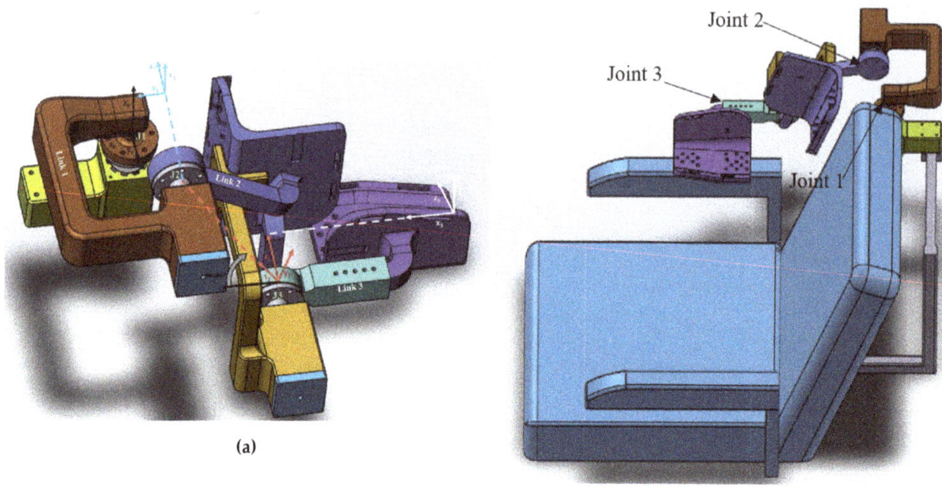

(a)

(b)

Figure 2. (**a**) Assembly of the exoskeleton, (**b**) assembly of the exoskeleton mounted on the wheelchair.

2.2. *Surface Electromyography Signal Acquisition and Hand Gestures*

The Myo armband was used to acquire surface-level muscle signals known as surface electromyography (sEMG) from the flexor carpi radialis, extensor carpi radialis longus and brevis, palmaris longus, pronator and supinator teres, flexor and extensor carpi ulnaris, brachioradialis, flexor digitorium superficialis, and flexor digitorum communis muscles in the forearm. These muscles generate electrical potentials during finger movements and various hand gestures. Another important decision, besides the features to extract for sEMG signals, was the hand gestures to use for the exoskeleton control. A relaxed open hand [14–17]; closed hand, or fist, or grasp [14,15,18]; flexion/extension or similar wave in/wave out [15–17]; radial and ulnar deviation/flexion [14–18] ; pronation/supination [18]; and non-relaxed open or fingers spread [15] are the hand and wrist gestures suggested by various past studies on robotic arm, prosthetic, and exoskeleton control and general EMG signal classification. Five hand gestures were chosen to obtain sEMG muscle signals for classification: wave in, wave out, radial deviation (angled up), ulnar deviation (angled down), and rest. However, 5 gestures alone are not sufficient to control a 3 DOF exoskeleton. Thus, to add to these five gestures without including any other gestures that involve making a fist or articulating the fingers in spread or individual configurations, it was initially decided to include forearm pronation and supination rotations, using a threshold for each direction. Unfortunately, it was found during initial testing that pronation/supination of the forearm was difficult to achieve while the user is strapped into the exoskeleton, so two more hand gestures had to be devised instead. All gestures needed to allow someone to grasp an object, such as a water bottle or the joystick to control a wheelchair. The first five gestures were chosen to perform this task. Other gestures such as fingers spread and a fist, recognized as poses by Thalmic Lab's software for the Myo armband (Myo Connect/Myo Armband Manager), were avoided due to the conflict of being able to make these poses while trying to grasp an object, as well as the possible confusion by the classification algorithm between the desired movement and the user simply wanting to grasp an object. Thus, the goal was to choose gestures that would support object grasping and other operations without accidentally signaling the movement of the exoskeleton.

A grasp taxonomy of 33 grasp types was identified, with top-level divisions focused on whether the grasp required more power—that is, "all movements of the object have to be evoked by the arm"—or precision, followed by the opposition type or direction of force needed to grasp the object in reference to the palm, namely palm (perpendicular), side (transverse), or pad (parallel) [19]. Interestingly, only for 4 of the 12 precision categories did grasping involve the pinky finger at all, while 2 of the 6 intermediate categories (between precision and power) used the pinky finger. The pinky finger seemed to be more necessary for power grasping, as it appeared to be used in 12 of the 15 grasp schemes shown in the GRASP taxonomy [19]. Hence, it could be concluded that the pinky finger may not be necessary for grasping the majority of the time. Thus, it is proposed that the pinky finger could be flexed and extended while grasping an object, so flexion and extension of the pinky finger were ultimately chosen as the other two hand gestures to control the robotic exoskeleton. This made it possible to have seven different classes of movement in total. Three more redundant classes were also added, making a total of 10 classes, but still with 7 movements, in order to help with pick and place tasks, such as the water bottle pick and place task of this research project. The three additional classes were holding a water bottle as an additional rest position, holding a water bottle while flexing the pinky finger, and holding a water bottle while extending the pinky finger backward. An illustration of the gestures for EMG is shown in Figure 3, with class labels (C1–C10) and gesture names (e.g., radial deviation) labeled on each gesture image. Feature extraction included the use of the RMS, median filter, and mean. Raw sEMG signals were processed through these features, followed by a normalization prior to artificial neural network classification.

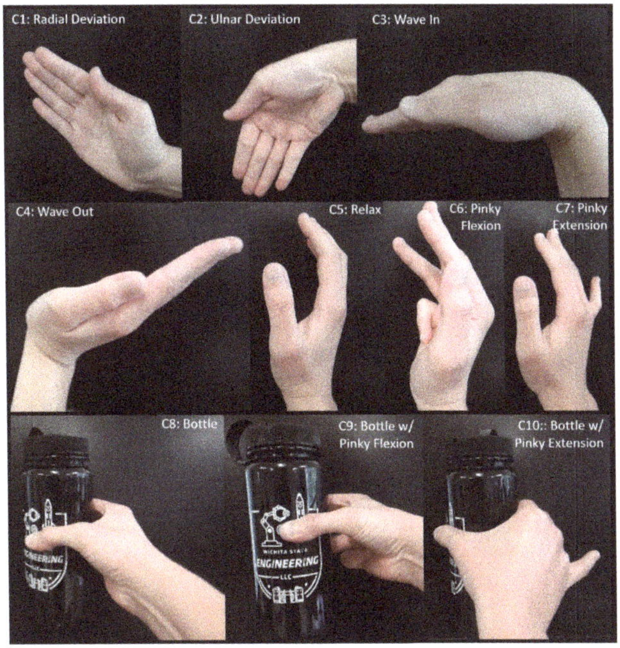

Figure 3. Hand gestures used for exoskeleton control.

Radial deviation, as shown, signals to the exoskeleton controller that the user desires to rotate the elbow joint in the flexion direction, while ulnar deviation signals desired assistance with elbow extension. Waving the hand inward corresponds to shoulder extension, while waving out corresponds to shoulder flexion. Shoulder horizontal adduction is signaled by curling the pinky finger inward towards the palm in a flexion motion without and with the water bottle, while horizontal abduction is signaled by stretching the pinky finger back in extension without and with the water bottle, as shown in the Figure 3. No movement of the exoskeleton is signaled by the relax gesture and the bottle grasping gesture.

2.3. *Adaptive Controller Mechanism*

MATLAB offers several toolboxes for machine learning, including three for neural networks specifically, namely the Neural Network Pattern Recognition Toolbox, the Neural Fitting app, and the Neural Clustering app. The first two were of interest for this particular study. The Neural Network Pattern Recognition Toolbox only uses scaled conjugate gradient backpropagation for training, while the Fitting app allows a choice between Levenberg–Marquardt, Bayesian regularization, and scaled conjugate gradient backpropagation. Initial testing was conducted to determine which would offer the highest accuracy for these features and number of classes, and it was determined that scaled conjugate gradient backpropagation had the highest accuracy; thus, the scaled conjugate-type neural network was used for all test subjects. The neural network architecture used for the real-time control of the exoskeleton is shown in Figure 4, including the number of input, hidden, and output layers.

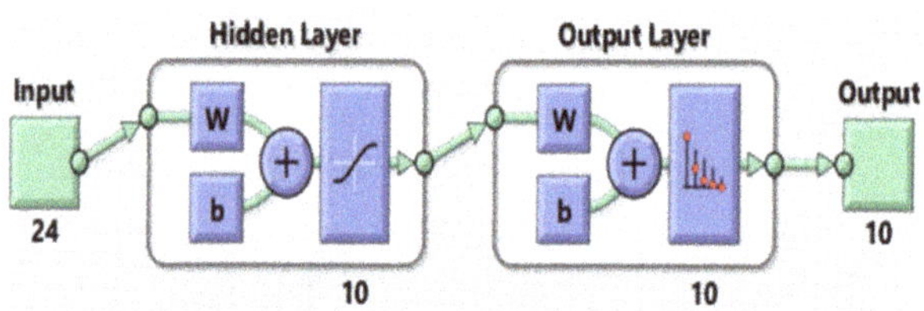

Figure 4. Artificial neural network structure.

The number 24 under the input block represents the 24 features used in this study, and the number 10 under the output block represents the 10 classes; hence, an input vector of length 24 is expected, as is an output vector of length 10. The figure is a basic representation of the 10 hidden layer neural network that was used for testing, with the W block representing an S by R weight matrix, with S referring to the neuron count in that layer and R representing the input vector elements count and the b block representing the bias vector of length S [20].

A trained scaled conjugate backpropagation neural network block with two layers (not to be confused with hidden layers)—one containing netsum and tansig functions and the other containing netsum and softmax functions—was generated. The netsum function is a summing function, while the tansig function is a transfer function, as shown in Equation (1), which calculates the hyperbolic tangent and fits the results between −1 and 1 [21].

$$c = tansig(n) = 2/(1 + exp(-2 * n)^{-1}) \tag{1}$$

The Simulink model for communicating with the Myo armband was supplied by the Myo SDK MATLAB MEX Wrapper [??], a free, open-access package of code and libraries downloadable from GitHub. Initially, the code was built in Simulink only using Arduino encoder reading blocks to read the encoders and Digital Output blocks with stair sequence signal inputs for motor control (as a sort of PWM signal), in addition to the EMG and IMU data acquisition blocks from the Myo SDK MATLAB MEX Wrapper to read the signals from the Myo Armband. This made sense because there was already pre-written Simulink and MATLAB MyoMex code for reading from the armband freely available on GitHub, and the neural pattern recognition toolbox allows for both MATLAB and Simulink. The Simulink blocks making up the tansig and softmax functions are shown in Figure 5.

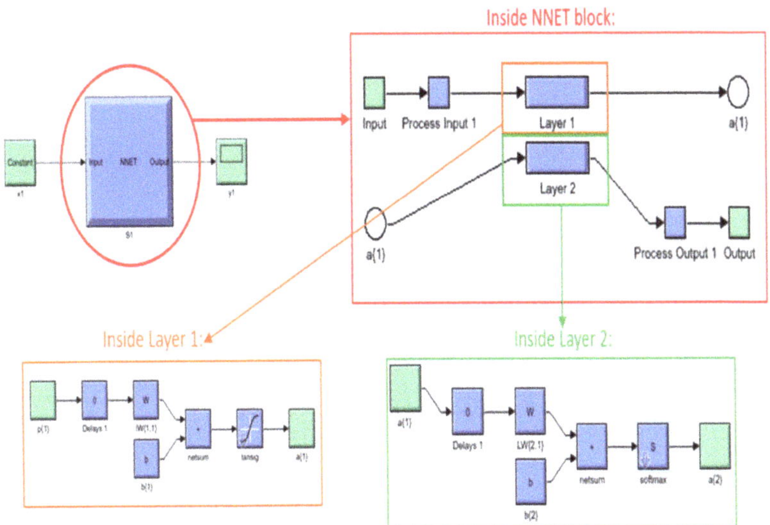

Figure 5. Scaled conjugate backpropagation neural network layers.

While MATLAB scripts could have been utilized for both the Myo armband and neural network, it was quickly discovered that the communication rates between MATLAB and any Arduino board were extremely slow, making MATLAB an impractical option for real-time control. However, it was also found that Simulink alone was not the best option, as when the EMG and IMU blocks were combined with the Arduino encoder blocks and for-loop blocks of stair sequence signals to produce the PWM needed for the stepper motors, this was ultimately too slow for real-time control. Finally, the solution to the timing issue was found by dividing tasks between two Arduino boards and two different software programs, essentially creating parallel computing. Simulink was still used to communicate with the Myo armband, classify gestures with a trained neural network for features of sEMG signals, and to output a total of seven digital values to seven different pins on one Arduino Mega 2560 board. Figure 6b shows a decision making flow chart of the proposed mechanism.

The seven active digital output pins of this first board were then wired to a second Arduino Mega 2560 board. This second Arduino microcontroller was programmed according to our previously published work on a force myography-based exoskeleton controller with some minor changes [23]. The first and most important change, of course, was to read in and use the seven digital inputs from the other board to determine which exoskeleton joint should move and in what direction. FSR values would still also be recorded and displayed, but only as data for post-processing, and perhaps later to add in another layer of safety by using it as an additional logical check. The second revision was to combine all the individual if statements used to allow multiple joints to move at the same time in FSR control into one if statement. This is because it was decided early on that the sEMG control of the exoskeleton would only allow the movement of one exoskeleton joint at the time, with each hand gesture or arm rotation determining the joint and direction of rotation.

(a)

(b)

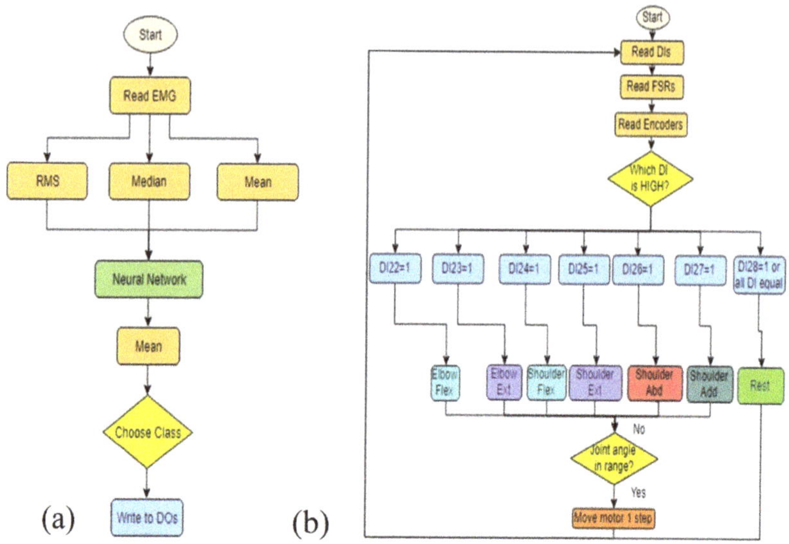

Figure 6. Algorithm flow chart for (**a**) Simulink Myo armband hand gesture recognition and (**b**) Arduino software controller for exoskeleton joints.

3. Results

3.1. Artificial Neural Network Training and Accuracy

Institutional Review Board (IRB) approval (# 4220) was acquired prior to testing human subjects at Wichita State University. This feasibility study recruited 8 participants between the ages of 18 to 60 years without any upper movement disabilities. Surface EMG signals of participants were recorded and used to train custom neural networks for each test subject. Based on our previous research on hand gesture-based powered wheelchair control and virtual robotic arm control, it was decided to use the scaled conjugate backpropagation type of neural network, due to its higher accuracy and shorter training time [24,25]. The MATLAB neural network pattern recognition toolbox offers several different graphical representations of the accuracy of neural networks, including confusion matrices and error histograms. Confusion matrices show both the count and percentage of false positives, or when the neural network algorithm falsely identifies the class, and true positives. The target class on a confusion matrix plot is the correctly-labeled class, while the output class is the actual labeling of the data by the neural network . A confusion matrix was plotted by the application after the training, validation, and testing stages of the artificial neural network. Test subject 3 had the highest training classification accuracy at 97.6%, while subject 8 had the lowest at 87.3%. For validation accuracies, subject 3 again had the highest at 97.5%, while subject 8 was the lowest at 88.6%. Testing accuracies ranged from 87.6% for subject 8 up to 96.7% for subject 3. For the overall confusion matrices, test subject 3 had the highest accuracy at 97.4%, while subject 8 had the lowest accuracy at 87.7%. For classification accuracy, scores around 85% and upward showed fairly good results for the actual use of the neural network in real-time testing. Therefore, since classification accuracies for testing accuracies of all these eight subjects were over 85%, their neural networks were deemed to be accurate enough to use in the sEMG controller for the exoskeleton. Table 1 shows a summary of the total classification accuracy of each subject's neural network during each phase of the neural network training.

Table 1. Artificial neural network classification accuracy.

Subject #	Training Accuracy %	Validation Accuracy %	Testing Accuracy %	Overall Accuracy %
1	92.3	91.5	91.1	92.0
2	90.2	90.4	89.9	90.2
3	97.6	97.5	96.7	97.4
4	96.4	96.2	95.7	96.3
5	90.9	90.0	90.1	90.8
6	89.3	87.5	89.0	89.1
7	89.9	91.1	89.9	89.9
8	87.8	88.6	87.6	87.7

3.2. *Human Subject Testing Results*

Each participant was given two trials to individually control each joint of the exoskeleton using the proposed hand gestures followed by a water bottle pick and place task. The exoskeleton setup is shown in Figure 7, where a participant was seated on a powered wheelchair and wore the actively-actuated exoskeleton with a Myo armband on the right forearm. Each participant's individual joint movements (Section 3.2.1) and water bottle pick and place (Section 3.2.2) tasks were recorded.

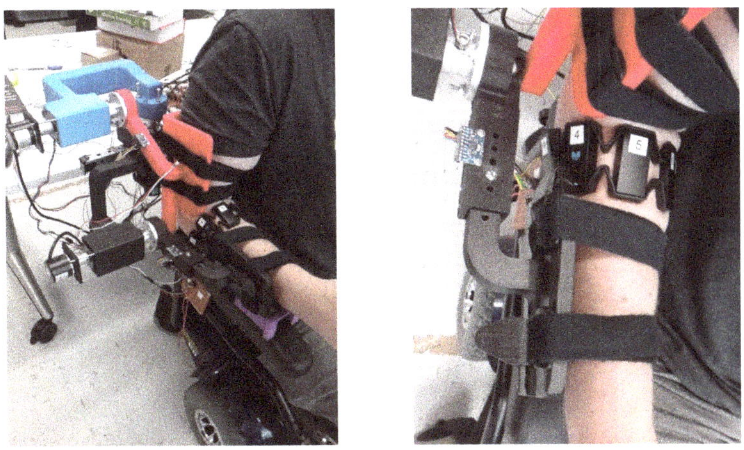

Figure 7. Human subject testing setup views (isometric and top).

3.2.1. Individual Joint Movements

To evaluate the overall efficacy of the control system in real time, all the participants data sets were recorded during the experimental testing of the exoskeleton. Median and mean values of different movement types for each participant were calculated and represented in box plots. Each participant was given two trials to control individual joints of the exoskeleton, and their second trial was recorded. The adaptive control mechanism limits any joint movement beyond 70°, which ranges from +35° to −35° at the rest condition. This plays an important role in human safety because it can override human intent and ensure the user's safety. Figure 8 shows the raw electromyography signals from two participants' forearms while controlling the exoskeleton's individual joints using hand gestures. These raw sEMG signals show distinct patterns thats clearly explain the need for mathematical features to be determined by an artificial neural network to accurately identify hand gestures. A similar strategy can also be applied in future research to test people with upper-arm muscle weakness.

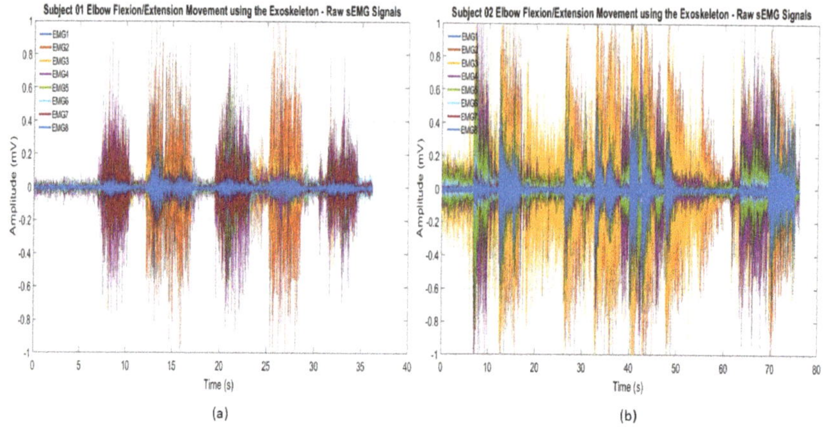

Figure 8. Raw sEMG signals recorded during the elbow flexion/extension movements of the exoskeleton: (**a**) subject 01 and (**b**) subject 02.

Each participant's trained artificial neural network was implemented into the controller unit to identify the number of occurrences for each joint while controlling each individual joint at a time, mimicking the natural movement of the arm. For example, if a user only used specific gestures to control elbow flexion/extension movement, there is a possibility that the user could perform other gestures associated with shoulder flexion/extension or shoulder abduction/adduction movements. Figure 9 demonstrates the number of occurrences of all three exoskeleton joints during the elbow flexion/extension movement. The highest median and mean values recorded for elbow flexion and extension were 11.5 and 14 and 8.5 and 10, respectively. These values are higher than the median and mean values of shoulder flexion (3 and 4.75) and extension (2 and 2.75) and shoulder horizontal abduction (5 and 9.125) and adduction (5.5 and 9.375) readings. This type of adaptability not only improves user acceptance but also allows more than one degree of freedom to mimic the natural movement of the arm. Average elbow flexion was recorded at 30%, while average elbow extension was 20%, yielding a combined total of 50% for the elbow flexion or extension movements.

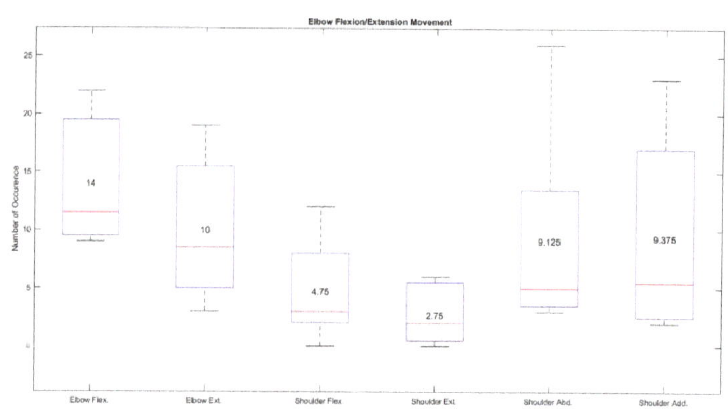

Figure 9. Average movement occurrence of all the joints during the specific elbow flexion/extension movement using hand gestures.

The shoulder flexion/extension movement of the exoskeleton using hand gestures seems to exhibit the best performance based on the recordings from all the participants. Figure 10 shows high median and mean values of 14.5 and 14.5 for shoulder flexion and 12 and 11.75 for shoulder extension respectively compared with the very low median and mean values recorded from elbow flexion (5.75 and 8), elbow extension (0 and 0.625), shoulder abduction (3.5 and 4.25), and shoulder adduction (2.5 and 2.75). These results show that the hand gestures (wave in and wave out) associated with this task made it very easy to control the specific joint compared with other elbow flexion/extension or shoulder horizontal abduction/adduction movements.

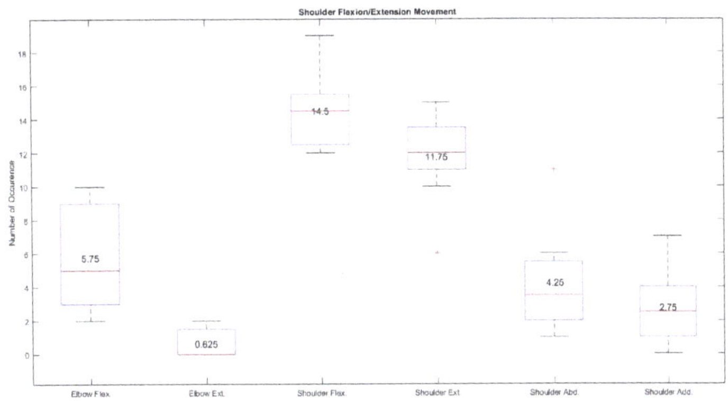

Figure 10. Average occurrence of all the joints during the specific shoulder flexion/extension movement using hand gestures.

Shoulder horizontal abduction/adduction movement of the exoskeleton was often mistaken by the neural network during the individual testing of the joint. Figure 11 shows higher occurrences in terms of median and mean values (21.5 and 19.125 for shoulder horizontal abduction and 16.5 and 19.75 shoulder horizontal adduction) during the shoulder horizontal abduction/adduction movement using the associated gestures (pinky finger flexion/extension). Despite having a higher occurrence rate using pinky finger movements for shoulder horizontal abduction and adduction movements, quick changes in hand gesture classification were observed where elbow flexion/extension movement falsely identified by the trained artificial neural network due to the use of the pinky finger in radial and ulnar deviations.

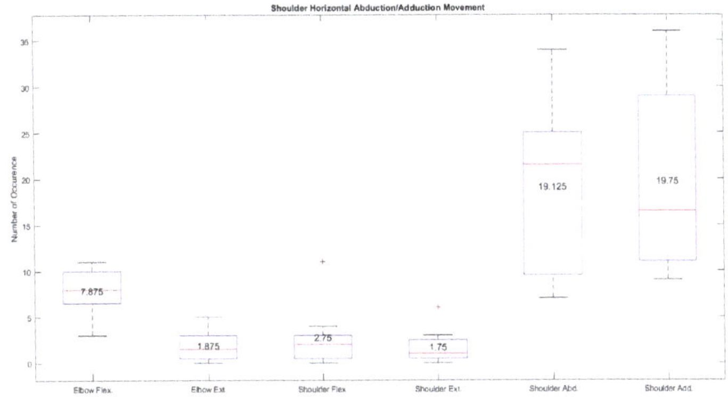

Figure 11. Average occurrence of all the joints during the specific shoulder horizontal abduction/adduction movement using hand gestures.

3.2.2. Water Bottle Pick and Place Task

All eight participants who completed the individual joint tests with the EMG controller also successfully completed the water bottle pick and place task. The shortest time for completing the water bottle pick and place task was 15.3 s by subject 5, while the longest time was 62.6 s by subject 3. A summary of the times for the hand gesture-based adaptive controller used in the water bottle pick and place task is shown in Table 2. The average completion time of the water bottle pick and place task for EMG control was 36.35 s.

Table 2. Water bottle pick and place time using the proposed hand gestured-based artificial neural network-trained adaptive wheelchair-mounted upper limb robotic exoskeleton.

Subject #	Completion Time (sec)
1	27.7
2	21.2
3	62.6
4	51.2
5	15.3
6	39.4
7	24.2
8	49.2

Figure 12 shows the number of occurrences in terms of median and mean values for each joint of the exoskeleton during a water bottle pick and place task. The median and mean values of shoulder horizontal abduction and adduction were 14.5 and 22.375 and 20.5 and 20, respectively. These values, compared with scores of 10 and 11.5 for elbow flexion, 3.5 and 7.125 for elbow extension, 11 and 12 for shoulder flexion, and 3.5 and 3.625 for shoulder extension, demonstrates the higher use of shoulder horizontal abduction and adduction during a water bottle pick and place task compared with the almost equally utilized elbow and shoulder flexion and extension movements.

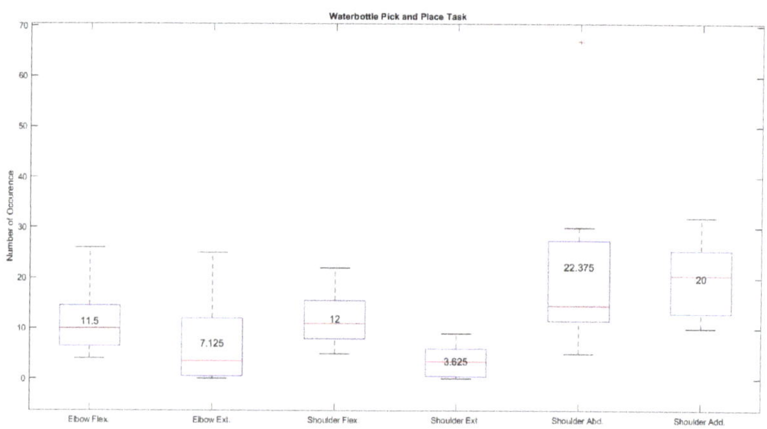

Figure 12. Number of occurrences during water bottle pick and place task using hand gestures.

The most used movements for the water bottle pick and place task with the proposed controller were, on average, shoulder adduction at 30% and abduction at 28%, while shoulder flexion was used 20% of the time on average and elbow flexion was used 15% of the time, followed by elbow extension at 7% and shoulder extension at only 5% of the time. The high use of shoulder adduction/adduction could be misleading if taken purely as intention. Visual observations of the tests showed that desired shoulder abduction and adduction movements were often confused with each other, which was also evident in the individual shoulder abduction/adduction joint test. A summary of these average joint movement occurrences for all eight participants, with the intended movements highlighted, is given in Table 3. The highlighted results prove that each participant's specific hand gesture associated with specific joint control of the exoskeleton was recognized by the trained artificial neural network followed by generating control signals for that specific joint. For example, first three columns of the Table 3 identify the maximum and minimum usage of the exoskeleton joint which clearly shows that the adaptive controller was able to maneuver a specific joint during the specific task. The fourth column of the Table 3 shows majority of the participants used shoulder horizontal abduction and adduction to perform water bottle pick and place compared with shoulder and elbow flexion and extension movements.

Table 3. Average joint movement occurrence during individual joint control and water bottle pick and place task using the wheelchair-mounted upper limb robotic exoskeleton.

Motion/Task	Elbow Flex/Ext %	Shoulder Flex/Ext %	Shoulder Horizontal Abd/Add %	Water Bottle Pick & Place %
Elbow Flex	30	14	16	15
Elbow Ext	20	2	3	7
Shoulder Flex	10	37	5	20
Shoulder Ext	5	31	3	5
Shoulder Horizontal Abd	17	10	35	28
Shoulder Horizontal Add	16	6	37	30

Table 4 shows subjective feedback on the ease of the proposed hand gestured-based controller for individual joint control. The average results are on a rating of 1–5, with 5 showing that the controller was easy to control and 1 representing that it was difficult for questions listed in the Table 4. The scores from all eight participants were averaged together. Shoulder inward/flexion movement received the highest preference, whereas the rotating inward shoulder joint movement was noted as being least preferred.

Table 4. Subjective feedback on exoskeleton joint control.

Subjective Feedback Questions	Average Score (0–5)
Bending exoskeleton elbow joint inward	3.75
Straightening out exoskeleton elbow joint	3.25
Bending exoskeleton shoulder joint inward	4.125
Straightening out exoskeleton shoulder joint	3.375
Rotating inward (towards body) exoskeleton shoulder joint	2.875
Rotating out (away from body) exoskeleton shoulder joint	3.125

4. Discussion

Experimental testing on human subjects without upper arm movement disability showed the need for improvements in the upper arm flexion/extension assembly angle for better fitting, comfort, and classification and improvements in registering hand gestures for classification.

4.1. Fitting and Comfort

Despite having a compact design of the exoskeleton, participants with a smaller upper arm radius felt a greater gap between the posterior side of the upper arm and the upper arm assembly. Due to this gap, participants with a small arm radius sometimes felt a minor slippage of the upper arm, specifically during shoulder flexion movement. This gap was filled with a plastic foam to restrict the upper arm from slipping during exoskeleton control. The shoulder's complex anatomy makes it very difficult to design an exoskeleton that mimics the rotation axis of the exoskeleton joint that can align with the human joint and offer all degrees of freedom when the exoskeleton robot is attached to the exterior of the human body [26,27]. Future designs of the prototype can include an angular joint assembly of the upper arm exoskeleton that can reduce the gap between the exoskeleton and upper arm assembly. A future design will include a locking hinge mechanism for the upper arm exoshell to adjust to the user's shoulder angle along the anatomical frontal plane, as well as the isolation of the links between shells and motor cases from the surfaces interfacing directly with the user. The interfacing surfaces can probably still continue to be fusion deposition modeling-printed PLA or NylonX materials, as well as the motor cases; however, the separated links will need to be manufactured out of a stronger material, such as aluminum. The exoshells in the future design will be made more rounded, maybe using the 3D scanning of an arm to gather further information about the geometry of the arm, although the current open square design allows for a larger range of arm sizes. One of the participants commented that flexion of the elbow with the Myo armband on the forearm could cause pinching because of its size and shape.This issue restricted wrist movements and was observed during the experimental system setup with one participant. The use of flexible sEMG sensors has the potential to reduce the pinching effect.

4.2. Hand Gestures and Classification

The use of a feed forward artificial neural network with 10 hidden layers achieved the highest overall accuracy of 97.4% and the lowest accuracy of 87.7%. Despite its very high accuracy, the overall system requires a computer with Matlab® software installed in order to implement the trained artificial neural network. The use of a Simulink-based model slows down the real-time control operation due to its limited data transfer rate. This issue also relates to the lower sampling rate of the Myo armband (200 Hz). The future work can utilize an on-board processor with a built-in trained network and an EMG system with a faster sampling rate to minimize delays in feature extraction and translation. Two out of eight participants reported issues regarding pinky finger gestures for the shoulder joint actuation of the exoskeleton while holding an object in their hand. This issue limits participants providing a higher gripping force, which will be important for lifting heavy objects in the near future. With regards to control, the data, survey results, and general observations show that the shoulder abduction and adduction control with the proposed hand gestures need to be improved, due to difficulties in signaling user intent to begin with and misclassification in recognition patterns from sEMG signals from the human forearm.

5. Conclusions

This paper successfully demonstrates an integration of artificial neural network-based hand gesture recognition with an adaptive upper limb exoskeleton device to extract 10 gestures in order to actuate a 3 DOF wheelchair-mounted upper limb robotic exoskeleton for activities of daily living. Overall, the trained scaled conjugate gradient neural network achieved an average of more than 93% accuracy using root mean square, medial filter, and mean features. The average movement occurrence shows a higher acceptance of the proposed technology for shoulder flexion/extension followed by elbow flexion/extension and shoulder horizontal abduction/adduction movements. The designed exoskeleton was printed using PLA and NylonX, weighs less than 6.61 lb, and can be easily mounted on the back of a powered wheelchair. All the participants were easily able to adapt to the proposed controller mechanism without any prior experience and performed individual joint movements followed by a water bottle pick and place task. The fastest response recorded while performing the water bottle pick and place task using the proposed technology was 15.3 s, which can be compared with the longest response (62.6 s), demonstrating the variability in user acceptance rates. Subjective feedback from each participant shows a higher acceptance rate (an average score of 4.125 out of 5) for bending the exoskeleton shoulder joint, whereas the lowest average score of 2.875 out of 5.0 was found for rotating the exoskeleton shoulder joint towards the body (shoulder horizontal adduction). Future work will include testing a force myography (FMG)-based controller mechanism and a comparison between EMG and FMG controller schemes based on real-time control and subjective feedback from participants, which will be beneficial for future work on people with muscular dystrophy.

Author Contributions: Conceptualization, J.D.; methodology, B.S. and J.D.; software, B.S. and J.D.; validation, B.S., J.D. and Y.Y.; formal analysis, B.S.; investigation, B.S. and J.D.; resources, J.D.; data curation, B.S. and J.D.; writing—original draft preparation, B.S., J.D. and Y.Y.; writing—review and editing, J.D. and Y.Y.; visualization, J.D.; supervision, J.D. and Y.Y.; project administration, J.D. All authors have read and agreed to the published version of the manuscript.

Funding: This research received no external funding.

Institutional Review Board Statement: The study was approved by the Institutional Review Board of Wichita State University (#4220—14 August 2018).

Informed Consent Statement: Informed consent was obtained from all subjects involved in the study.

Acknowledgments: The authors would like to thank Cerebral Palsy Research Foundation in Wichita, KS for donating a powered wheelchair.

Conflicts of Interest: The authors declare no conflict of interest.

References

1. Van der Heide, L.A.; Van Ninhuijs, B.; Bergsma, A.; Gelderblom, G.J.; Van Der Pijl, D.J.; De Witte, L.P. An Overview and Categorization of Dynamic Arm Supports for People with Decreased Arm Function. *Prosthetics Orthot. Int.* **2014**, *38*, 287–302. [CrossRef] [PubMed]
2. Schill, O.; Wiegand, R.; Schmitz, B.; Matthies, R.; Eck, U.; Pylatiuk, C.; Reischl, M.; Schulz, S.; Rupp, R. OrthoJacket: An active FES-hybrid orthosis for the paralysed upper extremity. *Biomed. Tech. Eng.* **2011**, *56*, 35–44. [CrossRef] [PubMed]
3. Pedrocchi, A.; Ferrante, S.; Ambrosini, E.; Gandolla, M.; Casellato, C.; Schauer, T.; Klauer, C.; Pascual, J.; Vidaurre, C.; Gföhler, M.; et al. MUNDUS project: MUltimodal Neuroprosthesis for daily Upper limb Support. *J. Neuroeng. Rehabil.* **2013**, *10*, 1–20. [CrossRef] [PubMed]
4. Tu, X.; He, J.; Wen, Y.; Huang, J.; Huang, X.; Huang, H.; Guo, M.; Yuan, Y. Cooperation of electrically stimulated muscle and pneumatic muscle to realize RUPERT bi-directional motion for grasping. In Proceedings of the 36th Annual International Conference of the IEEE Engineering in Medicine & Biology Society, Chicago, IL, USA, 26–30 August 2014.
5. Klein, J.; Spencer, S.; Bobrow, J.E.; Allington, J.; Reinkensmeyer, D.J. Optimization of a Parallel Shoulder Mechanism to Achieve a High-Force, Low-Mass, Robotic-Arm Exoskeleton. *IEEE Trans. Robot.* **2010**, *26*, 710–715. [CrossRef]
6. Johnson, G.; Carus, D.A.; Parini, G.; Marchese, S.S.; Valeggi, R. The design of a five-degree-of-freedom powered orthosis for the upper limb. *Proc. Inst. Mech. Eng.* **2001**, *215*, 275–284. [CrossRef] [PubMed]
7. Maciejasz, P.; Eschweiler, J.; Gerlach-Hahn, K.; Jansen-Troy, A.; Leonhardt, S. A survey on robotic devices for upper limb rehabilitation. *J. Neuroeng. Rehabil.* **2014**, *11*, 3. [CrossRef] [PubMed]
8. Stewart, A.M.; Pretty, C.G.; Adams, M.; Chen, X. Review of Upper Limb Hybrid Exoskeletons. *Int. Fed. Autom. Control. Pap.* **2017**, *50*, 15169–15178. [CrossRef]
9. Mulas, M.; Folgheraiter, M.; Gini, G. An EMG-Controlled Exoskeleton for Hand Rehabilitation. In Proceedings of the 2005 IEEE 9th International Conference on Rehabilitation Robotics, Chicago, IL, USA, 28 June–1 July 2005.
10. Kiguchi, K.; Hayashi, Y. An EMG-Based Control for an Upper-Limb Power-Assist Exoskeleton Robot. *IEEE Trans. Syst. Man Cybern. Part (Cybern.)* **2012**, *42*, 1064–1071. [CrossRef]
11. Gopura, R.C.; Kiguchi, K. EMG-Based Control of a 6DOF Upper-Limb Exoskeleton Robot. In Proceedings of the Robotics and Mechatronics Conference (ROBOMEC 2009), Fukuoka, Japan, 24–26 May 2009.
12. Li, Z.; Wang, B.; Sun, F.; Yang, C.; Xie, Q.; Zhang, W. sEMG-Based Joint Force Control for an Upper-Limb Power-Assist Exoskeleton Robot. *IEEE J. Biomed. Health Inform.* **2014**, *18*, 1043–1050. [PubMed]
13. Rahman, M.H.; Ochoa-Luna, C.; Saad, M. EMG Based Control of a Robotic Exoskeleton for Shoulder and Elbow Motion Assist. *J. Autom. Control. Eng.* **2015**, *3*, 270–276. [CrossRef]
14. Khezri, M.; Jahed, M. A Neuro–Fuzzy Inference System for sEMG-Based Identification of Hand Motion Commands. *IEEE Trans. Ind. Electron.* **2011**, *58*, 1952–1960. [CrossRef]
15. Allard, U.C.; Nougarou, F.; Fall, C.L.; Giguère, P.; Gosselin, C.; Laviolette, F.; Gosselin, B. A convolutional neural network for robotic arm guidance using sEMG based frequency-features. In Proceedings of the 2016 IEEE/RSJ International Conference on Intelligent Robots and Systems (IROS), Daejeon, Korea, 9–14 October 2016.
16. Oyama, T.; Mitsukura, Y.; Karungaru, S.G.; Tsuge, S.; Fukumi, M. Wrist EMG signals identification using neural network. In Proceedings of the 2009 35th Annual Conference of IEEE Industrial Electronics, Porto, Portugal, 3–5 November 2009.
17. Chu, J.U.; Moon, I.; Mun, M.-S. A Real-Time EMG Pattern Recognition System Based on Linear-Nonlinear Feature Projection for a Multifunction Myoelectric Hand. *IEEE Trans. Biomed. Eng.* **2006**, *53*, 2232–2239. [PubMed]
18. Shenoy, P.; Miller, K.J.; Crawford, B.; Rao, R.P.N. Online electromyographic control of a robotic prosthesis. *IEEE Trans. Biomed. Eng.* **2008**, *55*, 1128–1135. [CrossRef] [PubMed]
19. Feix, T.; Romero, J.; Schmiedmayer, H.-B.; Dollar, A.M.; Kragic, D. The GRASP Taxonomy of Human Grasp Types. *IEEE Trans. Hum. Mach. Syst.* **2016**, *46*, 66–77. [CrossRef]

20. Smith, J. *Machine Learning with Neural Networks Using MATLAB*; CreateSpace Independent Publishing Platform: Lexington, KY, USA, 2017.
21. The MathWorks, Inc. Tansig. 2019. Available online: https://www.mathworks.com/help/deeplearning/ref/tansig.html (accessed on 29 June 2019).
22. Tomaszewski, M. Myo SDK MATLAB MEX Wrapper. 18 October 2017. Available online: https://www.mathworks.com/matlabcentral/fileexchange/55817-myo-sdk-matlab-mex-wrapper (accessed on 16 March 2018).
23. Desai, J.; Schabron, B.; Yihun, Y. Force Myography Controlled Intelligent Assistive Wheelchair-Mounted Robotic Exoskeleton for Arm Movements. In Proceedings of the 2019 IEEE International Symposium on Measurement and Control in Robotics (ISMCR), Houston, TX, USA, 19–21 September 2019; pp. D2-5-1–D2-5-5.
24. Stroh, A.; Desai, J. Human-Centered Deep Learning Neural Network Trained Myoelectric Controller for a Powered Wheelchair. In Proceedings of the 2019 IEEE International Symposium on Measurement and Control in Robotics (ISMCR), Houston, TX, USA, 19–21 September 2019; pp. D2-4-1–D2-4-4.
25. Schabron, B.; Reust, A.; Desai, J.; Yihun, Y. Integration of Forearm sEMG Signals with IMU Sensors for Trajectory Planning and Control of Assistive Robotic Arm. In Proceedings of the 2019 41st Annual International Conference of the IEEE Engineering in Medicine and Biology Society (EMBC), Berlin, Germany, 23–27 July 2019; pp. 5274–5277.
26. Lee, H.; Kim, W.; Han, J.; Han, C. The Technical Trend of the Exoskeleton Robot System for Human Power Assistance. *Int. J. Precis. Eng. Manuf.* **2012**, *13*, 1491–1497. [CrossRef]
27. Delgado, P.; Sajja, A.; Majidirad, A.; Hakansson, N.A.; Desai, J.; Yihun, Y. Shoulder Kinematics Assessment towards Exoskeleton Development. *Appl. Sci.* **2020**, *10*, 6336. [CrossRef]

sensors

MDPI

Article

Effect of Fixed and sEMG-Based Adaptive Shared Steering Control on Distracted Driver Behavior

Zheng Wang [1,*], Satoshi Suga [2], Edric John Cruz Nacpil [3], Bo Yang [1,*] and Kimihiko Nakano [1]

1 Institute of Industrial Science, The University of Tokyo, Tokyo 153-8505, Japan; knakano@iis.u-tokyo.ac.jp
2 Department of Mechatronics Engineering, Technical University of Darmstadt, 64289 Darmstadt, Germany; satosuga95@gmail.com
3 Corpy & Co., Inc., Tokyo 113-0033, Japan; edric@corpy.co.jp
* Correspondence: z-wang@iis.u-tokyo.ac.jp (Z.W.); b-yang@iis.u-tokyo.ac.jp (B.Y.)

Abstract: Driver distraction is a well-known cause for traffic collisions worldwide. Studies have indicated that shared steering control, which actively provides haptic guidance torque on the steering wheel, effectively improves the performance of distracted drivers. Recently, adaptive shared steering control based on the forearm muscle activity of the driver has been developed, although its effect on distracted driver behavior remains unclear. To this end, a high-fidelity driving simulator experiment was conducted involving 18 participants performing double lane change tasks. The experimental conditions comprised two driver states: attentive and distracted. Under each condition, evaluations were performed on three types of haptic guidance: none (manual), fixed authority, and adaptive authority based on feedback from the forearm surface electromyography of the driver. Evaluation results indicated that, for both attentive and distracted drivers, haptic guidance with adaptive authority yielded lower driver workload and reduced lane departure risk than manual driving and fixed authority. Moreover, there was a tendency for distracted drivers to reduce grip strength on the steering wheel to follow the haptic guidance with fixed authority, resulting in a relatively shorter double lane change duration.

Keywords: driver–automation shared control; haptic guidance steering; adaptive automation design; surface electromyography; driver distraction

Citation: Wang, Z.; Suga, S.; Nacpil, E.J.C.; Yang, B.; Nakano, K. Effect of Fixed and sEMG-Based Adaptive Shared Steering Control on Distracted Driver Behavior. *Sensors* **2021**, *21*, 7691. https://doi.org/10.3390/s21227691

Academic Editor: Ernest N. Kamavuako

Received: 11 October 2021
Accepted: 17 November 2021
Published: 19 November 2021

1. Introduction

Steering a car necessitates continuous real-time visual information from the road ahead. Distractions that inhibit safe and timely driver responses to this information are detrimental to driving safety; thus, technology designed to monitor and assist distracted drivers is a crucial advancement [1,2]. Although fully autonomous driving is not likely to be realized in the near future, partial automation is becoming more readily available in the form of steering assistance systems and driver attention monitoring [3]. Shared steering control systems assist drivers with curve negotiation and lane changes by producing a proper haptic guidance torque on the steering wheel [4,5], particularly when a lack of attention to visual information results from driver distraction or fatigue [6,7]. Decreased performance due to the lack of visual information can be compensated for by haptic guidance [8,9].

Some attempts to design an adaptive shared steering control system have been conducted to improve driver–automation cooperation. One way to adjust the driver–automation control authority is to address vehicle–environment factors, including vehicle position, steering wheel angle, and yaw rate [10–13]. A haptic shared steering control has been designed with an adaptive level of authority based on time-to-line crossing, and driver-in-the-loop experiments demonstrate the effectiveness of the proposed system on decreasing conflict torques [10]. Shared steering control with adaptive authority based on driver input torque has been designed to achieve smooth authority transfer from automated driving to manual driving [11]. Based on the data from lateral offset and lateral velocity to the road center line, a human driver's driving intention and the desired maneuver have been recognized and used for designing a shared steering control system to avoid obstacles [12]. To further improve shared steering control performance, driver-in-the-loop model has been included in the optimal controller design [14–20]. A time-varying assistance factor has been developed to modulate the haptic steering torque, which is designed from an integrated driver-in-the-loop vehicle model, and the effectiveness of proposed haptic shared control method on driver–automation conflict management has been presented by driving tests conducted with high-fidelity simulations [14]. A Takagi–Sugeno fuzzy control approach has been proposed to deal with the time-varying driver activity parameter and vehicle speed under multiple system constraints to improve driver–automation shared steering performance [17]. A predictive control framework that uses a model of driver-in-the-loop steering dynamics to optimize the torque intervention with respect to the neuromuscular response of the driver has been presented. Results show the effectiveness of the proposed system in avoiding hazardous situations under different driver behaviors with driver-in-the-loop experiments [15]. Regarding the driver-in-the-loop control, a driver model that describes the driver steering motion as a sequence of identifiable motion primitives has been proposed for the calculation of the optimal torque [16]. In addition, a time-varying driver-in-the-loop path tracking model has been developed for shared steering control, where the preview behavior of the modelled driver is designed to mimic real human drivers [19]. By assuming that drivers can learn and incorporate the controller strategy into their internal model for predictive path following, a driver model has been proposed to include interactive steering behavior based on model predictive control for indirect shared control has been proposed [20].

Compared to adjusting the authority based on driver physiological status, the advantage of addressing vehicle–environment factors are apparent, as drivers do not need to wear bio-signal measuring equipment and driving performance data can be more easily recorded and processed in real time. However, the disadvantage is that it is difficult to establish an accurate mapping between driving performance data and individual driver behavior when compared with directly monitoring physiological status [21]. Moreover, it is sometimes not possible to establish an accurate driver-in-the-loop model for the controller design. An adaptive system has been designed to increase the automation authority, as driver control authority decreases owing to increased workload or decreased workload capacity [22], and this approach is preferable to an assistance system with fixed and rigid authority [23–25]. Driver workload can be estimated through physiological signal measurements such as eye and head movements [25] and surface electromyography (sEMG) signals [7]. For steering tasks, drivers can allocate the control authority with regard to the haptic guidance system by adjusting the arm admittance [4]. Previous research has suggested a relationship between arm admittance and grip strength on the steering wheel [26] as measured by sEMG signals from forearm muscles [27]. Inspired by research on the grip-force-based scheduling of guidance forces [28], our previous study adjusted the authority of shared steering control according to driver grip strength on the steering wheel to achieve better driver–automation cooperation performance [29]. We found a reduction in both lane departure risk and driver workload to be associated with sEMG-based adaptive authority compared to fixed authority [29].

However, the effect of haptic guidance with adaptive authority on distracted driver behavior remains unknown. As driver distraction is a well-known cause of traffic collisions worldwide, the effectiveness of the proposed shared steering control remains to be tested. Although adaptive automation relieves the driver of the task of engaging and disengaging the automation, it imposes an additional task of monitoring the time-varying automation level of the adaptive system with the possibility of increased workload [22]. Moreover, several experimental studies have indicated that the steering effort of drivers may be even higher for haptic guidance than for manual driving [4,5]. This situation may be more complicated when drivers become distracted. Furthermore, humans have a good ability for adaptation to take advantage of the shared steering control system. When the drivers are under a distracted state due to the secondary task, they may change their adaptive behavior.

Therefore, this study aimed to investigate the effect of sEMG-based adaptive shared control on distracted driver behavior by comparing with an attentive driver state. Fixed shared steering control and manual driving are also addressed for comparisons. It is hypothesized that the adaptive haptic guidance would provide more driving safety and comfort to distracted drivers, whereas fixed haptic guidance could more effectively reduce the duration of a lane change. This study is expected to especially benefit distracted drivers by contributing to safer and smoother steering performance in lane changing as well as lane following tasks under haptic shared control.

2. Materials and Methods

2.1. Participants

Eighteen healthy subjects (two women and 16 men) were recruited to participate in the experiment. Their ages ranged from 21 to 32 years (mean = 23.5 years, SD = 2.8 years). All subjects had a valid Japanese driver's license with driving experience (mean = 2.7 years, SD = 2.6 years). The experiment was approved by the Office for Life Science Research Ethics and Safety, Graduate School of Interdisciplinary Information Studies, University of Tokyo (No. 12 in 2017). Each subject provided written consent to the experimental protocol and received compensation for their participation.

2.2. Apparatus

A Myo armband (Thalmic Labs, Inc., Waterloo, Canada) acquired the sEMG from the dominant forearm of the driver (Figure 1). The sEMG was conditioned for further processing by calculating the root mean square (RMS) value of the activation signal from the stainless-steel armband sensors [30]. Driver grip strength was normalized with respect to the maximum sEMG for each participant ($sEMG_{REF}$). The steering wheel torque provided by the adaptive haptic guidance was real-time computed based on the normalized sEMG value.

Figure 1. Active steering system and armband worn by driver to enable sEMG-based operation of a high-fidelity driving simulator.

The experiment was conducted in a moving-based driving simulator with brake and accelerator pedals, an actuated steering wheel, and an instrument dashboard. The driving simulator included a 140° field-of-view and the moving platform had six degrees of freedom, which was considered to be high-fidelity. In order to emulate the feeling of on-road driving, high-frequency vibrations were also produced by the moving platform, engine sounds were provided by two speakers, and a self-aligning torque was generated by the actuated steering wheel. The electronic steering system was connected to the host computer of the driving simulator through a controller area network. The electronic steering system comprised a steering wheel, servo motor, and electronic control unit (ECU). The real-time haptic guidance torque calculated by the host computer served as the input to an ECU. A servomotor was subsequently actuated by the ECU to apply a haptic guidance torque to the steering wheel. The haptic guidance torque was calculated based on a model with two look-ahead points, whereas the magnitude and direction of the haptic guidance torque were determined by comparing the target vehicle trajectory based on the fifth-degree Bezier curve with the actual trajectory of the simulated vehicle [29].

2.3. Experimental Conditions and Scenario

The participants drove under six conditions as shown in Table 1. Two driver states were considered: attentive and distracted. For each state, there were three types of haptic guidance: manual steering or HG-Non, HG-Fixed (haptic guidance with fixed authority), and HG-Adaptive (haptic guidance with adaptive authority). Using a 6 × 6 Latin square, the sequence in which the experimental conditions were presented to the participants was partially counterbalanced. The three Latin squares partially counterbalanced the within-subject order of the conditions.

Table 1. Experimental conditions.

Condition	Driver State	Haptic Guidance
1	Attentive	Manual (HG-Non)
2	Attentive	HG-Fixed
3	Attentive	HG-Adaptive
4	Distracted	Manual (HG-Non)
5	Distracted	HG-Fixed
6	Distracted	HG-Adaptive

To induce driver distraction, a challenging secondary task, called the paced auditory serial addition task, was applied during the entire driving course. The subjects were given a number every 3 s and were asked to add the number they just heard with the number they heard before. The attentive experimental condition was a control session in which the driver was under normal driving conditions.

For manual driving, the overall gain of haptic guidance was set to 0, whereas the overall gain of the haptic guidance torque was held at 0.25 for HG-Fixed. The normalized torque was based on 25% of the gain for an automated double lane change (DLC). To manage the driver–automation conflict, the sEMG-based adaptive shared steering control was designed to reduce the authority of haptic guidance when the driver increased the grip strength to gain more manual control authority and vice versa. Specifically, the gain of the haptic guidance torque was computed in real time, and it decreased linearly from 1 (gain for automated DLC) to 0, when the driver's grip strength increased from 0 to $sEMG_{REF}$. Grip strength above $sEMG_{REF}$ during the DLC task prompted adjustment of the haptic guidance gain to 0.

The DLC task was composed of two stages of lane changing as illustrated with pylons in Figure 2. Before the DLC section, there was a straight lane 300 m long and, thus, the driver was asked to perform a lane keeping task. The speed was controlled by the driving simulator, and the driver did not need to operate the gas pedal. The initial speed was set at 0 km/h, and the acceleration rate was 1.8 m/s^2. After reaching 50 km/h at approximately 50 m from the starting point, the driving speed was kept constant. The aim was to motivate the driver to adapt their steering behavior, especially for driving with the haptic guidance. In order to determine the speed value, we conducted a pilot test to compare speeds of 40, 50, and 60 km/h given the same DLC task shown in Figure 2. According to the lane change performance evaluation, the speed of 40 km/h was found to be less challenging, whereas 60 km/h was overly challenging. Therefore, to improve the replicability of the experimental results with respect to steering behavior, a PID controller automatically maintained the simulated vehicle speed at 50 km/h.

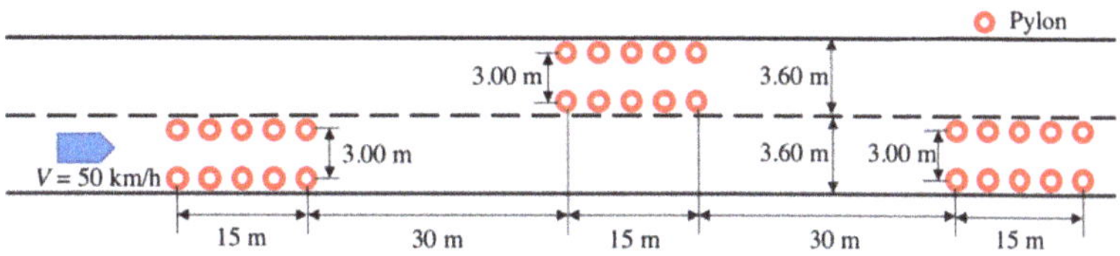

Figure 2. Experimental driving scenario performed in the driving simulator.

2.4. Experimental Procedure

Considering that different drivers have different usage habits with their left and right hands, each participant was asked about his/her dominant hand so that the armband was mounted on the dominant forearm. The sEMG normalization was realized by having each participant grip the driving simulator steering wheel in a "ten-and-two" position for 2 s with maximum grip strength. This procedure was repeated thrice with 10 s of rest between each repetition. The mean value across all repetitions for a given participant was used as the reference sEMG value, $sEMG_{REF}$, for normalization.

Each subject performed a practice task prior to the actual experiment so that they were accustomed to operating the simulator. Throughout the practice and experimental trials, drivers were trained to maintain both hands at the ten-and-two steering wheel position. The drivers repeated the DLC task five times for each of the six driving conditions and, thus, each driver performed a total of 30 trials for the actual experiment. The subjective task load was measured by having each participant complete a questionnaire after each trial.

2.5. Measured Variables

The DLC performance was evaluated based on driver steering behavior, lane departure risk, and subjective evaluation. The measured driver input torque, DLC duration, steering wheel angle (SWA), lateral acceleration, and normalized sEMG constituted the evaluation of driver steering behavior. The normalized sEMG RMS ($sEMG/sEMG_{REF}$) was calculated from the measured forearm signals. An increased normalized value indicated greater grip strength. The driver steering effort was determined by the calculated RMS value of the driver input torque. Calculating the RMS value of the SWA and peak SWA determined the magnitude of the steering control activity. The DLC process consisted of two parts: the first lane change part and the second lane change part. The peak value of SWA for each lane change part was calculated. The duration of DLC was calculated to express the timing of lane change maneuvers in a quantitative way. The starting point of the DLC was determined when the vehicle yaw angle was higher than 1.0 degree through a trial-and-error process. The duration of DLC was calculated from the starting point to the ending point. The driver normally made a small adjustment at the ending point to return to the centerline of the lane. Hence, the yaw rate was considered with yaw angle to determine the ending point. Furthermore, the ending point of the DLC was determined by trial-and-error process to correspond to a yaw angle and yaw rate both below 0.5 degrees.

At the conclusion of each lane change stage with the simulated vehicle driving parallel to the entered lane, the lateral error relative to the centerline of the lane was measured. The lane departure risk during DLC was evaluated using the lateral error.

Subjective preferences for the different types of tested haptic guidance paired with each driver state were recorded in conjunction with the NASA task load index (NASA-TLX) to conduct a subjective evaluation. The participants rated their workload according to the index after each driving condition. At the conclusion of all experimental trials, each participant selected the experimental condition with the highest degree of satisfaction. The preference score for a given condition equaled the number of times the condition was selected across all participants. Dividing the preference score by two yielded a relative preference score.

2.6. Analysis

In accordance with two-way repeated-measures analysis of variance (ANOVA), the extent of interaction between the state of the driver and haptic guidance that affected driver behavior was determined. Setting the level of significance to $p = 0.05$, Mauchly's test was executed before the repeated-measures ANOVA. Furthermore, Fisher's least significant difference for pairwise comparisons identified the main effects with a selected significance criterion of $p = 0.05$. Differences were considered statistically significant when the p-value ≤ 0.05, and a p-value ≤ 0.1 was interpreted as a tendency toward statistical significance.

3. Results and Discussion

The results in this section are described with regard to driver steering behavior and lane departure risk in addition to subjective evaluation. The p-values for two-way repeated-measures ANOVA related to driver behavior as well as the corresponding mean and standard deviations for experimental variables are listed in Table 2. This includes the main effect for the driver state, main effect for HG, and interaction effect between the drive state and HG.

The main effect was significant for HG in terms of the RMS of driver input torque ($p < 0.001$), lateral error at the end of the first lane change ($p < 0.001$), and relative score of pairwise preference ($p < 0.01$), although there was no significance for the main effect of driver state and interaction effect. As for the peak value of SWA in the first lane change, the main effect was significant for HG ($p < 0.05$) and driver state ($p < 0.05$). As for the peak value of SWA in the second lane change, there was a tendency toward significance for the HG ($p < 0.1$) and driver state ($p < 0.1$). The results of peak value for lateral acceleration had a similar tendency as the peak value of SWA. As for DLC duration, the main effect was significant for HG ($p < 0.01$), and there was a tendency toward significance for the interaction effect ($p < 0.1$). The main effect was significant for the driver state in terms of lateral error at the end of second lane change ($p < 0.05$) and overall workload according to the NASA-TLX ($p < 0.001$), whereas no significant difference was observed for the main effect of HG and the interaction effect.

3.1. Driver Steering Behavior

The results of driver input torque are shown in Table 2. From pairwise comparisons, the driver input torque for manual was significantly greater than that for HG-Fixed ($p < 0.001$), greater for manual than for HG-Adaptive ($p < 0.001$), and greater for HG-Fixed than for HG-Adaptive ($p < 0.001$). Therefore, our hypothesis was validated because haptic guidance significantly reduced driver steering effort, and HG-Adaptive was more effective.

The results of peak value for SWA in the first lane change are shown in Table 2 and Figure 3. The steering wheel angle and lateral acceleration data from Subject no. 9 were eliminated due to the fact of its extreme deviation from the data of other subjects when plotting the figure and conducting the statistical analysis. For the distracted state, the peak value of SWA was significantly lower for manual steering than for HG-Adaptive ($p < 0.01$) and lower for HG-Fixed than for HG-Adaptive ($p < 0.05$). Moreover, the peak value of SWA for manual steering was significantly higher with the attentive state than with the distracted state ($p < 0.05$). Hence, the distraction reduced the steering activity, and the haptic guidance system increased the steering activity to a level comparable to that of the attentive state. A similar tendency was found for the RMS of the SWA, although the difference was not statistically significant for HG. This may be due to the effect of HG and distracted state on SWA being less pronounced during the second lane change compared with first lane change as shown in Table 2. Thus, a more demanding secondary task or lane change task will be considered in future studies. Moreover, we found that there was a high correlation and similar tendency between the results of steering wheel angle and lateral acceleration. In this study, lateral error was mainly used to evaluate the lane departure risk. If there was no significant difference for lateral error, a lower lateral acceleration would indicate a smoother lane change with less steering control activity, thus implying a relatively lower risk of lane departure.

The results for the DLC duration are shown in Table 2 and Figure 4. From pairwise comparisons, for attentive, the duration of the DLC was significantly shorter with HG-Adaptive than with HG-Fixed with $p < 0.05$, whereas the DLC tended to be significantly shorter for HG-Adaptive than for manual steering, where $p < 0.1$. For distracted, the DLC duration was significantly shorter with HG-Fixed than with manual ($p < 0.001$) and shorter with HG-Adaptive than with manual ($p < 0.01$). Moreover, for HG-Fixed, there was a tendency for the DLC duration to be shorter with distracted than with attentive ($p < 0.1$). Therefore, HG-Fixed could more effectively reduce lane change duration for distracted drivers as predicted by our hypothesis.

The RMS results of sEMG are shown in Table 2 and Figure 5. Based on pairwise comparisons, for distracted, the RMS of sEMG was significantly lower with HG-Fixed than with HG-Adaptive ($p < 0.05$) and lower with HG-Fixed than with manual ($p < 0.1$), indicating that distracted drivers tended to give more control authority to the HG-Fixed by reducing grip strength. Consequently, the DLC duration was relatively shorter (Figure 4).

Table 2. Experimental conditions.

Variable	Attentive and Manual (1) M (SD)	Attentive and HG-Fixed (2) M (SD)	Attentive and HG-Adaptive (3) M (SD)	Distracted and Manual (4) M (SD)	Distracted and HG-Fixed (5) M (SD)	Distracted and HG-Adaptive (6) M (SD)	Driver State p-Value	HG p-Value	Interaction p-Value
RMS of driver input torque (N·m)	1.029 (0.064)	0.734 (0.072)	0.633 (0.106)	1.029 (0.058)	0.722 (0.081)	0.643 (0.149)	0.941	0.000 ***	0.454
RMS of SWA (degree)	18.897 (3.182)	19.122 (2.551)	19.403 (2.674)	17.976 (2.015)	18.592 (2.358)	18.555 (2.049)	0.059 +	0.198	0.732
Peak value of SWA at the 1st LC (degree)	28.497 (7.116)	28.040 (6.986)	29.468 (7.394)	25.274 (4.366)	26.323 (3.826)	28.274 (5.265)	0.046 *	0.024 *	0.203
Peak value of SWA at the 2nd LC (degree)	−33.815 (6.301)	−33.135 (5.184)	−32.867 (5.763)	−33.035 (4.758)	−31.296 (5.052)	−31.203 (3.811)	0.056 +	0.092 +	0.595
Peak value of lateral acceleration at the 1st LC (m/s^2)	1.702 (0.437)	1.670 (0.418)	1.761 (0.445)	1.501 (0.264)	1.571 (0.229)	1.684 (0.313)	0.046 *	0.026 *	0.180
Peak value of lateral acceleration at the 2nd LC (m/s^2)	−1.983 (0.381)	−1.944 (0.312)	−1.936 (0.356)	−1.929 (0.280)	−1.828 (0.189)	−1.833 (0.228)	0.047 *	0.147	0.619
Duration of double LC (s)	8.613 (1.044)	8.437 (0.689)	8.123 (0.595)	8.731 (0.788)	8.088 (0.626)	8.290 (0.811)	0.867	0.002 **	0.083 +
RMS of normalized sEMG (%)	7.116 (3.501)	6.746 (3.686)	6.696 (3.547)	6.947 (3.966)	6.309 (3.908)	7.120 (4.053)	0.822	0.099 +	0.208
Lateral error at the end of 1st LC (m)	0.432 (0.203)	0.354 (0.172)	0.366 (0.181)	0.430 (0.204)	0.364 (0.195)	0.319 (0.171)	0.391	0.000 ***	0.296
Lateral error at the end of 2nd LC (m)	0.193 (0.111)	0.214 (0.123)	0.169 (0.100)	0.241 (0.099)	0.215 (0.101)	0.224 (0.133)	0.033 *	0.650	0.114
NASA-TLX overall workload	43.685 (16.020)	43.019 (15.408)	40.593 (16.489)	71.407 (17.924)	68.130 (17.330)	66.241 (16.541)	0.000 ***	0.166	0.826
Relative score of pairwise preference	0.647 (0.786)	0.882 (0.781)	1.471 (0.624)	0.353 (0.702)	1.059 (0.748)	1.588 (0.507)	1	0.001 **	0.314

+ $p < 0.1$, * $p < 0.05$, ** $p < 0.01$, and *** $p < 0.001$. SWA: steering wheel angle; LC: lane change; HG: haptic guidance.

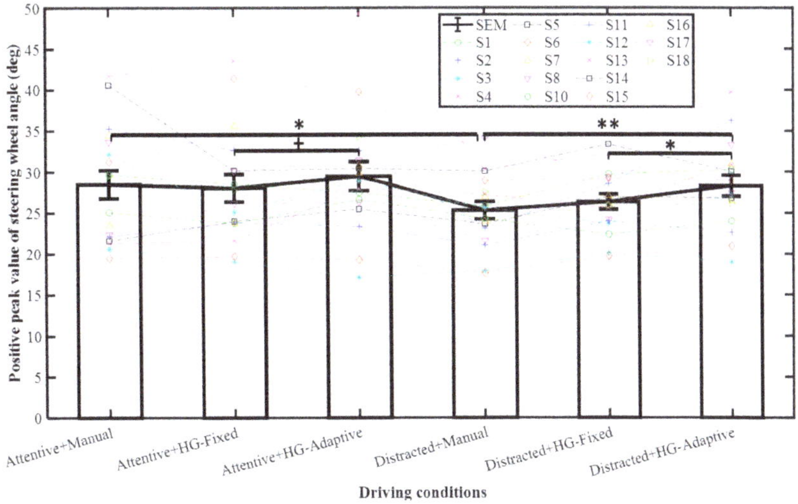

Figure 3. Peak value of SWA at the first lane change. Mean +/− SEM (standard error of mean) represented by error bars. $^{+}$ $p < 0.1$, * $p < 0.05$, and ** $p < 0.01$.

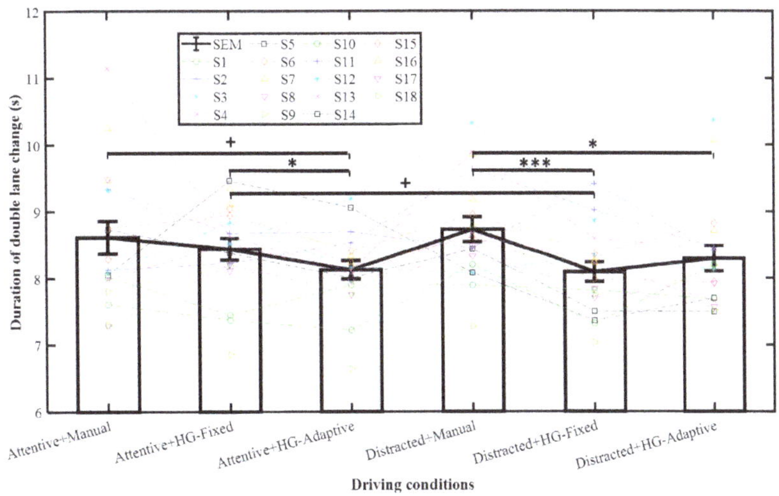

Figure 4. Double lane change duration. Mean +/− SEM represented by error bars. $^{+}$ $p < 0.1$, * $p < 0.05$, and *** $p < 0.001$.

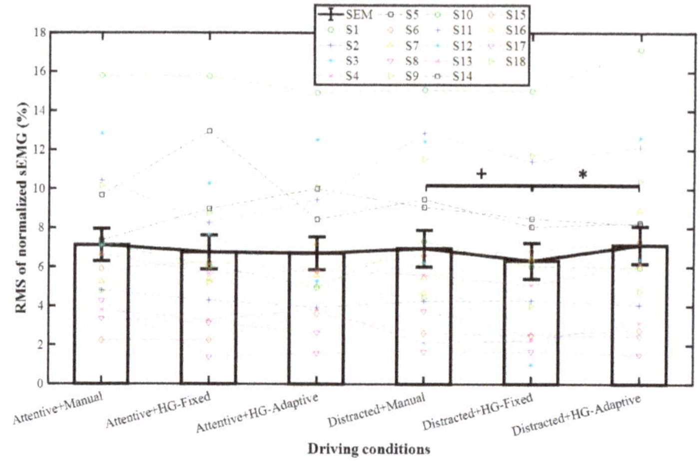

Figure 5. RMS of normalized sEMG (%). Mean +/− SEM represented by error bars. $^+ p < 0.1$ and $^* p < 0.05$.

3.2. Lane Departure Risk

Figure 6 and Table 2 show the results of the lateral error at the end of the first lane change. Pairwise comparisons indicated that for the attentive condition, the lateral error was significantly higher for manual steering than for HG-Fixed with $p < 0.05$ and tended to be higher for HG-Adaptive with $p < 0.1$. The lateral error for distracted was significantly lower in the case of HG-Fixed than in manual, where $p < 0.05$; HG-Adaptive was significantly lower than HG-Fixed ($p < 0.01$). Furthermore, HG-Adaptive tended to be significantly lower than HG-Fixed with $p < 0.1$. Thus, haptic guidance can reduce lane departure risk when the driver is attentive as confirmed by our previous study [29]. This outcome could be attributed to the human driver being limited by the response of the neuromuscular system, thereby making it difficult to complete the DLC accurately [31]. Furthermore, from this result, haptic guidance is also capable of reducing the lateral error for distracted drivers with HG-Adaptive being more effective.

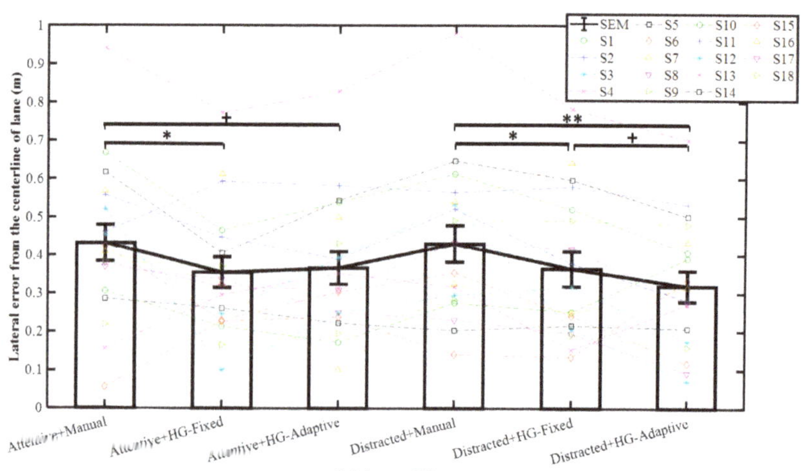

Figure 6. Lateral error with respected to lane centerline at end of first lane change. Mean +/− SEM represented by error bars. $^+ p < 0.1$, $^* p < 0.05$, and $^{**} p < 0.01$.

The results for lateral error in the case of the second lane change are listed in Table 2. The lateral error was significantly increased by driver distraction, but the effect of haptic guidance was insignificant. According to the peak SWA in the second lane change, no significance was observed with regard to the effect of HG. In accordance with our previous study [29], the lateral error at the end of first lane change was almost twice as large than at the end of the second lane change. The proposed haptic guidance system was more effective for the first lane change than the second one. This could be because there was more frequent overshoot during the first lane change, and haptic guidance was effective in diminishing the overshoot. Moreover, HG-Adaptive was more effective than HG-Fixed when the driver was distracted.

3.3. *Subjective Evaluation*

The results of the driver workload assessed by the NASA-TLX are shown in Table 2 and Figure 7. Taking into account the pairwise comparisons, the overall driver workload for distracted was significantly higher than for attentive ($p < 0.001$). Moreover, for distracted, HG-Fixed yielded a lower overall workload ($p < 0.1$), lower physical demand ($p < 0.1$), and lower effort ($p < 0.1$), and HG-Adaptive yielded a lower temporal demand ($p < 0.1$). This indicates that the driver workload increased during the secondary task, and the haptic guidance system effectively reduced the driver workload.

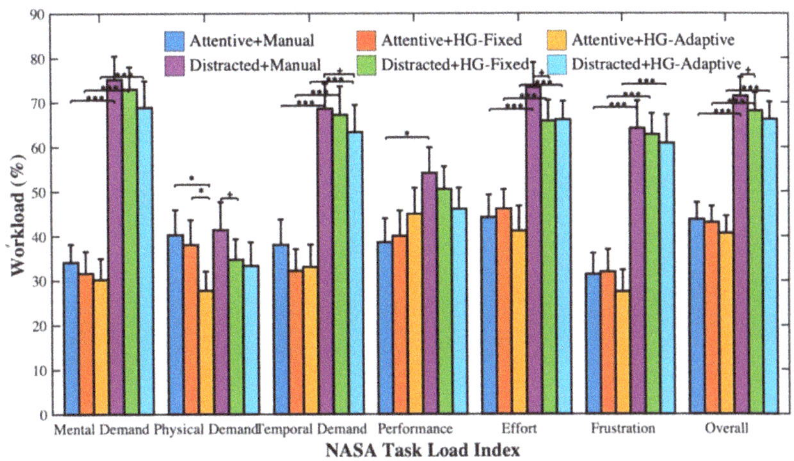

Figure 7. Mean scores on NASA-TLX. Data error bars represent the mean + SEM. $^{+}$ $p < 0.1$, * $p < 0.05$, and *** $p < 0.001$.

Table 2 shows the relative scores of pairwise driver preferences. For both attentive and distracted, drivers preferred HG-Fixed over manual, HG-Adaptive over manual, and HG-Adaptive over HG-Fixed. Moreover, there was a tendency for more drivers to prefer haptic guidance over the manual when they were distracted, which was expected, as haptic guidance reduces lane departure risk as well as driver workload.

3.4. *Limitations and Future Works*

One limitation of this study is that only a constant gain of 0.25 was considered for the HG-Fixed and linear law (i.e., linearly decreasing from 1 to 0 as the driver grip strength increased from 0 to $sEMG_{REF}$) for the HG-Adaptive. In the current study, the linear law was employed because it was assumed to be more intuitive to the drivers and could act as a benchmark and a reference for subsequent studies. Considering that drivers might not have the same behaviors when the fixed gain or adaptive strategies are changed, future studies could address other fixed gains and other laws (e.g., sigmoid law) for the haptic guidance design. Another limitation of the current work is that the driver distraction state was not measured. Considering that driver distraction could be estimated by monitoring eye and head movements in real time, an adaptive shared steering control based on multimodal sensory signal processing will be designed in our future work. Moreover, the driving speed was kept constant in the current study, resulting in less natural driving behavior. In the future study, more natural driving conditions with speed variability and a haptic shared controller design that takes into account speed variation will therefore be addressed. The current study only focused on lane change behavior on a relatively short driving course. A longer driving scenario with curved roads will be designed as future work, and the sEMG-based adaptive shared steering control system will be investigated for both lane change and lane following tasks. Given that the experimental results in the current study were based on driving simulations, future studies could consider evaluation of sense of presence in relation to virtual environments.

4. Conclusions

This driving simulator study focused on the effect of haptic guidance with adaptive authority on distracted driver behavior. DLC tasks were completed according to experimental conditions that were designed by combining two driver states, namely, attentive and distracted, along with three haptic guidance categories: HG-Fixed, HG-Adaptive, and Manual. HG-Adaptive relied on feedback from the real-time forearm sEMG of the driver.

For both attentive and distracted drivers, HG-Adaptive yielded a greater reduction in the driver workload and lane departure risk than that of HG-Fixed and manual steering. Moreover, drivers tended to reduce the steering wheel grip strength to provide admittance to the haptic guidance with fixed authority, resulting in the completion of the DLC in a relatively short period. As the current study indicated a small lane departure risk induced by driver distraction, we plan to address the possibility of greater risk by conducting a future study with a more demanding secondary task or lane change task. Although the current experiment was conducted in a high-fidelity driving simulator, future work will address how driver behavior would be in a more natural driving setting via real-vehicle experiments. Furthermore, the sample size of the present study was relatively small and biased toward male drivers and, thus, data from more female drivers should be collected in future work to further assess the effectiveness of the proposed system.

Author Contributions: Conceptualization, Z.W. and K.N.; methodology, Z.W., S.S. and B.Y.; software, S.S., E.J.C.N. and B.Y.; validation, Z.W. and S.S.; formal analysis, Z.W. and S.S.; resources, Z.W., B.Y. and K.N.; writing—original draft preparation, Z.W.; writing—review and editing, E.J.C.N., B.Y. and K.N.; supervision, K.N.; funding acquisition, Z.W. and K.N. All authors have read and agreed to the published version of the manuscript.

Funding: This research was funded by Grant-in-Aid for Early-Career Scientists No. 19K20318 and No. 21K17781 from the Japan Society for the Promotion of Science.

Institutional Review Board Statement: The study was approved by the Office for Life Science Research Ethics and Safety, Graduate School of Interdisciplinary Information Studies, University of Tokyo (protocol code No. 12 in 2017).

Informed Consent Statement: Informed consent was obtained from all subjects involved in the study.

Data Availability Statement: The data for this study are available in the IEEE DataPort: 10.21227/ckxa-8655 (accessed on 12 May 2021).

Conflicts of Interest: The authors declare no conflict of interest.

References

1. Koesdwiady, A.; Soua, R.; Karray, F.; Kamel, M.S. Recent trends in driver safety monitoring systems: State of the art and challenges. *IEEE Trans. Veh. Technol.* **2017**, *66*, 4550–4563. [CrossRef]
2. Xing, Y.; Lv, C.; Wang, H.; Cao, D.; Velenis, E.; Wang, F.Y. Driver activity recognition for intelligent vehicles: A deep learning approach. *IEEE Trans. Veh. Technol.* **2019**, *68*, 5379–5390. [CrossRef]

3. Jensen, M.J.; Tolbert, A.M.; Wagner, J.R.; Switzer, F.S.; Finn, J.W. A customizable automotive steering system with a haptic feedback control strategy for obstacle avoidance notification. *IEEE Trans. Veh. Technol.* **2011**, *60*, 4208–4216. [CrossRef]
4. Abbink, D.A.; Mulder, M.; Boer, E.R. Haptic shared control: Smoothly shifting control authority. *Cogn. Technol. Work* **2012**, *14*, 19–28. [CrossRef]
5. Tsoi, K.K.; Mulder, M.; Abbink, D.A. Balancing safety and support: Changing lanes with a haptic lane-keeping support system. In Proceedings of the 2010 IEEE International Conference on Systems, Man and Cybernetics, Istanbul, Turkey, 10–13 October 2010; pp. 1236–1243.
6. Griffiths, P.G.; Gillespie, R.B. Sharing control between humans and automation using haptic interface: Primary and secondary task performance benefits. *Hum. Factors J. Hum. Factors Ergonom. Soc.* **2005**, *47*, 574–590. [CrossRef] [PubMed]
7. Wang, Z.; Zheng, R.; Kaizuka, T.; Shimono, K.; Nakano, K. The effect of a haptic guidance steering system on fatigue-related driver behavior. *IEEE Trans. Hum.-Mach. Syst.* **2017**, *47*, 741–748. [CrossRef]
8. Mars, F.; Deroo, M.; Hoc, J.M. Analysis of human-machine cooperation when driving with different degrees of haptic shared control. *IEEE Trans. Haptics* **2014**, *7*, 324–333. [CrossRef]
9. Wang, Z.; Zheng, R.; Kaizuka, T.; Nakano, K. Influence of haptic guidance on driving behaviour under degraded visual feedback conditions. *IET Intell. Transp. Syst.* **2018**, *12*, 454–462. [CrossRef]
10. Zwaan, H.M.; Petermeijer, S.M.; Abbink, D.A. Haptic shared steering control with an adaptive level of authority based on time-to-line crossing. *IFAC-PapersOnLine* **2019**, *52*, 49–54. [CrossRef]
11. Saito, T.; Wada, T.; Sonoda, K. Control authority transfer method for automated-to-manual driving via a shared authority mode. *IEEE Trans. Veh. Technol.* **2018**, *3*, 198–207. [CrossRef]
12. Li, M.; Cao, H.; Song, X.; Huang, Y.; Wang, J.; Huang, Z. Shared control driver assistance system based on driving intention and situation assessment. *IEEE Trans. Ind. Inform.* **2018**, *14*, 4982–4994. [CrossRef]
13. Ji, X.; Yang, K.; Na, X.; Lv, C.; Liu, Y. Shared steering torque control for lane change assistance: A stochastic game-theoretic approach. *IEEE Trans. Ind. Electron.* **2019**, *66*, 3093–3105. [CrossRef]
14. Nguyen, A.-T.; Rath, J.J.; Lv, C.; Guerra, T.-M.; Lauber, J. Human-machine shared driving control for semi-autonomous vehicles using level of cooperativeness. *Sensors* **2021**, *21*, 4647. [CrossRef]
15. Ercan, Z.; Carvalho, A.; Tseng, H.E.; Gökaşan, M.; Borrelli, F. A predictive control framework for torque-based steering assistance to improve safety in highway driving. *Veh. Syst. Dyn.* **2018**, *56*, 810–831. [CrossRef]
16. Flad, M.; Fröhlich, L.; Hohmann, S. Cooperative shared control driver assistance systems based on motion primitives and differential games. *IEEE Trans. Hum.-Mach. Syst.* **2017**, *47*, 711–722. [CrossRef]
17. Nguyen, A.-T.; Sentouh, C.; Popieul, J.-C. Driver-automation cooperative approach for shared steering control under multiple system constraints: Design and experiments. *IEEE Trans. Ind. Electron.* **2016**, *64*, 3819–3830. [CrossRef]
18. Sentouh, C.; Nguyen, A.-T.; Benloucif, M.A.; Popieul, J.-C. Driver-automation cooperation oriented approach for shared control of lane keeping assist systems. *IEEE Trans. Control Syst. Technol.* **2018**, *27*, 1962–1978. [CrossRef]
19. Yang, K.; Liu, Y.; Na, X.; He, X.; Liu, Y.; Wu, J.; Nakano, S.; Ji, X. Preview-scheduled steering assistance control for co-piloting vehicle: A human-like methodology. *Veh. Syst. Dyn.* **2020**, *58*, 518–544. [CrossRef]
20. Li, R.; Li, Y.; Li, S.E.; Zhang, C.; Burdet, E.; Cheng, B. Indirect shared control for cooperative driving between driver and automation in steer-by-wire vehicles. *IEEE Trans. Intell. Transp. Syst.* **2020**, 1–11. [CrossRef]
21. Nacpil, E.J.C.; Wang, Z.; Nakano, K. Application of physiological sensors for personalization in semi-autonomous driving: A review. *IEEE Sens. J.* **2021**, *21*, 19662–19674. [CrossRef]
22. Kaber, D.B.; Riley, J.M.; Tan, K.W.; Endsley, M.R. On the design of adaptive automation for complex systems. *Int. J. Cogn. Ergonom.* **2001**, *5*, 37–57. [CrossRef]
23. Zhu, B.; Yan, S.; Zhao, J.; Deng, W. Personalized lane-change assistance system with driver behavior identification. *IEEE Trans. Veh. Technol.* **2018**, *67*, 10293–10306. [CrossRef]
24. Jiang, Y.; Deng, W.; Wu, J.; Zhang, S.; Jiang, H. Adaptive steering feedback torque design and control for driver–vehicle system considering driver handling properties. *IEEE Trans. Veh. Technol.* **2019**, *68*, 5391–5406. [CrossRef]
25. Benloucif, M.; Sentouh, C.; Floris, J.; Simon, P.; Popieul, J.-C. Online adaptation of the level of haptic authority in a lane keeping system considering the driver's state. *Transp. Res. Part F Traffic Psychol. Behav.* **2019**, *61*, 107–119. [CrossRef]
26. Nakamura, H.; Abbink, D.; Mulder, M. Is grip strength related to neuromuscular admittance during steering wheel control? In Proceedings of the 2011 IEEE International Conference on Systems, Man, and Cybernetics, Anchorage, AK, USA, 9–12 October 2011; pp. 1658–1663.
27. Wang, Z.; Kaizuka, T.; Nakano, K.; Zheng, R.; Shimono, K. Evaluation of driver steering performance with haptic guidance under passive fatigued situation. In Proceedings of the 2016 IEEE International Conference on Systems, Man, and Cybernetics (SMC), Budapest, Hungary, 9–12 October 2016; pp. 3334–3339.
28. Smisek, J.; Mugge, W.; Smeets, J.B.; Paassen, M.M.; Schiele, A. Haptic guidance on demand: A grip-force based scheduling of guidance forces. *IEEE Trans. Haptics* **2018**, *11*, 255–266. [CrossRef] [PubMed]

29. Wang, Z.; Suga, S.; Nacpil, E.J.C.; Yan, Z.; Nakano, K. Adaptive driver-automation shared steering control via forearm surface electromyography measurement. *IEEE Sens. J.* **2021**, *21*, 5444–5454. [CrossRef]
30. Tomaszewski, M. GitHub. Myo SDK MATLAB MEX Wrapper. Available online: https://www.github.com/mark-toma/MyoMex (accessed on 7 January 2020).
31. Chen, G.; Chen, S.; Langari, R.; Li, X.; Zhang, W. Driver-behavior-based adaptive steering robust nonlinear control of unmanned driving robotic vehicle with modeling uncertainties and disturbance observer. *IEEE Trans. Veh. Technol.* **2019**, *68*, 8183–8190. [CrossRef]

sensors

MDPI

Article

A Statistical Method for Exploratory Data Analysis Based on 2D and 3D Area under Curve Diagrams: Parkinson's Disease Investigation

Olga Sergeevna Sushkova [1,*], Alexei Alexandrovich Morozov [1], Alexandra Vasilievna Gabova [2], Alexei Vyacheslavovich Karabanov [3] and Sergey Nikolaevich Illarioshkin [3]

1 Kotel'nikov Institute of Radio Engineering and Electronics of RAS, Mokhovaya 11-7, 125009 Moscow, Russia; morozov@cplire.ru

2 Institute of Higher Nervous Activity and Neurophysiology of RAS, Butlerova 5A, 117485 Moscow, Russia; agabova@yandex.ru

3 FSBI "Research Center of Neurology", Volokolamskoe Shosse 80, 125367 Moscow, Russia; doctor.karabanov@mail.ru (A.V.K.); snillario@gmail.com (S.N.I.)

* Correspondence: o.sushkova@mail.ru

Abstract: A statistical method for exploratory data analysis based on 2D and 3D area under curve (AUC) diagrams was developed. The method was designed to analyze electroencephalogram (EEG), electromyogram (EMG), and tremorogram data collected from patients with Parkinson's disease. The idea of the method of wave train electrical activity analysis is that we consider the biomedical signal as a combination of the wave trains. The wave train is the increase in the power spectral density of the signal localized in time, frequency, and space. We detect the wave trains as the local maxima in the wavelet spectrograms. We do not consider wave trains as a special kind of signal. The wave train analysis method is different from standard signal analysis methods such as Fourier analysis and wavelet analysis in the following way. Existing methods for analyzing EEG, EMG, and tremor signals, such as wavelet analysis, focus on local time–frequency changes in the signal and therefore do not reveal the generalized properties of the signal. Other methods such as standard Fourier analysis ignore the local time–frequency changes in the characteristics of the signal and, consequently, lose a large amount of information that existed in the signal. The method of wave train electrical activity analysis resolves the contradiction between these two approaches because it addresses the generalized characteristics of the biomedical signal based on local time–frequency changes in the signal. We investigate the following wave train parameters: wave train central frequency, wave train maximal power spectral density, wave train duration in periods, and wave train bandwidth. We have developed special graphical diagrams, named AUC diagrams, to determine what wave trains are characteristic of neurodegenerative diseases. In this paper, we consider the following types of AUC diagrams: 2D and 3D diagrams. The technique of working with AUC diagrams is illustrated by examples of analysis of EMG in patients with Parkinson's disease and healthy volunteers. It is demonstrated that new regularities useful for the high-accuracy diagnosis of Parkinson's disease can be revealed using the method of analyzing the wave train electrical activity and AUC diagrams.

Keywords: electromyogram; EMG; exploratory data analysis; wave train electrical activity analysis method; wave trains; wavelets; signal processing; AUC diagrams; ROC analysis; Parkinson's disease

Citation: Sushkova, O.S.; Morozov, A.A.; Gabova, A.V.; Karabanov, A.V.; Illarioshkin, S.N. A Statistical Method for Exploratory Data Analysis Based on 2D and 3D Area Under Curve Diagrams: Parkinson's Disease Investigation. *Sensors* **2021**, *21*, 4700. https://doi.org/10.3390/s21144700

Academic Editor: Ernest N. Kamavuako

Received: 7 May 2021
Accepted: 6 July 2021
Published: 9 July 2021

1. Introduction

The paper provides a detailed description of the method used for analyzing the wave train electrical activity in biomedical signals. The method was developed to investigate electroencephalograms (EEG), electromyograms (EMG), and accelerometer signals (tremorograms) in patients with Parkinson's disease (PD) and identify regularities that are promising for the early diagnosis of this disease. Recently, many mathematical methods for analyzing EEG, EMG, and tremor signals have been developed. Historically, EMG analysis methods evolved from spectral analysis [1–7] and time-domain signal analysis methods such as morphological analysis [8], amplitude analysis [9], and autoregressive analysis [10,11] towards time–frequency domain analysis [12–16]. The state-of-the-art of EMG analysis methods is characterized by the active use of nonlinear data analysis methods [17], such as fractal analysis [18], phase analysis [19], recurrent quantification analysis [4,20,21], and the deep learning of neural networks [12,22–25]. According to the authors, the existing methods for analyzing EEG, EMG, and tremor signals, such as wavelet analysis [26–28], focus on local time–frequency changes in the signal and, therefore, do not reveal the generalized properties of the signal. By contrast, other methods, such as standard Fourier analysis, ignore local time–frequency changes in the signal and, therefore, lose a large amount of information that existed in the signal.

Let us consider the spectra of envelopes of EMG signals collected from PD patients and healthy volunteers (see Figure 1). On the left, an average spectrum of the tremor right hands of twelve PD patients and an average spectrum of the right hands of ten healthy volunteers are demonstrated. On the right, an average spectrum of the non-tremor left hands of the PD patients and an average spectrum of the left hands of the healthy volunteers are demonstrated. Three peaks are observed in the 4–10 Hz frequency range in the left figure. Two peaks are observed in this frequency range in the right figure. The Mann–Whitney statistical test discovers statistically significant differences in the spectra of the PD patients only in the tremor hands (see the left figure). These differences are well-known physiological regularity and are used for the diagnosis of PD [3]. However, the statistically significant differences between the spectra are not observed in the right figure. Therefore, the conventional spectral analysis does not reveal diagnostic features of PD in the non-tremor hands of the PD patients. In this paper, we will demonstrate that our method extracts much more information from the signals. In particular, statistically significant differences will be demonstrated between EMG signals collected from the non-tremor hands of the PD patients and healthy volunteers.

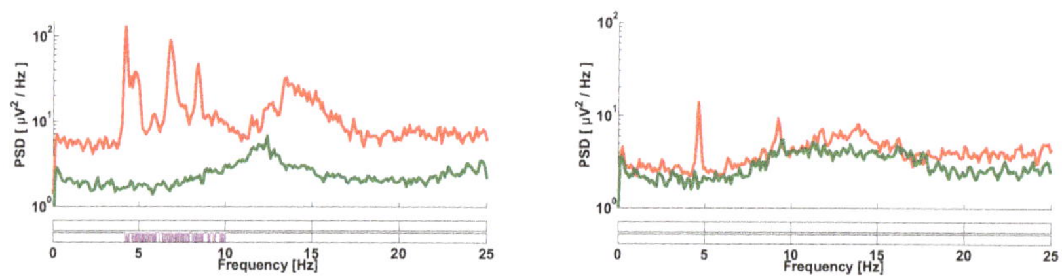

Figure 1. An example of the spectra of envelopes of EMG signals collected from PD patients and healthy volunteers. The spectra are computed using the standard Welch method. The Hann window was used, the window width was 10 s, and the window overlap was 7/8. The red curve indicates the PD patients. The green curve indicates the healthy volunteers. On the left, an average spectrum of the tremor right hands of twelve PD patients and an average spectrum of the right hands of ten healthy volunteers are demonstrated. On the right, an average spectrum of the non-tremor left hands of the PD patients and an average spectrum of the left hands of the healthy volunteers are demonstrated. The abscissa is the frequency. The ordinate is the power spectral density in the logarithmic scale. Two bars below the figure indicate the results of the Mann–Whitney statistical test. Statistically significant differences are indicated by the magenta color. The lower bar corresponds to the alpha level 0.05. The upper bar corresponds to the Bonferroni-corrected alpha level 0.0002. The significant differences are observed only in the tremor hands of the PD patients.

The wave train analysis method differs from standard signal analysis methods such as Fourier analysis and wavelet analysis in that it addresses the generalized characteristics of the biomedical signal based on local time–frequency changes in the signal. The idea of the method of wave train electrical activity analysis is to extract and analyze so-called wave trains in wavelet spectrograms. The wave train is the increase in the power spectral density

(PSD) of the signal localized in time, frequency, and space. The wave trains correspond to local maxima in the wavelet spectrograms. We investigate the following wave train parameters: wave train central frequency, wave train maximal PSD, wave train duration in periods, and wave train bandwidth. Note that in [9,29–33], for EMG analysis, the term "burst" is used. However, the meaning of this term is different. Usually, the term "burst" refers to the EMG signal areas characterized by a sharp increase in amplitude. In contrast to these papers, we investigate the wave trains in the time–frequency domain but not in the time domain. We consider the biomedical signal as a combination of wave trains, and we do not consider the wave train as a special type of signal.

We extract the wave trains in a wide frequency range using complex Morlet wavelets. We consider local maxima in the wavelet spectrograms as wave trains. A technique based on so-called area under curve (AUC) diagrams is used to identify regularities in signals in a wide frequency range. The AUC diagrams are specially designed graphical diagrams that help to determine the wave train parameters that are characteristic of a given neurodegenerative disease. We distinguish the Frequency AUC diagram, Power spectral density AUC diagram, Duration AUC diagram, and Bandwidth AUC diagram. These types of AUC diagrams will be discussed in Section 2.4. Moreover, we distinguish 2D and 3D AUC diagrams. The 2D AUC diagrams are useful for manually searching regularities in the wave train electrical activity. The 3D AUC diagrams are used for searching statistically significant differences between groups of subjects using AUC diagrams. The AUC diagram technique is illustrated by examples of analyzing the EMG data from patients with Parkinson's disease and healthy volunteers.

Note that the wavelets are not a critical issue of the method of wave train analysis. Generally speaking, similar data analysis can be carried out based on windowed Fourier transform. However, the wavelets have the following advantage: the time resolution of the wavelet changes automatically when different frequencies are investigated. Thus, the wavelets allow one to investigate wave trains simultaneously in high- and low-frequency bands. We use the Morlet wavelet because it is simple and people can easily understand the wavelet diagrams. Our method differs from other methods based on wavelets [15,16] in that the wave trains are considered and AUC diagrams are applied.

The problem of the early and differential diagnosis of PD is all too real [17,34–37]. It is difficult to identify the early features of PD because the disease develops over a long time without clear clinical manifestations. The first clinical stage of PD is characterized by the patient having a pathological tremor on only one side of the body. At the same time, another side of the body does not demonstrate the clinical manifestations of PD (has no trembling hyperkinesis [38,39]). O. E. Khutorskaya [1,2] suggested that the non-tremor side of the body of PD patients can be considered as a model of the preclinical (early) stage of PD. Therefore, it is important to investigate the non-tremor side of the body of first-stage PD patients. This paper demonstrates that wave train analysis can reveal new regularities in the non-tremor side of the PD patient body which are useful for the diagnosis of Parkinson's disease at the preclinical stage.

The method used for analyzing the wave train electrical activity of signals is discussed in Section 2. Section 3 describes the results of the group data analysis. A discussion of the data analysis results is given in Section 4.

2. Materials and Methods

The wave train is the increase in the signal PSD localized in space, time, and frequency. We applied wavelet spectrograms calculated using the complex Morlet wavelet to determine wave trains in signals. An adaptive two-dimensional Gaussian filter was used to smooth the wavelet spectrogram to eliminate artifacts arising in the process of calculating wavelets. Then, we detected the local maxima on the wavelet spectrogram. The attributes of the wave trains were calculated, such as the central frequency of the wave train, the maximal PSD of the wave train, the duration of the wave train in periods, and the bandwidth of the wave train.

2.1. Experimental Data

The object of our investigation is the electromyographic signals in PD patients at the first stage of the disease according to the classical Hoehn–Yahr scale. Approximately half of the patients had never taken antiparkinsonian drugs before, and the other patients had not taken antiparkinsonian drugs for one to two days before the investigation. Additionally, a group of healthy volunteers participated in the investigation. The average age of the patients was 56 years (the minimum age was 38 years; the maximum age was 69 years). The average age of the healthy volunteers was 51 years (the minimum age was 24 years; the maximum age was 71 years). There were no statistically significant differences between the ages of the patients and the healthy volunteers (the Mann–Whitney test was used). Note that the group of PD patients included patients with left-hand tremor (10 persons) and patients with right-hand tremor (12 persons), with 22 persons in total (see Figure 2). All the patients were examined at the FSBI Research Center of Neurology, and PD was diagnosed. The number of healthy volunteers was 10 persons. All patients and healthy volunteers were right-handed.

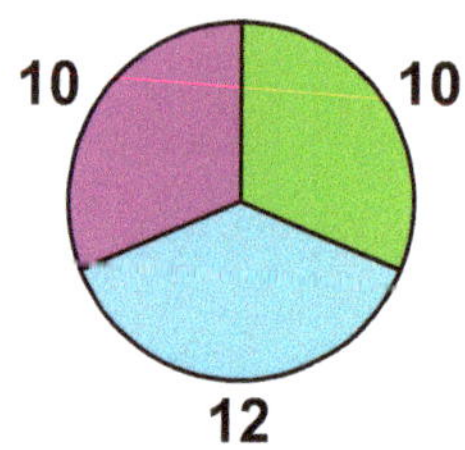

Figure 2. A diagram of the investigated groups of subjects. The left-hand tremor PD patients are indicated by the magenta color; the right-hand tremor PD patients are indicated by the cyan color; the healthy volunteers are indicated by the green color.

The subjects were sitting in a chair during the data acquisition. Arms were outstretched forward. The duration of the single recording was 1 min and 30 s. EMG electrodes were placed on both arms of the patient on the antagonist muscles of the wrist joint (extensor and flexor muscles: Musculus extensor carpi radialis longus and Musculus flexor carpi radialis). The eyes were closed during the measurement. The Neuron-Spectrum-5 multifunctional system for neurophysiological studies (Neurosoft Ltd.) was used for EMG recording. The sampling rate was 500 Hz. The Butterworth high-pass filter with a cut-off frequency 0.5 Hz and a 50 Hz notch filter were used during the data acquisition.

2.2. Signal Preprocessing

The preprocessing of EMG signals included the following stages:

1. The 50, 100, 150, and 200 Hz notch filters removed the power line interference.
2. The 60–240 Hz fourth-order Butterworth bandpass filter was applied to EMG in the forward and reverse directions.
3. The envelope of the EMG signal was calculated using the Hilbert transform. The signal envelope was used for tremor analysis according to the classical method [1,2].
4. The envelope of the signal was decimated; the decimation factor was 4.

2.3. Calculation of Local Maxima in the Wavelet Spectrogram

We used the wavelet spectrograms calculated using the complex Morlet wavelet (1) to determine wave trains in the signals:

$$\psi(x) = \frac{1}{\sqrt{\pi F_b}} exp(2\pi\imath F_c x) exp(\frac{-x^2}{F_b}) \tag{1}$$

where $F_b = 1$ and $F_c = 1$. We calculated the wavelet spectrogram in the frequency range from 0.1 to 50 Hz in the examples considered in the paper; the frequency step was 0.1 Hz.

The wavelet spectrogram was smoothed by an adaptive two-dimensional Gaussian filter to eliminate artifacts arising in the process of calculating the wavelets. The width of the Gaussian window in time and frequency depends on the width of the time and frequency windows of the wavelet at the considered frequency. We used a smoothing window width that was twice less than the time and frequency width of the wavelet window. The width of the smoothing window should be less than the width of the wavelet window to prevent the distortion of the wavelet spectrogram shape.

Let us consider an example of a wave train on a wavelet spectrogram of EMG signal in an extensor muscle of the non-tremor (right) arm of a patient with the left-side tremor of the body (Figure 3). The central frequency of the wave train is 15.2 Hz; the signal is clearly distinguished in the time–frequency space.

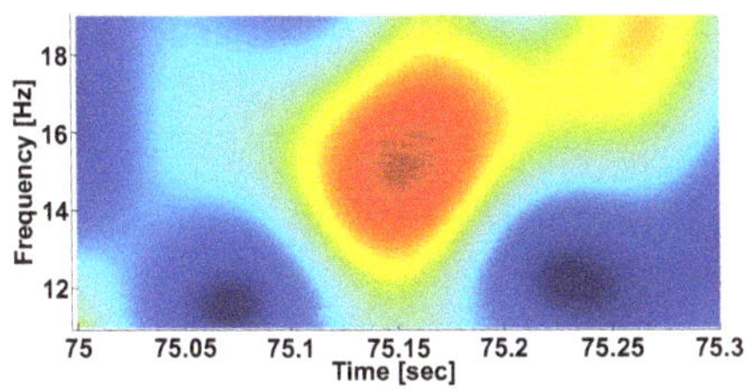

Figure 3. A wave train on the wavelet spectrogram of the EMG signal envelope. The abscissa axis indicates the time; the ordinate axis indicates the frequency.

The envelope of the EMG signal (Figure 3) is demonstrated in Figure 4 (on the left). One can see three periods of the wave train envelope in the figure. On the right, the source EMG signal is demonstrated. It is almost impossible to reveal the wave train considered in the source signal without special processing. Therefore, the standard methods for the morphological analysis [8] of signals are inapplicable for this signal.

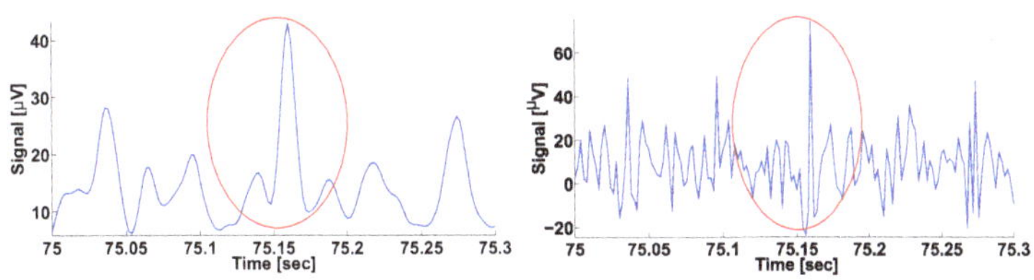

Figure 4. On the left is the envelope of the EMG signal considered in Figure 3. The abscissa axis indicates the time; the ordinate axis indicates the envelope of the signal in µV. On the right is the source EMG signal. The abscissa axis indicates the time; the ordinate axis indicates the amplitude of the signal in µV. The wave train is indicated by the red circle in both figures.

In Figure 5, other examples of the wave trains are demonstrated. On the left, the envelope of the EMG signal in the tremor left hand of a PD patient is demonstrated. On the right, the envelope of the EMG signal in the left hand of a healthy volunteer is demonstrated. The wave trains are very similar. The central frequency of both wave trains is ~6.5 Hz. In the framework of our method, we do not try to distinguish "normal" and "abnormal" wave trains. Instead, we use a statistical analysis based on the number of detected wave trains.

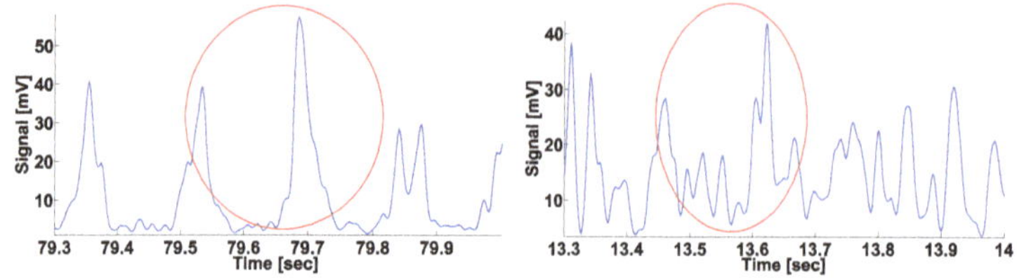

Figure 5. On the left: the envelope of the EMG signal in the tremor left hand of a PD patient. On the right: the envelope of the EMG signal in the left hand of a healthy volunteer. The wave trains are indicated by the red circles. The abscissa axis indicates the time; the ordinate axis indicates the envelope of the signal in μV. The wave trains are very similar. We do not try to distinguish "normal" and "abnormal" wave trains. Instead, we use a statistical analysis based on the number of detected wave trains.

Note that the computation of wavelet spectrograms and detection of wave trains are the most time-consuming data processing steps. The processing of the EMG data for the total group of subjects (32 persons) takes about 2 hours on a 2.30 GHz PC machine. We do not consider the wave trains in the wavelet spectrogram if the duration of the wave train is less than 1/10 of the signal period at the central frequency of the local maximum to increase the speed of computation.

The time duration and the frequency width of the local maximum are measured at the $1/\sqrt{2}$ maximum height of the local maximum. Figure 6 demonstrates time and frequency slices of the wave train wavelet spectrogram.

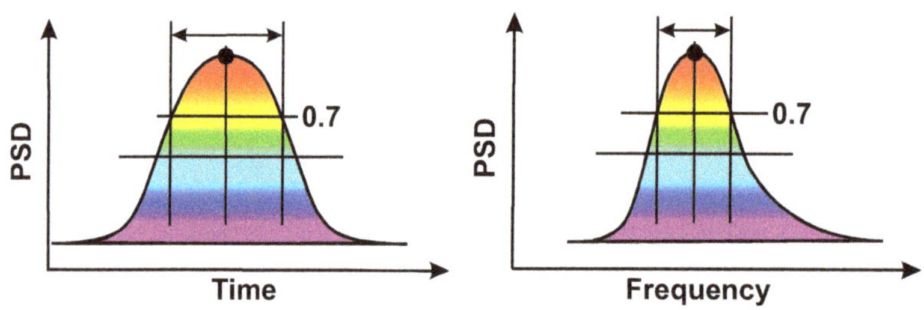

Figure 6. An example of a wave train spectrogram in the time–frequency domain. On the left is a time slice of the wavelet spectrogram. The abscissa is time and the ordinate is PSD. On the right is a frequency slice of the wavelet spectrogram. The abscissa is the frequency and the ordinate is PSD.

The wave train can be characterized by several parameters: the leading (central) frequency, the maximal PSD, the duration in periods (at $1/\sqrt{2}$ maximum height), and the bandwidth (at $1/\sqrt{2}$ maximum height). These parameters form a multidimensional space. The analysis aims to select a certain subspace in the given space where a difference between the groups of subjects is observed. The following notation is used below to denote the subspace bounds: *MinFreq* (the minimal wave train frequency), *MaxFreq* (the maximal wave train frequency), *MinPSD* (the minimal wave train PSD), *MaxPSD* (the maximal wave train PSD), *MinDurat* (the minimal wave train duration in periods), *MaxDurat* (the maximal wave train duration in periods), *MinBandwidth* (the minimal wave train bandwidth), and *MaxBandwidth* (the maximal wave train bandwidth).

The number of wave trains in the PD patients was compared with the number of wave trains in the healthy volunteers using ROC curves. The quality of the ROC curve is characterized by the area under the ROC curve (AUC). AUC values from 0 to 1 can be obtained when comparing groups of subjects. We were interested in values that significantly differed from 0.5—that is, values close to 0 and 1. These AUC values have the following interpretation. AUC > 0.5 means that the number of wave trains is higher in the patients than in the healthy volunteers. AUC < 0.5 means that the number of the wave trains is higher in the healthy volunteers. Both cases are of interest for the investigation and the diagnosis of PD.

We developed so-called AUC diagrams to search for regularities in the multidimensional space of the wave train parameters. We considered the following types of AUC diagrams: 2D and 3D diagrams.

2.4. 2D AUC Diagrams

The 2D AUC diagram demonstrates the AUC values corresponding to different ranges of the given wave train parameter. The range of the wave train parameter is characterized by the lower and upper bounds—namely, the minimal and maximal values of the parameter. The abscissa indicates the lower bound of the considered parameter, while the ordinate indicates the upper bound of the considered parameter. The AUC value is displayed using a colormap. The standard jet colormap is applied in the examples given in this paper. Low AUC values are displayed in blue and high AUC values are displayed in red in this colormap. AUC values close to 0.5 are displayed in green.

We considered AUC diagrams of different types—namely, Frequency AUC diagrams (see example in Figure 7), Power spectral density AUC diagrams (see example in Figure 8), Duration AUC diagrams (see example in Figure 9), and Bandwidth AUC diagrams (see example in Figure 10).

Let us consider the frequency range from 1 to 50 Hz and calculate the number of wave trains in the EMG signals of each PD patient and each healthy volunteer. The patients with the left-hand tremor and the patients with the right-hand tremor were investigated separately.

An example of a Frequency AUC diagram demonstrates the AUC values calculated for various ranges in the frequency interval from 1 to 50 Hz with 1 Hz steps (see Figure 7). Corresponding ROC curves compare the number of wave trains in the extensor muscle in the right non-tremor arm of the left-hand-tremor PD patients with the number of wave trains in the extensor muscle in the right arm of the healthy volunteers. The red color in the diagram indicates that the number of wave trains in the patients is greater than that in the healthy subjects. The blue color in the AUC diagram indicates that the number of the wave trains in the patients is lower than the number in the healthy subjects. The diagram has a triangular shape because the upper bound of the range is always bigger than the lower bound of the range.

Reading the diagram should be carried out in the following way. One should start by looking at the AUC values located on the diagonal line of the diagram. In the Frequency AUC diagram, the diagonal line corresponds to narrow frequency ranges $MinFreq \approx MaxFreq$, which allows one to accurately estimate the frequencies where differences appear between the patient group and the control group. These frequencies correspond to red and blue dots on the diagonal line. Next, one should consider the monochromatic areas adjacent to the diagonal line. The area must be of the same color as the red/blue dot on the diagonal line. The bigger the area is, the stronger the revealed difference between the groups of subjects is.

In Figure 7, two bright red areas are observed in the frequency range. The first red area is situated along the abscissa axis from 0 to 18 Hz; the y-coordinate is equal to approximately 20 Hz. The second area is situated along the ordinate axis from 17 Hz and above; the x-coordinate is equal to approximately 14 Hz. The brightest point has coordinates of 8 Hz on the abscissa axis and 20 Hz on the ordinate axis. This point corresponds to the frequency range from 8 to 20 Hz. The red color indicates that the PD patients have more wave trains than the healthy subjects in the human physiological tremor frequency area. The AUC value is approximately 0.88 in this frequency range; thus, the observed regularity can be used as a diagnostic criterion for Parkinson's disease.

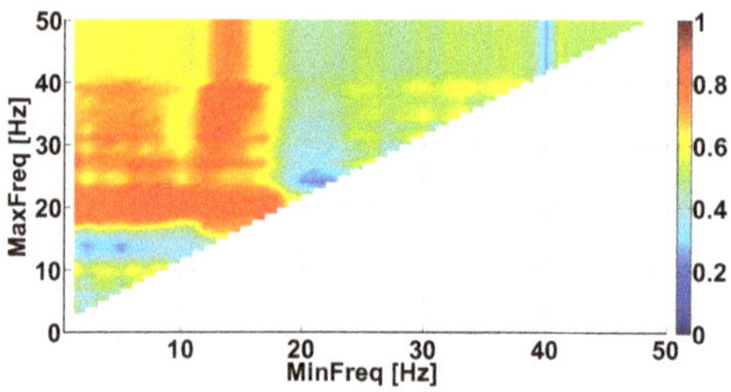

Figure 7. An example of the Frequency AUC diagram. The abscissa axis is the lower bound of the frequency range; the ordinate axis is the upper bound of the frequency range. Two bright red areas are observed. The first red area is situated along the abscissa axis from 0 to 18 Hz; the y-coordinate is equal to approximately 20 Hz. The second area is situated along the ordinate axis from 17 Hz and above; the x-coordinate is equal to approximately 14 Hz. The brightest point corresponds to the human physiological tremor frequency area from 8 to 20 Hz.

Let us consider a Power spectral density AUC diagram (Figure 8). The diagram is calculated on the same dataset. In contrast to the Frequency AUC diagram, the ranges of PSD are considered in the Power spectral density AUC diagram. The range of PSD is characterized by the lower and upper bounds. The abscissa indicates the lower bound of the PSD range, while the ordinate indicates the upper bound of the PSD range. The values of the wave train PSD are considered in the interval from 0 to 1000 $\mu V^2/$ Hz with 10 $\mu V^2/$ Hz steps. In the figure, a bright red area is observed along the ordinate axis above 70 $\mu V^2/$ Hz.

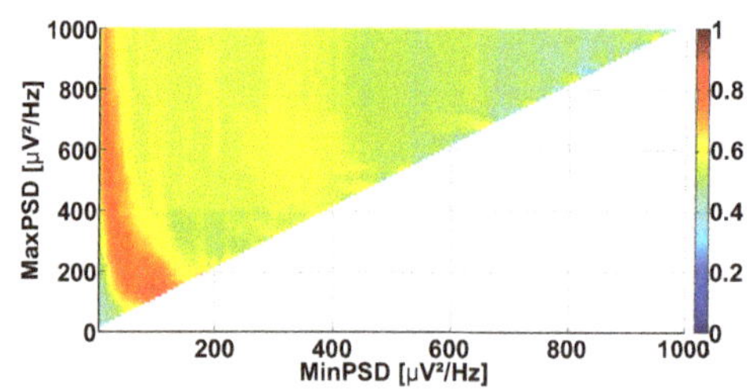

Figure 8. An example of the Power spectral density AUC diagram. The abscissa axis is the lower bound of the PSD range; the ordinate axis is the upper bound of the PSD range. A bright red region is observed along the ordinate axis above 70 $\mu V^2/$ Hz.

The third type of AUC diagrams is the Duration AUC diagram. Let us consider an example of the Duration AUC diagram (see Figure 9). The diagram is based on the same dataset as the previous diagrams; however, the ranges of wave train durations are considered. As in previous figures, the duration range is characterized by the lower and upper bounds of the range. The abscissa indicates the lower bound of the range, while the ordinate indicates the upper bound of the range. The duration of the wave trains is considered in the interval from 0 to 10 periods with 0.1 period steps.

Figure 9 demonstrates a bright yellow area with the x-coordinate of less than 3.8 periods. A narrow bright orange area is situated along the ordinate axis; the x-coordinate is equal to approximately 3.8 periods. This diagram can be explained in the following way: most wave trains have a duration of approximately 3.8 periods, but shorter wave trains are observed too.

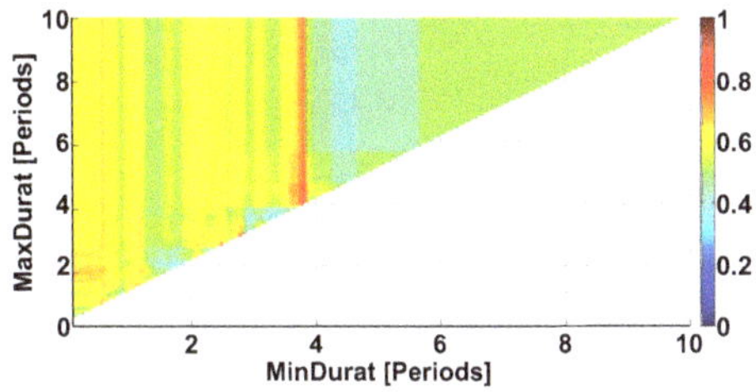

Figure 9. An example of the Duration AUC diagram. The abscissa axis is the lower bound of the duration range in periods; the ordinate axis is the upper bound of the duration range in periods. A narrow bright orange area is situated along the ordinate axis; the x-coordinate is equal to approximately 3.8 periods.

Let us consider an example of a Bandwidth AUC diagram (Figure 10). The diagram is based on the same dataset. In the Bandwidth AUC diagram, the ranges of bandwidth of the wave trains are considered. The bandwidth range is characterized by the lower and upper range bounds. The abscissa axis indicates the lower bound of the range, while the ordinate axis indicates the upper bound of the range. In this example, the frequency bands in the interval from 0 to 50 Hz are considered; the step size is 0.1 Hz.

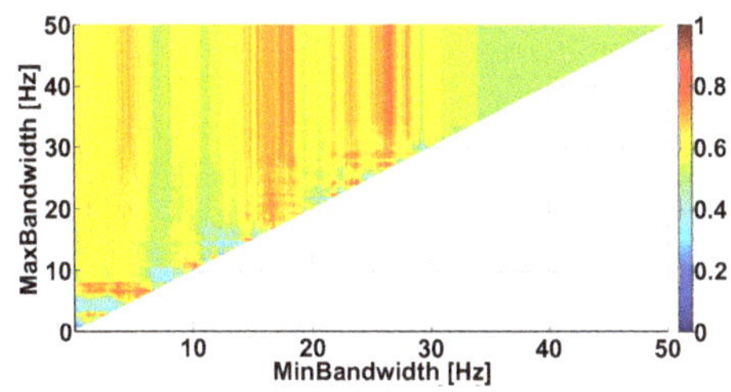

Figure 10. An example of the Bandwidth AUC diagram. The abscissa axis indicates the lower bound of the bandwidth; the ordinate axis indicates the upper bound of the bandwidth. The diagram demonstrates the multidirectional effects.

The frequency bandwidth of the signal characterizes the shape of the signal. The narrowband signal is close to the harmonic one; the wideband signal contains fragments of a complex shape. Figure 10 demonstrates several areas corresponding to multidirectional differences in the wave train bandwidth between the groups of subjects. In particular, a bright orange area is observed along the abscissa axis from 0 to 7 Hz; the y-coordinate is equal to approximately 7 Hz. In addition, a light blue area is observed along the abscissa axis from 0 to 4.5 Hz; the y-coordinate is equal to approximately 4.3 Hz. There are also vertical orange columns with x-coordinates equal to approximately 17 Hz and 26 Hz. Note that multidirectional effect diagrams are more difficult to interpret. In this example, we can only conclude that it is possible to obtain and investigate multidirectional differences between the groups of subjects by detailing the wave train bandwidth ranges. This will be carried out in the further steps of analysis.

The analysis of the wave train electrical activity begins with the calculation of AUC diagrams of all four types (see Figures 7–10). At the first stage of the analysis, one has to choose one out of four diagrams that demonstrates the most pronounced differences between the patients and healthy volunteers—that is, the diagram that contains the most prominent red or blue area with AUC values close to 0 or 1. The selected red/blue area corresponds to a certain range of the corresponding parameter. The calculation of all four diagrams is repeated in the next steps of the analysis. However, only the wave trains that correspond to the ranges of the wave train parameters selected in the previous steps are taken into account. It is possible to identify interesting ranges of all four parameters of the wave trains (the central frequency, PSD, duration in periods, and bandwidth) by iteratively repeating the described operations (see the flowchart in Figure 11).

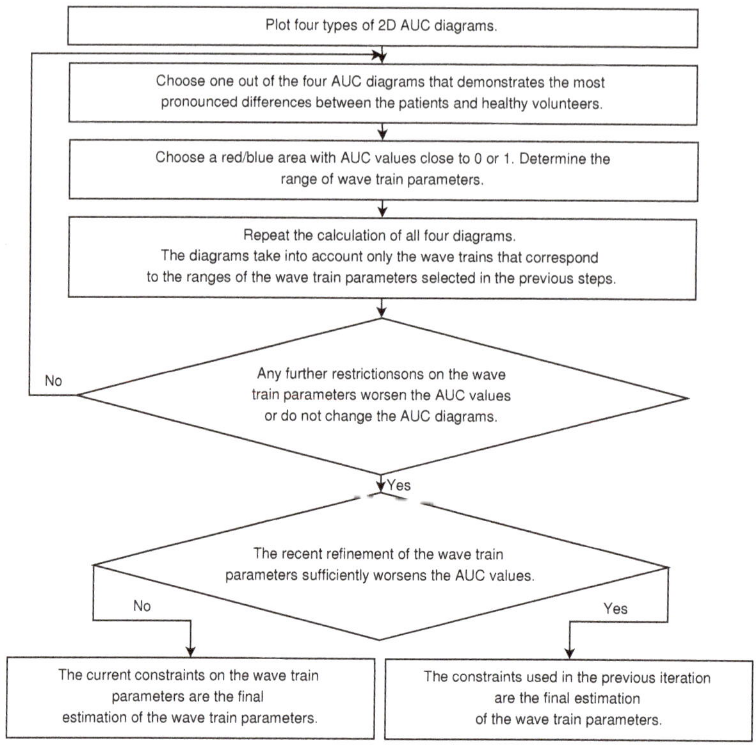

Figure 11. The flowchart of the method of analysis of the wave train electrical activity in EMG signals.

In the Frequency AUC diagram (Figure 7), we chose a red area with the following coordinates: frequency range from 8 to 20 Hz. Therefore, we will consider only the wave trains that belong to the frequency interval from 8 to 20 Hz in the further steps of the analysis. Let us recalculate the other three diagrams taking into account the chosen constraint.

Figure 12 demonstrates the Power spectral density AUC diagram for the wave trains that belong to the frequency interval from 8 to 20 Hz. The diagram differs from Figure 8 because the frequency band of the considered wave trains is narrowed. The colored areas in Figure 12 are brighter, but the size and position of the areas are approximately the same. This means that the applied frequency constraint allows the better recognition of the differences between the groups of subjects. The diagram demonstrates the most substantial differences between the patients and control subjects in the following point: the PSD range from 30 to 700 $\mu V^2/$ Hz (AUC = 0.96). The maximal PSD of the wave trains may be a hardware-dependent characteristic; thus, one can consider only the wave trains with a PSD above 30 $\mu V^2/$ Hz in the further steps of the analysis.

Let us consider the Duration AUC diagram for the wave trains that belong to the frequency interval from 8 to 20 Hz (see Figure 13). The diagram differs from Figure 9 because the frequency band is narrowed. The colored areas on the diagram became brighter. A red column appears along the ordinate; the x-coordinate is equal to approximately 1 period. This means that one can better distinguish the groups of subjects when the frequency band of the wave trains is narrowed. The most substantial differences between the patients and control subjects are observed in the wave train duration range from 0.7 to

2.6 periods (AUC = 0.88). Thus, we can narrow the interval of the wave train durations in the further step of the analysis. Only durations from 0.7 to 2.6 periods will be considered.

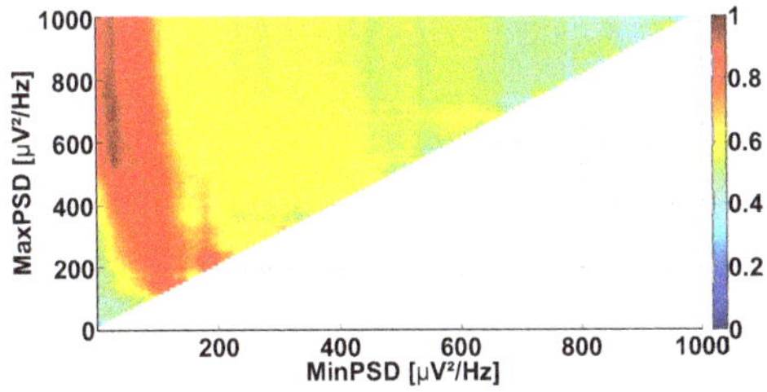

Figure 12. An example of the Power spectral density AUC diagram. The frequency band of the wave trains is constrained, and the frequency interval from 8 to 20 Hz is considered. The abscissa axis indicates the lower bound of the PSD range; the ordinate axis indicates the upper bound of the PSD range. A bright red area is observed along the ordinate axis; the x-coordinate is equal to approximately 30 $\mu V^2/$ Hz. The brightest point has the following coordinates: 30 $\mu V^2/$ Hz and 700 $\mu V^2/$ Hz. This point corresponds to the PSD range from 30 to 700 $\mu V^2/$ Hz.

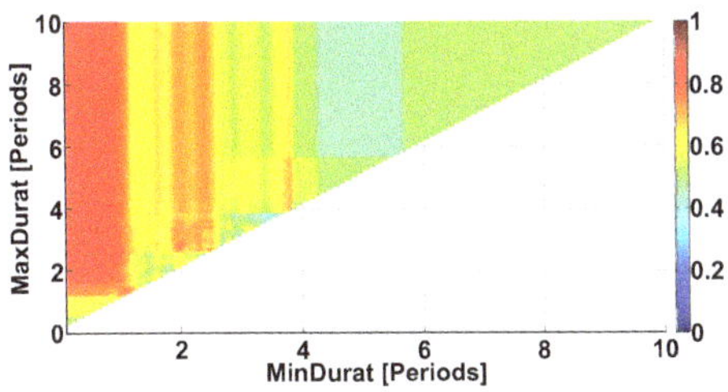

Figure 13. An example of the Duration AUC diagram. The frequency interval of considered wave trains is from 8 to 20 Hz. The abscissa axis indicates the lower bound of the range of durations in periods; the ordinate axis indicates the upper bound of the range of durations in periods. A bright red area is observed along the ordinate; the x-coordinate is equal to approximately 1 period. The brightest point has coordinates of 0.7 and 2.6 periods. This point corresponds to the range of durations from 0.7 to 2.6 periods.

Figure 14 demonstrates the Bandwidth AUC diagram. In the diagram, the wave trains belong to the frequency interval from 8 to 20 Hz. The diagram differs from Figure 10 because the frequency band is narrowed. The size and position of the red column have changed. A bright red area appears along the ordinate axis; the x-coordinate is equal to approximately 2 Hz. This means that the bandwidth of the wave trains characterizing PD differs sufficiently from the bandwidth of the other wave trains observed during the medical examination.

Now we are ready to implement the next iteration of the analysis. Once again, we have to choose which one of the four diagrams demonstrates the most striking regularities. A new constraint is applied to the wave train parameters based on this diagram. Let us choose the Power spectral density AUC diagram at this stage. We apply the following constraint based on this diagram: PSD no less than 30 $\mu V^2/$ Hz. Let us recalculate the other three AUC diagrams (frequency, duration, and bandwidth), taking into account two constraints: a frequency from 8 to 20 Hz and a PSD no less than 30 $\mu V^2/$ Hz.

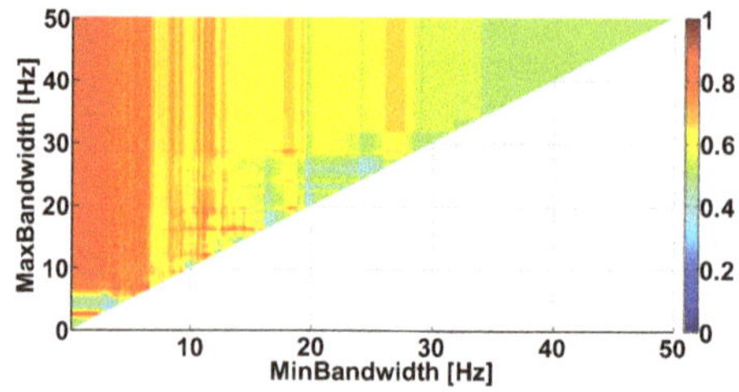

Figure 14. An example of the Bandwidth AUC diagram. The frequency interval of the considered wave trains is from 8 to 20 Hz. The abscissa axis indicates the lower bound of the frequency bandwidth; the ordinate axis indicates the upper bound of the frequency bandwidth. The bright red column corresponds to the frequency bandwidth values of the wave trains that are typical for PD patients.

Figure 15 demonstrates the Frequency AUC diagram with a PSD that is no less than 30 $\mu V^2 /$ Hz. Note that the frequency constraint needs to be refined using this diagram. The Frequency AUC diagram (see Figure 15) has changed substantially in comparison with that of Figure 7. The red areas in Figure 15 are brighter and larger. The blue areas have disappeared. Thus, the applied constraints made the differences between the groups of subjects more contrasting. The diagram demonstrates the strongest differences between the patients and control subjects in the frequency range from 8 to 20 Hz; the AUC value is equal to 0.93. Note that the disappearing blue area may correspond to another statistical regularity that differentiates the groups of subjects; however, this regularity is out of the scope of this paper.

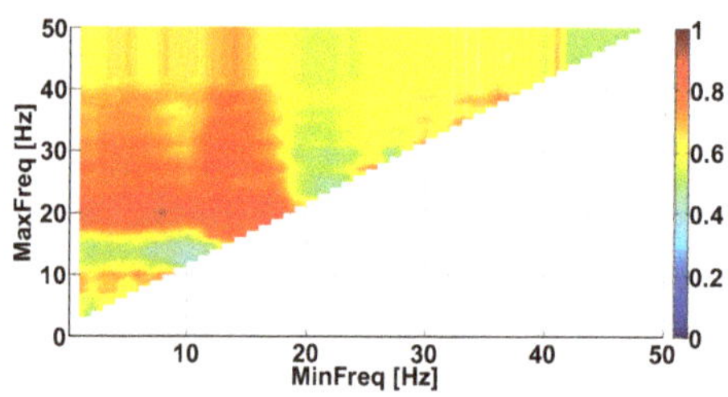

Figure 15. An example of the Frequency AUC diagram. The PSD of the considered wave trains is no less than 30 $\mu V^2 /$ Hz. The abscissa axis indicates the lower bound of the frequency range; the ordinate axis indicates the upper bound of the frequency range. A bright red area is observed in the frequency range along the abscissa; the y-coordinate is equal to approximately 20 Hz. The brightest red point has coordinates: 8 Hz and 20 Hz.

In Figure 16, the Duration AUC diagram is demonstrated. In this diagram, the following constraints are applied to the wave train parameters: a frequency from 8 to 20 Hz and a PSD no less than 30 $\mu V^2 /$ Hz. Figure 16 differs slightly from Figure 13. The red areas on the diagram became more intense. An intense red area appears inside the red region. The intense red area has the following coordinates: x-coordinates from 0 to 0.6 periods and y-coordinates from 3.6 periods and more. This diagram can be interpreted in the following way. The duration of most wave trains typical for the PD patients is equal to approximately 1 period. However, there are shorter and longer wave trains as well. Thus, the better recognition of PD patients is obtained when considering the wave trains in a wider range from 0.5 to 4 periods. The AUC value in the detected intense red area reaches 0.93, which is sufficient for the high-quality recognition of PD patients.

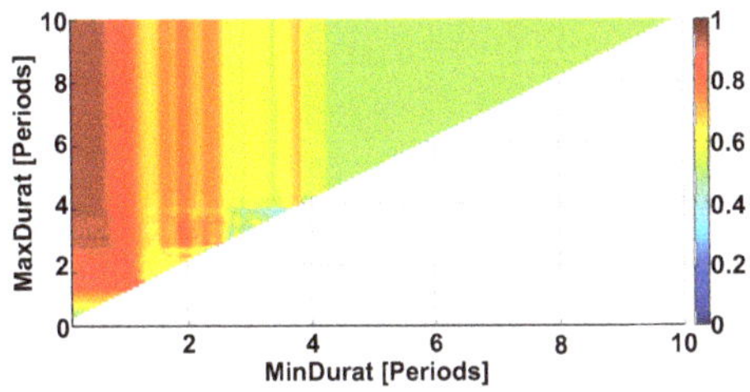

Figure 16. An example of the Duration AUC diagram. The following constraints are applied to the wave train parameters: a frequency from 8 to 20 Hz and a PSD no less than 30 $\mu V^2/$ Hz. The abscissa axis indicates the lower bound of the range of the duration in periods; the ordinate axis indicates the upper bound of the range of the duration in periods. An intense red area appears inside the red region. The intense red area has the following coordinates: x-coordinates from 0 to 0.6 periods and y-coordinates from 3.6 periods and more.

Figure 17 demonstrates the Bandwidth AUC diagram for wave trains that have a frequency from 8 to 20 Hz and a PSD no less than 30 $\mu V^2/$ Hz. The diagram differs from Figure 14 because of the constraints applied to the wave train parameters. The red areas on the diagram became more intense, while the multidirectional effects are still observed as the red and blue areas present on the diagram. This means that we cannot confidently determine the characteristic bandwidth of the wave trains typical for PD based on the available dataset. We can only conclude that the frequency bandwidth of the wave trains belongs to a wide interval; the value of the bandwidth can rise to ~28 Hz.

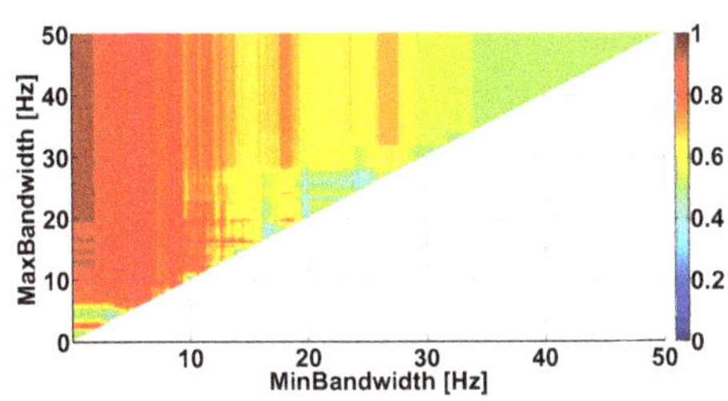

Figure 17. An example of the Bandwidth AUC diagram. The following constraints are applied to the wave train parameters: a frequency from 8 to 20 Hz and a PSD no less than 30 $\mu V^2/$ Hz. The abscissa axis indicates the lower bound of the bandwidth; the ordinate axis indicates the upper bound of the bandwidth. The diagram demonstrates multidirectional effects.

At the current step of the analysis, further iterations of the analysis do not change the Frequency AUC diagram (Figure 15), Power spectral density AUC diagram (Figure 12), Durations AUC diagram (Figure 16), and Bandwidth AUC diagram (Figure 17). Further detailing of the duration and bandwidth of the wave trains does not improve the AUC values. Thus, the iterative process of fitting the wave train characteristics typical for PD patients can be completed at this point. It was determined that the wave trains that help to distinguish the PD patients and healthy subjects have the following attributes: a frequency band from 8 to 20 Hz, a maximum PSD no less than 30 $\mu V^2/$ Hz, a duration from 0.5 to 4 periods, and a bandwidth from 1 to 28 Hz. The Mann–Whitney test confirms that a statistically significant difference between the numbers of wave trains with the given attributes is observed in PD patients and healthy volunteers ($p \leq 0.0011$).

In the general case, the iterative process of refinement of the wave train parameters is to be completed in the following situations:

1. Any further restrictions on the parameters of the wave trains do not change the AUC diagrams. This means that the further refinement of the wave train parameters makes no sense.
2. The refinement of the wave train parameters worsens the AUC values sufficiently in the AUC diagrams. This means that the investigated ranges of the wave train parameters became too narrowed; the number of wave trains considered in the AUC diagrams is too small. Theoretically speaking, in this case the refinement of the wave train parameters could be continued. However, the available dataset is not sufficient for this. The investigation of the wave train parameters could be continued if the number of subjects and/or the duration of EMG records are sufficiently increased.

Note that the initial AUC diagrams (Figures 7–10) had several red and blue areas that serve as a starting point for the iterative refinement of the wave train parameters. In the considered example, only one wave train type observed in the right non-tremor arm of the PD patients with a tremor in the left arm was investigated. The results of the analysis of other regularities observed in the dataset are given in Tables 1 and 2. In particular, we analyzed the EMG signals in the left arm of the left-hand-tremor PD patients, the EMG signals in the left non-tremor arm of the right-hand-tremor PD patients, and the EMG signals in the right arm of the right-hand-tremor PD patients. Thus, we analyzed both the non-tremor arms and arms with trembling hyperkinesis of the PD patients. Table 1 contains the results of the iterative analysis of the wave train parameters in the extensor muscles. Table 2 contains the results of the iterative analysis of the wave train parameters in the flexor muscles.

Table 1. The characteristics of the wave trains in the extensor muscles.

Investigated Regularity	**Frequency, Hz**	**PSD, $\mu V^2/Hz$**	**Duration, Periods**	**Bandwidth, Hz**	**AUC**	p
A red area. The right non-tremor arm in the left-hand-tremor PD patients.	8–20	≥30	0.5–4	1–28	0.93	0.0011
A red area. The left non-tremor arm in the right-hand-tremor PD patients.	2–9	any	0.8–2.3	any	0.87	0.0033
A blue area. The left tremor arm in the left-hand-tremor PD patients.	1–50	any	≥1	≥3	0	≤0.001
A blue area. The right tremor arm in the right-hand-tremor PD patients.	6–33	any	≥0.5	≥3.5	0.02	≤0.001
A red area. The left tremor arm in the left-hand-tremor PD patients.	3–7	≥11	≥1.5	any	1	≤0.001
A red area. The right tremor arm in the right-hand-tremor PD patients.	4–8	≥103	≥1.3	any	1	≤0.0001

2.5. 3D AUC Diagrams

Wave train electrical activity analysis based on 2D AUC diagrams requires considering a large number of combinations of the upper and lower bounds of parameter ranges. It is useful to check what positions of the 2D AUC diagram correspond to statistically significant differences between the numbers of wave trains in the groups of subjects compared. We implement this check using the Mann–Whitney nonparametric test; however, the multiple comparisons problem arises. The essence of the multiple comparisons problem is that the statistical test may give a sufficient first-type error when a large number of ranges of parameters are checked in the 2D AUC diagram. The simplest way to solve the multiple comparisons problem is to apply Bonferroni correction [40]. The value of the Bonferroni correction depends on the number of cells in the 2D AUC diagram. We call the number of rows/columns in the 2D AUC diagram the resolution R of the diagram. If the resolution of the AUC diagram is high, the value of the Bonferroni correction also becomes high. The Bonferroni correction changes (2) the alpha level of the Mann–Whitney test using the correction coefficient C:

$$\alpha_B = 1 - (1 - \alpha_0)^{1/C} \tag{2}$$

where $\alpha_0 = 0.05$. Let the C correction coefficient (3) be equal to the number of cells in the upper triangle of the 2D AUC diagram, including the number of cells on the diagonal of the AUC diagram:

$$C = R(R+1)/2 \tag{3}$$

Table 2. The characteristics of the wave trains in the flexor muscles.

Investigated Regularity	Frequency, Hz	PSD, $\mu V^2/Hz$	Duration, Periods	Bandwidth, Hz	AUC	p
A red area. The right non-tremor arm in the left-hand-tremor PD patients.	5–13	0–50	any	3.1–3.8	0.92	0.0017
A red area. The left non-tremor arm in the right-hand-tremor PD patients.	2–16	any	1.4–2.1	any	0.8	0.0161
A blue area. The left tremor arm in the left-hand-tremor PD patients.	1–39	any	$\geq$0.5	$\geq$2.5	0	$\leq$0.001
A blue area. The right tremor arm in the right-hand-tremor PD patients.	24–34	any	any	any	0.07	$\leq$0.001
A red area. The left tremor arm in the left-hand-tremor PD patients.	4–7	$\geq$4	$\geq$1.2	any	1	$\leq$0.001
A red area. The right tremor arm in the right-hand-tremor PD patients.	2–8	$\geq$2	$\geq$2.3	any	0.85	0.0037

The value of the correction coefficient depends on the R resolution according to the quadratic law. Therefore, the probability of detecting statistically significant differences in the 2D AUC diagram sufficiently decreases when the resolution is high. On the other hand, if the resolution R is low, the detailing of the 2D AUC diagram is reduced, and one can miss certain regularities present in the dataset. Thus, it is necessary to find a compromise between the level of detail in the 2D AUC diagram and the value of the Bonferroni correction to reveal interesting statistically significant differences between the groups of subjects. We developed a special type of AUC diagram, named a 3D AUC diagram, to solve this problem. The 3D AUC diagram is a generalization of the 2D AUC diagram. The abscissa and ordinate axes on the 3D AUC diagram indicate the values of the upper and lower bounds of the range of the considered wave train parameter, as in the 2D AUC diagram. However, the applicate axis indicates the R resolution of the AUC diagram. Thus, the horizontal slice of the 3D AUC diagram is a case of the 2D AUC diagram. In the 3D AUC diagrams, only the points that correspond to statistically significant differences between the numbers of wave trains in the groups of subjects are displayed; Bonferroni correction (2), which depends on the R resolution, is taken into account when the statistical significance is checked.

Let us consider an example of the 3D AUC diagram (see Figure 18). This diagram is a form of Frequency 3D AUC diagram; it demonstrates statistically significant differences between the numbers of wave trains in the groups of subjects when considering various ranges of frequencies. The number of wave trains in the left arm of the left-hand-tremor PD patients and the number of wave trains in the left arm of the healthy subjects are compared. The values of the wave train parameter ranges are given in Table 1, line 5. The 3D AUC diagram demonstrates a 3D isosurface that corresponds to various $p \leq \alpha_B$. The upper plateau of the isosurface corresponds to the resolution of 23. The coordinates of the plateau are from 3.1 to 3.4 Hz along the abscissa axis and from 6.4 to 7.5 Hz along the ordinate axis. The horizontal slice area of the isosurface grows and then decreases when the resolution decreases. This is because the Bonferroni correction is softened; however, the degree of detail in the diagram also decreases. The top point of the isosurface is of interest because it reveals statistically significant differences in the dataset with the greatest degree of detail. In the example being considered, the 3D AUC diagram confirms that there are statistically significant differences between the numbers of wave trains in the groups of subjects in the frequency range from about 3 to 7 Hz.

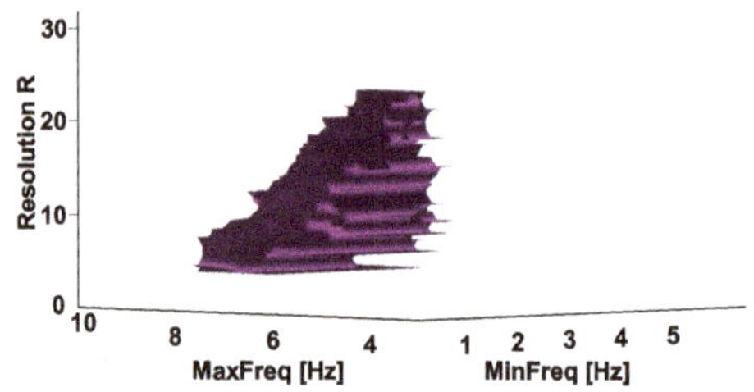

Figure 18. The isosurface $p \leq \alpha_B$. Left-hand-tremor PD patients; extensor muscle; frequencies from 1 to 10 Hz. The abscissa axis indicates the lower bound of the frequency range, the ordinate axis indicates the upper bound of the frequency range, and the applicate axis indicates the R resolution.

Figure 19 demonstrates the Frequency 2D AUC diagram that corresponds to the horizontal slice of the 3D AUC diagram (Figure 18) at the resolution R = 15. Figure 19 includes only the points that correspond to statistically significant differences between the groups of subjects. The diagram demonstrates that the slice has an irregular shape with coordinates from 2.28 to 4.2 Hz along the abscissa axis and from 6.12 to 8 Hz along the ordinate axis. The AUC values in the 2D AUC diagram are higher than 0.98; there is an area with high AUC values up to 1 in the central part of the diagram. The coordinates of this area are from 2.3 to 4 Hz along the abscissa axis and from 6.12 to 8 Hz along the ordinate axis. This example demonstrates that one can obtain the best AUC values when choosing the optimal level of detail in the AUC diagram; this allows high accuracy when diagnosing PD.

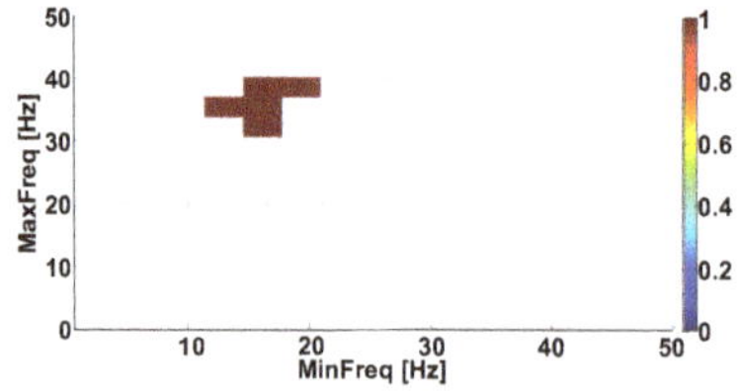

Figure 19. The Frequency 2D AUC diagram that corresponds to the horizontal slice of the 3D AUC diagram (see Figure 18) at the resolution R = 15. This Frequency 2D AUC diagram includes only the points that correspond to statistically significant differences between the groups of subjects. The abscissa axis indicates the lower bound of the frequency range; the ordinate axis indicates the upper bound of the frequency range.

3. Group Data Analysis

The analysis of 2D and 3D AUC diagrams that compare PD patients and healthy volunteers revealed several types of wave train electrical activity that are a distinctive feature of PD patients. Let us consider some regularities discovered in the dataset to clarify what neurophysiological mechanisms may control these types of electrical activity.

Let us consider the scatter plot (see Figure 20) that demonstrates the number of wave trains detected in two frequency intervals in the extensor muscle of the arms with trembling hyperkinesis in the PD patients: the *physiological* tremor frequency interval and the *Parkinsonian* tremor frequency interval. The abscissa axis corresponds to lines 3 and 4 in Table 1. The ordinate axis corresponds to lines 5 and 6 in Table 1. We have included the characteristics of the healthy subjects in the scatter plot for comparison. Each point in the scatter plot corresponds to one subject. The PD patients are indicated by the red color; the control subjects are indicated by the green color.

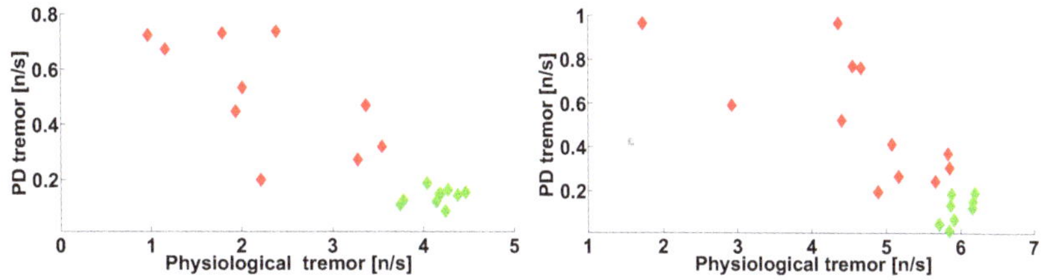

Figure 20. The scatter plot of the numbers of wave trains per second detected in the extensor muscle in the tremor arms of the PD patients. The abscissa axis indicates the wave train numbers corresponding to the physiological tremor; the ordinate axis indicates the Parkinsonian tremor. The PD patients are indicated by the red color. The control subjects are indicated by the green color. On the left, the left hand of the subjects are shown. On the right, the right hand of the subjects are shown.

The point cloud corresponding to the PD patients has an elongated shape (see Figure 20). The point cloud corresponding to the healthy subjects is situated in the lower right corner of the scatter plot. Note that the point clouds can be easily separated. The PD patient point cloud is perpendicular to the diagonal of the scatter plot, which is evidence of the negative correlation between the wave train numbers corresponding to physiological and Parkinsonian tremors.

The check of the correlation confirmed that the correlation between the number of wave trains corresponding to the physiological and Parkinsonian tremors is statistically significant in the right-hand-tremor PD patients (see Figure 20, on the right); the correlation coefficient is -0.6885, the first-type error probability is 0.0133, the Spearman's rank correlation coefficient is -0.7483, and the first-type error probability in the Spearman's nonparametric test is 0.0074.

The investigation of these types of wave trains in the left-hand-tremor PD patients reveals a statistical trend (see Figure 20, on the left) that confirms the regularity discovered in the right-hand-tremor PD patients (see Figure 20, on the right). Note that the point cloud corresponding to the left-hand-tremor PD patients has approximately the same shape as that of the right-hand-tremor PD patient point cloud; however, the correlation coefficient is -0.6349, the first-type error probability is 0.0486, the Spearman's rank correlation coefficient is -0.4909, and the first-type error probability in the Spearman's nonparametric test is 0.1544. A significant correlation is not detected in the healthy subject point clouds (see Figure 20).

Let us compare the number of wave trains detected in the *physiological* tremor frequency band in the extensor muscle of the *non-tremor* hand of the PD patients with the number of wave trains detected in the *Parkinsonian* tremor frequency band in the extensor muscle of the *tremor* hand of the PD patients (see Figure 21). The abscissa axis of the scatter plot corresponds to lines 1 and 2 in Table 1. The ordinate axis corresponds to lines 5 and

6 in Table 1. We included the characteristics of the healthy subjects in the scatter plot for comparison. Each point corresponds to one subject. The PD patients are indicated by the red color; the control subjects are indicated by the green color.

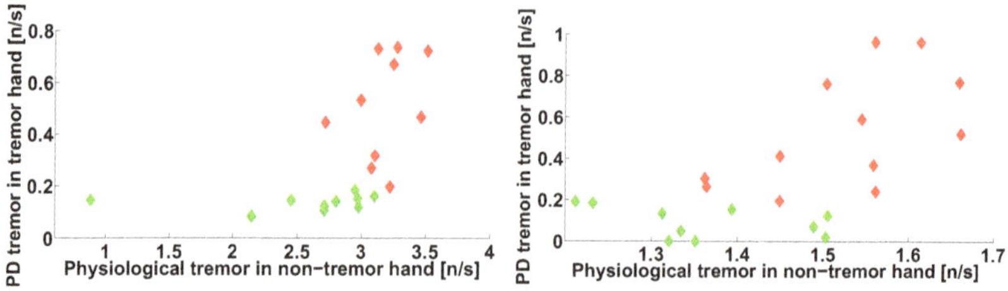

Figure 21. The scatter plot of the numbers of wave trains per second detected in the extensor muscle of the tremor and non-tremor arms of PD patients. The abscissa axis indicates the wave train numbers corresponding to the physiological tremor in the non-tremor hand; the ordinate axis indicates the wave train numbers corresponding to the Parkinsonian tremor in the tremor hand. The PD patients are indicated by the red color; the control subjects are indicated by the green color. On the left, the left-hand-tremor PD patients are shown. On the right, the right-hand-tremor PD patients are shown.

The scatter plot demonstrates that the point cloud corresponding to the healthy subjects is situated along the abscissa axis. The point cloud corresponding to the PD patients has an elongated shape and is located under the healthy subject point cloud (Figure 21). Note that the point clouds can be easily separated.

Let us investigate the correlation between the number of the wave trains detected in the physiological tremor frequency band in the extensor muscle of the left non-tremor hand of the PD patients with the number of the wave trains detected in the Parkinsonian tremor frequency band in the extensor muscle of the right tremor hand of the PD patients (Figure 21, on the right). The correlation coefficient is 0.5775, the first-type error probability is 0.0493, the Spearman's correlation coefficient is 0.5315, and the first-type error probability in the Spearman's nonparametric test is 0.0793. Thus, a statistical trend is observed.

Let us investigate the correlation between the numbers of wave trains detected in the left-hand-tremor PD patients (Figure 21, on the left). The correlation coefficient is 0.3105 and the probability of the first-type error is 0.3826—that is, the correlation is not significant. The Spearman's correlation coefficient is 0.4424 and the probability of the first-type error in the Spearman's nonparametric test is 0.2042. A significant correlation is also not detected in the healthy subject point clouds (see Figure 21).

Let us compare the number of wave trains detected in the extensor muscle in the physiological tremor frequency band of the tremor hand of the PD patients with the number of wave trains detected in the extensor muscle in the physiological tremor frequency band of the non-tremor hand of the PD patients (see Figure 22). The abscissa axis corresponds to lines 3 and 4 in Table 1. The ordinate axis corresponds to lines 1 and 2 in Table 1. We included the characteristics of the healthy subjects in the scatter plot for comparison. Each point in the scatter plot corresponds to one subject. The PD patients are indicated by the red color; the control subjects are indicated by the green color.

Figure 22 demonstrates that the point clouds corresponding to the PD patients and healthy subjects can be easily separated. The healthy subject point cloud is located to the right of the PD patient point cloud.

The correlation between the number of the wave trains in the tremor and non-tremor arms of the PD patients is not significant (Figure 22). However, a statistical trend is observed in the right-hand tremor PD patients (Figure 22, on the right). In the right-hand-tremor PD patients, the correlation coefficient is −0.4488, the first-type error probability is 0.1434, the Spearman's correlation coefficient is −0.5455, and the first-type error probability in the Spearman's nonparametric test is 0.0707.

In the left-hand-tremor PD patients (Figure 22, on the left), the correlation coefficient is −0.0872, the first-type error probability is 0.8107, the Spearman's correlation coefficient is −0.1515, and the first-type error probability in the Spearman's nonparametric test is 0.6818. No significant correlation was detected in the control subject point clouds (see Figure 22).

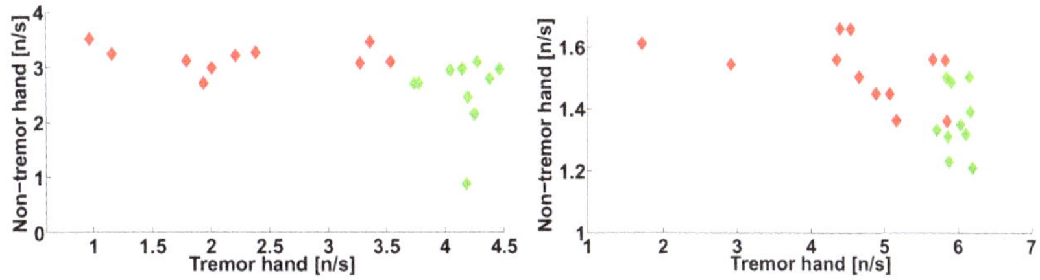

Figure 22. The scatter plot of the numbers of wave trains per second detected in the physiological tremor frequency band in the extensor muscle of the tremor and non-tremor arms of the PD patients. The abscissa axis indicates the wave train numbers corresponding to the physiological tremor in the tremor hand; the ordinate axis indicates the wave train numbers corresponding to the physiological tremor in the non-tremor hand. The PD patients are indicated by the red color; the control subjects are indicated by the green color. On the left, the left-hand-tremor PD patients are shown. On the right, the right-hand-tremor PD patients are shown.

The analysis of the correlation between the numbers of wave trains in the physiological tremor frequency band in the tremor arms and the age of the PD patients (see Figure 23) revealed a statistically significant correlation only in the left-hand-tremor PD patients (Figure 23, on the left). The correlation coefficient is −0.7246, the first-type error probability is 0.0178, the Spearman's correlation coefficient is −0.7356, and the first-type error probability in the Spearman's nonparametric test is 0.0153.

Note that the correlation is not observed in the right-hand-tremor PD patients (Figure 23, on the right). The correlation coefficient is −0.0512, the first-type error probability is 0.8745, the Spearman's correlation coefficient is −0.1399, and the first-type error probability in the Spearman's nonparametric test is 0.6672.

No significant correlation was found between the other wave train parameters in Table 1 and age. The correlation analysis of the wave train parameters of the flexor muscle (see Table 2) revealed no significant correlation.

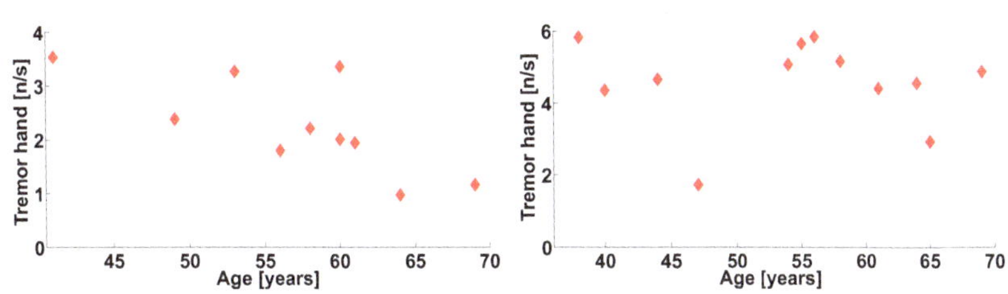

Figure 23. The scatter plot of the numbers of wave trains per second detected in the physiological tremor frequency band in the extensor muscle of the tremor arms of the PD patients. The scatter plot demonstrates the relation between the wave train number and the age of the patients. On the left, the left hands of the patients are shown. On the right, the right hands of the patients are shown.

4. Discussion

The obtained results can be explained by the neurophysiological mechanisms of the tremor maintenance known nowadays.

The negative correlation between the numbers of wave trains in the frequency ranges corresponding to the Parkinsonian and physiological tremors may indicate the mutual negative influence of some neurophysiological mechanisms underlying both tremor types

(Figure 20). It is known that the mechanism of physiological tremor (in the frequency range from 8 to 12 Hz) is maintained in the cerebello-thalamo-premotor-motor cortical network [41]. The Parkinsonian tremor mechanism is maintained in the cerebello-thalamic pathways [42]. Thus, the thalamus is involved in the mechanisms of both types of tremor. We can assume that the capabilities of the thalamus are limited and that both tremor mechanisms compete for the participation of the thalamus. We propose that Parkinsonian tremor is more successful than physiological one in this competition. Therefore, as Parkinsonian tremor intensifies, physiological tremor is suppressed.

We did not observe a correlation between the Parkinsonian tremor in the tremor arms and the physiological tremor in the non-tremor arms (Figure 21). However, there is a statistical trend in the right-hand-tremor PD patients. Thus, we cannot prove or deny the existence of a relation between the tremor in the tremor arms and the non-tremor arms. The relationship between the Parkinsonian tremor in the tremor arms and the physiological tremor in the non-tremor arms requires more investigations to be carried out on a larger group of corresponding patients.

No correlation was observed between the numbers of wave trains in the frequency range of the physiological tremor in the tremor arms and the numbers of the wave trains in the frequency range of physiological tremor in the non-tremor arms in the PD patients (see Figure 22). Note that there is a separation of the wave train clouds in the arms of the healthy subjects and the PD patients (Figure 22). This separation of the clouds can be explained by the fact that the physiological tremor in the healthy subjects and the physiological tremor in the PD patients have different mechanisms. It is known that the physiological tremor in healthy subjects occurs mainly due to homeostatic peripheral movements of muscles and joints to maintain posture or the movement of the limbs [43]. The resting tremor in the PD patients is associated with increased activity in the cerebello-thalamic pathways [44]. In PD patients, the increased activity of the cerebello-thalamic pathways contributes to the physiological tremor of the non-tremor arm. Thus, the Parkinsonian tremor contributes to the movement of the non-tremor arm. The cumulative tremor in the non-tremor arms of the PD patients can be named the preclinical tremor. Different characteristics of the preclinical tremor in the non-tremor limbs of the PD patients and physiological tremor in the healthy subject limbs observed during the medical examination can be used for the early diagnosis of PD.

The analysis of the correlation between the numbers of wave trains in the frequency range of the physiological tremor and the age of the PD patients revealed substantial differences between the patients with the right- and left-sided-debut of PD (Figure 23). PD usually progresses with age. Increasing the age can be considered to be a factor contributing to the progression of PD [45]. It has been suggested that the dominant and non-dominant arms may have different specializations [46,47]. Motor lateralization hypothesis [48] suggests that when the right hand is dominant, it is specialized in the predictive control of the dynamics of smooth and efficient movements. In contrast, the non-dominant left hand is specialized in the resilience to unforeseen disturbances [49,50]. Thus, there is a piece of evidence that different mechanisms control the dominant and non-dominant hands. As the non-dominant hand specializes in resistance to unforeseen disturbances, we assume that this mechanism increases with PD progression. It is possible that, in the preclinical stage, the enhancement of the stability function is more substantial than the process associated with the development of PD and the increase in the tremor.

The idea of the 3D AUC diagrams described in this paper is based on multiscale data analysis. We changed the resolution of the diagrams to look for Bonferroni correction that enables to observe statistically significant differences between the subject groups. The investigation of the multiple comparisons problem [40,51–54] is currently an important topic in biomedicine. A large number of methods have been developed to solve the multiple comparisons problem. These methods are based on the analysis of the family-wise error rate (FWE), the false discovery rate (FDR), the application of random field theory (RFT) [54–56], the permutation method [52], etc. Unfortunately, most of these methods do not consider the connection between the multiple comparisons problem and the problem of the multiscale analysis of biomedical data. Meanwhile, these two problems are closely related both from the point of view of the mathematical apparatus and from the point of view of the practical application of the methods. The problem with multiscale analysis is that biomedical data may contain patterns that appear on various, previously unknown scales of images/signals. It is necessary to investigate data in a certain space of scales to detect such patterns. However, different scales correspond to different numbers of multiple comparisons and, therefore, suggest a different level of statistical correction for the number of multiple comparisons. Moreover, the investigation of the data on the multiple scales implies multiple comparisons and, therefore, may require the application of additional correction for the multiple comparisons. This problem, in particular, was considered in the random field theory, and an approach to analyzing the data based on the space of scales (the scale space approach) was proposed [57]. Thus, a possible direction in the development of the method for the analysis of wave train electrical activity is the use of more accurate corrections for the multiple comparisons instead of the standard Bonferroni correction.

5. Conclusions

The method developed for analyzing the wave train electrical activity is a universal method for exploratory data analysis and can be applied to other types of biomedical signals [58–70]. In particular, we demonstrated that the statistical analysis of some characteristics of wave trains in EEG can identify features of the preclinical stage of PD [58–61]. It was found that the number of wave trains in wavelet spectrograms in the beta frequency range in first-stage PD patients was significantly reduced in comparison with the control subjects [71–73]. Statistically significant differences in the signals of the accelerometer [68] and EMG [63] in patients with PD, essential tremor (ET), and healthy volunteers in the 0.5–4 Hz low-frequency range were found using 2D AUC diagrams. Note that this frequency range has remained unexplored for a long time. The revealed regularities can be used for the differential diagnostics of PD and ET. The problem of the early and differential diagnostics of PD and ET by means of wave train analysis was considered in paper [69]. The source code of the Matlab program used for the analysis of EMG data has been published in GitHub [74]. The method used for the differential diagnostics of the essential tremor disease and early and first stages of Parkinson's disease based on the wave train electrical activity analysis was patented [75].

6. Patents

Sushkova O.S., Morozov A.A., Gabova A.V., Karabanov A.V. Patent number RU 2741233 C1. Russian Federation. Method for differential diagnosing of essential tremor and early and first stages of Parkinson's disease using wave train activity analysis of muscles. Published: 22.01.2021 Bul. No. 3. Application: 2020118098, 24.04.2020. Starting date of the patent validity period: 04/24/2020. Date of registration: 01/22/2021. Application date: 04/24/2020. Address for correspondence: 125009, Moscow, Mokhovaya Street, 11/7, Kotelnikov IRE RAS, Patent Department.

Author Contributions: Conceptualization, O.S.S. and A.A.M.; methodology, O.S.S., A.A.M., and A.V.G.; software, O.S.S. and A.A.M.; validation, O.S.S., A.A.M., A.V.G., and A.V.K.; formal analysis, O.S.S. and A.A.M.; investigation, O.S.S., A.A.M., A.V.G., and A.V.K.; resources, A.V.K.; data acquisition, O.S.S. and A.V.G.; data curation, O.S.S.; writing—original draft preparation, O.S.S., A.A.M., and A.V.G.; writing—review and editing, O.S.S., A.A.M., A.V.G., and A.V.K.; visualization, O.S.S.; supervision, O.S.S., A.A.M., and S.N.I.; project administration, O.S.S.; funding acquisition, O.S.S. All authors have read and agreed to the published version of the manuscript.

Funding: The work was carried out within the framework of the state task.

Institutional Review Board Statement: The study was conducted according to the guidelines of the Declaration of Helsinki, and approved by the Ethics Committee of FSBI "Research Center of Neurology".

Informed Consent Statement: Informed consent was obtained from all subjects involved in the study.

Data Availability Statement: The data presented in this study are available on request from the corresponding author. The clinical data are not publicly available due to the ethical policy of the institute.

Acknowledgments: The authors are grateful to O. E. Khutorskaya, whose ideas were a starting point for this research. We are grateful to Yu. V. Obukhov, who proposed the idea of the analysis of the local maxima in the wavelet spectrograms. We are grateful to I. A. Kershner for the help in the data acquisition and his remarks on the initial version of the paper. We are grateful to E. M. Timanin, who is the principal developer of the wonderful accelerometer used in our research.

Conflicts of Interest: The authors declare no conflicts of interest. The funders had no role in the design of the study; in the collection, analysis, or interpretation of data; in the writing of the manuscript; or in the decision to publish the results.

References

1. Andreeva, Y.; Khutorskaya, O. EMGs spectral analysis method for the objective diagnosis of different clinical forms of Parkinson's disease. *J. Electromyogr. Clin. Neurophysiol.* **1996**, *36*, 187–192.
2. Khutorskaya, O.E. Method for Early and Differential Electromyographic Diagnosis of the Main Symptoms of Parkinson's Disease. Russian Federation Patent Application No. 2626557, 28 July 2017.
3. Lukhanina, E.; Kapoustina, M.; Karaban, I. A quantitative surface electromyogram analysis for diagnosis and therapy control in Parkinson's disease. *Park. Relat. Disord.* **2000**, *6*, 77–86. [CrossRef]
4. Fattorini, L.; Felici, F.; Filligoi, G.; Traballesi, M.; Farina, D. Influence of high motor unit synchronization levels on non-linear and spectral variables of the surface EMG. *J. Neurosci. Methods* **2005**, *143*, 133–139. [CrossRef] [PubMed]
5. Robichaud, J.A.; Pfann, K.D.; Vaillancourt, D.E.; Comella, C.L.; Corcos, D.M. Force control and disease severity in Parkinson's disease. *Mov. Disord.* **2005**, *20*, 441–450. [CrossRef] [PubMed]
6. Yang, W.; Yang, D.; Liu, Y.; Liu, H. EMG pattern recognition using convolutional neural network with different scale signal/spectra input. *Int. J. Humanoid Robot.* **2019**, *16*, 1950013. [CrossRef]
7. Sung, P.S.; Zurcher, U.; Kaufman, M. Comparison of spectral and entropic measures for surface electromyography time series: A pilot study. *J. Rehabil. Res. Dev.* **2007**, *44*, 599–609. [CrossRef] [PubMed]
8. Rissanen, S.; Kankaanpää, M.; Tarvainen, M.P.; Nuutinen, J.; Tarkka, I.M.; Airaksinen, O.; Karjalainen, P.A. Analysis of surface EMG signal morphology in Parkinson's disease. *Physiol. Meas.* **2007**, *28*, 1507. [CrossRef]
9. Pfann, K.D.; Buchman, A.S.; Comella, C.L.; Corcos, D.M. Control of movement distance in Parkinson's disease. *Mov. Disord.* **2001**, *16*, 1048–1065. [CrossRef]
10. Zeng, Y.; Yang, J.; Peng, C.; Yin, Y. Evolving Gaussian process autoregression based learning of human motion intent using improved energy kernel method of EMG. *IEEE Trans. Biomed. Eng.* **2019**, *66*, 2556–2565. [CrossRef]
11. Al-Timemy, A.H.; Bugmann, G.; Escudero, J.; Outram, N. Classification of finger movements for the dexterous hand prosthesis control with surface electromyography. *IEEE J. Biomed. Health Inform.* **2013**, *17*, 608–618. [CrossRef]
12. Xiong, D.; Zhang, D.; Zhao, X.; Zhao, Y. Deep learning for EMG-based human-machine interaction: A review. *IEEE/CAA J. Autom. Sin.* **2021**, *8*, 512–533. [CrossRef]
13. Parajuli, N.; Sreenivasan, N.; Bifulco, P.; Cesarelli, M.; Savino, S.; Niola, V.; Esposito, D.; Hamilton, T.J.; Naik, G.R.; Gunawardana, U.; et al. Real-time EMG based pattern recognition control for hand prostheses: A review on existing methods, challenges and future implementation. *Sensors* **2019**, *19*, 4596. [CrossRef]
14. Reaz, M.B.I.; Hussain, M.S.; Mohd-Yasin, F. Techniques of EMG signal analysis: Detection, processing, classification and applications. *Biol. Proced. Online* **2006**, *8*, 11–35. [CrossRef] [PubMed]
15. Dogan, S.; Tuncer, T. A novel statistical decimal pattern-based surface electromyogram signal classification method using tunable q-factor wavelet transform. *Soft Comput.* **2021**, *25*, 1085–1098. [CrossRef]
16. Subba, R.; Bhoi, A.K. Feature Extraction and Classification Between Control and Parkinson's Using EMG Signal. In *Cognitive Informatics and Soft Computing*; Springer: Berlin/Heidelberg, Germany, 2020; pp. 45–52.
17. Meigal, A.; Rissanen, S.; Tarvainen, M.; Airaksinen, O.; Kankaanpää, M.; Karjalainen, P. Non-linear EMG parameters for differential and early diagnostics of Parkinson's disease Front. *Neurol* **2013**, *4*, 135.
18. Namazi, H. Decoding of hand gestures by fractal analysis of electromyography (EMG) signal. *Fractals* **2019**, *27*, 1950022. [CrossRef]
19. Ivanova, E.; Fedin, P.; Brutyan, A.; Ivanova-Smolenskaya, I.; Illarioshkin, S. Analysis of tremor activity of antagonist muscles in essential tremor and Parkinson's diseases. *Nevrol. Zhurnal* **2014**, *19*, 11–18. [CrossRef]
20. Webber, C.; Marwan, N. *Recurrence Quantification Analysis–Theory and Best Practices*; Springer: Berlin/Heidelberg, Germany, 2015.
21. Farina, D.; Fattorini, L.; Felici, F.; Filligoi, G. Nonlinear surface EMG analysis to detect changes of motor unit conduction velocity and synchronization. *J. Appl. Physiol.* **2002**, *93*, 1753–1763. [CrossRef]
22. Buongiorno, D.; Cascarano, G.D.; De Feudis, I.; Brunetti, A.; Carnimeo, L.; Dimauro, G.; Bevilacqua, V. Deep learning for processing electromyographic signals: A taxonomy-based survey. *Neurocomputing* **2020**, *452*, 549–565. [CrossRef]
23. Rim, B.; Sung, N.J.; Min, S.; Hong, M. Deep learning in physiological signal data: A survey. *Sensors* **2020**, *20*, 969. [CrossRef] [PubMed]

24. Mahmud, M.; Kaiser, M.S.; Hussain, A.; Vassanelli, S. Applications of deep learning and reinforcement learning to biological data. *IEEE Trans. Neural Netw. Learn. Syst.* **2018**, *29*, 2063–2079. [CrossRef]
25. Mahmud, M.; Kaiser, M.S.; McGinnity, T.M.; Hussain, A. Deep learning in mining biological data. *Cogn. Comput.* **2021**, *13*, 1–33. [CrossRef] [PubMed]
26. Phinyomark, A.; Limsakul, C.; Phukpattaranont, P. Application of wavelet analysis in EMG feature extraction for pattern classification. *Meas. Sci. Rev.* **2011**, *11*, 45–52. [CrossRef]
27. Englehart, K.; Hudgin, B.; Parker, P.A. A wavelet-based continuous classification scheme for multifunction myoelectric control. *IEEE Trans. Biomed. Eng.* **2001**, *48*, 302–311. [CrossRef] [PubMed]
28. De Michele, G.; Sello, S.; Carboncini, M.C.; Rossi, B.; Strambi, S.K. Cross-correlation time-frequency analysis for multiple EMG signals in Parkinson's disease: A wavelet approach. *Med. Eng. Phys.* **2003**, *25*, 361–369. [CrossRef]
29. Valls-Solé, J.; Valldeoriola, F. Neurophysiological correlate of clinical signs in Parkinson's disease. *Clin. Neurophysiol.* **2002**, *113*, 792–805. [CrossRef]
30. Hallett, M.; Shahani, B.T.; Young, R.R. EMG analysis of stereotyped voluntary movements in man. *J. Neurol. Neurosurg. Psychiatry* **1975**, *38*, 1154–1162. [CrossRef]
31. Flament, D.; Vaillancourt, D.; Kempf, T.; Shannon, K.; Corcos, D. EMG remains fractionated in Parkinson's disease, despite practice-related improvements in performance. *Clin. Neurophysiol.* **2003**, *114*, 2385–2396. [CrossRef]
32. Robichaud, J.A.; Pfann, K.D.; Comella, C.L.; Corcos, D.M. Effect of medication on EMG patterns in individuals with Parkinson's disease. *Mov. Disord.* **2002**, *17*, 950–960. [CrossRef]
33. Robichaud, J.A.; Pfann, K.D.; Comella, C.L.; Brandabur, M.; Corcos, D.M. Greater impairment of extension movements as compared to flexion movements in Parkinson's disease. *Exp. Brain Res.* **2004**, *156*, 240–254.
34. Meigal, A.I.; Rissanen, S.; Tarvainen, M.; Karjalainen, P.; Iudina-Vassel, I.; Airaksinen, O.; Kankaanpää, M. Novel parameters of surface EMG in patients with Parkinson's disease and healthy young and old controls. *J. Electromyogr. Kinesiol.* **2009**, *19*, e206–e213. [CrossRef] [PubMed]
35. Rissanen, S.M.; Kankaanpää, M.; Meigal, A.; Tarvainen, M.P.; Nuutinen, J.; Tarkka, I.M.; Airaksinen, O.; Karjalainen, P.A. Surface EMG and acceleration signals in Parkinson's disease: Feature extraction and cluster analysis. *Med. Biol. Eng. Comput.* **2008**, *46*, 849–858. [CrossRef] [PubMed]
36. Oktay, A.B.; Kocer, A. Differential diagnosis of Parkinson and essential tremor with convolutional LSTM networks. *Biomed. Signal Process. Control* **2020**, *56*, 101683. [CrossRef]
37. Kim, H.B.; Lee, W.W.; Kim, A.; Lee, H.J.; Park, H.Y.; Jeon, H.S.; Kim, S.K.; Jeon, B.; Park, K.S. Wrist sensor-based tremor severity quantification in Parkinson's disease using convolutional neural network. *Comput. Biol. Med.* **2018**, *95*, 140–146. [CrossRef]
38. Ivanova, E.; Fedin, P.; Brutyan, A.; Ivanova-Smolenskaya, I.; Illarioshkin, S. Clinical and electrophysiological analysis of tremor in patients with essential tremor and Parkinson's disease. *Nevrol. Zhurnal* **2013**, *5*, 21–26. Available online: https://cyberleninka.ru/article/n/kliniko-elektrofiziologicheskiy-analiz-drozhatelnogo-giperkineza-pri-essentsialnom-tremore-i-bolezni-parkinsona/viewer (accessed on 7 July 2021).
39. Illarioshkin, S.N.; Ivanova-Smolenskaya, I.A. *Trembling Hyperkinesis*; Atmosfera: Moscow, Russia, 2011. (In Russian). Available online: https://www.ozon.ru/product/drozhatelnye-giperkinezy-rukovodstvo-dlya-vrachey-162350391/ (accessed on 7 July 2021).
40. Shaffer, J.P. Multiple hypothesis testing: A review. *Annu. Rev. Psychol.* **1995**, *46*, 561–584. [CrossRef]
41. Schnitzler, A.; Timmermann, L.; Gross, J. Physiological and pathological oscillatory networks in the human motor system. *J. Physiol. Paris* **2006**, *99*, 3–7. [CrossRef]
42. Deiber, M.P.; Pollak, P.; Passingham, R.; Landais, P.; Gervason, C.; Cinotti, L.; Friston, K.; Frackowiak, R.; Mauguière, F.; Benabid, A.L. Thalamic stimulation and suppression of parkinsonian tremor: Evidence of a cerebellar deactivation using positron emission tomography. *Brain* **1993**, *116*, 267–279. [CrossRef]
43. Elias, W.J.; Shah, B.B. Tremor. *JAMA* **2014**, *311*, 948–954. [CrossRef]
44. Helmich, R.C.; Janssen, M.J.; Oyen, W.J.; Bloem, B.R.; Toni, I. Pallidal dysfunction drives a cerebellothalamic circuit into Parkinson tremor. *Ann. Neurol.* **2011**, *69*, 269–281. [CrossRef]
45. Poewe, W. The natural history of Parkinson's disease. *J. Neurol.* **2006**, *253*, vii2–vii6. [CrossRef]
46. Woytowicz, E.J.; Westlake, K.P.; Whitall, J.; Sainburg, R.L. Handedness results from complementary hemispheric dominance, not global hemispheric dominance: Evidence from mechanically coupled bilateral movements. *J. Neurophysiol.* **2018**, *120*, 729–740. [CrossRef]
47. Woytowicz, E.J.; Sainburg, R.L.; Westlake, K.P.; Whitall, J. Competition for limited neural resources in older adults leads to greater asymmetry of bilateral movements than in young adults. *J. Neurophysiol.* **2020**, *123*, 1295–1304. [CrossRef] [PubMed]
48. Sainburg, R.L. Evidence for a dynamic-dominance hypothesis of handedness. *Exp. Brain Res.* **2002**, *142*, 241–258. [CrossRef] [PubMed]
49. Sainburg, R.L. Handedness: Differential specializations for control of trajectory and position. *Exerc. Sport Sci. Rev.* **2005**, *33*, 206–213. [CrossRef] [PubMed]
50. Sainburg, R.L. Convergent models of handedness and brain lateralization. *Front. Psychol.* **2014**, *5*, 1092. [CrossRef] [PubMed]
51. Pigeot, I. Basic concepts of multiple tests—A survey. *Stat. Pap.* **2000**, *41*, 3–36. [CrossRef]
52. Nichols, T.E. Multiple testing corrections, nonparametric methods, and random field theory. *Neuroimage* **2012**, *62*, 811–815. [CrossRef]

53. Austin, S.R.; Dialsingh, I.; Altman, N. Multiple hypothesis testing: A review. *J. Indian Soc. Agric. Stat.* **2014**, *68*, 303–314.
54. Petersson, K.M.; Nichols, T.E.; Poline, J.B.; Holmes, A.P. Statistical limitations in functional neuroimaging II. Signal detection and statistical inference. *Philos. Trans. R. Soc. Lond. B Biol. Sci.* **1999**, *354*, 1261–1281. [CrossRef]
55. Brett, M.; Penny, W.; Kiebel, S. Chapter 44-Introduction to random field theory. In *Human Brain Function*, 2nd ed.; Frackowiak, R.S., Friston, K.J., Frith, C.D., Dolan, R.J., Price, C.J., Zeki, S., Ashburner, J.T., Penny, W.D., Eds.; Academic Press: Burlington, MA, USA, 2004; pp. 867–879.
56. Rohani, F. Nonparametric Random Fields with Applications in Functional Imaging. Ph.D. Thesis, McGill University, Montréal, QC, Canada, 2009.
57. Worsley, K.J.; Wolforth, M.; Evans, A.C. Scale space searches for a periodic signal in fMRI data with spatially varying hemodynamic response. In *Proceedings of BrainMap*; Citeseer: Princeton, NJ, USA, 1997; Volume 96. Available online: https://citeseerx.ist.psu.edu/viewdoc/download?doi=10.1.1.34.7019&rep=rep1&type=pdf (accessed on 7 July 2021).
58. Sushkova, O.S.; Morozov, A.A.; Gabova, A.V. Data mining in EEG wave trains in early stages of Parkinson's disease. In *Advances in Soft Computing*; MICAI 2016. Lecture Notes in Computer Science; Springer: Cham, Switzerland, 2017; Volume 10062, pp. 403–412.
59. Sushkova, O.S.; Morozov, A.A.; Gabova, A.V.; Karabanov, A.V. Application of brain electrical activity burst analysis method for detection of EEG characteristics in the early stage of Parkinson's disease. *S.S. Korsakov J. Neurol. Psychiatry* **2018**, *118*, 45–48. [CrossRef] [PubMed]
60. Sushkova, O.S.; Morozov, A.A.; Gabova, A.V. Investigation of specificity of Parkinson's disease features obtained using the method of cerebral cortex electrical activity analysis based on wave trains. In Proceedings of the 2017 13th International Conference on Signal-Image Technology & Internet-Based Systems (SITIS), Jaipur, India, 4–7 December 2017; pp. 168–172.
61. Sushkova, O.S.; Morozov, A.A.; Gabova, A.V.; Karabanov, A.V. An investigation of the specificity of features of early stages of Parkinson's disease obtained using the method of cortex electrical activity analysis based on wave trains. *J. Phys. Conf. Ser.* **2018**, *1096*, 012078. [CrossRef]
62. Sushkova, O.S.; Morozov, A.A.; Gabova, A.V.; Karabanov, A.V. Investigation of surface EMG and acceleration signals of limbs' tremor in Parkinson's disease patients using the method of electrical activity analysis based on wave trains. In *Advances in Artificial Intelligence*; Simari, G., Eduardo, F., Gutierrez, S.F., Melquiades, J.R., Eds.; Springer International Publishing: Cham, Switzerland, 2018; pp. 253–264.
63. Sushkova, O.S.; Morozov, A.A.; Gabova, A.V.; Karabanov, A.V.; Chigaleychik, L.A. Investigation of the 0.5–4 Hz low-frequency range in the wave train electrical activity of muscles in patients with Parkinson's disease and essential tremor. *Radioelectron. Nanosyst. Inf. Technol.* **2019**, *11*, 225–236. [CrossRef]
64. Sushkova, O.; Morozov, A.; Gabova, A.; Karabanov, A. Investigation of the multiple comparisons problem in the analysis of the wave train electrical activity of muscles in Parkinson's disease patients. *J. Phys. Conf. Ser.* **2019**, *1368*, 052004. Available online: https://iopscience.iop.org/article/10.1088/1742-6596/1368/5/052004/meta (accessed on 7 July 2021). [CrossRef]
65. Sushkova, O.; Morozov, A.; Gabova, A.; Karabanov, A. Investigation of the Multiple Comparisons Problem in the Wave Train Electrical Activity Analysis in Parkinson's Disease Patients. *Basic Clin. Pharmacol. Toxicol.* **2019**, *124*, 24–25. [CrossRef]
66. Sushkova, O.; Gabova, A.; Karabanov, A.; Kershner, I.; Obukhov, K.Y.; Obukhov, Y.V. Time–frequency analysis of simultaneous measurements of electroencephalograms, electromyograms, and mechanical tremor under Parkinson disease. *J. Commun. Technol. Electron.* **2015**, *60*, 1109–1116. [CrossRef]
67. Obukhov, Y.V.; Gabova, A.; Zaljalova, Z.; Illarioshkin, S.; Karabanov, A.; Korolev, M.; Kuznetsova, G.; Morozov, A.; Nigmatullina, R.; Obukhov, K.; et al. Electroencephalograms features of the early stage Parkinson's disease. *Pattern Recognit. Image Anal.* **2014**, *24*, 593–604. [CrossRef]
68. Sushkova, O.S.; Morozov, A.A.; Gabova, A.V.; Karabanov, A.V.; Chigaleychik, L.A. An investigation of accelerometer signals in the 0.5–4 Hz range in Parkinson's disease and essential tremor patients. In *Proceedings of International Conference on Frontiers in Computing and Systems*; Springer: Berlin/Heidelberg, Germany, 2021; pp. 455–462.
69. Sushkova, O.; Morozov, A.; Gabova, A.; Karabanov, A. Development of a method for early and differential diagnosis of Parkinson's disease and essential tremor based on analysis of wave train electrical activity of muscles. In Proceedings of the 2020 International Conference on Information Technology and Nanotechnology (ITNT), Samara, Russia, 26–29 May 2020; pp. 1–5.
70. Sushkova, O.S.; Morozov, A.A.; Kershner, I.A.; Petrova, N.G.; Gabova, A.V.; Chigaleychik, L.A.; Karabanov, A.V. Investigation of distribution laws of the phase difference of the envelopes of electromyograms of antagonist muscles in Parkinson's disease and essential tremor patients. *RENSIT* **2020**, *12*, 415–428. [CrossRef]
71. Sushkova, O.S.; Morozov, A.A.; Gabova, A.V. A method of analysis of EEG wave trains in early stages of Parkinson's disease. In Proceedings of the International Conference on Bioinformatics and Systems Biology (BSB-2016), Allahabad, India, 4–6 March 2016; pp. 1–4.
72. Sushkova, O.S.; Morozov, A.A.; Gabova, A.V. Development of a method of analysis of EEG wave packets in early stages of Parkinson's disease. In Proceedings of the International Conference Information Technology and Nanotechnology ITNT 2016, Samara, Russia, 17–19 May 2016; CEUR: Samara, Russia, 2016; pp. 681–690.
73. Sushkova, O.S.; Morozov, A.A.; Gabova, A.V. EEG beta wave trains are not the second harmonic of mu wave trains in Parkinson's disease patients. In Proceedings of the International conference Information Technology and Nanotechnology ITNT 2017, Samara, Russia, 25–25 April 2017; CEUR: Samara, Russia, 2017; Volume 1901, pp. 226–234.

74. Sushkova, O.S.; Morozov, A.A. GitHub Repository. Wave Train Analysis of EMG Signals. AUC Diagrams. 2021. Available online: https://github.com/OlgaSushkova/Wave-Train-Analysis (accessed on 7 July 2021).
75. Sushkova, O.S.; Morozov, A.A.; Gabova, A.V.; Karabanov, A.V. Method for Differential Diagnosing of Essential Tremor and Early and First Stages of Parkinson's Disease Using Wave Train Activity Analysis of Muscles. Russian Federation Patent Application No. 2741233, 22 January 2021.

Short Biography of Authors

Olga S. Sushkova defended her Ph.D. thesis at Bauman Moscow State Technical University in 2017. She is a senior research associate of Kotel'nikov Institute of Radio Engineering and Electronics of Russian Academy of Sciences. She is a 2019 Moscow Government Award laureate. Her research interests include biomedical signal analysis, Parkinson's disease, statistical analysis of biomedical signals, and development of methods for differential diagnosing of neurodegenerative diseases.

Alexei A. Morozov graduated from Bauman Moscow State Technical University in 1991. In 1998, he defended his Ph.D. thesis on logic programming. He is a senior research associate of Kotel'nikov Institute of Radio Engineering and Electronics of Russian Academy of Sciences. His research interests include object-oriented logic programming, intelligent video surveillance, terahertz video surveillance, and biomedical signal analysis (EEG, EMG, MEG). He is a member of the International Association for Logic Programming (ALP). He is a principal developer of the Actor Prolog language (www.fullvision.ru).

Alexandra V. Gabova graduated from Moscow Region State University in 1979. In 1990, she defended her Ph.D. thesis. She is a senior research associate of Institute of Higher Nervous Activity and Neurophysiology of Russian Academy of Sciences. Her research interests include neurophysiology, EEG analysis, wavelet analysis, neurodegenerative diseases, Parkinson's disease, and absence epilepsy.

Alexei V. Karabanov is a medical doctor of the highest category. In 1986, he graduated from Sechenov First Moscow Institute of Medicine. In 2000, he defended his Ph.D. thesis. He is a neurologist of the scientific advisory department of FSBI Research Center of Neurology. His research interests include experimental neurology, neurodegenerative diseases, Parkinson's disease, essential tremor, Atypical Parkinsonism (MSA), EEG, and EMG.

Sergei N. Illarioshkin is the Corresponding Member of the Russian Academy of Sciences. He is a medical doctor of the highest category. In 1986, he graduated from Sechenov First Moscow Institute of Medicine. In 1992, he defended his Ph.D. thesis. In 1997, he defended his Doctor of Science thesis. In 2003, he became a professor. Since 2003, he is a Deputy Director for Research of FSBI Research Center of Neurology. Since 2006, he is the Head of the Brain Research Department. His research interests include neurodegenerative diseases, Parkinson's disease, gene and cell therapy, brain aging mechanisms, and experimental neurology.

sensors

MDPI

Review

A Systematic Review of EMG Applications for the Characterization of Forearm and Hand Muscle Activity during Activities of Daily Living: Results, Challenges, and Open Issues

Néstor J. Jarque-Bou *, Joaquín L. Sancho-Bru and Margarita Vergara

Department of Mechanical Engineering and Construction, Universitat Jaume I, E12071 Castellón, Spain; sancho@uji.es (J.L.S.-B.); vergara@uji.es (M.V.)
* Correspondence: jarque@uji.es; Tel.: +34-964-728-125

Abstract: The role of the hand is crucial for the performance of activities of daily living, thereby ensuring a full and autonomous life. Its motion is controlled by a complex musculoskeletal system of approximately 38 muscles. Therefore, measuring and interpreting the muscle activation signals that drive hand motion is of great importance in many scientific domains, such as neuroscience, rehabilitation, physiotherapy, robotics, prosthetics, and biomechanics. Electromyography (EMG) can be used to carry out the neuromuscular characterization, but it is cumbersome because of the complexity of the musculoskeletal system of the forearm and hand. This paper reviews the main studies in which EMG has been applied to characterize the muscle activity of the forearm and hand during activities of daily living, with special attention to muscle synergies, which are thought to be used by the nervous system to simplify the control of the numerous muscles by actuating them in task-relevant subgroups. The state of the art of the current results are presented, which may help to guide and foster progress in many scientific domains. Furthermore, the most important challenges and open issues are identified in order to achieve a better understanding of human hand behavior, improve rehabilitation protocols, more intuitive control of prostheses, and more realistic biomechanical models.

Keywords: ADL; EMG; forearm muscles; muscles role; synergies; muscle coordination

Citation: Jarque-Bou, N.J.; Sancho-Bru, J.L.; Vergara, M. A Systematic Review of EMG Applications for the Characterization of Forearm and Hand Muscle Activity during Activities of Daily Living: Results, Challenges, and Open Issues. *Sensors* **2021**, *21*, 3035. https://doi.org/10.3390/s21093035

Academic Editor: Ernest N. Kamavuako

Received: 1 April 2021
Accepted: 24 April 2021
Published: 26 April 2021

1. Introduction

The ability to carry out activities of daily living (ADL) is critical to ensure a full and autonomous life [1], and has been established by the WHO as the main factor for classifying the degree of disability [2]. ADLs refers to those elementary tasks that allow anyone to function with minimal autonomy and independence, including any daily activity that we perform for self-care, work, housework, and leisure. The ability of the hands to grasp and manipulate is fundamental in the performance of ADL [3] but also for working life [4]. This ability is achieved thanks to a complex musculoskeletal system, with 25 degrees of freedom that are controlled by approximately 38 muscles located in the forearm and hand [5]. These muscles can be divided into two groups: extrinsic and intrinsic muscles. The extrinsic muscles are located in the anterior and posterior compartments of the forearm whereas intrinsic muscles are located within the hand itself. Broadly, the extrinsic muscles are considered to control crude movements of the hand and produce forceful grip, while the intrinsic muscles would be responsible for the fine motor functions of the hand [6–8]. However, both the specific role of the different muscles in ADL and how the Central Nervous System dares with this redundant and complex muscular system are still unknown [9]. This information is essential for determining the impact on functionality when a given muscle is compromised because of an accident or pathology.

The measurement and interpretation of the hand kinematics and the associated muscle activation signals is complex but of great importance to deepen the knowledge of the role of the muscles in ADL [8]. This knowledge is not only important to rate disability but also to improve rehabilitation processes [10,11] or to help in decision-making during surgical planning, among others [12]. Another important field of application is in the control of hand prostheses [13–16]. Myoelectric hand prostheses use the electrical action potential of the residual muscles in the limb emitted during muscular contractions. These emissions are measured on the skin surface, picked up by electrodes, and are amplified to be used as control signals for the functional elements of the prosthesis. Therefore, deepening the knowledge of the role of the forearm muscles in ADL may help in the selection of the muscles to control this type of prostheses.

Electromyography (EMG) emerged as a diagnostic procedure to assess the health of Muscles and the nerve cells that control them (motor neurons). The electrodes receive the electrical signals transmitted by the motor neurons that cause muscle contraction. However, these EMG signals acquired from muscles require advanced methods for their detection, decomposition, processing, and classification [17–19] that a specialist interprets. There are two basic types of electrodes to acquire these signals: surface and intramuscular electrodes. Surface electrodes are placed on the skin directly over the muscles, recording the signal from all the fibers under the two electrodes. Intramuscular electrodes can be indwelling (also known as needle) or fine wire electrodes (Fw-EMG), and they are inserted through the skin directly into the muscle [20], thus recording the signals from only few fibers. The general advantage of surface electrodes is that they are non-invasive and easy to apply. Their use, however, is limited to superficial muscles that are large enough to support electrode mounting on the skin surface, and crosstalk is particularly problematic for smaller muscles within a complex mechanical arrangement, such as the forearm [21]. Indwelling electrodes need significantly more training for their proficient use in comparison to surface electrodes. Although they are ideal for recording the activity of deep muscles, correct placement requires a detailed knowledge of musculoskeletal anatomy. Furthermore, the invasiveness of inserting a needle into the muscles, as well as the associated pain, is the major disadvantage of intramuscular electrodes [20].

EMG has been incorporated as a diagnostic technique for the detection of pathologies that affect nervous and muscular structures, and for the spatial location of the origin of the injury. Examination with needle EMG allows motor unit action potentials (MUAPs) to be evaluated. High density EMG grids allow also the identification of the MUAPs, but only for those motor units whose innervation zone was close to the surface of the muscle [22]. The morphology (duration, amplitude, and number of phases) and recruitment pattern of MUAPs are the key element for diagnosing pathologies using needle EMG. MUAPs are analyzed per muscle and the results compared with those normally expected for that particular muscle. For this reason, due to the need for a normal pattern, evaluation of MUAPs is not useful to gain deeper knowledge of the role of the muscles in ADL. Parameters, such as time-domain, time-frequency domain, or intensity of muscle activation, could be more useful for studying tasks. EMG data for these purposes are commonly normalized to a reference value to avoid variability arising from electrode placement, participants, or even the day of the experiment. The most popular method is to normalize EMG data to the maximum voluntary contraction (MVC) of the muscle of interest [23], which, besides making data comparable, also informs about how active muscles are relative to their maximum capabilities. Surface EMG is applied in many fields, such as motor control of human movement, myoelectric control of prosthetic and orthotic devices, and rehabilitation [24–27]. Some studies have performed EMG analyses for intrinsic and/or extrinsic hand muscles in specific situations: while grasping objects [28–40], during working postures [41–49], and for the design and improvement of sports equipment, as well as for the study of the role of Muscles in sports performance [50–55].

The concept of synergy has been used in the field of control of myoelectric hand prostheses in an attempt to simplify the study of the complex kinematics and muscular action of the hand [56–59]. There are some studies describing muscle patterns or muscular synergies during some postures [60], grasps [61,62], or hand movements [60,63], and during particular actions [64–68]. In that research, different activation patterns have been obtained, revealing coordination between certain intrinsic and extrinsic hand muscles. Thus, EMG patterns have been studied as a way to control signals [69]. However, the usability of myoelectric prostheses is still challenged because of issues, such as the effect of electrode location or changes in EMG patterns over time, which can lead to long training processes [70]. A small number of studies have investigated the existence of hand muscle synergies in ADLs, which could help in the selection of Muscles to control myoelectric prostheses.

In this work, a review of the studies in which EMG has been used to record the muscular activity of hand and forearm muscles during ADL is presented, which may help to identify the role of these muscles in ADLs. In addition, studies examining EMG patterns or muscular synergies between the muscles of the hand and forearm are also presented, in order to simplify the study of the muscular action of the hand. The contents are organized in two sections: muscle activation in different activities and hand muscle synergies (dimensional reduction of EMG).

2. Materials and Methods

The literature review consisted in examining research studies that recorded EMG of hand and forearm muscles regarding the activation of these muscles in ADLs, and the dimensional reduction of the muscular action of the hand. A systematic literature search was conducted in the Scopus and PubMed databases until March 2021. Figure 1 shows the flowchart followed. The search was restricted to papers published in English and containing the terms ("Electromyography" and "muscles") and ("thumb" or "finger" or "hand" or "forearm") in the title, abstract, or keywords. Then, a refined search was conducted including different keywords in the title, abstract, or keywords (see Figure 1). Finally, a manual screening was carried out to remove duplicates and reject non-relevant articles.

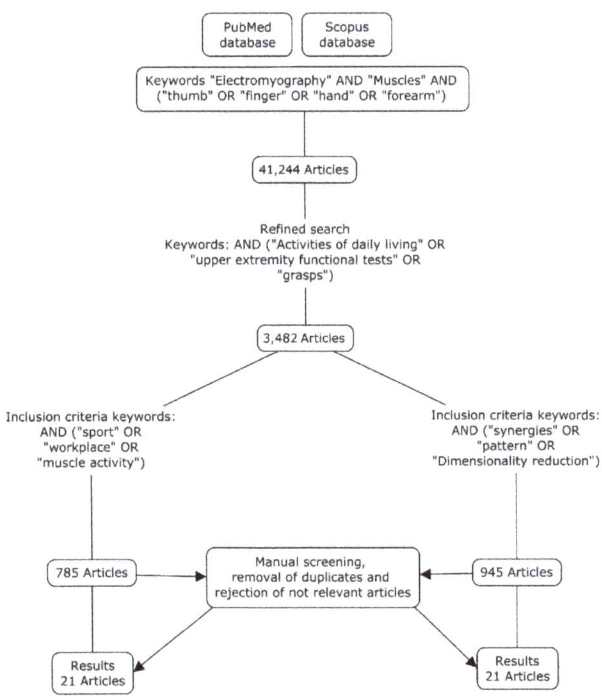

Figure 1. Methodology followed for the literature search.

After the manual screening, 21 articles related to muscle activity during different ADLs were selected, and 21 articles related to dimensional reduction of EMG of the hand were identified. Altogether, 42 articles (including two reviews) were selected for inclusion in the current review. The articles selected are discussed in the following sections.

3. Results

3.1. Muscle Activation in Different Activities

This section includes a review of studies that have characterized hand and forearm muscle activity while performing specific activities, such as grasps, ADL, work activities, and sports. Table 2 summarizes the most relevant information of the 21 papers related to muscle activity during different ADL found in the literature.

Several studies have analyzed the activation of hand and forearm muscles (both extrinsic and intrinsic) during certain types of grasps. Regarding extrinsic muscles, they found that in power grasps both the flexor and extensor groups of Muscles (extrinsic muscles) were activated, although the extensor part underwent greater fatigue [47]. As regards the intrinsic muscles, they found that during precision grasps intrinsic muscles play a major role in finely graded force generation, since fine movements require less stabilization and counterforce to the long flexor action [36].

Many studies have focused specifically on the thumb muscle activation through EMG while performing different grasps [30,37–40,71], especially during the opposition movement, due to its great importance in precision grasps. In general, these studies found a need for a cooperation of thumb muscles to accomplish the tasks performed [39], with the exception of the Extensor Policis Longus and Flexor Policis Longus (EPL, FPL), which could be activated separately from the other muscles [40]. Another study [30] explored and demonstrated the importance of the opposition of the thumb during stable and unstable lateral grasps. They observed that instability affects some thumb muscles with greater activation of Abductor Policis Longus (APL) and EPL in the unstable tasks. Similarly, Kaufman et al. [71] recorded the EMG activity of 7 thumb muscles and their contribution at the carpometacarpal (CMC) joint during voluntary isometric contractions. They found that:

- Thumb CMC flexion is supported by Flexor Policis Brevis (FPB), Abductor Policis Brevis (APB), thumb Adductor (ADD), and FPL.
- CMC extension by APB, APL, and EPL; CMC abduction by FPB, APB, APL, and EPL; and CMC adduction by FPB, APB, EPL, and FPL.
- The Opponens Policis (OPP) was active in all motions.

However, the studies in the literature all focused on small sets of very controlled and simple activities (a few grasps or simple finger movements). Additionally, they were limited to very specific muscles, or specific fingers or joints, especially for the role of the thumb.

EMG has also been used to study the effect of different kinds of work activities on the forearm muscles, evaluating the influence of different factors on fatigue during repetitive tasks [41,43,49], such as typing, keying, writing, reading and mousing, and on pulling and pushing tasks [45,46]. These studies have focused on evaluating and comparing different forearm and hand positions. Nevertheless, the relationship between force production and EMG is not well understood, and there are factors that influence the forces generated and therefore prevent the direct quantification of muscle force from EMG signals. These factors include variations in the location of the recording electrodes, crosstalk, the involvement of synergistic muscles, properties of Muscles, tendons, ligaments, etc. Consequently, the EMG-force relationship differs for each muscle and for each situation [72]. Many studies in the literature have focused on examining muscle activation of both intrinsic and extrinsic muscles in writing activities [31,32,35]. One of these studies [35] compared two different typewriting tasks and the results suggested that the major function of the Extensor Carpi Radialis (ECR) muscle as a stabilizer of the wrist joint is maintained during handwriting. It is also suggested that the increased use of extrinsic muscles could result in a diminished role of intrinsic hand muscles. In that research, the authors showed that EMG of hand and arm muscles may be converted into handwriting patterns. However, the results of those studies were focused on specific activities and in many cases with a low number of subjects and activities, lacking representativeness in ADL.

EMG has also been used in the design and improvement of sports equipment, as well as for the study of the role of Muscles in sports performance. Some researchers have studied the regions that are activated, thereby making the main movement possible, in sports, such as golf [50] and tennis [51–54]. Other studies have focused on examining the effect of different features of sports equipment, such as the size of the handle of rackets [53] or of a golf club [55].

Some authors [50,52] have observed that there is considerable diversity in the protocol design used for sEMG recording. For example, most of the studies did not specify the electrode placement, so it is not clear which locations were used to acquire the EMG data, thus making it difficult to compare values. A recent study [73] recorded sEMG activity from 30 spots distributed over the skin of the whole forearm of six subjects while performing 21 representative ADL from the Sollerman Hand Function Test (SHFT). As a result, they proposed that the number of sEMG sensors could be reduced from 30 to 7 without losing any relevant information, using them as representative spots of the muscular activity of the forearm in ADL.

There are few EMG analyses of upper extremities that examine muscle function during daily tasks, and they use little variability and a limited number of tasks (no more than 10) [74–76]. A wide variety of clinical tests (such as the Jebsen Taylor Hand Function (JTHF) test [77], Chedoke Arm and Hand Activity Inventory (CAHAI) [78], or SHFT [79]) are often used to evaluate and track functional recovery of the upper extremity simulating ADL. In these cases, EMG recordings may provide a window into the central nervous system to evaluate muscle recruitment and coordination. In this sense, Peters and collaborators [75] evaluated the recruitment and coordination between several upper-limb muscles during some of these clinical tests (JTHF, CAHAI, and Block and box test (BBT)). Specifically, they recorded sEMG from eight upper-extremity muscles (Anterior and Posterior Deltoid (AD and PD), Biceps Brachii (BB), Triceps lateral head (TriB), Brachioradialis (Br), ECR, Flexor Carpi Ulnaris (FCU), and Extensor Digitorum (EDC)), and evaluated which muscles were used to execute each task and whether activation and co-contraction levels were similar across tasks. As results, they found that co-contraction levels were similar across tests and EDC was found to have the greatest activation levels across all tasks, thereby denoting its importance for common tasks. However, this study has several limitations: they evaluated a small set of forearm muscles (only four forearm and finger muscles).

Summing up, most research found in the literature presents gaps that require further investigation, as, in many cases, the studies are focused on small sets of very controlled and simple activities. Few of them characterize the EMG activity of all the hand muscles while performing representative actions, either by carrying out all possible grasp types required in ADL or by performing a representative and conveniently standardized set of ADL. Furthermore, the lack of a methodology and a standardized protocol hinders the comparison of EMG results between tasks and subjects. Indeed, at the hand level there are few specific recommendations to help in this decision, although results from a recent study [73] could assist in this task, as 7 specific spots were identified as being representative of the muscular activity of the forearm in ADL. In addition, more studies are required to improve the knowledge about the relationship between force production and EMG.

Table 1. Summary of the studies included in the systematic literature review (I). Relevant information contains subjects, type of EMG used, and muscles recorded in the studies.

Study	Relevant Information	Description of the Task	Observations about Role of Muscles
Cooney et al. [38]	8 healthy subjects Fw-EMG Extrinsic muscles: FPL, APL, EPL, EPB Intrinsic muscles: ADD, APB, OPP	Isometric F/E and Abd/Ad thumb movements, pinch and power grasps	Extensor muscles (EPL, EPB, and APL) were primary and contributed nearly equally to the extension. In flexion, only the FPL was primary. ADD and APB are primary in adduction but the EPL (adduction) and OPP (abduction) contribute significantly. Three muscles appear to be primary in pinch and power grasp: the ADD, OPP, and FPL.
Kilbreath and Gandevia [80]	7 healthy subjects needle EMG Extrinsic muscles: FPL, FDP Intrinsic muscles: None	Dynamic movements: Lift a weight by flexion of one digit	There is a coactivation of other deep digital flexor muscles and this coactivation increases when the digit flexes at a greater velocity or through a larger angle.
Maier and Hepp-Reymond [36]	6 healthy subjects needle EMG Extrinsic muscles FDP, FDS, APB, FPL, EPL EPB, APL, EDC, EI Intrinsic muscles: FDI, FPI, FPB, FLUM ADD, OPP	Isometric forces with thumb and index finger	The intrinsic muscles (FDI, FPI, and FLUM) and the long flexors (FDP, FDS) of the index finger, as well as two intrinsic muscles of the thumb (ADD FPB), increase their activity according to the load. The other thenar muscles (OPP, APB) and the extrinsic muscles of the thumb (FPL, EPL, EPB, and APL) become active only at higher loads and may serve to stabilize joints. The long extensors of the index finger (EDC, EI) were classified as antagonistic, and only act to balance the applied load and maintain joint equilibrium. The intrinsic muscles play a major role in finely graded force generation since less stabilization and counterforce to the long flexor action are needed, thus releasing the intrinsic muscles for precise force regulation.
Hägg and Milerad [47]	9 healthy subjects sEMG Extrinsic muscles: FCU, FDS, ECR Longus, ECR Brevis, EDC Intrinsic muscles: None	Simulations of grips in industrial work	Fatigue effects are generally larger on the extensor side although none of the regimes studied were acceptable from point of view of EMG fatigue.

Table 1. *Cont.*

Study	Relevant Information	Description of the Task	Observations about Role of Muscles
Chow et al. [51]	7 healthy and skilled tennis subjects sEMG Extrinsic muscles: FCR, ECR Intrinsic muscles: None	Tennis volley	In general, the ECR was more active than the FCR during the volley, suggesting the presence of wrist extension and abduction.
Kaufman et al. [71]	5 healthy subjects Fw-EMG Extrinsic muscles: APL, FPL, EPL Intrinsic muscles: APB, OPP, FPB, ADD	Isometric thumb motions in F/E and Abd/Ad	The FPB was most active in the range from flexion to abduction with about 50% activity or less in extension and adduction. The OPP displayed activity in all directions of motion. The APB displayed maximal activity in abduction and abduction-flexion directions. The ADD was active during flexion. The APL was most active in abduction and/or extension. The EPL showed the highest activity during extension in combination with abduction/adduction functions. The FPL was the most active in flexion and/or adduction functions.
Johanson et al. [30]	7 healthy subjects Fw-EMG Extrinsic muscles: FPL, EPL, EPB, APL Intrinsic muscles: APB, ADD, FDI, FPB OPP	Key and opposition pinch postures between stable and unstable tasks	Activation patterns are different between key and opposition pinch posture and between stable and unstable pinch tasks. APB and EPL muscles are necessary to accurately direct thumb-tip forces in a functional pinch, not just to position the thumb, independently of pinch force magnitude. In all unstable conditions, APB and EPL were among the most activated muscles and could provide the task with directional accuracy.
Van Galen et al. [41]	20 subjects sEMG Extrinsic muscles: BB, TriB, FCU, ECR Intrinsic muscles: None	Fast movements with an electronic pen along the surface of a digitizer	For the forearm muscle movers, our findings show that the participants reacted with a substantial increase in static muscle activity, i.e., of antagonistic activation. For the wrist stabilization musculature, however, the effect was reversed.

Table 1. *Cont.*

Study	Relevant Information	Description of the Task	Observations about Role of Muscles
Danion and Galléa [37]	7 healthy subjects sEMG Extrinsic muscles: EPL Intrinsic muscles: FPB	Constant force matching task during precision grasp	Muscle co-contraction is not a critical factor for force steadiness during a precision grasp task. Muscle co-contraction and grip force steadiness depend on grip force magnitude, but grip force steadiness does not depend on muscle co-contraction.
Hatch et al. [53]	16 healthy tennis players Fw-EMG Extrinsic muscles: EDC, ECR Longus, ECR Brevis, PT, FCR Intrinsic muscles: None	Back-hand tennis stroke	There was a progressive increase in ECRL and FCR activity from early acceleration through ball impact. There was a progressive increase in EDC activity through the early and late acceleration phases. At ball impact, all muscle activity tended to peak, and then gradually decreased in the early follow-through phase.
Ertan [67]	20 healthy subjects sEMG Extrinsic muscles: EDC, FDS Intrinsic muscles: None	Dynamic archery shooting	Elite archers relax their finger flexors so as not to grip the bow-handle, and contract the extensors to avoid holding/gripping the handle throughout the whole shot. The main difference between elite and beginner archers was that the elite archers had a greater activation of the EDC.
Linderman et al. [32]	6 healthy subjects sEMG Extrinsic muscles: FCR, EDC, ECU, ECR Intrinsic muscles: OPP, APB, FDI	Writing numeric characters	It is feasible to recreate handwriting solely from EMG signals thanks to the existence of muscle patterns during writing.
Di Dominizio and Keir [46]	12 healthy subjects sEMG Extrinsic muscles: FCR, FCU, FDS, ECR, ECU, EDC, AD, PD Intrinsic muscles: None	Grips with pull and push tasks	Flexor muscle activity tended to be lower when performing push with grip tasks and pull with grip tasks than extensor muscle activity. The highest wrist and finger extensor (ECR, ECU, and EDC) activity was elicited when performing grip tasks in a pronated posture. ECU was found to be the most sensitive to postural changes.

Table 1. *Cont.*

Study	Relevant Information	Description of the Task	Observations about Role of Muscles
Szeto and Lin [49]	17 healthy subjects and 9 symptomatic office workers sEMG Extrinsic muscles: ECR, FCU, ECU, FCR Intrinsic muscles: None	Performing mouse-clicking tasks under different speed and precision conditions	Higher EMG amplitudes in the Control Group over Case Group mostly in the ECU and ECR muscles and more so in the more stressful condition. ECR muscle recorded significant group differences in both precision and speed condition analyses, and FCU in speed condition analysis.
Marta et al. [50]	Review paper about amateur and professional golfers sEMG/Fw-EMG Extrinsic muscles: ECR Brevis, PT, FCR, FCU, Intrinsic muscles: None	Different phases of the golf swing	Higher peak activity in the leading PT during the acceleration phase and just after the impact in professional golfers compared to amateur players who showed peak activation in the ECRB. This study also reported considerably higher levels of activity in the ECRB in amateurs during all swing phases. Some studies did not specify the electrode placement, so it is not clear which locations were used to acquire the EMG data, thus making it difficult to compare values.
Almeida et al. [35]	24 healthy subjects sEMG Extrinsic muscles: Tr, BB, ECR Brevis, FDS Intrinsic muscles: None	Writing a word five times	The major function of the ECRB muscle as a stabilizer of the wrist joint is maintained during handwriting tasks and the increased use of extrinsic muscles could result in a diminished role of intrinsic hand muscles during handwriting.
Birdwell et al. [40]	7 healthy subjects Fw-EMG Extrinsic muscles: APL, EPB, EPL EDC, FDP, FDS, FPL Intrinsic muscles: None	Activating each specific muscle during 3 s of MVC	Only two extrinsic thumb muscles, EPL and FPL, were capable of sustaining individual activations from the other thumb muscles. Activation of EPB elicited coactivity levels from EPL and APL.
Park [31]	36 healthy subjects sEMG Extrinsic muscles: FCU, FCR, ECU, ECR, Upper Tr Intrinsic muscles: None	Writing subtests of the JHFT	ECU was the most active muscle during writing in both left- and right-handers.

Table 1. *Cont.*

Study	Relevant Information	Description of the Task	Observations about Role of Muscles
Alizadehkhaiyat and Frostick [52]	Review paper: 198 healthy (controls) 18 Lateral-epicondylitis patients sEMG/Fw-EMG Extrinsic muscles: FCU, FCR, ECR, ECU, PT Intrinsic muscles: None	Different tennis strokes	Increase in the activity of wrist extensors including ECR Brevis and ECR Longus in multiple phases of forehand, serve, and backhand strokes with the activity of wrist flexors remaining fairly constant. Higher EMG activity of ECR during repetitive pre- and post-impact in the presence of unchanged FCR activity has been suggested as predisposing players to injury or delaying the recovery process. Finally, an earlier, longer, and greater activation of ECR Brevis during backhand volleys at combined conditions of velocity and racket-head impact locations has been reported in LE patients compared to non-injured players. There was considerable diversity in the protocol design used for EMG recording.
Kerkhof et al. [39]	10 healthy subjects Fw-EMG Extrinsic muscles: FPL, EPB, EPL, APL Intrinsic muscles: ADD, APB, FPB OPP	Isometric contractions in a lateral key pinch, a power grasp, and a jar twist task	Extrinsic thumb muscles were significantly more active than intrinsic muscles in all tasks. The thumb muscles display a high variability in muscle activity during functional tasks of daily life. To produce a substantial amount of force, a co-contraction between the intrinsic and extrinsic thumb muscles is necessary.
Peters et al. [75]	20 healthy subjects sEMG Extrinsic muscles: AD, PD, BB, TriB, BB, ECR Longus, FCU, EDC Intrinsic muscles: None	Clinical tests of upper extremity function	Minimal muscle force is required to perform these tests. Co-contraction levels were similar across tests. EDC has the greatest activation levels across all tasks. The results suggest that healthy participants used different strategies to execute the tests.

Table 1. *Cont.*

Study	Relevant Information	Description of the Task	Observations about Role of Muscles
Jarque-Bou et al. [73]	6 healthy subjects sEMG Extrinsic muscles: Muscles recorded from all the forearm Intrinsic muscles: None	21 ADL selected and adapted from SHFT	The EMG sensors needed to record relevant information about forearm muscle activity could be reduced to 7. The signals from these seven spots would be related to seven different movements: 1. wrist flexion and ulnar deviation 2. wrist flexion and radial deviation 3. digit flexion 4. thumb extension and abduction/adduction 5. finger extension 6. wrist extension and ulnar deviation 7. wrist extension and radial deviation
Jarque-Bou et al. [81]	22 healthy subjects sEMG Extrinsic muscles: Seven spots representative of all available muscle activity of the whole forearm Intrinsic muscles: None	26 representative ADL	Minimal muscle force is required to perform ADL. Greater activity is shown during reaching (to place the hand to grasp) than during manipulation. Finger and wrist extensors were the most active muscles while performing ADL. Muscle activity presented some variability among subjects, highlighting the different possibilities that each subject may have to carry out the same activities.

Abbreviations: FLUM—First Lumbrical; FDI—First Dorsal Interosseous; FPI—First Palmar Interosseous; ECU—Extensor Carpi Ulnaris; PT—Pronator Teres; FCR—Flexor Carpi Radialis; FDS—Flexor Digitorum Superficialis; FDP—Flexor digitorum Profundus.

3.2. Hand Muscle Synergies

This section includes a review of studies that have characterized hand and forearm muscle activity by studying EMG patterns or muscular synergies between the muscles of the hand and forearm in order to simplify the study of muscular action of the hand. Table 2 summarizes the most relevant information in the 21 papers related to hand muscle synergies that were selected.

The human hand has a complex biomechanical structure, controlled by a neural structure that is still not completely understood. In the analysis of the biomechanical and behavioral aspects of the hand, one of the most striking is the high redundancy of its structure, seemingly having many more muscles than are actually required [56]. Synergies are thought to be used by the nervous system to simplify the control of these numerous muscles by actuating them in task-relevant subgroups. There are studies both for and against muscle synergies [82]. Many researchers seek to detect and describe such simplifying functional muscle groups and how to interpret them in order to reveal the underlying control strategy used by the brain to coordinate muscles [56,60]. Others point out the importance of the ability of the brain to break and dissolve such patterns of neural synchrony. This might happen to enable flexible and individuated control of hand muscles [83–85], thus indicating that muscles are recruited flexibly in accordance with their mechanical action, rather than in fixed groupings. In both cases, EMG of hand muscle activity has been extensively used to infer the control strategies underlying the complex coordination of muscle activity within and across digits and as a tool to study the spatial and temporal coordination of multiple muscles. In fact, this technique has been used to examine the organization of these muscle synergies in healthy and neurologically impaired individuals [83,86]. In addition, several studies have examined the covariations in EMG amplitudes across muscle pairs [36] and among multiple muscles [87,88] related to hand function.

Previous studies reinforce the idea of hand muscle synergies, and their results concerning the features of hand muscle synergies can be summarized as: muscle synergy occurs primarily across muscles with similar mechanical features [89]; the coactivity of some muscles is a way to adapt the limb to different environmental conditions [90,91]; and the whole set of hand and forearm muscles may be approximated with relatively few adequately scaled and synchronized muscle synergies [60,92–96].

The basic approach of these studies involves four steps:

(1). Measuring sEMG from a large number of Muscles during a complex behavior (or more than one behavior). Note that surface electrodes are the most widely used method, since they are non-invasive and a high number of Muscles on the forearm need to be recorded.
(2). Using a computational analysis, such as non-negative matrix factorization, to identify a set of synergies from the sEMG recorded. Different factorization methods have been used to extract muscle synergies from muscle activation patterns during dynamic tasks. The two most commonly used factorization methods reported in the literature are non-negative matrix factorization and principal component analysis [97]. Similar results are obtained in both cases, but the non-negative matrix factorization method is the most suitable when recording a high number of channels [98].
(3). Evaluating whether the sEMG observed can be well described as the combination of these synergies.
(4). Relating the muscle synergies identified to task-relevant variables [82].

As a result of the application of this procedure, two different types of synergies are described in the literature: synchronous synergies and time-varying synergies. A synchronous synergy is a vector of weighting coefficients that specify the relative involvement (strength of membership) of each muscle in the group. In contrast, a time-varying synergy is a collection of EMG bursts in various muscles.

Several studies describe muscle patterns or muscular synergies during certain specific postures or grasps [99] or during the whole-hand grasping performance [60,62]. Some synergies during specific tasks are also described, such as during finger spelling [60,95], or the preparatory muscle activation response when a fall occurs [100]. Weiss and Flanders [60] recorded the EMG activity of 6 hand and forearm muscles (APB, FPB, FDI, EDC, ADM, and FDS) in four subjects while they held the hand statically, shaping around 26 grasped objects and forming the 26 letter shapes of a manual alphabet. They found that a single muscle can be a member of more than one muscle synergy [60,101]. Klein Breteler et al. [95] expanded the synergy analysis from static synergies to time-varying synergies in order to explore the timing of muscle activations during finger spelling using a manual alphabet. They recorded FDI, APB, FPB, ADM, FDS, and EDC and concluded that four time-varying synergies could account for 80% of the temporal EMG patterns observed, with the first two synergies accounting for about 60%. In addition, they showed that the first component displayed a consistent pattern, the first and second component waveforms showed similarities across subjects, and higher order components were far more variable across subjects. The first component was a pattern where the EDC and the thumb muscles (APB and FPB) were active early on, and the other muscles were active later. Recently, Scano and collaborators [62] extracted muscle synergies from 20 hand grasps with an array of 8 equally spaced electrodes on the forearm, two electrodes on finger flexors and extensors, and another two on BB and TriB. The synergies they found were characterized by two temporal activation patterns: a strong coactivation corresponding to the grasp/hold phase, and two minor coactivating patterns related to hand opening (visible in the pre-shaping and release phase).

Synergistic finger patterns have also been described during dynamic free movements of the wrist and single fingers [102–104]. Tanzarella et al. [104] performed isometric contractions with each finger and with three combinations of fingers in opposition with the thumb. They observed a low dimensional control of motor neurons across multiple intrinsic and extrinsic muscles. Gazzoni et al. [102] identified distinct areas of sEMG activity on the forearm for different fingers during hand and finger movements. In the same way, Hu et al. [103] revealed distinct activation patterns during individual finger extensions, especially for the index and middle fingers. Nevertheless, the detailed location of the recording electrodes was not reported in most of the studies, which makes comparison between subjects and activities difficult.

However, few studies have assessed muscle patterns during complex tasks in which manipulation is the most relevant phase, such as in ADL [61,68], or during particular actions, such as playing a piano [63], archery [65,67], catching a ball [64], or while performing a karate punch [66]. Winges et al. [63] studied the muscle activation patterns of 10 pianists, suggesting that amateur pianists use the same balance as professionals. Nevertheless, in other research [65,67], the authors found different patterns between elite and beginner archers, where the main difference was that the expert archers had a greater activation of the ED. In this sense, in a study on karate punches [66], expert and non-expert karatekas presented distinct kinematic and EMG patterns. Regarding a more complex activity, such as catching a ball [64], the authors recorded sEMG data from 16 shoulder and elbow muscles, but only one forearm muscle (Br), in six subjects. They found that the variation in the muscle patterns was captured by two time-varying muscle synergies, modulated in amplitude and shifted in time according to the height at which the ball arrived and the flight duration. The initial muscular response, captured by the first synergy, allowed the subject's hand to reach the interception zone. The following component of the muscle pattern, captured by the second synergy, guided the hand to the interception. Zariffa et al. [61] characterized what muscle synergies were present while using different types of hand grips (gripping a block, a cylinder, a ball, a key, and rotating a disk 180 degrees) extracted from clinical tests. sEMG data was recorded from FDI, FCU, FCR, FDS, ECR, EDC, EIP, and the thenar eminence muscle group. Two main synergies were found: the first between EDC and EIP, and the second between FDS and FCU. However, they had some limitations due to crosstalk, the small number of Muscles recorded, and the little variability of the ADLs chosen. Ricci et al. [68] recorded data from shoulder and elbow muscles along with FDS, FCU, ECRLB, and ECU while subjects poured water. In the transport phase, characterized by weight bearing, handgrip and displacement of the arm in space, a higher activity of almost all muscles was found. Furthermore, they found that ECR seems to play a key role in maintaining optimal wrist posture and function regardless of task demand. That stabilization could be provided by a delicate balance of co-contraction of forearm muscles to keep the hand in the proper posture to grasp or produce handgrip force [105]. However, few forearm muscles were measured and for very specific actions, and therefore further studies should be conducted to evaluate more forearm muscle patterns in a wide range of ADL. Likewise, hand kinematics should be recorded in order to relate muscular and kinematic hand synergies during representative ADL.

Moving on to the assessment of pathologies, as mentioned above, sEMG has recently been used for the evaluation of patients with neuromuscular disorders by using muscle synergies. Muscle synergies have been investigated in acute, subacute, and chronic stroke, showing abnormalities compared to healthy people [84,106–109], as well as in patients with dystonia [110] and sclerosis [111] or after spinal cord injury [61]. The results illustrate that muscle synergy patterns contain rich information in their spatial components and temporal profiles. Comparing pathological synergies of patients with the baseline synergy can reveal deficits in the underlying neuromuscular coordination and control. The analysis of task-specific muscle synergies should offer both researchers and clinicians new insights into the impairments in the neural organization of motor control. Yet, in these studies, a considerable diversity in the protocol design was used for sEMG recording, and it is not clear which locations were used to acquire the sEMG data, resulting in difficulty when it comes to comparing values.

Summarizing, EMG has been widely used to detect muscle patterns, although a small number of studies have investigated the muscular synergies in the hand in greater depth. There are some gaps that need to be studied in more detail. First, muscular synergies seem to be task-dependent, and a single muscle can be a member of more than one muscle synergy. In the literature, researchers have generally investigated the presence of synergies during some specific hand movements or grasps, but few studies have analyzed the different coordination and muscular patterns or synergies during the performance of a representative set of ADL. Second, little has been studied about kinematics and muscular synergies of the forearm and hand relationship. Consequently, little is known about the role of the muscles linked to the joint movement of the hand during ADL.

Table 2. Summary of the studies included in the systematic literature review (II). Relevant information contains subjects, type of EMG used, and muscles recorded in the studies.

Study	Relevant Information	Description of the Task	Observations about Muscles Role
Valero-Cuevas et al. [87]	8 healthy subjects Fw-EMG Extrinsic muscles: FDP, FDS, EI, EDC Intrinsic muscles: FLUM, FDI, FPI	Static force in five directions	CNS is implementing the predicted mechanically advantageous strategies, and scaling them down to produce less than maximal forces. Palmar force used flexors, extensors, and FDI. Dorsal force used all muscles. Distal force used all muscles except for extensors. Medial and lateral forces used all muscles including significant co-excitation of FDI.
Valero-Cuevas [88]	8 healthy subjects Fw-EMG Extrinsic muscles: FDP, FDS, EI, EDC Intrinsic muscles: FLUM, FDI, FPI	Different levels of fingertip forces while maintaining their forefinger in a static posture	Significant muscle coordination patterns similar to those previously reported for 100% of maximal fingertip forces were found for 50% of maximal voluntary force.
Weiss and Flanders [60]	4 healthy subjects sEMG Extrinsic muscles: EDC, FDS Intrinsic muscles: ADM, APB, FPB, FDI	Static postures for 26 objects and 26 letter shapes of a manual alphabet	Single muscles may be a member of more than one muscle synergy.
Klein Breteler et al. [95]	9 healthy subjects sEMG Extrinsic muscles: EDC, FDS Intrinsic muscles: FDI, APB, FPB, ADM	Finger spell words, presented on a computer screen	The first synergy represented the main temporal synergy, accounting for more of the EMG variance (up to 40%). This main synergy began with a burst in the EDC and a silent period in the flexors. There were then progressively later and shorter bursts in the APB, FPB, ADM, and, finally, the FDS.

Table 2. *Cont.*

Study	Relevant Information	Description of the Task	Observations about Muscles Role
Martelloni et al. [96]	6 healthy subjects sEMG Extrinsic muscles: TriB, Deltoid, Trapezius, FCR, ECR, BB Intrinsic muscles: None	Performing reach-to-grasp movements for different objects placed in different locations	Activation of proximal muscles can be statistically different for different grip types. Proximal and distal muscles are simultaneously controlled during reaching and grasping. Patterns of EMG activation in arm muscles can provide a reliable representation of motor behavior during reaching and grasping of different objects.
Valero-Cuevas et al. [85]	8 healthy subjects Fw-EMG Extrinsic muscles: EDC, EI, FDP, FDS Intrinsic muscles: FDI, FPI, FLUM	Vertical fingertip force vectors of prescribed constant or time-varying magnitudes	Evidence for preferential control of task-relevant parameters that strongly suggest the use of a neural control strategy compatible with the principle of minimal intervention. Only one synergy accounting for >40% of the variance with positive correlation among all muscles (coactivation). There was no reduction in dimensionality because each of the seven principal components explains a nontrivial amount of variance.
Marc H. Schieber et al. [83]	10 stroke subjects sEMG Extrinsic muscles: None Intrinsic muscles: APB, FDI, ADM	Cyclical F/E or Ab/Ad movements of each digit	FDI in the control hand was active only when the index finger was abducting. FDI in the affected hand was also active during movement of the thumb or the ring finger. These inappropriate contractions of FDI in the affected hand would cause the index finger to move when the subject attempted to move only the thumb or the ring finger. Muscle synergies of the stroke-affected arm were strikingly similar to those of the unaffected arm despite marked differences in motor performance between the arms. In subjects with severe motor impairment, there was far less resemblance between the synergies of the two arms.

Table 2. Cont.

Study	Relevant Information	Description of the Task	Observations about Muscles Role
VencesBrito et al. [66]	18 karatekas and 19 non-karatekas sEMG Extrinsic muscles: BB, Br, Deltoid, Pectoralis, PT, Infraspinatus Intrinsic muscles: None	Analysis of a karate punching movement (choku-zuki) on a fixed target	The two groups presented distinct EMG patterns. The first muscles to be activated were the agonists of the arm flexion and internal rotation. This was followed by an initial activation of the forearm flexor and pronator muscles. The forearm extensor muscle initiates its activity slightly later, followed by the second activation moment of forearm pronator muscle.
Cheung et al. [107]	31 stroke subjects sEMG Extrinsic muscles: infraspinatus; rhomboid major; Trapezius; pectoralis major; Deltoid; TriB; BB; brachialis, Br; supinator; PT Intrinsic muscles: None	Tasks and reaching movements with shoulder and forearm motions	Muscle synergies of the stroke-affected arm were strikingly similar to those of the unaffected arm despite marked differences in motor performance between the arms. In subjects with severe motor impairment, there was much less resemblance between the synergies of the two arms.
Zariffa et al. [61]	10 healthy subjects 6 Spinal cord-injured subjects sEMG Extrinsic muscles: FDS, FCR, FCU, ECR, EDC Intrinsic muscles: EI, FDI, Thenar eminence	7 functional tasks using grasp types relevant to ADLs	The synergies found were: (1) EDC and EIP, and (2) FDS and FCU. Many tasks involving finger extension tasks can be expected to recruit both EDC and EIP. The FDS and FCU synergy suggests that a wrist flexion was often used to position the hand during a grasping action, though this may be a product of the specific set of tasks employed in this study. The most common synergy in SCI subjects was FCR and ECR, which was also one of the average able-bodied synergies. FDI and Thenar eminence were common in both groups, possibly because of the need for independent fine thumb and index finger movements in many dextrous tasks. In subjects with SCI, similar synergies were observed, but in different proportions.

Table 2. *Cont.*

Study	Relevant Information	Description of the Task	Observations about Muscles Role
Burkhart and Andrews [100]	20 healthy subjects sEMG Extrinsic muscles: BB, Br, Anconaeus, FCR, ECR Intrinsic muscles: None	Impacts occurred to the hand from two heights	Individuals are capable of selecting an upper extremity posture that allows them to minimize the effects of an impact and the presence of a preparatory muscle activation response has been confirmed.
Castellini and Van Der Smagt [99]	6 healthy subjects sEMG Extrinsic muscles: Two bands surrounding the forearm Intrinsic muscles: None	Five static grasps: flat grasp, pinch grip, tripodal grip, small power grasp, and large power grasp	Three main synergies were found: uniform activation, activation of the dorsal muscles near the radius, and activation of the flexors near the radius.
Lee et al. [109]	4 healthy subjects 14 subjects with chronic hemiparesis sEMG Extrinsic muscles: FDS, EDC, FCR, FCU, ECR, ECU Intrinsic muscles: Thenar muscles, FDI, hypothenar muscles	Wrist F/E finger extension, lateral pinch, power grip, and tip pinch	The first synergy, containing mainly thenar and FDI activity, was largely active in the three grip tasks. The second synergy, consisting of EDC, ECR, and ECU, was heavily weighted during finger/wrist extension. The third synergy, involving coactivation of the wrist and finger muscles. The fourth synergy, with FCR, FCU, and EDC activity, was employed during wrist flexion. For stroke survivors, the composition of these modules was generally similar to those of subjects with no impairment.
Winges et al. [63]	10 healthy subjects sEMG Extrinsic muscles: FDS (2 portions), EDC Intrinsic muscles: ADM, APB, FPB, FDI	Piano dynamic movements: playing 14 selected pieces with the right hand at a uniform tempo	Phasic coactivation was evident between extensor and flexor muscles during piano playing. For the thumb sequence, PC1 first synergy was dominated by bursts of activity in the APB and the FPB with activity in the four-finger ED muscle. For the index finger sequence, the central burst of the first synergy included activity in two to three flexors of the index finger. Higher PC synergies were variable across subjects.

Table 2. *Cont.*

Study	Relevant Information	Description of the Task	Observations about Muscles Role
Hu et al. [103]	10 healthy subjects sEMG Extrinsic muscles: surface grid centered on the EDC Intrinsic muscles: None	Static and dynamic finger movements: To extend MCP joints individually	When the four fingers were extended simultaneously, the entire EDC was active. When individual fingers were extended separately, distinct regions of the EDC were selectively activated, with the index finger in the most distal region, the middle finger in the most proximal region, and the ring and little fingers in between. Index and middle fingers have a greater degree of individuation in comparison to the little and ring fingers.
Ricci et al. [68]	25 healthy subjects sEMG Extrinsic muscles: Trapezius, Serratus, Deltoid, Pectoralis, BB, TriB, FDS, FCU, ECR, ECU Intrinsic muscles: None	Pouring water task belonging to the Elui Functional Test of the Upper Extremity	In the reaching phase, the main movements observed were shoulder flexion and elbow and wrist extension, accompanied by significant higher activity of S, D, and TriB. The sequence of movements in this phase ended up with the subjects grasping the pitcher, which could be related to the late coactivation between ECU and FCU. Transport phase was mainly characterized by higher muscle activity of all muscles, except for Pectoralis. There were almost no significant differences in muscle activity within the release phase. ECR is a key muscle for wrist posture and function regardless of the task demand. Activation of FCU and ECRLB were identified as the main control strategy performed to maintain optimal grasping.
Roh et al. [108]	6 healthy subjects 16 post-stroke subjects sEMG Extrinsic muscles: Br, BB, TriB, Deltoid, and pectoralis Intrinsic muscles: None	Grasping the MACARM's gimbaled handle	EMG spatial patterns were well explained by task-dependent combinations of only a few (typically 4) muscle synergies. Elbow-related synergies were conserved across stroke survivors, regardless of level of impairment. Alterations in the shoulder muscle synergies underlying isometric force generation appear prominently.

Table 2. *Cont.*

Study	Relevant Information	Description of the Task	Observations about Muscles Role
Hesam-Shariati et al. [110]	24 post-stroke subjects sEMG Extrinsic muscles: Trapezius, Deltoid medius, BB, ECR, FCR, FDI Intrinsic muscles: None	14-day program focused on the more- affected upper limb	The profile of coordinated muscle activation varied by the level of residual motor-function in chronic stroke. The number of synergies used increased (although not significantly) with therapy for patients with low and moderate motor-function. The distribution of muscle weightings within synergies changed as a consequence of therapy.
Lunardini et al. [111]	9 dystonia subjects 9 healthy subjects sEMG Extrinsic muscles: FCU, ECR, BB, TriB, Deltoid, Supraspinatus Intrinsic muscles: None	Writing task	Synergy analysis revealed no difference in the number of synergies between children with and without dystonia. Two synergies primarily involved upper limb distal muscles (distal synergies). Distal synergies were different depending on the task. The other two synergies mainly included proximal muscles (proximal synergies). Proximal synergies were very similar across groups and tasks: Synergy 3 involved shoulder flexors (D), while synergy 4 mainly comprised shoulder extensors (D and supraspinatus).
Pellegrino et al. [112]	11 healthy subjects 11 subjects with multiple sclerosis sEMG Extrinsic muscles: 15 upper limb muscles with only two forearm muscles (Br, PT) Intrinsic muscles: None	Reaching tasks: subjects grasped the handle	For both populations, the analysis identified three primary synergies that involved the distal muscles, another synergy that involved proximal muscles, and the last synergy included shoulder muscles. Muscle synergy analysis detected aspects related to muscle coordination that were not evident from the analysis of single muscle activity.

Table 2. *Cont.*

Study	Relevant Information	Description of the Task	Observations about Muscles Role
Scano et al. [62]	28 healthy subjects sEMG Extrinsic muscles: One band of 8 electrodes surrounding the forearm + BB, TriB, finger flexor and extensor Intrinsic muscles: None	Performance of 20 grasps	Ten spatial motor modules, properly elicited in time, are enough to describe the whole dataset with good accuracy, generalizing across subjects. The coactivating group composed of forearm electrodes is very often grouped together, especially in the hold phase. Two activation patterns are recognizable: a strong coactivation, often (but not always) corresponding to the grasp/hold phase, and two minor coactivating patterns in the pre-shaping and release phases that are often grouped in a single synergy. BB is activated during the reaching phase, thereby confirming that it is indeed an active reaching component that is active in the pre-shaping and release phase.

Abbreviations: ADM—Abductor Digiti Minimi; FLUM—First Lumbrical; FDI—First Dorsal Interosseous; FPI—First Palmar Interosseous; ECU—Extensor Carpi Ulnaris; PT—Pronator Teres; FCR—Flexor Carpi Radialis; FDS—Flexor Digitorum Superficialis; FDP—Flexor Digitorum Profundus.

4. Discussion

This literature review found 42 papers that matched the defined search criteria: 21 papers regarding the role played by hand and forearm muscles, and 21 dealing with hand muscle synergies.

First, studies focused on the role of specific small sets of forearm/hand muscles during some common tasks and grasps, sport activities, and working tasks were analyzed. Both extrinsic and intrinsic forearm muscles are required to accomplish these tasks, the extrinsic ones being responsible for the gross movements and the intrinsic ones in command of the fine movements, but they also complement each other. Some specific muscles show a high level of activation across all the tasks, such as EDC, while others seem to have a specific role, such as ECR as a wrist stabilizer. The tasks performed in ADL seem to require moderate levels of co-contraction of forearm muscles, needing the cooperation between different groups of Muscles, this cooperation being non task-dependent [75]. Thumb muscles, such as EPL and EPB, are able to activate separately from the flexors and extensors of the other fingers, and their important role in grasps has been widely demonstrated in the literature.

Second, the muscle synergies reflecting the relationship between muscles provide information in two domains: co-contractions and timing of activation. Therefore, studying muscle synergies can help to reach a better interpretation of the role of the muscles during the execution of different movements/tasks. The idea is consistent with the concept that the central nervous system may embed a modular structure that relies on a limited number of synergies at hand level. Non-negative matrix factorization and principal component analysis methods are the most used and present similar results in terms of coordination patterns. However, the non-matrix factorization method is the most preferred when a high number of sEMG are recorded [98]. The studies reviewed have demonstrated that a small subset of synergies could be generalized across tasks, representing basic building blocks underlying natural human hand motions/actions. Therefore, muscle synergy analysis could also be useful for comparing different therapies and evaluating the function recovery of subjects regarding ADL performance. It has been hypothesized that patients' functional deficit may be identified by regularly assessing their muscle synergy profile, which might be used to track the results of rehabilitation, and to adjust treatments [113]. Synergies have been suggested as being useful for clinicians to treat motor dysfunctions more effectively by organizing patients into subclasses and tailoring the treatment to each patient's specific deficit [113].

However, some important gaps have also been identified, which should be addressed in further studies. One of the main gaps found in the literature is the considerable diversity in the protocol design used to record sEMG from forearm muscles. Most of the studies do not specify the electrode placement, so it is not clear which locations were used to acquire the sEMG data. This makes it difficult to compare values or may affect the crosstalk level, which will depend on the longitudinal level of the muscle where it has been placed. Therefore, it could be useful to define a method of placing the sEMG electrodes that is comparable between subjects and that considers all the muscles involved in wrist and hand movements. It has been seen that this could be achieved by identifying the most representative forearm areas for ADL performance in terms of EMG activity [73].

The second main gap concerns the lack of representativeness of the tasks used for the EMG characterization. Most studies found in the literature are focused on studying the role of specific muscles during simple tasks (hand postures or free finger movements), or during single activities (such as writing or typing), or during small sets of very controlled activities (a few grasps, sport movement, etc.). However, only a few studies have dealt with the analysis of the forearm and hand muscles during ADL, and none of them consider a wide representative set of ADL. Therefore, defining a selection of a limited set of representative tasks would improve the current methodology, given the wide variety of ADL that can be performed by humans. Furthermore, standardization of the tasks would allow for comparison between subjects and sessions (important for tracking function recovery). The use of standardized tasks is especially important considering that each different individual may perform the same activity using several different strategies. Standardization would help in the comparison of muscular patterns and the identification of different strategies, by distinguishing between the different task phases [114].

In addition, to go further into synergies, simultaneous measurement of hand kinematics is not usually performed, and, when it is measured, it is used only to segment the different phases of the movement. Therefore, linked EMG-kinematic datasets, at the hand level, are very limited [81]. Such synchronized datasets are needed if we want to analyze how hand movements are produced and controlled. This could be helpful in some fields, like rehabilitation (to help choose the most suitable approaches) or prosthetics (to find a more reliable and natural control of hand prosthetics).

The review performed provides a basis of knowledge about the role of hand/forearm muscles, but the lack of a clear methodology introduces some limitations. These methodological inconsistencies add additional difficulty for an effective interpretation of findings and to draw any decisive conclusions.

Author Contributions: Writing—review and editing, N.J.J.-B.; Revision and supervision, M.V. and J.L.S.-B. All authors have read and agreed to the published version of the manuscript.

Funding: This research was funded by the Spanish Ministry of Science, Innovation and Universities (project PGC2018-095606-B-C21), and under a grant from the Spanish MINECO and the European Union (FEDER funds) through project DPI2014-52095-P and FPI grant BES-2015-072480.

Institutional Review Board Statement: Not applicable.

Informed Consent Statement: Not applicable.

Data Availability Statement: The study does not report any data.

Conflicts of Interest: The authors declare no conflict of interest.

Abbreviations

Ab/Ad	Abduction/adduction
AD	Anterior Deltoid
ADL	Activities of daily living
ADM	Abductor Digiti Minimi
APL	Abductor Policis Longus
APB	Abductor Policis Brevis
ADD	Thumb Adductor
BB	Biceps Brachii
Br	Brachioradialis
CMC	Carpometacarpal
ECR	Extensor Carpi Radialis
ECRB	Extensor Carpi Radialis Brevis
ECRL	Extensor Carpi Radialis Longus
ECU	Extensor Carpi Ulnaris
EDC	Extensor Digitorum Communis
EI	Extensor Indicis
EMG	Electromyography
EPB	Extensor Policis Brevis
EPL	Extensor Policis Longus
F/E	Flexion/extension
FCR	Flexor Carpi Radialis
FCU	Flexor Carpi Ulnaris
FDP	Flexor Digitorum Profundus
FDS	Flexor Digitorum Superficialis
FDI	First Dorsal Interosseous
FLUM	First Lumbrical
FPI	First Palmar Interosseous
Fw-EMG	Fine-wire EMG
iEMG	Intramuscular Electromyography
ICF	International Classification of Functioning, Disability and Health
IP	Interphalangeal
MCP	Metacarpophalangeal
MUAPs	Motor unit action potentials
MVC	Maximum Voluntary Contractions
OPP	Opponens policis
PD	Posterior Deltoid
PIP	Proximal interphalangeal
PL	Palmaris Longus
PT	Pronator Teres
sEMG	Surface Electromyography
SENIAM	Surface electromyography for the non-invasive assessment of Muscles
SHFT	Sollerman Hand Function Test
Tr	Trapezius
TriB	Triceps Brachii
WHO	World Health Organization
WMSD	Work-related musculoskeletal disorders

References

1. Vergara, M.; Gracia-Ibáñez, V.; Sancho-Bru, J.-L.L. Evaluation of Hand Functionality during Activities of Daily Living (ADL): A Review. In *Activities of Daily Living, Adl: Cultural Differences, Impacts of Disease and Long-Term Health Effects*, Lively, S.T., Ed.; Nova Science Pub Inc.: New York, NY, USA, 2015; pp. 103–132. ISBN 978-1-63463-941-5.
2. WHO. *International Classification of Functioning, Disability and Health (ICF)*; WHO: Geneva, Switzerland, 2001.
3. Vergara, M.; Sancho-Bru, J.L.; Gracia-Ibáñez, V.; Pérez-González, A. An introductory study of common grasps used by adults during performance of activities of daily living. *J. Hand Ther.* **2014**, *27*, 1–28. [CrossRef]

4. Bullock, I.M.; Zheng, J.Z.; Rosa, S.D.L.; Guertler, C.; Dollar, A.M. Grasp Frequency and Usage in Daily Household and Machine Shop Tasks. *IEEE Trans. Haptics* **2013**, *6*, 296–308. [CrossRef] [PubMed]
5. Yu, H.-L.; Chase, R.A.; Robert, A.; Strauch, B. *Atlas of Hand Anatomy and Clinical Implications*; Mosby: St. Louis, MO, USA, 2004; ISBN 0815179278.
6. Kapandji, A.I. Fisiologia Articular. In *Miembro Superior -6a ed. Madrid: 2006–2007*; Editorial Médica Panamericana: Madrid, Spain, 1996.
7. Brand, P.W.; Hollister, A.M. *Clinical Mechanics of the Hand*, 3rd ed.; Mosby Publishing: St. Louis, MO, USA, 1999.
8. Lee, K.-S.; Jung, M.-C. Ergonomic evaluation of biomechanical hand function. *Saf. Health Work* **2015**, *6*, 9–17. [CrossRef] [PubMed]
9. Oatis, C.A. *Kinesiology: The Mechanics and Pathomechanics of Human Movement*; Lippincott Williams & Wilkins: Philadelphia, PA, USA, 2009; ISBN 0781774225.
10. Lum, P.S.; Mulroy, S.; Amdur, R.L.; Requejo, P.; Prilutsky, B.I.; Dromerick, A.W. Gains in Upper Extremity Function After Stroke via Recovery or Compensation: Potential Differential Effects on Amount of Real-World Limb Use. *Top. Stroke Rehabil.* **2009**, *16*, 237–253. [CrossRef]
11. Dietz, V.; Schrafl-Altermatt, M. Control of functional movements in healthy and post-stroke subjects: Role of neural interlimb coupling. *Clin. Neurophysiol.* **2016**, *127*, 2286–2293. [CrossRef] [PubMed]
12. Elkwood, A.I.; Kaufman, M.; Schneider, L.F. (Eds.) *Rehabilitative Surgery*; Springer International Publishing: Cham, Switzerland, 2017; ISBN 978-3-319-41404-1.
13. Scheme, E.; Englehart, K. Electromyogram pattern recognition for control of powered upper-limb prostheses: State of the art and challenges for clinical use. *J. Rehabil. Res. Dev.* **2011**, *48*, 643. [CrossRef] [PubMed]
14. Hahne, J.M.; Wilke, M.A.; Koppe, M.; Farina, D.; Schilling, A.F. Longitudinal Case Study of Regression-Based Hand Prosthesis Control in Daily Life. *Front. Neurosci.* **2020**. [CrossRef]
15. Roche, A.D.; Lakey, B.; Mendez, I.; Vujaklija, I.; Farina, D.; Aszmann, O.C. Clinical Perspectives in Upper Limb Prostheses: An Update. *Curr. Surg. Rep.* **2019**, *7*, 5. [CrossRef]
16. Parker, P.; Englehart, K.; Hudgins, B. Myoelectric signal processing for control of powered limb prostheses. *J. Electromyogr. Kinesiol.* **2006**, *16*, 541–548. [CrossRef]
17. Reaz, M.B.I.; Hussain, M.S.; Mohd-Yasin, F. Techniques of EMG signal analysis: Detection, processing, classification and applications (Correction). *Biol. Proced. Online* **2006**, *8*, 163. [CrossRef]
18. Chowdhury, R.H.; Reaz, M.B.I.; Ali, M.A.B.M.; Bakar, A.A.A.; Chellappan, K.; Chang, T.G. Surface Electromyography Signal Processing and Classification Techniques. *Sensors* **2013**, *13*, 12431–12466. [CrossRef]
19. Khokhar, Z.O.; Xiao, Z.G.; Menon, C. Surface EMG pattern recognition for real-time control of a wrist exoskeleton. *Biomed. Eng. Online* **2010**, *9*, 41. [CrossRef]
20. Kamen, G.; Gabriel, D.A. *Essentials of Electromyography*; Human Kinetics: Champaign, IL, USA, 2010; ISBN 9780736085502.
21. Mogk, J.P.M.; Keir, P.J. Crosstalk in surface electromyography of the proximal forearm during gripping tasks. *J. Electromyogr. Kinesiol.* **2003**, *13*, 63–71. [CrossRef]
22. Holobar, A.; Farina, D.; Gazzoni, M.; Merletti, R.; Zazula, D. Estimating motor unit discharge patterns from high-density surface electromyogram. *Clin. Neurophysiol.* **2009**. [CrossRef] [PubMed]
23. Konrad, P. *The ABC of EMG A Practical Introduction to Kinesiological Electromyography*; Noraxon: Scottsdale, AZ, USA, 2005; pp. 1–60.
24. Kumar, D.; Poosapadi Arjunan, S.; Singh, V. Towards identification of finger flexions using single channel surface electromyography—Able bodied and amputee subjects. *J. Neuroeng. Rehabil.* **2013**, *10*, 50. [CrossRef] [PubMed]
25. Rojas-Martínez, M.; Mañanas, M.A.; Alonso, J.F. High-density surface EMG maps from upper-arm and forearm muscles. *J. Neuroeng. Rehabil.* **2012**, *9*, 85. [CrossRef] [PubMed]
26. Castellini, C.; Fiorilla, A.E.; Sandini, G. Multi-subject/daily-life activity EMG-based control of mechanical hands. *J. Neuroeng. Rehabil.* **2009**, *6*, 41. [CrossRef]
27. Hogrel, J.-Y. Clinical applications of surface electromyography in neuromuscular disorders. *Neurophysiol. Clin. Neurophysiol.* **2005**, *35*, 59–71. [CrossRef]
28. Bonnefoy, A.; Louis, N.; Gorce, P. Muscle activation during a reach-to-grasp movement in sitting position: Influence of the distance. *J. Electromyogr. Kinesiol.* **2009**, *19*, 269–275. [CrossRef]
29. Hoozemans, M.J.M.; Van Dieën, J.H. Prediction of handgrip forces using surface EMG of forearm muscles. *J. Electromyogr. Kinesiol.* **2005**, *15*, 358–366. [CrossRef]
30. Johanson, M.E.; Valero-Cuevas, F.J.; Hentz, V.R. Activation patterns of the thumb muscles during stable and unstable pinch tasks. *J. Hand Surg. Am.* **2001**, *26*, 698–705. [CrossRef] [PubMed]
31. Park, S. Comparison of Muscle Activation during Dominant Hand Wrist Flexion when Writing. *J. Phys. Ther. Sci.* **2013**, *25*, 1529–1531. [CrossRef] [PubMed]
32. Linderman, M.; Lebedev, M.A.; Erlichman, J.S. Recognition of handwriting from electromyography. *PLoS ONE* **2009**, *4*. [CrossRef]
33. Duque, J.; Masset, D.; Malchaire, J. Evaluation of handgrip force from EMG measurements. *Appl. Ergon.* **1995**, *26*, 61–66. [CrossRef]
34. Kamavuako, E.N.; Farina, D.; Yoshida, K.; Jensen, W. Relationship between grasping force and features of single-channel intramuscular EMG signals. *J. Neurosci. Methods* **2009**, *185*, 143–150. [CrossRef]

35. De Almeida, P.H.T.Q.; da Cruz, D.M.C.; Magna, L.A.; Ferrigno, I.S.V. An electromyographic analysis of two handwriting grasp patterns. *J. Electromyogr. Kinesiol.* **2013**, *23*, 838–843. [CrossRef]
36. Maier, M.A.; Hepp-Reymond, M.-C. EMG activation patterns during force production in precision grip. *Exp. Brain Res.* **1995**, *103*, 108–122. [CrossRef]
37. Danion, F.; Galléa, C. The relation between force magnitude, force steadiness, and muscle co-contraction in the thumb during precision grip. *Neurosci. Lett.* **2004**, *368*, 176–180. [CrossRef] [PubMed]
38. Cooney, W.P.; An, K.-N.; Daube, J.R.; Askew, L.J. Electromyographic analysis of the thumb: A study of isometric forces in pinch and grasp. *J. Hand Surg. Am.* **1985**, *10*, 202–210. [CrossRef]
39. Kerkhof, F.D.; Deleu, G.; D'Agostino, P.; Vereecke, E.E. Subject-specific thumb muscle activity during functional tasks of daily life. *J. Electromyogr. Kinesiol.* **2016**, *30*, 131–136. [CrossRef] [PubMed]
40. Birdwell, J.A.; Hargrove, L.J.; Kuiken, T.A.; Weir, R.F. ff Activation of individual extrinsic thumb muscles and compartments of extrinsic finger muscles. *J. Neurophysiol.* **2013**, *110*, 1385–1392. [CrossRef]
41. Van Galen, G.P.; Müller, M.L.T.M.; Meulenbroek, R.G.J.; Van Gemmert, A.W.A. Forearm EMG response activity during motor performance in individuals prone to increased stress reactivity. *Am. J. Ind. Med.* **2002**, *41*, 406–419. [CrossRef]
42. Cifrek, M.; Medved, V.; Tonković, S.; Ostojić, S. Surface EMG Based Muscle Fatigue Evaluation in Biomechanics. *Clin. Biomech.* **2009**, *24*, 327–340. [CrossRef] [PubMed]
43. Taib, M.F.M.; Bahn, S.; Yun, M.H. The effect of psychosocial stress on muscle activity during computer work: Comparative study between desktop computer and mobile computing products. *Work* **2016**, *54*, 543–555. [CrossRef] [PubMed]
44. Straker, L.; Burgess-Limerick, R.; Pollock, C.; Maslen, B. The influence of desk and display design on posture and muscle activity variability whilst performing information technology tasks. *Appl. Ergon.* **2009**, *40*, 852–859. [CrossRef] [PubMed]
45. Kao, H.C.; Lin, C.J.; Lee, Y.H.; Chen, S.H. The Effects of Direction of Exertion, Path, and Load Placement in Nursing Cart Pushing and Pulling Tasks: An Electromyographical Study. *PLoS ONE* **2015**, *10*, e0140792. [CrossRef] [PubMed]
46. Di Domizio, J.; Keir, P.J. Forearm posture and grip effects during push and pull tasks. *Ergonomics* **2010**, *53*, 336–343. [CrossRef]
47. Hägg, G.M.; Milerad, E. Forearm extensor and flexor muscle exertion during simulated gripping work—An electromyographic study. *Clin. Biomech.* **1997**, *12*, 39–43. [CrossRef]
48. Sako, S.; Sugiura, H.; Tanoue, H.; Kojima, M.; Kono, M.; Inaba, R. Electromyographic analysis of relevant muscle groups during completion of computer tasks using different computer mouse positions. *Int. J. Occup. Saf. Ergon.* **2017**, *23*, 267–273. [CrossRef]
49. Szeto, G.P.Y.; Lin, J.K.M. A study of forearm muscle activity and wrist kinematics in symptomatic office workers performing mouse-clicking tasks with different precision and speed demands. *J. Electromyogr. Kinesiol.* **2011**, *21*, 59–66. [CrossRef]
50. Marta, S.; Silva, L.; Castro, M.A.; Pezarat-Correia, P.; Cabri, J. Electromyography variables during the golf swing: A literature review. *J. Electromyogr. Kinesiol.* **2012**, *22*, 803–813. [CrossRef]
51. Chow, J.W.; Carlton, L.G.; Lim, Y.T.; Shim, J.H.; Chae, W.S.; Kuenster, A.F. Muscle activation during the tennis volley. *Med. Sci. Sports Exerc.* **1999**, *31*, 846–854. [CrossRef] [PubMed]
52. Alizadehkhaiyat, O.; Frostick, S.P. Electromyographic assessment of forearm muscle function in tennis players with and without Lateral Epicondylitis. *J. Electromyogr. Kinesiol.* **2015**, *25*, 876–886. [CrossRef] [PubMed]
53. Hatch, G.F.; Pink, M.M.; Mohr, K.J.; Sethi, P.M.; Jobe, F.W. The effect of tennis racket grip size on forearm muscle firing patterns. *Am. J. Sports Med.* **2006**, *34*, 1977–1983. [CrossRef] [PubMed]
54. Rota, S.; Morel, B.; Saboul, D.; Rogowski, I.; Hautier, C. Influence of fatigue on upper limb muscle activity and performance in tennis. *J. Electromyogr. Kinesiol.* **2014**, *24*, 90–97. [CrossRef]
55. Sorbie, G.G.; Hunter, H.H.; Grace, F.M.; Gu, Y.; Baker, J.S.; Ugbolue, U.C. An electromyographic study of the effect of hand grip sizes on forearm muscle activity and golf performance. *Res. Sport. Med.* **2016**, *24*, 207–218. [CrossRef] [PubMed]
56. Santello, M.; Baud-Bovy, G.; Jörntell, H. Neural bases of hand synergies. *Front. Comput. Neurosci.* **2013**, *7*, 23. [CrossRef]
57. Santello, M.; Flanders, M.; Soechting, J.F. Postural hand synergies for tool use. *J. Neurosci.* **1998**, *18*, 10105–10115. [CrossRef] [PubMed]
58. Gracia-Ibáñez, V.; Sancho-Bru, J.L.; Vergara, M.; Jarque-Bou, N.J.; Roda-Sales, A. Sharing of hand kinematic synergies across subjects in daily living activities. *Sci. Rep.* **2020**. [CrossRef]
59. Jarrassé, N.; Ribeiro, A.T.; Sahbani, A.; Bachta, W.; Roby-Brami, A. Analysis of hand synergies in healthy subjects during bimanual manipulation of various objects. *J. Neuroeng. Rehabil.* **2014**. [CrossRef]
60. Weiss, E.J.; Flanders, M. Muscular and Postural Synergies of the Human Hand. *J. Neurophysiol.* **2004**, *92*, 523–535. [CrossRef] [PubMed]
61. Zariffa, J.; Steeves, J.; Pai, D.K. Changes in hand muscle synergies in subjects with spinal cord injury: Characterization and functional implications. *J. Spinal Cord Med.* **2012**, *35*, 310–318. [CrossRef] [PubMed]
62. Scano, A.; Chiavenna, A.; Molinari Tosatti, L.; Müller, H.; Atzori, M. Muscle Synergy Analysis of a Hand-Grasp Dataset: A Limited Subset of Motor Modules May Underlie a Large Variety of Grasps. *Front. Neurorobot.* **2018**, *12*, 57. [CrossRef] [PubMed]
63. Winges, S.A.; Furuya, S.; Faber, N.J.; Flanders, M. Patterns of muscle activity for digital coarticulation. *J. Neurophysiol.* **2013**, *110*, 230–242. [CrossRef]
64. D'Andola, M.; Cesqui, B.; Portone, A.; Fernandez, L.; Lacquaniti, F.; d'Avella, A. Spatiotemporal characteristics of muscle patterns for ball catching. *Front. Comput. Neurosci.* **2013**, *7*, 107. [CrossRef]

65. Ertan, H.; Kentel, B.; Tümer, S.T.; Korkusuz, F. Activation patterns in forearm muscles during archery shooting. *Hum. Mov. Sci.* **2003**, *22*, 37–45. [CrossRef]
66. VencesBrito, A.M.; Rodrigues Ferreira, M.A.; Cortes, N.; Fernandes, O.; Pezarat-Correia, P. Kinematic and electromyographic analyses of a karate punch. *J. Electromyogr. Kinesiol.* **2011**, *21*, 1023–1029. [CrossRef] [PubMed]
67. Ertan, H. Muscular activation patterns of the bow arm in recurve archery. *J. Sci. Med. Sport* **2009**, *12*, 357–360. [CrossRef]
68. Ricci, F.P.F.M.; Santiago, P.R.P.; Zampar, A.C.; Pinola, L.N.; de Fonseca, M.C.R. Upper extremity coordination strategies depending on task demand during a basic daily activity. *Gait Posture* **2015**, *42*, 472–478. [CrossRef] [PubMed]
69. Geethanjali, P. Myoelectric control of prosthetic hands: State-of-the-art review. *Med. Devices Evid. Res.* **2016**, *9*, 247–255. [CrossRef]
70. Johnson, S.S.; Mansfield, E. Prosthetic training: Upper limb. *Phys. Med. Rehabil. Clin. N. Am.* **2014**, *25*, 133–151. [CrossRef] [PubMed]
71. Kaufman, K.R.; An, K.N.; Litchy, W.J.; Cooney, W.P.; Chao, E.Y. In-vivo function of the thumb muscles. *Clin. Biomech.* **1999**, *14*, 141–150. [CrossRef]
72. Disselhorst-Klug, C.; Schmitz-Rode, T.; Rau, G. Surface electromyography and muscle force: Limits in sEMG–force relationship and new approaches for applications. *Clin. Biomech.* **2009**, *24*, 225–235. [CrossRef]
73. Jarque-Bou, N.J.; Vergara, M.; Sancho-Bru, J.L.; Roda-Sales, A.; Gracia-Ibáñez, V. Identification of forearm skin zones with similar muscle activation patterns during activities of daily living. *J. Neuroeng. Rehabil.* **2018**, *15*. [CrossRef]
74. Jakobi, J.M.; Edwards, D.L.; Connelly, D.M. Utility of Portable Electromyography for Quantifying Muscle Activity during Daily Use. *Gerontology* **2008**, *54*, 324–331. [CrossRef] [PubMed]
75. Peters, K.M.; Kelly, V.E.; Chang, T.; Weismann, M.C.; Westcott McCoy, S.; Steele, K.M. Muscle recruitment and coordination during upper-extremity functional tests. *J. Electromyogr. Kinesiol.* **2018**, *38*, 143–150. [CrossRef] [PubMed]
76. Kern, D.S.; Semmler, J.G.; Enoka, R.M. Long-term activity in upper- and lower-limb muscles of humans. *J. Appl. Physiol.* **2001**, *91*, 2224–2232. [CrossRef]
77. Jebsen, R.H.; Taylor, N.; Trieschmann, R.B.; Trotter, M.J.; Howard, L.A. An objective and standardized test of hand function. *Arch. Phys. Med. Rehabil.* **1969**, *50*, 311–319.
78. Barreca, S.; Kelly Gowland, C.; Stratford, P.; Huijbregts, M.; Griffiths, J.; Torresin, W.; Dunkley, M.; Miller, P.; Masters, L. Development of the Chedoke Arm and Hand Activity Inventory: Theoretical Constructs, Item Generation, and Selection. *Top. Stroke Rehabil.* **2004**, *11*, 31–42. [CrossRef]
79. Brogardh, C.; Persson, A.L.; Sjolund, B.H. Intra- and inter-rater reliability of the Sollerman hand function test in patients with chronic stroke. *Disabil. Rehabil.* **2007**, *29*, 145–154. [CrossRef] [PubMed]
80. Kilbreath, S.L.; Gandevia, S.C. Limited independent flexion of the thumb and fingers in human subjects. *J. Physiol.* **1994**, *479 Pt 3*, 487–497. [CrossRef]
81. Jarque-Bou, N.J.; Vergara, M.; Sancho-Bru, J.L.; Gracia-Ibáñez, V.; Roda-Sales, A. A calibrated database of kinematics and EMG of the forearm and hand during activities of daily living. *Sci. Data* **2019**, *6*, 270. [CrossRef] [PubMed]
82. Tresch, M.C.; Jarc, A. The case for and against muscle synergies. *Curr. Opin. Neurobiol.* **2009**, *19*, 601–607. [CrossRef]
83. Schieber, M.H.; Lang, C.E.; Reilly, K.T.; McNulty, P.; Sirigu, A. Selective activation of human finger muscles after stroke or amputation. *Adv. Exp. Med. Biol.* **2009**, *629*, 559–575. [CrossRef] [PubMed]
84. Kutch, J.J.; Kuo, A.D.; Bloch, A.M.; Rymer, W.Z. Endpoint force fluctuations reveal flexible rather than synergistic patterns of muscle cooperation. *J. Neurophysiol.* **2008**, *100*, 2455–2471. [CrossRef] [PubMed]
85. Valero-Cuevas, F.J.; Venkadesan, M.; Todorov, E. Structured variability of muscle activations supports the minimal intervention principle of motor control. *J. Neurophysiol.* **2009**, *102*, 59–68. [CrossRef] [PubMed]
86. Jo, H.J.; Maenza, C.; Good, D.C.; Huang, X.; Park, J.; Sainburg, R.L.; Latash, M.L. Effects of unilateral stroke on multi-finger synergies and their feed-forward adjustments. *Neuroscience* **2016**, *319*, 194–205. [CrossRef] [PubMed]
87. Valero-Cuevas, F.J.; ZAJAC, F.E.; Burgar, C.G. Large index-fingertip forces are produced by subject-independent patterns of muscle excitation. *J. Biomech.* **1998**, *31*, 693–703. [CrossRef]
88. Valero-Cuevas, F.J. Predictive modulation of muscle coordination pattern magnitude scales fingertip force magnitude over the voluntary range. *J. Neurophysiol.* **2000**, *83*, 1469–1479. [CrossRef]
89. Johnston, J.A.; Winges, S.A.; Santello, M. Neural control of hand muscles during prehension. *Adv. Exp. Med. Biol.* **2009**, *629*, 577–596.
90. Gribble, P.L.; Mullin, L.I.; Cothros, N.; Mattar, A. Role of Cocontraction in Arm Movement Accuracy. *J. Neurophysiol.* **2003**, *89*, 2396–2405. [CrossRef]
91. Milner, T.E.; Cloutier, C. Compensation for mechanically unstable loading in voluntary wrist movement. *Exp. Brain Res.* **1993**, *94*, 522–532. [CrossRef]
92. D'Avella, A.; Fernandez, L.; Portone, A.; Lacquaniti, F. Modulation of phasic and tonic muscle synergies with reaching direction and speed. *J. Neurophysiol.* **2008**, *100*, 1433–1454. [CrossRef] [PubMed]
93. d'Avella, A.; Lacquaniti, F. Control of reaching movements by muscle synergy combinations. *Front. Comput. Neurosci.* **2013**, *7*, 42. [CrossRef] [PubMed]
94. Smeets, J.B.J.; Brenner, E. Synergies in Grasping. In *Advances in Experimental Medicine and Biology*; Springer: Cham, Switzerland, 2016; Volume 957, pp. 21–34.

95. Klein Breteler, M.D.; Simura, K.J.; Flanders, M. Timing of muscle activation in a hand movement sequence. *Cereb. Cortex* **2007**, *17*, 803–815. [CrossRef] [PubMed]
96. Martelloni, C.; Carpaneto, J.; Micera, S. Characterization of EMG patterns from proximal arm muscles during object- and orientation-specific grasps. *IEEE Trans. Biomed. Eng.* **2009**, *56*, 2529–2536. [CrossRef] [PubMed]
97. Rabbi, M.F.; Pizzolato, C.; Lloyd, D.G.; Carty, C.P.; Devaprakash, D.; Diamond, L.E. Non-negative matrix factorisation is the most appropriate method for extraction of muscle synergies in walking and running. *Sci. Rep.* **2020**. [CrossRef]
98. Ebied, A.; Kinney-Lang, E.; Spyrou, L.; Escudero, J. Evaluation of matrix factorisation approaches for muscle synergy extraction. *Med. Eng. Phys.* **2018**. [CrossRef] [PubMed]
99. Castellini, C.; Van Der Smagt, P. Evidence of muscle synergies during human grasping. *Biol. Cybern.* **2013**, *107*, 233–245. [CrossRef]
100. Burkhart, T.A.; Andrews, D.M. Kinematics, kinetics and muscle activation patterns of the upper extremity during simulated forward falls. *J. Electromyogr. Kinesiol.* **2013**, *23*, 688–695. [CrossRef]
101. d'Avella, A.; Saltiel, P.; Bizzi, E. Combinations of muscle synergies in the construction of a natural motor behavior. *Nat. Neurosci.* **2003**, *6*, 300–308. [CrossRef]
102. Gazzoni, M.; Celadon, N.; Mastrapasqua, D.; Paleari, M.; Margaria, V.; Ariano, P. Quantifying forearm muscle activity during wrist and finger movements by means of multi-channel electromyography. *PLoS ONE* **2014**, *9*, e109943. [CrossRef]
103. Hu, X.; Suresh, N.L.; Xue, C.; Rymer, W.Z. Extracting extensor digitorum communis activation patterns using high-density surface electromyography. *Front. Physiol.* **2015**, *6*. [CrossRef]
104. Tanzarella, S.; Muceli, S.; Del Vecchio, A.; Casolo, A.; Farina, D. Non-invasive analysis of motor neurons controlling the intrinsic and extrinsic muscles of the hand. *J. Neural Eng.* **2020**. [CrossRef]
105. Hagert, E. Proprioception of the Wrist Joint: A Review of Current Concepts and Possible Implications on the Rehabilitation of the Wrist. *J. Hand Ther.* **2010**, *23*, 2–17. [CrossRef] [PubMed]
106. Hsiao, L.-P.; Cho, C.-Y. The effect of aging on muscle activation and postural control pattern for young and older computer users. *Appl. Ergon.* **2012**, *43*, 926–932. [CrossRef]
107. Cheung, V.C.K.; Turolla, A.; Agostini, M.; Silvoni, S.; Bennis, C.; Kasi, P.; Paganoni, S.; Bonato, P.; Bizzi, E. Muscle synergy patterns as physiological markers of motor cortical damage. *Proc. Natl. Acad. Sci. USA* **2012**, *109*, 14652–14656. [CrossRef]
108. Roh, J.; Rymer, W.Z.; Beer, R.F. Evidence for altered upper extremity muscle synergies in chronic stroke survivors with mild and moderate impairment. *Front. Hum. Neurosci.* **2015**, *9*, 1–14. [CrossRef] [PubMed]
109. Lee, S.W.; Triandafilou, K.; Lock, B.A.; Kamper, D.G. Impairment in task-specific modulation of muscle coordination correlates with the severity of hand impairment following stroke. *PLoS ONE* **2013**, *8*, e68745. [CrossRef]
110. Hesam-Shariati, N.; Trinh, T.; Thompson-Butel, A.G.; Shiner, C.T.; McNulty, P.A. A Longitudinal Electromyography Study of Complex Movements in Poststroke Therapy. 1: Heterogeneous Changes Despite Consistent Improvements in Clinical Assessments. *Front. Neurol.* **2017**, *8*, 340. [CrossRef] [PubMed]
111. Lunardini, F.; Casellato, C.; Bertucco, M.; Sanger, T.D.; Pedrocchi, A. Children With and Without Dystonia Share Common Muscle Synergies While Performing Writing Tasks. *Ann. Biomed. Eng.* **2017**, *45*, 1949–1962. [CrossRef] [PubMed]
112. Pellegrino, L.; Coscia, M.; Muller, M.; Solaro, C.; Casadio, M. Evaluating upper limb impairments in multiple sclerosis by exposure to different mechanical environments. *Sci. Rep.* **2018**, *8*, 2110. [CrossRef]
113. Safavynia, S.A.; Torres-Oviedo, G.; Ting, L.H. Muscle Synergies: Implications for Clinical Evaluation and Rehabilitation of Movement. *Top. Spinal Cord Inj. Rehabil.* **2011**, *17*, 16–24. [CrossRef] [PubMed]
114. Hebert, J.S.; Lewicke, J.; Williams, T.R.; Vette, A.H. Normative data for modified Box and Blocks test measuring upper-limb function via motion capture. *J. Rehabil. Res. Dev.* **2014**, *51*, 918–932. [CrossRef] [PubMed]

sensors MDPI

Review

Control Strategies and Performance Assessment of Upper-Limb TMR Prostheses: A Review

Federico Mereu [1,†], Francesca Leone [1,†], Cosimo Gentile [1], Francesca Cordella [1], Emanuele Gruppioni [2] and Loredana Zollo [1,*]

1 Unit of Advanced Robotics and Human-Centred Technologies, Università Campus Bio-Medico di Roma, 00128 Rome, Italy; f.mereu@unicampus.it (F.M.); f.leone@unicampus.it (F.L.); c.gentile@unicampus.it (C.G.); f.cordella@unicampus.it (F.C.)
2 INAIL Prosthetic Center, 40054 Vigorso di Budrio, Italy; e.gruppioni@inail.it
* Correspondence: l.zollo@unicampus.it
† These authors contributed equally to this work.

Abstract: The evolution of technological and surgical techniques has made it possible to obtain an even more intuitive control of multiple joints using advanced prosthetic systems. Targeted Muscle Reinnervation (TMR) is considered to be an innovative and relevant surgical technique for improving the prosthetic control for people with different amputation levels of the limb. Indeed, TMR surgery makes it possible to obtain reinnervated areas that act as biological amplifiers of the motor control. On the technological side, a great deal of research has been conducted in order to evaluate various types of myoelectric prosthetic control strategies, whether direct control or pattern recognition-based control. In the literature, different control performance metrics, which have been evaluated on TMR subjects, have been introduced, but no accepted reference standard defines the better strategy for evaluating the prosthetic control. Indeed, the presence of several evaluation tests that are based on different metrics makes it difficult the definition of standard guidelines for comprehending the potentiality of the proposed control systems. Additionally, there is a lack of evidence about the comparison of different evaluation approaches or the presence of guidelines on the most suitable test to proceed for a TMR patients case study. Thus, this review aims at identifying these limitations by examining the several studies in the literature on TMR subjects, with different amputation levels, and proposing a standard method for evaluating the control performance metrics.

Keywords: Targeted Muscle Reinnervation (TMR); upper limb amputee; prosthesis; prosthetic control; multi-DoF control; pattern recognition

Citation: Mereu, F.; Leone, F.; Gentile, C.; Cordella, F.; Gruppioni, E.; Zollo, L. Control Strategies and Performance Assessment of Upper-Limb TMR Prostheses: A Review. *Sensors* **2021**, *21*, 1953. https://doi.org/10.3390/s21061953

Academic Editor: Ernest N. Kamavuako

Received: 29 January 2021
Accepted: 5 March 2021
Published: 10 March 2021

1. Introduction

The amputation of the upper limb causes a huge decrease in dexterity, with a significant reduction in patients' quality of life. People who have had an upper-limb amputation need a prosthesis that replaces the lost arm functionality. It is very difficult to find epidemiology data on amputations of the upper limb. Over the world, the population of amputees was estimated as 10 million, 30% of whom are upper limb amputees [1]. Focusing on European countries, in Italy, there were 2720 upper-limb amputations in 2018, being equal to 18% of total amputations [2]; in 2003, in the UK, there were 5767 new amputations, 5% were upper limb amputees [3]. Between 2004 and 2013, only in the adult hand emergency clinic of the Nancy University Hospital (France), 2247 patients suffered an upper limb amputation (partial and pediatric amputation excluded) that was traumatic in 76.32% of cases [4]. In the USA, approximately 340,000 people have also suffered the loss of a limb and every year there are 10,000 new upper limb amputations, as reported by the National Center for Health Statistics [5]. The relevance of the upper limb loss has pushed international research to seek new prosthetic solutions [6,7].

Prosthesis technology ranges from passive or cosmetic typologies on one end to active or functional types on the other. Cosmetic prostheses are used to only restore the aesthetic aspect [8], while active ones are used to restore, as far as possible, the functionality of the lost arm. Active prostheses can be further classified into body powered, that exploit cables to control the device with the more proximal joints, and externally powered (electric or pneumatic) [9], which allow for the movement of the motors of the joints making up the prosthesis [10]. Externally powered prostheses require a control system in order to associate an input signal (generated by a sensor and/or a button) to an output action. One of the most used control systems is the myoelectric one, which exploits the electromyographic (EMG) signals of a specific muscle to provide discrete movement and of an antagonist muscle group to make complementary movements. EMG signals have been used to control prostheses since 1948 [11] and, over the years, various control strategies have been identified: among these, the control strategies that directly associate a movement of the prosthetic limb to an EMG input signal are usually referred to as Direct Myoelectric Control or simply Direct Control (DC). The conventional techniques, to control from one Degree of Freedom (DoF) to multiple DoFs [12], are [13]: the on/off strategy, which is typically used to control one DoF and allowing the performance of two opposite movements based on the exceeding of a preset threshold by the EMG amplitude of two residual antagonist muscles; the proportional control strategy that considers, instead, the voltage applied to the motor proportional to the contraction level/intensity of EMG signals.

Such control strategies are generally associated with a method for selecting the joint to be controlled. The co-contraction method is the first one, which allows the patient to change from one joint to another by simultaneously contracting the muscles used to control the joint; however, the principal limit of this technique is that it is possible to only control one joint at a time. The second is the simultaneous method that is used to control multi-DoF prostheses, handling more than one joint at the same time. However, in this case, the number of controllable DoFs depends on the number of independent EMG control sites [6] Figure 1.

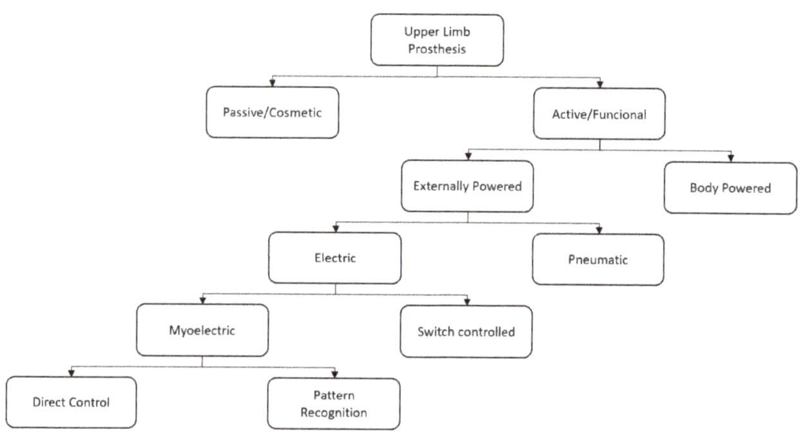

Figure 1. A Block diagram describing the types of Upper Limb Prostheses and control approaches.

The Pattern Recognition (PR) based myoelectric control has been proposed to reach a more intuitive and adaptable control of the prosthesis in order to overcome these limits and eliminate the need for mode switching; these PR strategies do not require independent muscle sites, but they consider muscular activation patterns of different muscle sites to classify several motion classes. In spite of the progress made in this field based on EMG-PR [14], the main reasons for the abandonment of the prosthetic device by many users can be imputed to comfort and function [15]. Regarding the function, technological factors, as the type of prosthetic control and the presence or absence of sensory feedback [16], have a key role in avoiding the decision of prosthesis abandonment. An indispensable requirement is that the control system must be simple, direct, and user-friendly [17].

To reach these requirements, a major advancement in the field of upper-limb prosthetics has been reached with the Targeted Muscle Reinnervation (TMR) surgery, which was developed by Prof. Todd Kuiken and his team at the Rehabilitation Institute of Chicago [18]. The idea behind TMR was that reinnervating the residual nerves of the amputated limb to new target muscles may allow users to more intuitively control the prosthesis and simply perform Activities of Daily Living (ADL). This is because the patient's intention, manifested with the phantom limb, can be sent, as neural information, to the reinnervated muscle, which amplifies the EMG signal that is used to control the prosthetic device.

Several control strategies were proposed in the literature for making myoelectric prostheses control easy, reliable, efficient, and, therefore, for lowering the users' cognitive burden for TMR patients. However, the optimal control system that allows users to control a multiple DoF prosthesis with dexterity, and by using intuitive interfaces between the user and the device, has not yet been developed [19].

2. Targeted Muscle Reinnervation

After an upper-extremity amputation, the employment of TMR allows for improving the functionality of myoelectric prostheses: the reinnervation of residual muscles creates additional myoelectric control sites available for obtaining the multi-DoF prosthetic control, without the need of switching between modalities available on the device [20]. In 1995, Kuiken examined muscle recovery and related changes in the motor unit population of "hyper-reinnervated" rats [18]. Only in 2004, the first TMR surgery was performed on one human subject with bilateral shoulder disarticulation amputation [21] Figure 2.

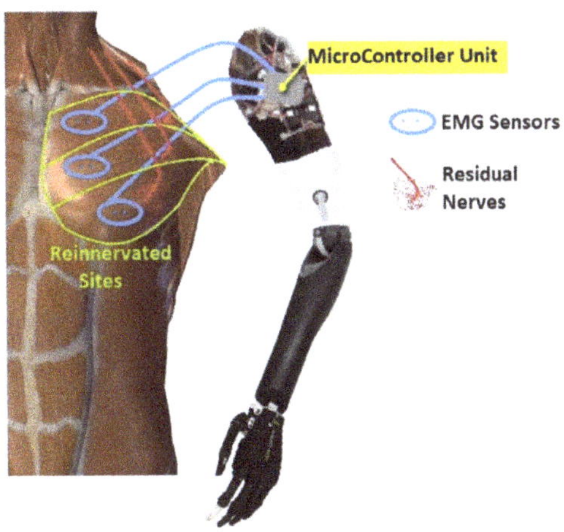

Figure 2. An example of a prosthesis control system after Targeted Muscle Reinnervation (TMR) surgery: the electromyographic (EMG) sensors collected from the reinnervated sites the EMG signals and communicated to the MicroController Unit (MCU) the user's intention to translate into arm and hand movements.

In 2006, Kuiken introduced the following requirements to make TMR surgery successful: (i) separate regions of muscles and skin must be reinnervated by multiple donor nerves; (ii) EMG signals must be acquired from each target area; and, (iii) the prosthesis must be able to receive numerous EMG input signals and control several motors [22]. TMR can be performed for three different levels of amputation: shoulder disarticulation, transhumeral, and transradial amputation. The innervation strategies depend on the type of amputation [23]. For the shoulder disarticulated patients Figure 3B, pectoralis muscles are usually denervated and then reinnervated with residual arm peripheral nerves [22]. Afterward, back muscles (if possible) are also reinnervated to have more active sites. For the transhumeral amputees Figure 3A, the median nerve is transferred to the short head of the biceps motor branch to restore the function of hand closing or pronation; the ulnar nerve is transferred to a residual brachialis motor branch to have additional control sites for hand closing; finally, the radial nerve is reinnervated to the lateral head of the triceps motor branch in order to control hand opening or supination [24]. For transradial amputees, the control of multifunctional prosthetic hands can be reached by using additional Targeted Muscle Reinnervation signals for improving the function of intrinsic finger and thumb muscles: the distal median nerve is transferred to the flexor digitorum superficialis, while the ulnar nerve is reinnervated to the flexor carpi ulnaris [25]. When the muscles usually chosen cannot be reinnervated, as in [26], three bundles of the anterior tight muscle are used to obtain three active sites for the prosthetic control. The TMR is also an emerging technique for the treatment and reduction of the phantom limb pain (PLP) and neuroma pain [27], for the osseointegrated prostheses [28], and for the targeted sensory Reinnervation [29] of bidirectional neuroprosthetic devices. Finally, another important outcome is the use of TMR in the oncologic population, due to the potential to reduce pain without the use of opioids [30].

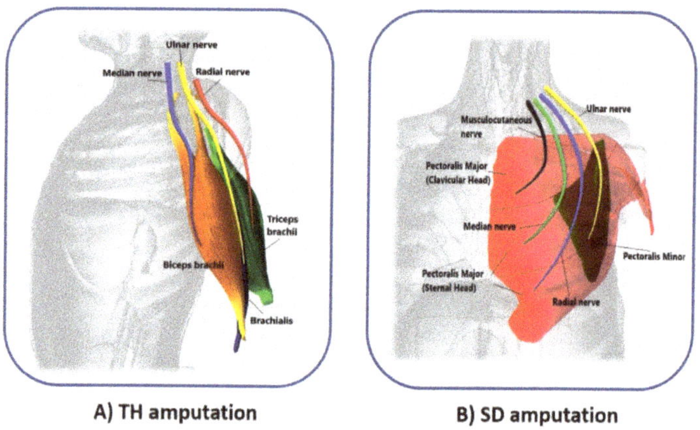

Figure 3. Scheme of the reinnervated sites for different levels of amputation. (**A**) Median (blue), ulnar (yellow), and radial (red) nerves transfer on biceps brachii (orange), brachialis (violet), and triceps brachii (green) muscles of transhumeral (TH) amputees; (**B**) Musculocutaneous (black), median (light green), radial (blue), and ulnar (yellow) nerves transfer on pectoralis major (clavicular and sternal head, in red), and pectoralis minor muscles (dark green) of shoulder disarticulation (SD) amputees.

3. Aim of the Study

This paper proposes an in-depth study of the literature on control strategies for prostheses that were developed for amputee subjects who underwent TMR procedure. The scope is to consolidate the current knowledge in this field and delineate the limits of these strategies that, up to now, do not yet allow for natural and simultaneous control of the prosthetic arm DoFs.

Nowadays, according to the literature search, the related review papers on TMR contributed to defining the advantages of using this surgery technique from a medical perspective [23,30] without analyzing the most suitable control strategies and performance evaluation tests that allow exploiting the additional targeted muscle reinnervation sites to improve the multi-DoF prosthesis control.

This work has the twofold purpose of (i) identifying the main issues and advantages of the control strategies that were proposed in the literature in order to address the future research towards the development of prostheses that are functional and able to mimic the lost upper limb behavior, replicating the performance of the human arm, for amputee subjects who have undergone the TMR procedure; and, (ii) suggesting a unified protocol test for the validation of these control strategies and, in the case of PR, in both offline and online mode. The expected added value provided by this work is to complete the current knowledge on the control strategies with more recent papers, by critically evaluating and comparing (when possible) the available results and pointing out inconsistencies and neglected aspects.

The paper is organized, as follows. Section 3 describes the methods that were used to select the reviewed articles. Section 4 introduces the benefits of TMR. Section 5 describes the control strategies (both DC and PR) that were used in the analyzed papers for TMR patients. In Section 6, the performance evaluation methods are reported. Section 7 underlines the principal limits of the current control strategies and suggests a unified protocol for the control performance evaluation. Finally, conclusions are drawn in Section 8.

4. Materials and Methods

A wide search was conducted through the following databases: PubMed and Google Scholar. The search terms included the following keywords and their combinations: Targeted Muscle Reinnervation (TMR), upper limb, prosthesis, amputation level, prosthetic control, real-time and offline performance, multi-DoF control, and pattern recognition. Only studies that were published between 2004 and February 2020 were selected.

All of the found articles were in English and available in full text on peer-reviewed journals or in conference proceedings. Some additional papers extracted from the references of the examined articles have been included.

The authors reviewed all articles fulfilling the following inclusion criteria:

1. Be a study on upper limb prosthesis users that underwent TMR surgery.
2. Concern control techniques in upper limb prosthesis.
3. Involve both direct and pattern recognition control strategies.
4. Use methods for evaluating the performance of the prosthetic control.
5. Be a full-length publication in a peer-reviewed journal or in conference proceedings.

The search strategy was based on the PRISMA (Preferred Reporting Items for Systematics reviews and Meta-Analyses) statement (2009), as shown in Figure 4.

A total of 136 papers was analyzed by using the previously mentioned search method. After considering the titles and abstracts, 108 articles were excluded from the initial 136, because they did not meet the inclusion criteria. The remaining 28 articles have been carefully analyzed. Eight of them were further excluded because the reported data were not significant or were repetitive for the purpose of this work. This review discusses the remaining 20 articles.

The selected studies have been classified into two main groups according to the used control strategy (DC or PR); one more group was devoted to papers on comparison among strategies:

1. Direct Control strategy—six papers: Kuiken et al. (2004) [21], Kuiken et al. (2005) [31], Kuiken et al. (2007) [24], Miller et al. (2008) [32], O'Shaughnessy et al. (2008) [33], and Miller et al. (2008-b) [34]
2. Pattern Recognition strategy—10 papers: Mastinu et al. (2018) [28], Kuiken et al. (2009) [35], Smith et al. (2013) [36], Huang et al. (2008) [37], Zhou et al. (2007) [38], Batzianoulis et al. (2019) [39], Batzianoulis et al. (2018) [40], Xu et al. (2018) [41], Hargrove et al. (2018) [42], and Tkach et al. (2014) [43].
3. Comparison of different types of control—four papers: Hargrove et al. (2013) [44], Wurth and Hargrove (2014) [45], Hargrove et al. (2017) [46], and Young et al. (2014) [47]

The following information has been extracted from the studies and reported in Table 1:

- the number of the enrolled patients;
- the amputation level: bilateral shoulder disarticulation (BSD), shoulder disarticulation (SD), and transhumeral (TH);

- the number of reinnervated sites/control sites after the TMR surgery;
- the use of prostheses/virtual reality (VR);
- the number of controllable DoF/motion classes; and,
- the adopted performance evaluation methods.

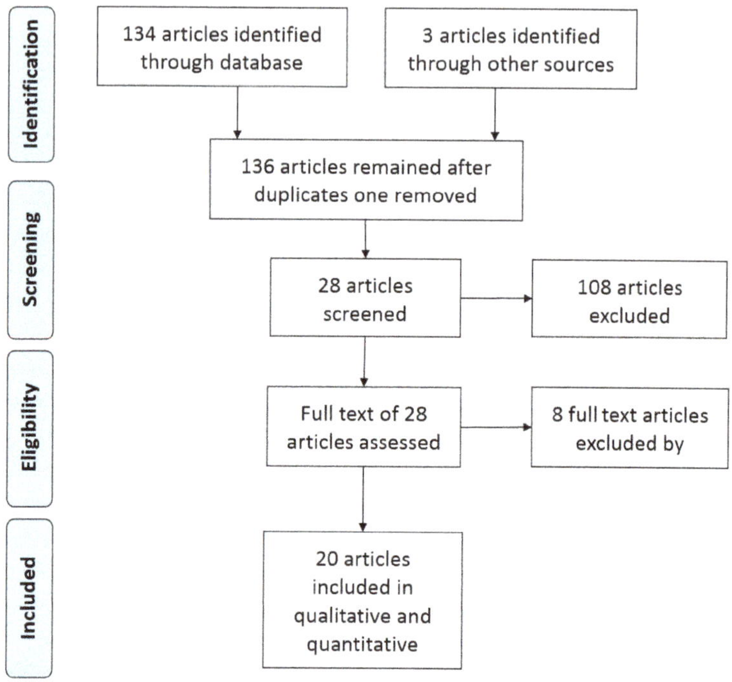

Figure 4. Flow diagram of the search and inclusion process.

Table 1. Summary of the reported analysis.

Study	No. of Patients	Amp. Level	No. of Reinnervated Sites/Control Sites	Prostheses/ Virtual Reality	DoF/ Motion Classes	Performance Evaluation Methods
Kuiken et al. [21]	1	BSD	4 reinnervated sites/ 3 muscle control sites	Prosthesis—DC	2	BBT, CRT
Kuiken et al. [31]	2	BSD, TH	4 reinnervated sites/ 3 muscle control sites (BSD) 2 muscle sites (TH)	Prosthesis—DC	2	BBT, CRT, WMFT, AMPS
Kuiken et al. [24]	1	TH	4 muscle sites and 2 sensory sites	Prosthesis—DC	2	BBT, AMPS, light touch, graded pressure, texture, edge detection, and thermal feedback
Miller et al. [32]	1	BSD	4 reinnervated sites	Prosthesis—DC	3	BBT, CRT, Cubbies, Cups
O'Shaughnessy et al. [33]	3	TH	2 reinnervated sites/ 4 control sites	Prosthesis—DC	2	BBT, CRT, AMPS
Miller et al. [34]	6	SD, TH	2 reinnervated sites (TH), 4 reinnervated sites (SD)	Prosthesis—DC	2	BBT, CRT, AMPS
Mastinu et al. [28]	2	TH	2 reinnervated sites	PR without prosthesis	4 discrete hand and elbow motions	accuracy offline, classification error rate of LDA with 4 time domain features (MAV, WL, ZC, SSC)
Kuiken et al. [35]	5	SD, TH	4 reinnervated sites, 4 control sites	PR without prosthesis—VR	10 discrete elbow, hand and wrist motions	accuracy offline, motion selection time, motion completion time, and motion completion rate of LDA with TD features [38]

Table 1. *Cont.*

Study	No. of Patients	Amp. Level	No. of Reinnervated Sites/Control Sites	Prostheses/ Virtual Reality	DoF/ Motion Classes	Performance Evaluation Methods
Smith et al. [36]	5	SD, TH	3–4 reinnervated sites (SD1, SD2), 2 reinnervated sites (TH)	PR without prosthesis	9 discrete elbow, hand and wrist motions	classification error rate of LDA with TD features [48]
Haung et al. [37]	3	BSD, TH	4 reinnervated sites (BSD), 4–2 reinnervated sites (STH, LTH)	PR without prosthesis	15 discrete elbow, hand and wrist motions	offline accuracy of LDA classifier with TD features (MAV, ZC, SSC, WL)
Zhou et al. [38]	4	BSD, STH, LTH	4 reinnervated sites (BSD), 4–2 reinnervated sites (STH, LTH)	PR without prosthesis	16 discrete movements of the arm, hand, and finger/thumb	offline accuracy of LDA classifier with TD feature set, and a combination of AR-RMS
Batzianoulis et al. [40]	2	TR	TMR surgery for the neuroma pain, not for control sites	PR without prosthesis	5 grasp types (prismatic-2 fingers, precision disk, palm pinch, lateral, prismatic-4 fingers)	offline accuracy, standard errors of LDA, two SVMs, and ESN Network
Batzianoulis et al. [39]	2	TR	TMR surgery for the neuroma pain, not for control sites	PR without prosthesis	3 grasp types (precision disk, lateral, and palm pinch)	offline accuracy of LDA classifier with TD feature
Xu et al. [41]	1	TH	3 reinnervated sites/ 5 control sites	Prosthesis—PR	6 discrete elbow, wrist and hand motions	offline accuracy, ARAT, LDA classifier with TD features (MAV, WL, ZC, SSC)
Hargrove et al. [42]	9	TH	not described	Prosthesis and VR—PR	6 discrete elbow, wrist and hand motions	SHAP, JTHFT, CRT, BBT, ACMC, the classification error rate, completion time, failure rate of LDA classifier with TDAR
Tkach et al. [43]	4	SD, TH	4 reinnervated sites (TH), 2 reinnervated sites (SD)	PR without prosthesis—VR	8 discrete and combined elbow, wrist and hand motions	offline accuracy of the LDA classifier with AR feature set
Hargrove et al. [44]	4	SD, TH	4–5 reinnervated control sites	Prosthesis—DC and PR	2 DoFs (sequentially PR system)	BBT, BST, CRT, classification error rates
Wurth et al. [45]	1	TH	4 control sites	PR and DC without prosthesis—VR	2 DoFs (sequentially and simultaneously PR systems)	FTAT, throughput (bits/second), path efficiency (%), completion rate (%)
Hargrove et al. [46]	8	TH	4 control sites	Prosthesis—DC and PR	2 DoFs	ACMC, SHAP, BBT, CRT
Young et al. [47]	3	SD, TH	2 reinnervated sites/ 4 control sites	Prosthesis—DC and PR	2 DoFs (sequentially and simultaneously PR systems)	TAC test (completion time, completion rate, length error), offline classification error

Acronyms of Table 1: BSD: Bilateral Shoulder Disarticulation; SD: Shoulder Disarticulation; TH: Transhumeral; LTH-STH: Long (L)–Short (S) Transhumeral; TR: Transradial; DC: Direct Control; PR: Pattern Recognition; VR: Virtual Reality; BBT: Box and Block Test CRT: Clothespin Relocation Test; WMFT: Wolf Motor Functions Tests; AMPS: Assessment of Motor and Process Skills; LDA: Linear Discriminant Analysis; MAV: Mean Absolute Value; WL: Waveform Length; ZC: Zero Crossing; SSC: Slope Sign Changes; TD: Time Domain; AR-RMS: Auto Regressive-Root Mean Square; ESN: Echo State Network; SVM: Support Vector Machine; TD-AR: Time Domain and Auto Regressive; ARAT: Action Research Arm Test; BST: Block stacking test; FTAT: Fitts' Target Acquisition Task; SHAP: Southampton Hand Assessment Procedure; JTHFT: Jebsen-Taylor test of Hand Function.

5. Control Strategies

5.1. Direct Control

The control strategies where EMG signals are directly associated with a specific movement are named direct control strategies, as mentioned in Section 1. Among them, the most used are on/off and proportional techniques. Multiples control techniques can be combined with the joint selection method to control multi-DoF prostheses. Figure 5 shows the DC approach.

In detail, the control techniques indicate the relationship between the value of the input signal and the value of the output. In the on/off technique there are only two possible output signals: a predefined speed value (on) and zero (off). The input signal must exceed the preset threshold in order to generate the output value that is used for motor control. Instead, with the proportional technique, it is possible to create a proportional link between the motor speed (output) and the amplitude of the EMG signal (input). Of course, there is also a threshold below which the output signal is zero; in some cases, there is another threshold, above which the output signal is the maximum possible. Regardless of which control technique is used, when the EMG signals are fewer than the DoFs to be activated, the user can employ muscle co-contraction to choose the DoFs to be controlled with the same EMG signal, as mentioned in Section 1. The introduction of TMR surgery made possible to have more muscle sites to uniquely associate a sEMG signal with a movement and simultaneously control more than one DoF. This procedure is resumed in Figure 6.

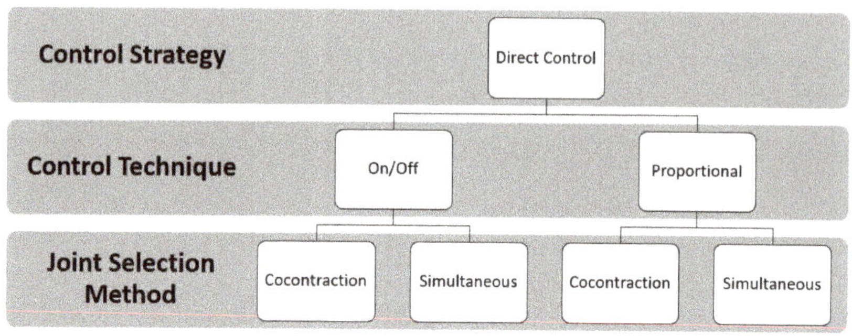

Figure 5. Direct Control approach: the EMG signals are the input to the controller unit. Two control techniques (the on/off and the proportional) defined the speed necessary to move the joint when the EMG signal is above a predefined threshold. The joint selection methods allow for the user to switch joints with muscle co-contraction or to select them simultaneously.

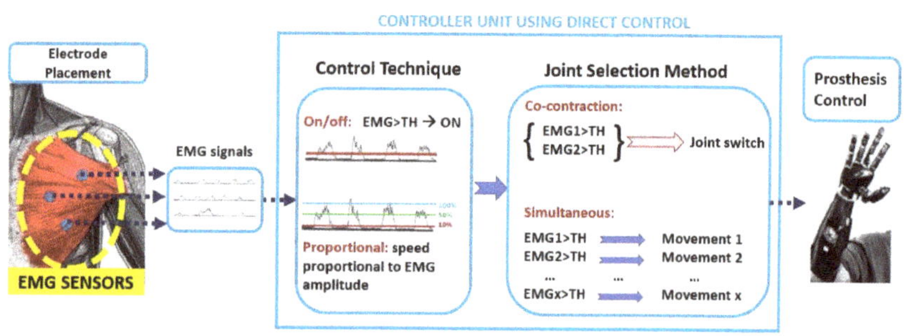

Figure 6. Schematic diagram of direct myoelectric control techniques and joint selection methods.

The following six articles make a clear reference to the use of direct control following a TMR intervention. Particular attention was paid to information regarding the level of amputation, the number of sites reinnervated and adopted for the control, and the used prosthetic devices (when indicated are reported in Figure 7), in order to critically evaluate and compare the performance of each method, and to point out the most functional prosthesis control method.

In Kuiken et al. [21], it was demonstrated that a man with a bilateral shoulder disarticulation (BSD), who underwent for the first time TMR, on the left side, was able to control a 3 DoFs prosthesis by using sEMG sensors placed on the three muscle sites successfully reinnervated. Prosthesis was composed of open/close (O/C) hand: Greifer Terminal (Ottobock); wrist prono/supination (P/S): Wrist rotator (Ottobock); elbow flexion/extension (F/E): Boston digital arm (included forearm); and, shoulder: LTI Collier shoulder joint. The proportional control was employed, with a simultaneous joint selection strategy for the hand and the elbow, and with co-contraction strategy to switch between hand and wrist. In details, the patient was able to pass to the control of the wrist from the hand by using the co-contraction of the hand open/close signals; while the elbow was controlled directly with the remaining active site, by modulating the contraction (from weak to strong) for flexing the elbow slightly or completely, respectively; therefore, the elbow extension was possible by relaxing the contraction.

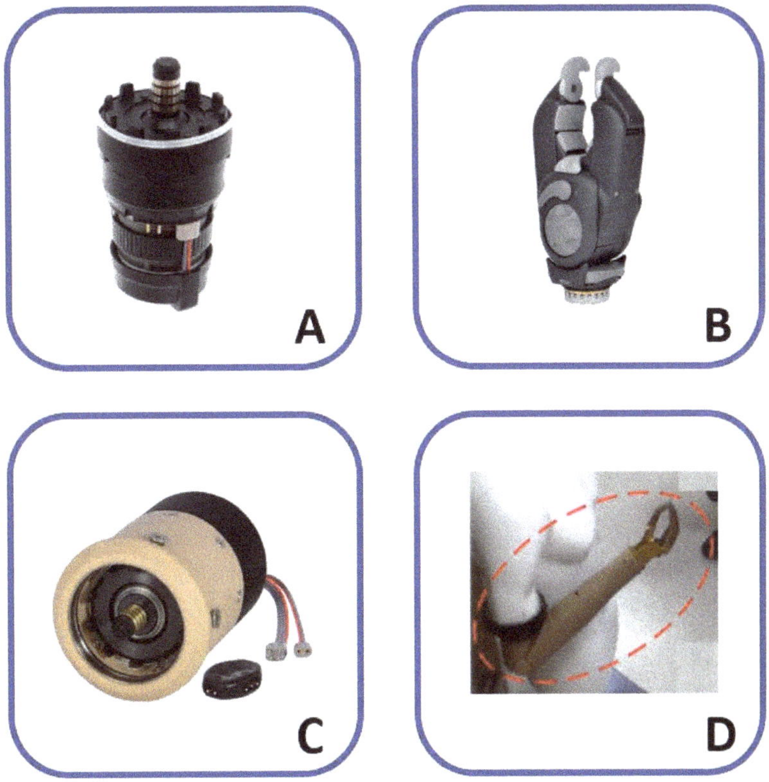

Figure 7. Prosthetic systems used in the analyzed papers on the TMR subjects: (**A**) Wrist rotator (Ottobock) used in [21,24,31,34]—Photo courtesy of Ottobock; (**B**) Greifer Terminal (Ottobock) used in [21,31,34,46]—Photo courtesy of Ottobock; (**C**) Wrist rotator (Motion Control Inc.) used in [42,46]—Photo courtesy of Motion Control, Inc.; (**D**) Danyang Prostheses Co. (Ltd, Danyang City, China) used in [41]— (http://creativecommons.org/licenses/by/4.0/, accessed on 5 March 2021), this image has been cut from [41].

The same author, in [31], also extended the reported outcomes in [21] on a patient with TH amputation, who received the TMR for only two muscle sites. In this case, a powered hand and the same elbow and wrist prostheses presented in [21] were used to simultaneously control three DoFs, by also considering the wrist rotation with shoulder motion.

In Kuiken et al. [24], a case study of a woman with TH amputation was reported to understand if (i) TMR can improve the prosthesis control and (ii) Targeted Sensory Reinnervation (TSR) can provide a region with "sensory perception", by reinnervating four muscle sites and two sensory sites, respectively. The proportional control with simultaneous selection strategy was used to control a prosthetic device after evaluating, with a grid of 128 monopolar electrodes, the most suitable placements of the sEMG sensors for recording the hand and elbow signals. In this way, the patient was finally able to control a three DoFs prosthesis composed of: a passive shoulder components, motorized elbow (F/E) with a computerised arm controller (LTI), motorized wrist rotator (P/S) (Otto Bock), and motorized hand (O/C) (Otto Bock). To control the wrist, two pressure-sensitive pads were mounted in the patient's socket, but she rarely operated them because the cognitive load of simultaneously controlling all three joints was high.

In O'Shaughnessy et al. [33], the proportional control technique was used in order to allow a simultaneous control of the prosthesis composed of the elbow F/E and the hand O/C joints. Three TMR patients with TH amputation were enrolled: only two of them had successful nerve transfers and were able to drive the experimental myoelectric prosthesis. For both of the patients, a total of four control sites was used for prosthetic control: the two reinnervated sites to control the hand and two other residual sites for the elbow.

In Miller et al. [34], a case report of three patients with SD and three with TH amputation are presented. The subjects with TH amputation underwent the reinnervation of two muscles, while, for the SD patient, four muscles were reinnervated. All of the patients were equipped with a prosthesis, including Boston Digital Arm, Ottobock device wrist rotator, and electric terminal device. The only difference among the subjects was the use of a prosthetic hand or a hook as a terminal organ. The prosthesis was equipped with proportional control of the elbow and hand joints through the four reinnervated muscle sites in the case of the SD and the two reinnervated sites plus two residual sites in the case of the TH patients. The wrist joint was controlled by the signal from one or two FSR sensors or a potentiometer. At the end of the trials, all of the subjects appreciated the ability to simultaneously control elbow and hand joints, without changing control with respect to conventional control, thanks to the TMR.

To summarize, in all the papers analyzed in this sub-section, the proportional control was employed, combined with both simultaneous and co-contraction joint selection methods. Prosthesis was composed of three DoFs, which were actuated by the different modules that are shown in Figure 5. Only in [33]; the prosthesis was composed of 2 DoFs.

5.2. Control via Pattern Recognition

Generally, the pattern recognition strategies applied to the prosthetic control associated the several inputs based on sEMG signals of different movements to several outputs, as limb motions related to specific myoelectric patterns [49].

These PR algorithms consist of a first step that is based on feature extraction, in the time and frequency domain [50], to enhance information about EMG contraction in selected time windows. Subsequently, in the sequential control technique, a single classifier is trained that is based on linear or non-linear decision boundaries; instead, in the simultaneous control technique, multiple classifiers are trained to control multiple joints simultaneously or a single classifier is trained by considering discrete and combined movements as separate classes, as shown in Figure 8.

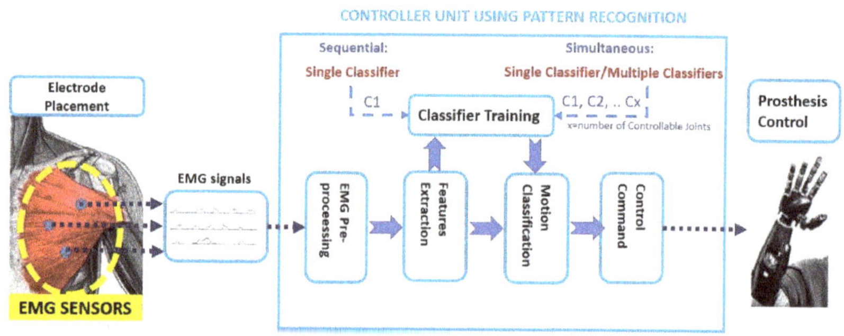

Figure 8. Pattern Recognition approach: the EMG signals are the input to the controller unit. Firstly the pre-processing step is done; then, in the features extraction step, the time and frequency domain features are used as input to train a single classifier or multiple classifiers. The classification output is the motion class to send as the command control to the prosthesis.

For instance, an extensive analysis can be found in [13] and in [51], where a comparative analysis among Non-linear Logistic Regression (NLR), Multi-Layer Perceptron (MLP), Support Vector Machine (SVM), and Linear Discriminant Analysis (LDA) is proposed: the main difference between these algorithms is the linear and nonlinear shape of the decision boundary; straight line or plane for the LDA algorithm; curved line, or surface, for the NLR, MLP, and SVM algorithms. Additionally, the robustness and reliability of the proposed algorithms are key factors for the online control of the prosthetic device and they depend on their offline performance, complexity, and computational time. In the case of trans-radial amputees, the LDA and NLR obtained statistically similar values in terms of F1 Score performance and computational burden in [51] .

To sum up, these strategies used machine learning techniques (Figure 9) in order to increase the amputee's ability to control the prosthesis, in a more natural way, by adding the number of controllable DoFs, because they do not require independent EMG sites for classifying motion classes of different joints [52].

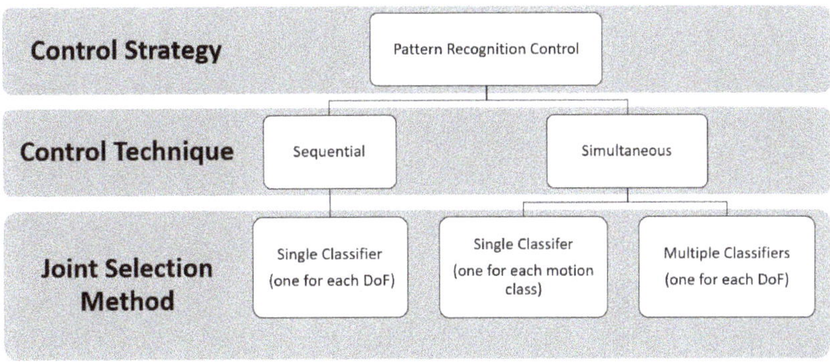

Figure 9. Schematic diagram of pattern recognition-based myoelectric control techniques and joint selection methods.

The TMR is considered to be very promising for improving the simultaneous control of multiple arm functions for many (ADLs) ([34,53]). This surgical technique, combined with PR-based systems, represents an opportunity, especially for SD and TH amputees, to overcome the limited number of independent EMG sites that are available for controlling a multi-DoF prosthetic systems [35]. Indeed, the advanced EMG-based pattern recognition strategies have the potential to perform, in a more natural way, the simultaneous control of multiple DoFs with respect to the conventional myoelectric control methods [54], because they do not require independently control sites or mode-switching to activate multiple joints like elbow, wrist, and hand. The following 10 articles have been found in the literature, in which pattern recognition algorithms have been employed in TMR patients:

In this section, for each study, we will analyze the most used strategies introduced to improve the prosthesis myoelectric control for TMR patients with PR control systems, to critically evaluate and compare the performance evaluation results of each method, and to point out the most functional control method and prostheses that replicate the behavior of the human arm.

In Mastinu et al. [28], the monitoring of TMR myoelectric signals of two TH amputee subjects, with TMR surgery and an e-OPRA, has been analyzed for 48 weeks after surgery in order to understand the potentiality as compared to conventional surface electrodes. The TMR-radial and TMR-ulnar sites were used for hand opening and closure, respectively, while the triceps and biceps muscles for the flexion and extension of the elbow. The LDA classifier was used with four TD features: the summation of absolute value of EMG signals, defined as mean absolute value (MAV); the cumulative length of the EMG signal waveform defined as waveform length (WL); the zero crossing (ZC) that measures how many times two consecutive samples have different sign (when the EMG signal crosses zero) in order to detect the onset of movement during the procedure of data segmentation; and, the slope sign changes (SSC), which represents the number of times the slope of EMG signal changes sign. Four discrete motions of elbow and hand were recorded with the Artificial Limb Controller, a prosthetic device that was designed for patients with e-OPRA implants [55].

In Kuiken et al. [35], five TMR patients with SD and TH amputations were able to perform, with a virtual prosthetic arm, 10 different motions that were related to different joints, like elbow, wrist, and hand (elbow F/E, wrist F/E, wrist P/S, hand opening, three types of hand grasps -3 jaw chuck, fine pinch, tool grip, and no movement). For each subject, 12 self-adhesive bipolar EMG electrodes were placed over the reinnervated sites: in detail, four electrodes were placed according to clinical evaluation, while eight additional sites were chosen by an electrode-placement optimization algorithm that allowed to select, from high density (HD) EMG recordings, a reduced number of electrodes necessary to preserve sufficient neural control information for the accurate classification of user's intention [56]. The proposed PR algorithm was based on an LDA classifier with four TD features (MAV, ZC, WL, SSC). The LDA classifier was used to produce, in real-time, a new prediction every 100 ms. In details, the performance metricsm such as motion selection time (mst), motion completion time (mct), and motion completion rate (mcr), were introduced for assessing the functionality, in real-time, of a virtual multifunction prosthesis.

In Smith et al. [36], the potentiality of PR myoelectric control was investigated when using wireless implantable devices. Five TMR subjects (three with SD and two with TH) were employed for evaluating the capability of performing nine motion classes (rest state, elbow F/E, wrist P/S, F/E, and hand O/C). However, two motion classes (hand open and wrist extension) were excluded for all subjects, because two subjects (one with SD, one with TH) did not have a successful fine-wire insertion into sites. In particular, for two SD subjects, the number of reinnervated muscle sites was equal to 3, while, for one SD subject, was equal to 4. Both intramuscular EMG signals (imEMG) and sEMG signals were acquired by locating bipolar fine-wire electrodes and adhesive bipolar surface electrodes, respectively, on TMR sites. One subject with SD was excluded from pattern classification, because he had the sEMG signals corrupted by a 60 Hz noise.

In Huang et al. [37], different spatial filters were tested in order to enhance the spatial selectivity of EMG recordings and the performance of EMG pattern classification by applying spatial filtering to high-density EMG recordings. Three subjects with TMR were recruited: the first one had a BSD amputation with four reinnervated muscle sites; the second one had a very short TH with four reinnervated muscle sites; and, the last TMR subject had a long TH amputation with two reinnervated muscle sites, and two natively innervated muscle sites. High-density surface EMG signals were recorded from the above mentioned muscle sites, which had been clinically selected. The following fifteen different movements were acquired: elbow F/E, wrist F/E, P/S, ulnar and radial deviation, two hand opening patterns (that included finger abduction and finger adduction), and five functional hand-closing patterns (power grip, prehensile (3-jaw chuck) grip, fine pinch grip, key grip, and trigger grip). The LDA classifier was used to classify the EMG signal with TD features (MAV, ZC, SSC, and WL) and the surface EMG signals were processed by various high pass spatial filters, including one-dimensional and two-dimensional filters.

In Zhou et al. [38], 16 movements of the arm, hand, and finger/thumb, with eight degrees of freedom, were discriminated with an LDA classifier with the TD feature set, and a combination of AR coefficients and RMS (AR-Root Mean Square) of the signals. The recordings were made by using monopolar electrode configuration and three bipolar electrodes in three different directions: transversal, longitudinal, and diagonal. Four TMR subjects were recruited: the first one with a BSD with four reinnervated muscle sites, the second one with a very short TH and four reinnervated muscle sites, and two other subjects with long TH amputations with two reinnervated muscle sites and two natively innervated sites for elbow flexion/extension.

In Batzianoulis et al. [40], three different classification systems that were based on LDA, SVMs (with linear and non-linear kernel), and an Echo State Network (ESN) were evaluated by considering, for each proposed strategy, the classification performance on three phases of dynamic reach-to-grasp motions: acceleration (first phase), deceleration (second phase), and rest (third phase). Eight able-bodied control subjects and four TR amputees, two of which underwent TMR surgery for the neuroma pain, were enrolled. These TMR patients did not have additional muscle sites for improving myoelectric control. The EMG muscle activity was recorded with 12 sEMG sensors from seven muscles of the upper arm and five muscles of the forearm. For LDA and SVM, three features (i.e., average activation of each time window, waveform length, and number of slope changes) for each window of 150 ms have been extracted. Five grasp types (prismatic-2 fingers, precision disk, palm pinch, lateral, and prismatic-4 fingers) were discriminated. In their most recent study [39], the same two TMR transradial amputees that were presented in [40] were employed to extend the previous results by addressing more insights on the LDA potentiality and introducing the use of the Hellinger distance to quantify the similarity between motion classes. In this case, the subjects were asked to perform a bimanual task by only considering three grasp types as the precision disk, lateral, and palm pinch motions. Different from [40], only the performance of an LDA classifier was evaluated in terms of classification accuracy when it was trained for each phase and over all motion phases. To train the classifier, the EMG signals of five muscles of the residual arm were recorded: Flexor Digitorum Superficialis, Extensor Digitorum Communis, Flexor Carpi Ulnaris, Extensor Carpi Ulnaris, and Flexor Carpi Radialis.

In Xu et al. [41], the authors investigated how the rehabilitation training improved the separability of some channels of sEMG signals that remained still coupled over TMR. A TMR TH patient with five targeted muscles with coupled sEMG signals has been engaged. Five bipolar EMG electrodes have been placed on targeted muscles that are associated with the following movements: hand C/O, wrist P/S, and elbow F/E. A new approach that was based on pattern recognition control with MAV-based threshold switches was introduced to improve the classification performance of an LDA classifier, based on Bayesian decision, with TD features (MAV, WL, ZC, and SSC). Subsequently, the obtained classification parameters have been used for allowing the patient to control a commercial prosthesis (Danyang Prostheses Co. Ltd, Danyang City, China); a subset of the modified ARAT test was proposed to compare the online performance of the prosthetic operation.

The LDA classifier with TD-AR (time-domain and auto-regressive) features [57] was introduced for classifying elbow F/E, wrist S/P, and hand O/C. A grid of stainless steel electrodes was placed over specific muscles. However, the exact sites of reinnervated muscles have been not described in detail. The outcome measures, which were obtained with both a virtual reality and a physical prosthetic system, were introduced to evaluate the improvements in terms of offline classification errors. For obtaining physical outcomes, all nine subjects used the following custom-fabricated prosthesis composed of: Boston Digital Elbow (Liberating Technologies Inc.), wrist Rotator (Motion Control Inc.), and single DoF terminal device.

In Hargrove et al. [42], the outcome measures, which were obtained with both virtual reality and a physical prosthetic system, were introduced to evaluate the improvements in terms of offline classification errors of nine transhumeral TMR subjects, when using prosthesis after a six-week home trial. Three blocks of the Target Achievement Control (TAC) test [58] were used to evaluate the performance of the LDA classifier with TD-AR (time-domain and auto-regressive) features [57]. For obtaining physical outcomes, all nine subjects used the following custom-fabricated prosthesis that was composed of: Boston Digital Elbow (Liberating Technologies Inc.), wrist Rotator (Motion Control Inc.), and single DoF terminal device.

In Tkach et al. [43], it was demonstrated that a generic grid arrangement of electrodes performed equivalently or better than the control site (specific site for electrode placement). Four TMR amputee subjects were employed: two TH subjects had four reinnervated muscle sites; two SD subjects presented, instead, only two reinnervated muscles sites. EMG signals were acquired by using 15 bipolar pairs of EMG electrodes placed according to two conditions: in the "Control Site" condition, the electrodes were placed over muscle control sites, after clinical palpation; in the "Grid" condition, electrodes were positioned in a grid configuration, around the residual limb and the surface of the chest, for the TH and SD subjects, respectively. The LDA algorithm was used with the AR feature set, including the six coefficients of a 6th order autoregressive model.

To sum up, all 10 studies presented in this sub-section take the pattern recognition strategy based on LDA classifier with different features set into account: TD features (MAV, WL, ZC, and SSC) [28,35–38,41]; TD-AR features [57]; AR-RMS [38]; the AR feature set [43]; Hellinger distance [39]; and, the average activation of each time window, the waveform length, and the number of slope changes [40]. In Batzianoulis et al. [40], the SVMs (with linear and non-linear kernel), and an Echo State Network (ESN) PR-based strategies were also evaluated by considering, for each proposed strategy, the classification performance on five reach-to-grasp motions. The minimum number of discriminated classes was equals to four discrete motions related to the elbow and hand [28] or only the hand [39,40]. While, for the other seven studies, the elbow, wrist, and hand joints were always considered by including from nine up to 29 motion classes [43] (for both discrete and simultaneous movements).

5.3. Comparison between DC and PR Strategies

The following four papers presented a comparison between direct control and pattern recognition based strategies.

Particular attention was paid to information regarding the level of amputation, the number of sites reinnervated and used for the control, the various DC strategies and PR algorithms used, and the prosthetic device used (when indicated).

The first study that directly compared the performance of pattern recognition systems to direct control systems using a physical prosthesis with TMR patients is Hargrove et al. [44]. Four patients (one male with SD, two males and one female with TH amputation) had at least four reinnervated control sites (five in one case) used for direct control of the elbow F/E and hand O/C joints. The P/S of the wrist joint was controlled and selected in different ways by the various patients, in a manner similar to that used with their old prostheses. For the PR-based control system, four pairs of bipolar electrodes have been added to the four pairs that were used for direct control. The PR control system was composed of a LDA classifier with TD features and AR coefficients. The velocity of the desired movement was computed while using a simple proportional control algorithm. Section 6.3 presents the performance achieved by all patients experts in the daily use of the myoelectric prosthesis with DC control and with experience in the laboratory use of the prosthesis controlled with PR. All of the subjects said that they preferred PR-based control, because it was more intuitive. However, the authors pointed out that direct control allowed the simultaneous movement of two joints, while the PR-based control was limited to sequential control, even when tasks required multiple DoFs.

In Wurth et al. [45], a real-time comparison between DC and PR-based control strategies was carried out to control a multi-DoF myoelectric prosthesis. Only one TH amputee among the enrolled subjects underwent the TMR procedure, with four independent control sites. The others were nine healthy control subjects and one TR amputee. For the DC control, the MAV EMG signals of the wrist flexors and extensors muscles were recorded from able bodied and TR subjects by using pre-gelled adhesive bipolar Ag-AgCl electrodes. Instead, for the TH amputee subject, four bipolar electrodes were placed on the flexor and extensor muscles, in order to simultaneously control more than one DoF. For the PR-control, the LDA classifier was used with four TD features (MAV, ZC, SSC, and WL) and six AR coefficients. In particular, the able bodied and TH subjects were asked to perform hand O/C, wrist F/E, and no motion. Instead, for the TH subject, the elbow F/E was replaced by wrist F/E, because this DoF was considered to be more intuitive and relevant to be controlled for this level of amputation.

In Hargrove et al. [46], a clinical study was reported on eight TH patients, with different levels of amputation and prosthetic solution composed of motorized Boston Digital Elbow (LTI), Motion Control Wrist Rotator (Motion Control Inc.), and terminal device (seven hook from Greifer or EDT and one hand). All of the subjects used prosthesis in both the controlled (laboratory) and uncontrolled (home) environment. The eight patients were randomly divided into two groups of four subjects, each of which completed the home-trial while initially using a prosthesis with a different control strategy, according to the group that they belonged to. The two configurations were used for six weeks each. The electrode sites were identified with different methods, depending on the different strategy adopted: when using direct control, the muscle sites were manually udentified, using a combination of surgical notes when available, palpation, and myoelectric signal testing. As for the PR control, linear electrode locations were not targeted over specific muscles, rather a grid of electrodes was used. The algorithm that was used for PR-based control was the LDA described in [57]. For the DC control, dual-site differential DC system employed antagonistic muscle pair in order to control elbow F/E and terminal device O/C (hand or hook). In addition, mode switches were configured, for each subject, to control the wrist P/S DoF according to their previous device use.

In Young et al. [47], three different control strategies (direct control with proportional strategy, sequential PR control—one DoF at time, and simultaneous PR control—two DoFs at time) were analyzed in order to evaluate the ability of four amputees (two TH and two SD), who underwent TMR surgery, to simultaneously control up to two DoFs with a virtual prosthesis. TH patients had two reinnervated muscle sites used for controlling hand O/C movements, and two natively innervated muscles (the biceps and triceps brachii) used for elbow F/E motions. The SD subjects had four reinnervated sites for controlling hand O/C and elbow F/E movements. Four pairs of self-adhesive Ag/AgCl bipolar surface electrodes were placed in the same muscle sites that were used for the conventional prostheses control. Other pairs of electrodes were placed near the primary sites where muscle activity could be palpated. The following eight discrete and combined motions were acquired: elbow F/E, hand O/C, and elbow F/E combined with hand O/C. The TMR amputees controlled discrete motions using their four independent muscle sites. For the PR-strategy, an LDA algorithm with four TD features (MAV, ZC, SSC, and WL) and six AR coefficients of a sixth-order were used for the classification. As for the sequential control condition, the same methods that were introduced in [48] were used. Instead, for the simultaneous control strategy, the authors used the system that was tested on able-bodied subjects in [59].

To summarize, in all of the articles presented this section, a physical prosthetic device with two [45] or three DoFs [44,46] was employed, except for the [47], in which a virtual prosthesis was used instead of the physical one. Regarding the direct control strategy, in all of the studies, the simultaneous joint selection methods were employed. Only in Young et al. [47] was the control technique specified, i.e., the proportional control technique. Regarding the reviewed papers on PR-based control, the LDA was always adopted with TD and AR features.

6. Performance Evaluation Methods

In this section, the performance evaluation methods introduced in the analyzed papers are reported. These methods allow for evaluating the functional effectiveness of the prosthesis as a whole with quantitative indicators that highlight the potential and advantages of the proposed systems.

The reported studies have shown the use of many performance evaluation criteria, although there is no standard procedure in place so far. The lack of a common standard is highlighted by the presence of multiple criteria that are necessary for the performance evaluation in the analyzed papers. In Table 2, for each method, the following information are reported: the percentage of the number of patients enrolled with respect to the total number of the patients considered in all the papers (i.e., 67, which is the sum of column 2 of Table 1) and, the total number of articles in which the performance evaluation method was employed.

Table 2. Performance evaluation methods.

Test Used	Percentage of Enrolled Patients (%)	N° of Articles
Qualitative/subjective evaluation (questionnaire)	20	7
Box and Block	50	9
Clothespin Relocation	49	8
Wolf Motor Functions—WMFT	1	1
Assessment of Motor and Process Skills—AMPS	14	3
Cubbies	1	1
Cups	1	1
Target Achievement Control—TAC	19	2
Southampton Hand Assessment Procedure—SHAP	25	2
Accuracy offline	35	8
Classification Error rate	35	5
Assessment of Capacity for Myoelectric Control—ACMC	25	2
Real Time Virtual Test	14	3
Action Research Arm Test—ARAT	1	1
Jebsen-Taylor Test of Hand Function—JTHFT	13	1
Block stacking	5	1
Fitts' Target Acquisition Task—FTAT	1	1

6.1. Direct Control

Regarding systems with DC strategies, the following performance evaluation methods were taken into account: in the Box and Block Test (BBT) [60], the patient had to move the higher number of standard size cubes from one side of the box to the other in a maximum time of one minute; however, Kuiken modified the test extending the time limit up to 2 min., to use it with amputee patients who had undergone the TMR procedure. In the Clothespin Relocation Test (CRT) [21], the patient is prompted to move three clips from the horizontal bar to the vertical bar; the execution time that is required to complete the task was measured. The test was repeated three times. Instead, the Cubbies test [32] is a cubicle reach and retrieve test (Cubbies) composed of 15 cubicles, containing 15 one-inch (2.5 cm) cubes placed on an adjustable height table; in this test, the subject, without moving his feet, has to reach, grasp, and place the 1-inch cubes on the table from as many cubicles as he is able to reach successfully; the total time to grasp a 1-inch cube, place it on the table, and activate a buzzer is recorded. The score consists of the average time per successful reach and retrieval of all blocks in the work-space. In the Cups test [32], 11 plastic cups must be individually retrieved from an inverted stack and positioned in a prescribed pyramid arrangement; four upside-down cups on the bottom row, until one on the top. The final eleventh cup has to be placed upright on top of the top upside-down cup of the pyramid. The time that is needed to stack and unstack the cups is recorded. Finally, the Miscellaneous ADL consisted of a series of the following activities: cutting meat with a knife and fork; place three objects onto a tray and then transport the tray; place 3 1-lb cans into a bag with handles; open and close a jar of peanut butter; stir a spatula in a big bowl; open an envelope with a tool; wrap a package; pull on both socks; and, remove and put on a long-sleeved shirt.

The following six papers used the BBT (modified up to 2 min.) [21,24,31–34] to evaluate the performance of the proposed control strategy. In detail, in Kuiken et al. [21], the comparison of the prosthetic control before and after the TMR surgery has shown an improvement of the new procedure. An increase in performance was present: +8.3 blocks, on average, over three tests. In Kuiken et al. [31], for the patient with TH amputation, the results were: +322% blocks moved with the experimental prosthesis. In Kuiken et al. [24], there was the evidence of an increase in the performance between pre-surgery and post-surgery: the average score over three trials changed from 4.0 ± 1.0 to 15.6 ± 1.5 blocks. In Miller et al. [32,34] and O'Shaughnessy et al. [33] the BBT (modified up to 2 min) was employed. In Miller et al. [32], the patient achieved better performance with the three DoFs arm: he moved, on average, 15.0 ± 0.1 blocks, while wearing the six DoFs prosthesis he only moved 13.7 ± 2.5 blocks. In Miller et al. [34], the improvements ranged from 95% to 271% (with an average of 198%) of the number of blocks moved. In O'Shaughnessy et al. [33], both of the patients demonstrated an increase in the number of blocks (on average, over three tests) of 611% and 150% for patient 1 and patient 2, respectively.

The CRT was also employed in the following six studies [21,22,31–34]. The task execution time of CRT equals -36 seconds in Kuiken et al. [21,22] and −54.24% in Kuiken et al. [31]. In Miller et al. [32], the subject was required to flex the shoulder forward once, and then sequentially activate terminal device, elbow and wrist rotations. This test showed that the subject was faster with the six DoF arm (58.0 ± 9.2 s of task execution time) than the three DoF arm (79.2 ± 14.3 s of task execution time). In Miller et al. [34], the improvement ranged from 31% to 55% (with an average of 45%) reduction in the time to complete the task. In O'Shaughnessy et al. [33], there was a decrease of the time (in seconds) that is required to move three clothespins from one beam to another: 55% and 41% for patient 1 and for patient 2, respectively.

The AMPS test was employed in the following four papers [24,31,33,34]. In Kuiken et al. [24], it went from the score of 0.30 and 0.90 for motor and process, respectively, to the score of 1.98 in both cases. In O'Shaughnessy et al. [33], a computer-tabulated score was reported, reflecting motor and processing function concerning activities of daily living. For patient 1, there was an increase in the score from 0.5 to 1 regarding the motor function and from 0.3 to 1.1 in the processing function. For patient 2, there was an increase in the score from 0.9 to 1.56 in the motor function and from 1.09 to 1.43 in the processing function.

In Kuiken et al. [31], the Wolf Motor Functions Tests (WMFT) [61] was also performed in addition to the Assessment of Motor and Process Skills (AMPS) [62]. The patient with TH amputation only had an increase in the score of +0.54 and +0.75, respectively. Thus, for the TH patient, a markedly greater increase in performance has been demonstrated than that obtained by the BSD patient. In Miller et al. [34], the tasks included cooking, cleaning, housework, garden work, and home maintenance; all of the TH subjects and two of the three SD amputees performed the AMPS test. For the SD amputees, an average increase in the motor score of +0.80 occurred; instead, for the TH subjects, the average scores was equal to +0.77; for the process score, there was an average score increase of +0.5 for the SD amputees, and +0.57 for TH ones. However, it should be pointed out that, as the current tests may not adequately measure improvements in control or prosthetic design due to the ceiling effect (the task is too easy to be performed by most subjects) or to the floor effect (the task is too difficult, almost none of the subjects can perform it).

Only in O'Shaughnessy et al. [33], the Miscellaneous ADL was evaluated: both of the patients used less time to complete the various activities (except wrapping a package) when using the experimental prosthesis.

Finally, in Miller et al. [32], the Range Of Motion (ROM), the Cubbies, and Cup test were also employed. In detail, an increase of the ROM of the shoulder flexion: it went from a range 0–90° for the passive shoulder with the old prosthesis to a range of 200°, from −15° to +185°, with the new experimental prosthesis. Regarding the Cubbies test, the

ROM was greater with the six DoF device due to the shoulder and wrist motors: with the three DoF prosthesis, the subject was only able to access 12 of the cubbies, whereas, with the 6 DoF device, the patient was able to access all 15 cubbies. Even with the increased work-space, the time per cubby was approximately the same: 18.6 ± 2.5 s with the 3 DoF arm and 15.3 ± 9.5 s. In the Cups test, the subject showed an increase in the amount of time necessary to stack all 11 cups in an inverted pyramid using the 6 DoFs (357 ± 36 s), compared to the 3 DoFs prosthesis (169 ± 58 s). To unstacking cups, the difference was lower: (83 ± 13 s) with the three DoF prosthesis and (89 ± 29 s) with the six DoFs one. Interestingly, the six DoF prosthesis was not always better than the three DoF prosthesis, on the contrary, it was sometimes worse due to some reasons as: the considerable weight (about 5.75 kg) of the six DoF prosthesis and the patient's cognitive effort forcing him to perform the tasks more slowly. Thus, to make a six DoF limb prosthesis clinically usable, it is necessary to develop a simpler and more intuitive control system, avoiding using the shoulder to control wrist movements.

6.2. Control via Pattern Recognition

Instead, in systems with PR control strategies, the following performance evaluation methods were taken into account: the offline classification accuracy [28,35,37,38], the offline classification error [28,36,42,43], the mst, mct, and mcr [35], the mean path length error percentages [43], and the score of six tasks rhat were chosen from Action Research Arm Test (ARAT) [41].

In detail, in Kuiken et al. [35] the performance metrics as the mst, mct and mcr were introduced for evaluating the control of a virtual arm in real-time, even if that of a real prosthesis is more challenging. In detail, mst was defined as the time from the onset to the first correct classification (i.e., the time that is taken to successfully select a target movement); mct was the time from movement onset to the 10th correct classification (i.e., the time from the onset to the completion of the intended movement); finally, mcr ("success" rate) was the percentage of successfully completed motions out of the total attempted motions.

In Xu et al. [41], the following six grasping tasks from the ARAT test were performed for assessing the recovery of upper limb function [63]: the grasping of a Block of 2.5, 5.0, and 7.5 cm^3, of a cricket ball, and a sharpening stone; to pour water from one glass to another; to displace 1 and 2.25 cm alloy tubes from one side of the table to another.

Only in Hargrove et al. [42], the following TAC, SHAP, JTHFT, and ACMC tests and the virtual outcome metrics were employed in order to evaluate the performance of the PR system as following. In particular, three blocks of Target Achievement Control (TAC) test [58] consisted of moving, within a 15-s time frame, a flesh-colored virtual limb to match the 12 postures of a translucent grey-colored virtual target limb % in real-time. For the virtual outcome metrics, the number of postures successfully acquired within their allotted 15-s time frames, and the median completion time required to match the set of postures in a block, were analyzed. Instead, for evaluating the differences between the control of a virtual prosthesis in a virtual environment with respect to that of the physical prosthesis, the following parameters were introduced as physical outcomes: the classification error rate, the completion time, and the failure rate. The classification error rate is defined as the number of incorrect decisions that were divided by the total number of decisions; the completion time was the time from the trial start to the target posture achievement, while the failure rate measured the percentage of trials that were unsuccessfully completed during the TAC tests. These physical outcomes were evaluated by using the following promising tests for assessing the functional effectiveness of a prosthetic system [64]: the Southampton Hand Assessment Procedure (SHAP) test [65], which consists of 12 abstract objects and 14 ADLs, and each task is timed by the participant in order to avoid reliability on the reaction times; Jebsen–Taylor test of Hand Function (JTHFT) [66], which is made up of seven subsets that are writing, simulated page-turning, lifting small objects, simulated feeding, stacking, and lifting large, lightweight, and heavy objects; the Assessment for

Capacity of Myoelectric Control (ACMC) [67], which consists of 22 items that are related to capacity for myoelectric control: the need for external support, grip force, coordination of both hands, different positions and in motion (timing), repetitive grasp and release, and the need for visual feedback. It is the only test validated for a good test-retest reliability and interpretation guidelines for evaluating the functionality of upper-limb prostheses [68]; and, BBT and CRT (as described in Section 5).

The following six papers used the classification errors [28,35,37–40] to evaluate the performance of the proposed PR-based control strategy.

In Mastinu et al. [28], four movements that were related to elbow and hand were discriminated with a sequential pattern recognition strategy based on a LDA classifier with TD features. The offline classification accuracy was over 97% since the last follow-up (week 48). Their results showed, for the first time, the evolution and quality of the TMR signals when using intramuscular electrodes instead of the conventional skin surface electrodes. In Kuiken et al. [35], ten different motions of elbow, wrist, and hand were classified with a sequential PR strategy that is based on a single LDA classifier with TD features. The mean classification accuracy was 88 ± 7% for TMR patients. In Huang et al. [37], fifteen different discrete movements were discriminated by using a single LDA classifier with TD features (MAV, ZC, SSC, and WL). The surface EMG signals were processed by various high pass spatial filters, including one-dimensional and two-dimensional filters. The use of high-density EMG recordings combined with a single differential filter in transverse direction (BipT), and a single differential filter in longitudinal direction (BipL), or higher order filters, allowed for reaching 95% classification accuracy for 15 movements for SD and patients with a long transhumeral amputation, and above 85% for patient with a short transhumeral amputation. However, when only considering 12 EMG signals, the double differential filters obtained 5–15% higher classification accuracies than the filters with a lower spatial resolution and comparable accuracies to the filters with higher spatial resolution. Thus, the use of double differential EMG recordings can improve the TMR-based neural interface for the control of artificial arms. In Zhou et al. [38], a single LDA classifier with the TD feature set, and a combination of AR coefficients and RMS (AR-RMS) of the signals were used to classify 16 movements of the arm, hand and fingers. The performance of the LDA classifier was reported in terms of classification accuracy for the various electrode configurations: with the monopolar channels, the average overall classification accuracy was equal to 90.5 ± 6.3% for TD feature sets and 90.0 ± 7.3% for AR-RMS feature sets. The accuracy of classification consistently improved to an average of 96.0 ± 3.9% with TD and to 95.0 ± 5.2% with AR-RMS features, for bipolar electrode configurations. Thus, TMR combined with the LDA classifier was able to extract motor control information from the reinnervated sites, by using high-density surface EMG recordings. In Batzianoulis et al. [40], a comparison of offline classification accuracy of four different classification systems based on LDA, two SVMs, and an Echo State Network (ESN) was presented. By applying an analysis of variance (ANOVA) (with a significance level of 5%) to the classifiers' performance, a significant difference for all the proposed classifiers in terms of offline accuracy values, for all 5, 4 and 3 classes of movements was not found. Namely, in the first and third phase, the average classification accuracies are similar for both non-TMR and TMR patients: 68.6 ± 8.8% and 64 ± 14.4% for the first phase and 87.6 ± 3.4% and 83.6 ± 4% for the third phase, respectively. However, the accuracy of TMR subjects in the second phase was better than that of non-TMR subjects (90.2 ± 4.6% and 77.8 ± 10.9%, respectively). Finally, an on-line evaluation of 20 reach-to-grasp motion with RIC hand [69] was tested only for a non-TMR subject. In Batzianoulis et al. [39], the EMG signals were analyzed by applying a sliding time window of 150 ms with an increment of 50 ms and three features (MAV, SSC, WL) were extracted from each time window. The classification system was composed by three LDA classifiers, one for each phase: the accuracies values were equal to 42.7 ± 8.2%, 57.8 ± 14.4%, and 74.2 ± 14% in the first, second, and third phase, respectively. Instead, the single LDA classifier, which was trained with all the phases, obtained an accuracy of 33.6 ± 12.5%, 51 ± 15.4%, and 66.2 ± 11% for each phase.

The obtained results underlined that the arm extension towards a specific direction during a reach-to-grasp motion affected the classification performance. Thus, the introduction of segmentation into motion phases revealed that higher accuracy values can be obtained when considering all of the reach-to-grasp motion phases.

The following four papers used the classification errors [28,36,42,43] to evaluate the performance of the proposed PR-based control strategy.

In Mastinu et al. [28], the evolution of the classification error over time was considered: for the first subject, the mean error decreased from 10.8% (week 4) to 1.7% (week 48), while, for the second subject, it remained stable below 5%. In Smith et al. [36], two LDA classifiers with TD [48] and autoregressive (AR) features [56] were implemented, by using sEMG and imEMG signals as input, respectively. The use of imEMG instead of sEMG produced a decrease of 1.39 ± 6.45% (90% confidence interval) of the average error rate, which was equal to 5.52% for sEMG and 4.13% for imEMG. The proposed results showed that, despite the variability of imEMG signals, the performance of the LDA classifier did not decrease. Thus, the imEMG signals can also be used for the PR myoelectric control, with the benefits of the reduction of EMG crosstalk, the placement on deeper muscles, and the overcoming of electrode shifting. However, the presence of sparse motor units at the reinnervated sites increased the MAV, based on the amplitude of the imEMG signals, and, consequently, this feature could not be used for estimating the proportional velocity. In Hargrove et al. [42], the average classification error across subjects decreased from 13.4 to 8.3% after home trials. In Tkach et al. [43], the classification errors of five classifiers, based on the LDA algorithm with the AR feature set, were evaluated when two different electrode placements were considered: the "Control Site" and Grid "configurations" (as described in Section 5.2). The five classifiers differ from the output classes: the "Seq Only" PR strategy was based on a single classifier based on 9 motion classes; the "Seq. Elbow+Hand", the "Seq. Elbow+Wrist", the "Seq. Hand+Wrist", and "All" employed a single classifier that was able to classify 13,17,17 and 29 motion classes, respectively. In detail, the eight discrete and twenty combined motions were the following: elbow F/E, wrist F/E, wrist P/S, and hand O/C; elbow F/E + hand O/C; elbow F/E + wrist F/E; elbow F/E + wrist P/S; wrist F/E + hand O/C; wrist P/S + hand O/C; no motion condition. The discrepancies of the classification error between the Control Site and Grid conditions were equals to 11.47% for the "All" classifier (29 motion classes) and 1.69% for the "Seq Only" classifier. The "Seq. Elbow+Wrist" and the "Seq. Hand+Wrist" classifiers, which discriminated 17 motion classes, both had similar classification errors of 11.5 ± 1.1% and 11.2 ± 1.2% for the Grid analysis condition. Regarding the Control Site condition, the two classifiers yielded higher errors of 17.9 ± 1.3% and 21.1 ± 1.7%, respectively. These results showed that the grid-like arrangement of electrodes can outperform the specific electrode placement on targeted muscle sites when considering classifiers with a greater number of motion classes.

For the real-time evaluation, in Kuken et al. [35], the LDA classifier was also used to produce, in real-time, a new prediction every 100 ms. The mean motion completion rate for the elbow and wrist movements was high (96.3% ± 3.8%) and it was lower for hand grasp movements (86.9% ± 13.9%) that were considered to be more challenging by some patients. Thus, the relevance of this study was to have assessed, for the first time, a protocol to evaluate, in real-time, the PR performance for controlling multi-DoF artificial arms, in patients with TMR.

Only in Xu et al. [41], comparisons have been made between the PR algorithms with no post-processing (Control), majority vote, and MAV-based threshold switches, while using six tasks that were chosen from Action Research Arm Test (ARAT). An LDA classifier based on Bayesian decision has been adopted with TD features (MAV, WL, ZC, SSC). The final scores were reported for the three methods: for PR control, they were equal to 11, 11, and 13 with majority vote lengths equal to 3, 5, and 10, respectively; for MAV-based threshold switches, three threshold values were defined: the standard threshold (ST) was equal to 0.2 mV; the lower threshold (LT) was equal to the 80% of ST; and, the higher threshold (HT) was equal to the 120% of ST. The final score values were equal to 14.7 (with lower threshold), 16 (with medium threshold), and 14.7 (with higher threshold). Finally, the improved PR control with MAV-based threshold switches turned out to be the best configuration for obtaining a robust control of the prosthesis.

Only in Hargrove et al. [42] were the following results about these performance evaluation tests reported: the TAC test performance metrics improved significantly from 19.9 to 3.7% after home trials: the failure rate improved from 19.9 to 3.7% ($p = 0.001$), and the completion time decreased from 7.5 to 5.5 s ($p = 0.007$). In the virtual test, the median completion time correlated significantly with the Southampton Hand Assessment Procedure ($p = 0.05$, R = −0.86), Box and Blocks Test ($p = 0.007$, R = −0.82), Jebsen–Taylor Test ($p = 0.003$, R = 0.87), and the Assessment of Capacity for Myoelectric Control ($p = 0.005$, R = −0.85). The classification error performance only had a strong correlation with the Clothespin Relocation Test ($p = 0.018$, R = 0.76). However, only the SHAP ($p = 0.001$) and the BBT ($p = 0.03$) have showed statistically significant improvements after a six-week home trial. Additionally, the physical outcomes that were related to the use of the physical prosthesis improved after the home trial. Thus, when considering all metrics as the classification error rates, the outcome metrics associated with both the virtual TAC test and the physical prosthesis, the home trial is the best solution for make subjects able to control the device.

Finally, in Tkach et al. [43], the mean path length error percentages and the mean offline classification errors of classifiers, with lower complexity than those reported above for the offline evaluation, were also considered with a virtual limb, for both experimental conditions ("Control Site" and "Grid"): the sequential real-time classifier ("SeqRT") was trained to predict four single-joint motions (elbow F/E and hand O/C, and no motion class); the simultaneous real-time classifier ("SimRT") was trained to predict the no motion class, the four single-joint motions of "SeqRT", and four combined motions classes (elbow F + hand O; elbow F + hand C; elbow E + hand O; and, elbow E + hand C). In particular, as regards the "SeqRT" classifier, the mean path length error percentages for "Control Sites" and "Grid" conditions were equal to 68.25% and 68.99%, while, for the "SimRT" classifier, they equaled 22.48% and 25.25%, respectively. Instead, the mean offline classification errors were equal to 1.6% ("Control Sites") and 1.3% ("Grid") for the "SeqRT" classifier; as regards the "SimRT" classifier, the mean offline classification errors were equals to 19.2% ("Control Sites") and 17.1% ("Grid"). The real-time results demonstrated that the simultaneous PR control of multiple DoFs perform equivalent or slightly better by using either a grid arrangement of electrodes or site-specific electrode placement.

6.3. Comparison between DC and PR Strategies

Finally, in the papers that reported a comparison between DC and PR, the performance evaluation methods were the following.

In Hargrove et al. [44], the performance of the proportional DC control and LDA classifier with TD features and AR coefficients was compared. The patients completed three different real-time performance tests with each system: BBT (modified up to 2 min) and block stacking test [70] (this test involves stacking the largest number of 1-inch cubes on top of each other in three minutes), CRT. In the BBT, the patients achieved an average 40% increase in the number of blocks passing from direct to PR control; in the block stacking test, the stacked towers were 59% higher while using the pattern recognition control system. The clothespin task was completed in 25% less time when using the pattern recognition control system. The average classification error rate for the pattern recognition systems was 16.3% (±1.6%).

In Wurth et al. [45], the real-time comparison between DC and PR-based control strategies have led to the following performance outcomes. In detail, the authors developed the Fitts' target acquisition task (FTAT) test, based on Fitts' law, which consisted of moving a cursor in two-dimensional Cartesian space from the center of the screen to a circular target appearing within a given radius at a given position. It was used in real time for assessing three EMG-based control strategies in virtual environments: the clinical standard of care (DC), a conventional, sequential PR (seqPR) strategy, and a simultaneous PR (simPR) strategy. The simPR approach considered the classification of each DoF independently. The parallel strategy that was introduced by [71] was adopted for the simultaneous classification of both DoFs. In order to comprehensively compare the three strategies, the following principal functional performance metrics were introduced: throughput (bits/second) and path efficiency (%). In detail, the throughput values for discrete motions (1 DoF) were equal to 2.64 ± 0.24, 3.67 ± 0.23, and 2.11 ± 0.18 for the conventional, seqPR, and simPR, respectively. Instead, for the combined motions (two DoFs), the throughput values were equal to 1.24 ± 0.04, 1.32 ± 0.03, and 1.63 ± 0.05 for the conventional, seqPR, and simPR strategies, respectively. Regarding the path efficiency values, they were equal for discrete motions (1 DoF) to 90.1 ± 0.23, 97.0 ± 0.96, and 96.3 ± 1.12 for the conventional, seqPR, and simPR strategies, respectively. Instead, for the combined motions (two DoFs), the path efficiency values were equal to 71.3 ± 0.8, 71.6 ± 0.76, and 87.7 ± 0.7 for the conventional, seqPR, and simPR strategies, respectively. Regarding discrete one DoF motions, the functional performance metrics of simultaneous PR were slightly lower than the sequential pattern recognition strategy that was revealed to be more precise and robust. Additionally, a qualitative evaluation of the control strategies was performed through a questionnaire demonstrating that both pattern recognition control strategies outperformed the amplitude-based DC control when considering two DoF tasks. In fact, the DC control was felt as unnatural and cumbersome in operating two 2 DoF control. It was only considered to be efficient for discrete 1 DoF tasks.

For the clinical trial reported by Hargrove et al. [46], the performance evaluation was carried out, for each type of control (PR and DC), by carrying out a series of tests before and after the six weeks of the home-trial: ACMC, SHAP, BBT (modified up to 2 min), and CRT. At the end of the trial, there were no significant differences in the ACMC test scores between pattern recognition (47.3 ± 3.9) and direct control (44.4 ± 3.4). As for the SHAP test, a significant improvement (p = 0.041) was noted in the performance that was achieved by using the PR-based control as compared to the direct control. The authors also noted that there was a difference between the performance in the pre and post trial cases (p = 0.038). In the BBT test, there was no particular difference between the pre and post trial tests; however, it was noted that, at the end of the home trial, the subjects moved 13.4 ± 2.6 blocks using pattern recognition control and 15.6 ± 2.7 blocks using direct control. Finally, in the CRT test, the subjects obtained significantly better results (p = 0.024) (i.e., the patients needed less time for concluding the test) using the PR-based control (90.2 ± 39.6 s) than using the direct control (137 ± 60.2 s). There were no statistically significant changes between pre- and post-home tests, nor was there any significant interaction between the pre- and post-home tests and the control strategy used. Quantitative statistics on household use have been carried out, as measured by the control system: on average, users cumulatively wore the prosthesis 138.7 ± 34.6 h during the direct control portion of the home trial and 147.7 ± 45.3 h during the PR phase of the home trial. The subjects chose to re-calibrate their control on 32.6 ± 8.2 occasions over the duration of the home trial. At the end of the trial, seven out of eight patients said that they preferred PR-based control over DC. As for DC, many had found pulse control not intuitive; they also found it difficult to control a single DoF when desired, having the simultaneous activation of two DoFs available. On the other hand, in the PR-based control, the subjects reported that the prosthesis was sometimes activated unwantedly.

In Young et al. [47], the performance of the conventional amplitude-based myoelectric control, the sequential (one DoF at time) PR control, and the simultaneous PR control (two DoFs at time) were reported. That three strategies were evaluated with a virtual prosthesis in a virtual environment using the TAC Test that considered the completion time, completion rate, and length error as performance metrics. The statistical ANOVA test ($p < 0.05$) was conducted for the comparison of the reported control strategies. For two DoFs tasks, the simultaneous PR system performed the best, with the lowest average completion times, completion rates, and length error when compared to the other control strategies. In particular, for the 2 DoFs tasks, the amputees chose to perform simultaneous movements in 78% of cases with simultaneous PR and in 64% of cases with conventional control. Furthermore, overall offline classification errors for the PR control strategies were compared with the ANOVA test ($p < 0.05$). The average classification error for sequential control was equal to 11.1% (±5.8 standard error of the mean—SEM), while the errors for simultaneous PR control were 23.1% (±10.3 SEM) and 33.19% (±11.3 SEM) for discrete and combined movement classification, respectively. Thus, finally, the authors have demonstrated that the simultaneous PR system had slightly lower performance with respect to the sequential PR system that was used for 1 DoF tasks (that required one DoF motions) and performed better than conventional control on 1 DoF tasks, while it had the best performance with simultaneous PR control on 2 DoFs tasks when compared to both conventional or sequential control.

Table 3 summarizes the results of the comparison papers between DC and PR. In most cases, the systems that are based on PR strategy allow for achieving better performance than DC one.

Table 3. Evaluation of the performance obtained with the DC and Pattern Recognition (PR) systems reported in the papers presented in Section 6.3.

Performance Evaluation Method	Metric Indicators	DC	PR	Study
BBT	Number of 1-inch blocks moved over a barrier in two min (average value on 4 TMR patients)	10.7 ± 4.3	15 ± 3	[44]
	Number of 1-inch blocks moved over a barrier in two min (average value on 8 TMR patients)	15.6 ± 2.7	13.4 ± 2.6	[46]
CRT	Time (s) to move three clothespins (average value on 4 TMR patients)	60 ± 15	45 ± 11	[44]
	Time (s) to move three clothespins (average value on 8 TMR patients)	137 ± 60.2	90.2 ± 39.6	[46]
ACMC	Test score (average value on 8 TMR patients)	44.4 ± 3.4	47.3 ± 3.9	[46]
SHAP	Index of function (average value on 8 TMR patients)	18 ± 5	31 ± 3	[46]
TAC	Completion Time (s) 1 DoF (average value on 4 TMR patients)	2.651.66	1.4 ± 0.3 (Seq) 2.03 ± 0.83 (Sim)	[47]
	Completion Time (s) 2 DoF (average value on 4 TMR patients)	3.55 ± 1.66	3.6 ± 0.42 (Seq) 1.93 ± 0.82 (Sim)	
	Completion rate (%) 1 DoF (average value on 4 TMR patients	86.25	100 (Seq) 93.75 (Sim)	
	Completion rate (%) 2 DoF (average value on 4 TMR patients)	81.25	92.5 (Seq) 98.75 (Sim)	
	Length error (%) 1 DoF (average value on 4 TMR patients	97.4 ± 81	13.96 ± 4.65 (Seq) 32 ± 29.52 (Sim)	
	Length error (%) 2 DoF (average value on 4 TMR patients)	86.65 ± 52.77	67.88 ± 11.8 (Seq) 21.88 ± 25.08 (Sim)	
FTAT	Throughput (bit/s) (1 DoF)	2.64 ± 0.24	3.67 ± 0.23 (Seq) 2.11 ± 0.18 (Sim)	[45]
	Throughput (bit/s) (2 DoF)	1.24 ± 0.04	1.32 ± 0.03 (Seq) 1.63 ± 0.05 (Sim)	
	Path efficiency (%) (1 DoF)	90.1 ± 0.23	97.00 ± 0.96 (Seq) 96.3 ± 1.12 (Sim)	
	Path efficiency (%) (2 DoF)	71.3 ± 0.80	71.60 ± 0.76 (Seq) 87.7 ± 0.7 (Sim)	

7. Results and Discussion

A comprehensive literature analysis on the most used prosthetic control strategies for TMR patients was carried out, when considering the amputation level of the enrolled patients, the number of signals of reinnervated sites, the number of controlled DoFs and, when available, the methods for validating the control of a prosthetic device, and the obtained performance. When the prosthesis is controlled with DC strategies (proportional strategy, usually), it is possible to use the simultaneous method for the selection of the joint to be controlled. The number of joints that can be controlled simultaneously depends on the number of input signals available. The reinnervated sites used to acquire EMG signals are usually three [21] or four [31,32,34,44,47] for SD patient, while it can be two [31,33,34,47] or four [24,44,45] for TH amputees. Especially when there are less then two reinnervated sites, additional signals from the residual muscles [21,31,33] can be used to actuate the prosthesis with the myoelectric control. However, patients simultaneously control, at most, two DoFs with the only use of EMG signals, whether from reinnervated sites or residual muscle. Almost all patients have demonstrated that they can control a three DoF prosthesis with F/E elbow, P/S wrist, and O/C hand movements, by switching with the co-contraction joint selection strategy from one of the two DoFs simultaneously controlled (usually elbow or hand) to the third one (wrist) and vice-versa. For patients with more than four reinnervated sites, as one of the TH amputees in [44], the P/S of the wrist can also be controlled simultaneously. In some studies, additional inputs were added to EMG signals, such as those from FSR sensors [32,34] or from switches [24]; these additional inputs allow for simultaneously controlling more than two DoFs, with a hybrid control. However, prostheses with more than three DoFs, as in [32], resulted in being difficult and less intuitive to control.

With respect to the conventional myoelectric control methods, which only consider the EMG amplitude at specific myoelectric control sites [54], the PR control systems that are based on both sequential and simultaneous strategies allow for the control of up to 2 DoFs, in a natural way, as asserted in [44,46], and [45,47]. Regarding the PR strategies, most studies considered the following discrete motion classes most clinically used for different amputation levels. In the case of BSD, SD, and TH amputees, from eight to 16 motion classes were mainly considered: elbow F/E, wrist F/E, wrist S/P, wrist ulnar and radial deviation, hand opening, two hand opening patterns, including finger abduction and finger adduction, a selection of various types of hand grasps (as 3-jaw chuck, power grip, fine pinch, key grip, trigger grip, and tool grip), and a rest state [35–38]. Instead, for the TR TMR amputees, as considered in [39,40], from three to a maximum number of five grasp types were taken into account: prismatic-2 fingers, precision disk, palm pinch, lateral, and prismatic-4 fingers. The majority of these studies [36–40] reported only the offline performance of the LDA classifier with TD, AR, AR-RMS, and TDAR, except [35], which introduced useful performance metrics as motions selection time, motion completion time, and motion completion rate for evaluating the control of a virtual arm in real-time. When only considering that the offline accuracy can be a relevant limitation for evaluating the prosthetic control, since many studies [72,73] have shown that offline accuracy does not necessarily correspond to real-time performance. To overcome these limitations, other studies, such as [41,42], introduced for TH amputees a physical prosthesis and a virtual limb for controlling, using an LDA classifier with time-domain feature set, the following discrete movements: elbow F/E, wrist rotation, and hand C/O. It is worth noticing that, in [42], the differences between the performance achieved with a virtual prosthesis with respect to that obtained with a physical prosthesis were also evaluated. It is shown that the TAC test completion time correlated significantly with all physical outcome measures except the CRT. These results support the importance of reporting also the online performance metrics rather than only the offline classification error analysis [43,74].

The following studies [44–47] also reported a comparison of real-time performance between DC and PR based control strategies for SD and TH amputees: [44–46] only considered a physical prosthesis, while [47] presented a virtual prosthesis for controlling up to 2 DoFs. In most cases, the patients preferred either the more intuitive sequential or simultaneous pattern recognition control to the DC control, because this latter appeared to be unnatural and especially cumbersome for two DoFs tasks.

It is worth noticing that, with respect to the traditional PR control strategies, which were limited to sequentially controlling one DoF at the time, [43,45,47,59,71] also introduced the simultaneous PR control strategy, considered more promising because it showed improvements in throughputs and path efficiencies when compared to direct control or sequential PR. Moreover, the simultaneous PR control allowed for amputees to perform tasks with more than two DoFs with the lowest average completion times, completion rates and length error compared to the other control strategies, even if [43] argues that the simultaneous PR system has slightly lower performance with respect to the sequential PR system, but performs better than conventional DC control.

Clinical Applications

In this section, the principal clinical outcomes of this review are discussed by focusing on the most relevant control strategies that allow a natural and simultaneous control of the prosthetic arm DoFs for TMR patients with different upper-limb amputations (i.e., BSD, SD, TH, and TR).

Firstly, for both the DC and PR strategies, the benefits of TMR surgery on the resulting control of different joints (as the elbow, wrist, and hand) have been introduced. As reported in almost all the analyzed papers [24,31,33,34,42,46,53], the patients can achieve better performance compared to the pre-TMR situation; furthermore, they also increase the ability to perform some ADLs, improving their quality of life.

These results confirmed that TMR provides new target muscle sites that are physiologically linked to the effective movements of the prosthetic device. However, in some studies [21,31,41], not all of the reinnervated sites are used as control sites due to the overlapping of EMG signals. In detail, the number of reinnervated sites is closely related to the number of DoFs that can be simultaneously controlled when using DC, whereas PR methods do not require mode-switching and independent EMG signals, even if almost all of the proposed classifiers only provide sequential control of multiple DoFs [13].

In order to examine the clinical robustness of myoelectric prosthetic control with TMR, the reported real-time performance metrics were taken into account to test whether the reinnervated muscles, after TMR surgery, can improve the myoelectric signals for real-time control of multifunction prostheses. Because of the lack of standard criteria, this review suggested a unified protocol test for the validation of these control strategies, by defining which tests are most suitable for the evaluation of prosthetic control for TMR patients with different amputation level.

Only the ACMC can be considered a standardized test for evaluating myoelectric control, because it has an established test-retest, inter-rater, and intra-rater reliability and clinical interpretation guidelines, as claimed by [68]. Additionally, as reported in [42], the following tests were also considered to be promising for assessing the functionality of a prosthesis: BBT, SHAP test, CRT, and JTHFT. Indeed, it can be noted that, in the analyzed works, the evaluation methods most commonly used are: BBT (modified up to 2 min) used in nine papers for a total of 34 subjects and CRT used in eight papers for a total of 33 subjects. Both of these tests have several advantages: they evaluate patient performance without differences among amputation levels or the number of reinnervated sites. Moreover, both tests are applicable for both DC and PR control, so that they were also used in the studies in which the two types of myoelectric control are compared. Furthermore, they can also be used on non-TMR patients to evaluate the performance of the prosthetic control, allow comparing the performance in pre and post-surgery situation, as in [21,24,31,33,34], to evaluate pre-and post-trial performance, as in [42], or to evaluate performance when different control strategies were used [44,46]. Lastly, these tests were used to compare performance in the control of prostheses with three or six DoFs [32]. Among the disadvantages, there is the fact that the above tests cannot be used in the absence of a prosthetic arm, therefore they cannot be used to evaluate the patient's performance while using virtual reality. In [42], it is demonstrated that the TAC test completion time is a measure that significantly correlates with all the physical outcome measures, except the CRT.

For these reasons, we recommend, if possible, the use of Box and Block and Clothespin Relocation tests to evaluate the performance of the functional effectiveness of the prosthesis (with both DC and PR control) as a whole. In addition, this literature analysis also highlighted the need to have more quantitative information and to use instrumental indicators to be associated with these tests. For instance, when virtual reality is used with PR, real-time performance metrics that include motion selection time, motion completion time, and motion completion ("success") rate can be considered for evaluating virtual arm movements.

Finally, regarding some clinical trials, we have also reported that the initial use of TMR was to prevent or treat phantom limb pain (PLP) and neuroma pain [25,27]. Other clinical investigations have focused on the role of the osseointegration [28] and the Targeted Sensory Reinnervation (TSR) [29] on TMR patients. Regarding the use of the osseointegration combined with TMR surgery, a direct interface that links the implant to the bone can provide even more stability when an external prosthesis is worn by amputees [19]. The TSR, instead, allows for controlling, in a bidirectional way, neuroprosthetic devices, thanks to the presence of a region with "sensory perception".

8. Conclusions

This paper has provided an overview of the main advancements of the state of the art regarding prosthetic control techniques of the upper limb and performance evaluation methods for patients who have undergone TMR surgery.

Twenty papers were analyzed, highlighting that the most commonly used prosthetic control techniques are: in the context of direct Control, the proportional strategy and the method of simultaneous joint selection with co-contraction; in the context of PR methods, the LDA algorithm with various feature selection sets.

The most common performance evaluation methods, both for DC and PR, are BBT and CRT. In the case of prostheses controlled with PR, there is always the offline analysis of accuracy.

The most commonly used myoelectric prostheses are composed of 3 DoFs, for elbow F/E, wrist P/S, and hand O/C.

This work further highlighted the presence of a variety of tests that were used for the functional performance evaluation (Table 2), but there is a lack of standard criteria allowing to define which tests are the most suitable for the evaluation of prosthetic control for TMR patients with a different amputation level. In order to fill this gap, both the Box and Blocks and the Clothespin Relocation seem to be the most promising tests for evaluating the performance of the prosthetic systems.

In addition, we believe that virtual reality can be used to further explore the potentiality of the proposed control approaches, before considering them on a physical prosthesis. In fact, we proposed extending the use of virtual reality performance indexes, defined for PR, like motion selection time, motion completion time, and motion completion ("success") rate also to the DC control in this way: the motion selection time and motion completion time can be modified by considering the time that is required from EMG onset to remain above the threshold, while the success rate does not need to be modified. In this way, a comparative analysis between DC and PR systems can also be done when using a virtual reality system.

It has to be noted that only few articles presented results regarding the simultaneous PR control strategy showing improvements in throughputs and path efficiencies when compared to direct control or sequential PR [43,45,47,59,71]. Thus, the possibility of simultaneously controlling the prosthesis still can be improved with simultaneous PR-based controllers.

In conclusion, despite the great progresses in the field of advanced prosthetic control, this paper highlights the necessity to still identify the best PR/DC-based system allowing for robust control when considering more than 2 DoFs and of defining standard evaluation methods of the real-time control strategy performance.

Author Contributions: F.M. and F.L. designed the paper, analyzed the literature and wrote the paper; C.G. and F.C. designed the study, supervised the writing and wrote the paper; L.Z. and E.G. designed the paper and supervised the writing. All authors have read and agreed to the published version of the manuscript.

Funding: This work was supported partly by the Italian Institute for Labour Accidents (INAIL) prosthetic center with WiFi Myo-Hand (CUP: E59E19001460005) project, partly by ReGiveMeFive (CUP: E59E19001460005), partly by funding from the European Union's Horizon 2020 research and innovation programme under grant agreement No 899822, SOMA project.

Conflicts of Interest: The authors declare no conflict of interest. The founders had no role in the design of the study; in the collection, analyses, or interpretation of data; in the writing of the manuscript, or in the decision to publish the results.

Abbreviations

The following abbreviations, reported in alphabetic order, are used in this manuscript:

ACMC	Assessment for Capacity of Myoelectric Control
ADL	Activities of Daily Living
AMPS	Assessment of Motor and Process Skills
AR	Auto Regressive
ARAT	Action Research Arm Test
BBT	Box and Block Test
BSD	Bilateral Shoulder Disarticulation
CRT	Clothespin Relocation Test
DC	Direct Control
DoF	Degree of Freedom
EMG	ElectroMyoGraphy
ESN	Echo State Network
F/E	Flexion/Extension
FSR	Force Sensor Resistor
HD	High Density
HT	Higher Threshold
I/E	Intra/Extra
imEMG	intramuscolarEMG
JTHFT	Jebsen-Taylor test of Hand Function
LDA	Linear Discriminant Analysis
MAV	Mean Absolute Value
mcr	motion completion rate
mct	motion completion time
MCU	MicroController Unit
MLP	Multi Layer Perceptron
mst	motion selection time
NLR	Nonlinear Logistic Regression
O/C	Open/Close
P/S	Pronation/Supination
PLP	Phantom Limb Pain
PR	Pattern Recognition
PRISMA	Preferred Reporting Items for Systematics reviews and Meta-Analyses
RMS	Root Mean Square
ROM	Range Of Motion
SD	Shoulder Disarticulation

sEMG	surfaceEMG
seqPR	sequential PR
SHAP	Southampton Hand Assessment Procedure
simPR	simultaneous PR
SSC	Slope Sign Changes
ST	Standard Threshold
SVM	Support Vector Machine
TAC	Target Achievement Control
TD	Time Domain
TD-AR	Time Domain and Auto Regressive
TH	Trans-Humeral
TMR	Targeted Muscle Reinnervation
TSR	Targeted Sensory Reinnervation
VR	Virtual Reality
WL	Waveform Length
WMFT	Wolf Motor Functions Tests
ZC	Zero Crossing

References

1. LeBlanc, M. Give Hope-Give a Hand. The LN-4 Prosthetic Hand. 2008. Available online: https://web.stanford.edu/class/engr110/2011/LeBlanc-03a.pdf (accessed on 10 May 2011).
2. Technical Report: Bionic Technologies and Disabilities: The Challenges of Limb Prosthetics. Available online: https://www.inail.it/cs/internet/docs/alg-evento-workshop-dcod-sfide-protesica-2019.pdf (accessed on 18 November 2019).
3. Montague, R. *Amputee and Prosthetic Rehabilitation—Standards and Guidelines*, 2nd ed.; British Society of Rehabilitation Medicine: London, UK, 2003; p. 13. Available online: https://www.bsrm.org.uk/downloads/ars-gfinaltext.pdf (accessed on 21 October 2003).
4. Pomares, G.; Coudane, H.; Dap, F.; Dautel, G. Epidemiology of traumatic upper limb amputations. *Orthop. Traumatol. Surg. Res.* **2018**, *104*, 273–276. [CrossRef]
5. Ziegler-Graham, K.; MacKenzie, E.J.; Ephraim, P.L.; Travison, T.G.; Brookmeyer, R. Estimating the prevalence of limb loss in the United States: 2005 to 2050. *Arch. Phys. Med. Rehabil.* **2008**, *89*, 422–429. [CrossRef] [PubMed]
6. Ciancio, A.L.; Cordella, F.; Barone, R.; Romeo, R.A.; Bellingegni, A.D.; Sacchetti, R.; Davalli, A.; Di Pino, G.; Ranieri, F.; Di Lazzaro, V.; et al. Control of prosthetic hands via the peripheral nervous system. *Front. Neurosci.* **2016**, *10*, 116. [CrossRef] [PubMed]
7. Cordella, F.; Di Corato, F.; Loianno, G.; Siciliano, B.; Zollo, L. Robust pose estimation algorithm for wrist motion tracking. In Proceedings of the 2013 IEEE/RSJ International Conference on Intelligent Robots and Systems, Tokyo, Japan, 3–7 November 2013; pp. 3746–3751.
8. Bruce, C. Cosmetic Prosthesis and Methods for Making the Same. U.S. Patent Application 10/539,433, 3 August 2006.
9. Geethanjali, P. Myoelectric control of prosthetic hands: State-of-the-art review. *Med. Devices (Auckl. NZ)* **2016**, *9*, 247. [CrossRef] [PubMed]
10. Sears, H.H.; Iversen, E.K.; Hays, K.B.; Dyck, A.D. Method and Apparatus for Controlling an Externally Powered Prosthesis. U.S. Patent 5,888,213, 30 March 1999.
11. Reiter, R. Eine neue elektrokunsthand. *Grenzgebiete der Medizin* **1948**, *1*, 133–135. [PubMed]
12. Roche, A.D.; Rehbaum, H.; Farina, D.; Aszmann, O.C. Prosthetic myoelectric control strategies: A clinical perspective. *Curr. Surg. Rep.* **2014**, *2*, 44. [CrossRef]
13. Scheme, E.; Englehart, K. Electromyogram pattern recognition for control of powered upper-limb prostheses: State of the art and challenges for clinical use. *J. Rehabil. Res. Dev.* **2011**, *48*, 643–659.
14. Parajuli, N.; Sreenivasan, N.; Bifulco, P.; Cesarelli, M.; Savino, S.; Niola, V.; Esposito, D.; Hamilton, T.J.; Naik, G.R.; Gunawardana, U.; et al. Real-time EMG based pattern recognition control for hand prostheses: A review on existing methods, challenges and future implementation. *Sensors* **2019**, *19*, 4596. [CrossRef]
15. Smail, L.C.; Neal, C.; Wilkins, C.; Packham, T.L. Comfort and function remain key factors in upper limb prosthetic abandonment: Findings of a scoping review. *Disabil. Rehabil. Assist. Technol.* **2020**, 1–10. [CrossRef]
16. Biddiss, E.; Chau, T. Upper-limb prosthetics: Critical factors in device abandonment. *Am. J. Phys. Med. Rehabil.* **2007**, *86*, 977–987. [CrossRef]
17. Cipriani, C.; Zaccone, F.; Micera, S.; Carrozza, M.C. On the shared control of an EMG-controlled prosthetic hand: Analysis of user–prosthesis interaction. *IEEE Trans. Robot.* **2008**, *24*, 170–184. [CrossRef]

18. Kuiken, T.A.; Childress, D.S.; Rymer, W.Z. The hyper-reinnervation of rat skeletal muscle. *Brain Res.* **1995**, *676*, 113–123. [CrossRef]
19. Chi, A.; Smith, S.; Womack, I.; Armiger, R. The evolution of man and machine—A review of current surgical techniques and cutting technologies after upper extremity amputation. *Curr. Trauma Rep.* **2018**, *4*, 339–347. [CrossRef]
20. Myers, H.; Lu, D.; Gray, S.J.; Bruscino-Raiola, F. Targeted muscle reinnervation to improve electromyography signals for advanced myoelectric prosthetic limbs: A series of seven patients. *ANZ J. Surg.* **2020**, *90*, 591–596. [CrossRef]
21. Kuiken, T.A.; Dumanian, G.A.; Lipschutz, R.D.; Miller, L.A.; Stubblefield, K. The use of targeted muscle reinnervation for improved myoelectric prosthesis control in a bilateral shoulder disarticulation amputee. *Prosthet. Orthot. Int.* **2004**, *28*, 245–253. [CrossRef]
22. Kuiken, T. Targeted reinnervation for improved prosthetic function. *Phys. Med. Rehabil. Clin.* **2006**, *17*, 1–13. [CrossRef]
23. Oh, C.; Carlsen, B.T. New innovations in targeted muscle reinnervation: A critical analysis review. *JBJS Rev.* **2019**, *7*, e3. [CrossRef] [PubMed]
24. Kuiken, T.A.; Miller, L.A.; Lipschutz, R.D.; Lock, B.A.; Stubblefield, K.; Marasco, P.D.; Zhou, P.; Dumanian, G.A. Targeted reinnervation for enhanced prosthetic arm function in a woman with a proximal amputation: A case study. *Lancet* **2007**, *369*, 371–380. [CrossRef]
25. Kuiken, T.A.; Barlow, A.K.; Hargrove, L.; Dumanian, G.A. Targeted muscle reinnervation for the upper and lower extremity. *Tech. Orthop. (Rockv. MD)* **2017**, *32*, 109. [CrossRef]
26. Bueno Jr, R.A.; French, B.; Cooney, D.; Neumeister, M.W. Targeted muscle reinnervation of a muscle-free flap for improved prosthetic control in a shoulder amputee: Case report. *J. Hand Surg.* **2011**, *36*, 890–893. [CrossRef]
27. Souza, J.M.; Cheesborough, J.E.; Ko, J.H.; Cho, M.S.; Kuiken, T.A.; Dumanian, G.A. Targeted muscle reinnervation: A novel approach to postamputation neuroma pain. *Clin. Orthop. Relat. Res.* **2014**, *472*, 2984–2990. [CrossRef] [PubMed]
28. Mastinu, E.; Brånemark, R.; Aszmann, O.; Ortiz-Catalan, M. Myoelectric signals and pattern recognition from implanted electrodes in two TMR subjects with an osseointegrated communication interface. In Proceedings of the 2018 40th Annual International Conference of the IEEE Engineering in Medicine and Biology Society (EMBC), Honolulu, HI, USA, 18–21 July 2018; pp. 5174–5177.
29. Hebert, J.S.; Chan, K.M.; Dawson, M.R. Cutaneous sensory outcomes from three transhumeral targeted reinnervation cases. *Prosthet. Orthot. Int.* **2016**, *40*, 303–310. [CrossRef]
30. Roubaud, M.S. Targeted Muscle Reinnervation in the Oncologic Population: A Literature Review and Current Practice. *Curr. Surg. Rep.* **2020**, *8*, 1–7. [CrossRef]
31. Kuiken, T.; Dumanian, G.; Lipschutz, R.; Miller, L.; Stubblefield, K. Targeted muscle reinnervation for improved myoelectric prosthesis control. In Proceedings of the 2nd International IEEE EMBS Conference on Neural Engineering, Arlington, VA, USA, 16–19 March 2005; pp. 396–399.
32. Miller, L.A.; Lipschutz, R.D.; Stubblefield, K.A.; Lock, B.A.; Huang, H.; Williams, T.W., III; Weir, R.F.; Kuiken, T.A. Control of a six degree of freedom prosthetic arm after targeted muscle reinnervation surgery. *Arch. Phys. Med. Rehabil.* **2008**, *89*, 2057–2065. [CrossRef] [PubMed]
33. O'Shaughnessy, K.D.; Dumanian, G.A.; Lipschutz, R.D.; Miller, L.A.; Stubblefield, K.; Kuiken, T.A. Targeted reinnervation to improve prosthesis control in transhumeral amputees: A report of three cases. *JBJS* **2008**, *90*, 393–400. [CrossRef] [PubMed]
34. Miller, L.A.; Stubblefield, K.A.; Lipschutz, R.D.; Lock, B.A.; Kuiken, T.A. Improved myoelectric prosthesis control using targeted reinnervation surgery: A case series. *IEEE Trans. Neural Syst. Rehabil. Eng.* **2008**, *16*, 46–50. [CrossRef]
35. Kuiken, T.A.; Li, G.; Lock, B.A.; Lipschutz, R.D.; Miller, L.A.; Stubblefield, K.A.; Englehart, K.B. Targeted muscle reinnervation for real-time myoelectric control of multifunction artificial arms. *JAMA* **2009**, *301*, 619–628. [CrossRef]
36. Smith, L.H.; Hargrove, L.J. Intramuscular EMG after targeted muscle reinnervation for pattern recognition control of myoelectric prostheses. In Proceedings of the 2013 6th International IEEE/EMBS Conference on Neural Engineering (NER), San Diego, CA, USA, 6–8 November 2013; pp. 1155–1158.
37. Huang, H.; Zhou, P.; Li, G.; Kuiken, T. Spatial filtering improves EMG classification accuracy following targeted muscle reinnervation. *Ann. Biomed. Eng.* **2009**, *37*, 1849–1857. [CrossRef] [PubMed]
38. Zhou, P.; Lowery, M.M.; Englehart, K.B.; Huang, H.; Li, G.; Hargrove, L.; Dewald, J.P.; Kuiken, T.A. Decoding a new neural–machine interface for control of artificial limbs. *J. Neurophysiol.* **2007**, *98*, 2974–2982. [CrossRef] [PubMed]
39. Batzianoulis, I.; Simon, A.; Hargrove, L.; Billard, A. Reach-to-grasp motions: Towards a dynamic classification approach for upper-limp prosthesis. In Proceedings of the 2019 9th International IEEE/EMBS Conference on Neural Engineering, San Francisco, CA, USA, 20–23 March 2019; pp. 287–290.
40. Batzianoulis, I.; Krausz, N.E.; Simon, A.M.; Hargrove, L.; Billard, A. Decoding the grasping intention from electromyography during reaching motions. *J. Neuroeng. Rehabil.* **2018**, *15*, 57. [CrossRef] [PubMed]
41. Xu, Y.; Zhang, D.; Wang, Y.; Feng, J.; Xu, W. Two ways to improve myoelectric control for a transhumeral amputee after targeted muscle reinnervation: A case study. *J. Neuroeng. Rehabil.* **2018**, *15*, 37. [CrossRef] [PubMed]

42. Hargrove, L.; Miller, L.; Turner, K.; Kuiken, T. Control within a virtual environment is correlated to functional outcomes when using a physical prosthesis. *J. Neuroeng. Rehabil.* **2018**, *15*, 60. [CrossRef] [PubMed]
43. Tkach, D.C.; Young, A.J.; Smith, L.H.; Rouse, E.J.; Hargrove, L.J. Real-time and offline performance of pattern recognition myoelectric control using a generic electrode grid with targeted muscle reinnervation patients. *IEEE Trans. Neural Syst. Rehabil. Eng.* **2014**, *22*, 727–734. [CrossRef]
44. Hargrove, L.J.; Lock, B.A.; Simon, A.M. Pattern recognition control outperforms conventional myoelectric control in upper limb patients with targeted muscle reinnervation. In Proceedings of the 2013 35th Annual International Conference of the IEEE Engineering in Medicine and Biology Society (EMBC), Osaka, Japan, 3–7 July 2013; pp. 1599–1602.
45. Wurth, S.M.; Hargrove, L.J. A real-time comparison between direct control, sequential pattern recognition control and simultaneous pattern recognition control using a Fitts' law style assessment procedure. *J. Neuroeng. Rehabil.* **2014**, *11*, 91. [CrossRef]
46. Hargrove, L.J.; Miller, L.A.; Turner, K.; Kuiken, T.A. Myoelectric pattern recognition outperforms direct control for transhumeral amputees with targeted muscle reinnervation: A randomized clinical trial. *Sci. Rep.* **2017**, *7*, 1–9.
47. Young, A.J.; Smith, L.H.; Rouse, E.J.; Hargrove, L.J. A comparison of the real-time controllability of pattern recognition to conventional myoelectric control for discrete and simultaneous movements. *J. Neuroeng. Rehabil.* **2014**, *11*, 5. [CrossRef] [PubMed]
48. Englehart, K.; Hudgins, B. A robust, real-time control scheme for multifunction myoelectric control. *IEEE Trans. Biomed. Eng.* **2003**, *50*, 848–854. [CrossRef]
49. Ortiz-Catalan, M.; Håkansson, B.; Brånemark, R. Real-time and simultaneous control of artificial limbs based on pattern recognition algorithms. *IEEE Trans. Neural Syst. Rehabil. Eng.* **2014**, *22*, 756–764. [CrossRef]
50. Tkach, D.; Huang, H.; Kuiken, T.A. Study of stability of time-domain features for electromyographic pattern recognition. *J. Neuroeng. Rehabil.* **2010**, *7*, 21. [CrossRef] [PubMed]
51. Bellingegni, A.D.; Gruppioni, E.; Colazzo, G.; Davalli, A.; Sacchetti, R.; Guglielmelli, E.; Zollo, L. NLR, MLP, SVM, and LDA: A comparative analysis on EMG data from people with trans-radial amputation. *J. Neuroeng. Rehabil.* **2017**, *14*, 82. [CrossRef]
52. Leone, F.; Gentile, C.; Ciancio, A.L.; Gruppioni, E.; Davalli, A.; Sacchetti, R.; Guglielmelli, E.; Zollo, L. Simultaneous sEMG classification of wrist/hand gestures and forces. *Front. Neurorobot.* **2019**, *13*, 42. [CrossRef]
53. Hijjawi, J.B.; Kuiken, T.A.; Lipschutz, R.D.; Miller, L.A.; Stubblefield, K.A.; Dumanian, G.A. Improved myoelectric prosthesis control accomplished using multiple nerve transfers. *Plastic Reconstr. Surg.* **2006**, *118*, 1573–1578. [CrossRef]
54. Cheesborough, J.E.; Smith, L.H.; Kuiken, T.A.; Dumanian, G.A. Targeted muscle reinnervation and advanced prosthetic arms. In *Seminars in Plastic Surgery*; Thieme Medical Publishers: New York, NY, USA, 2015; Volume 29, pp. 062–072.
55. Mastinu, E. Embedded Controller for Artificial Limbs. Ph.D. Thesis, Department of Signals and Systems, Chalmers University of Technology, Gothenburg, Sweden, 2017.
56. Huang, H.; Zhou, P.; Li, G.; Kuiken, T.A. An analysis of EMG electrode configuration for targeted muscle reinnervation based neural machine interface. *IEEE Trans. Neural Syst. Rehabil. Eng.* **2008**, *16*, 37–45. [CrossRef]
57. Kuiken, T.A.; Miller, L.A.; Turner, K.; Hargrove, L.J. A comparison of pattern recognition control and direct control of a multiple degree-of-freedom transradial prosthesis. *IEEE J. Transl. Eng. Health Med.* **2016**, *4*, 1–8. [CrossRef]
58. Simon, A.M.; Hargrove, L.J.; Lock, B.A.; Kuiken, T.A. The target achievement control test: Evaluating real-time myoelectric pattern recognition control of a multifunctional upper-limb prosthesis. *J. Rehabil. Res. Dev.* **2011**, *48*, 619. [CrossRef]
59. Young, A.J.; Smith, L.H.; Rouse, E.J.; Hargrove, L.J. A new hierarchical approach for simultaneous control of multi-joint powered prostheses. In Proceedings of the 2012 4th IEEE RAS & EMBS International Conference on Biomedical Robotics and Biomechatronics (BioRob), Rome, Italy, 24–27 June 2012; pp. 514–520.
60. Mathiowetz, V.; Weber, K.; Kashman, N.; Volland, G. Adult norms for the nine hole peg test of finger dexterity. *Occup. Ther. J. Res.* **1985**, *5*, 24–38. [CrossRef]
61. Wolf, S.L.; Catlin, P.A.; Ellis, M.; Archer, A.L.; Morgan, B.; Piacentino, A. Assessing Wolf motor function test as outcome measure for research in patients after stroke. *Stroke* **2001**, *32*, 1635–1639. [CrossRef]
62. Assessment of Motor and Process Skills. 2019. Available online: https://www.sralab.org/rehabilitation-measures/assessment-motor-and-process-skills# (accessed on 26 April 2019).
63. Yozbatiran, N.; Der-Yeghiaian, L.; Cramer, S.C. A standardized approach to performing the action research arm test. *Neurorehabilit. Neural Repair* **2008**, *22*, 78–90. [CrossRef]
64. Hill, W.; Kyberd, P.; Hermansson, L.N.; Hubbard, S.; Stavdahl, Ø.; Swanson, S. Upper limb prosthetic outcome measures (ULPOM): A working group and their findings. *JPO J. Prosthet. Orthot.* **2009**, *21*, P69–P82. [CrossRef]
65. Light, C.M.; Chappell, P.H.; Kyberd, P.J. Establishing a standardized clinical assessment tool of pathologic and prosthetic hand function: Normative data, reliability, and validity. *Arch. Phys. Med. Rehabil.* **2002**, *83*, 776–783. [CrossRef]
66. Jebsen, R.H. An objective and standardized test of hand function. *Arch. Phys. Med. Rehabil.* **1969**, *50*, 311–319. [PubMed]
67. Lindner, H.; Langius-Eklof, A.; Hermansson, L. Test-retest reliability and rater agreements of Assessment of Capacity for Myoelectric Control version 2.0. *J. Rehabil. Res. Dev.* **2014**, *51*, 635–644. [CrossRef]
68. Hermansson, L.M.; Bodin, L.; Eliasson, A.C. Intra-and inter-rater reliability of the assessment of capacity for myoelectric control. *J. Rehabil. Med.* **2006**, *38*, 118–123. [CrossRef]
69. Lenzi, T.; Lipsey, J.; Sensinger, J.W. The RIC Arm—A small anthropomorphic transhumeral prosthesis. *IEEE/ASME Trans. Mechatronics* **2016**, *21*, 2660–2671. [CrossRef]

70. Simon, A.M.; Hargrove, L.J.; Lock, B.A.; Kuiken, T.A. A decision-based velocity ramp for minimizing the effect of misclassifications during real-time pattern recognition control. *IEEE Trans. Biomed. Eng.* **2011**, *58*, 2360–2368. [CrossRef]
71. Young, A.J.; Smith, L.H.; Rouse, E.J.; Hargrove, L.J. Classification of simultaneous movements using surface EMG pattern recognition. *IEEE Trans. Biomed. Eng.* **2012**, *60*, 1250–1258. [CrossRef]
72. Lock, B.; Englehart, K.; Hudgins, B. Real-time myoelectric control in a virtual environment to relate usability vs. accuracy. In Proceedings of the 2005 MyoElectric Controls/Powered Prosthetics Symposium, Fredericton, NB, Canada, 17–19 August 2005; pp. 122–127.
73. Li, G.; Schultz, A.E.; Kuiken, T.A. Quantifying pattern recognition—Based myoelectric control of multifunctional transradial prostheses. *IEEE Trans. Neural Syst. Rehabil. Eng.* **2010**, *18*, 185–192.
74. Hargrove, L.J.; Scheme, E.J.; Englehart, K.B.; Hudgins, B.S. Multiple binary classifications via linear discriminant analysis for improved controllability of a powered prosthesis. *IEEE Trans. Neural Syst. Rehabil. Eng.* **2010**, *18*, 49–57. [CrossRef] [PubMed]

MDPI
St. Alban-Anlage 66
4052 Basel
Switzerland
Tel. +41 61 683 77 34
Fax +41 61 302 89 18
www.mdpi.com

Sensors Editorial Office
E-mail: sensors@mdpi.com
www.mdpi.com/journal/sensors

www.ingramcontent.com/pod-product-compliance
Lightning Source LLC
Chambersburg PA
CBHW061559220926
51850CB00005B/373